D0161130

Modeling and Simulation

Modeling and Simulation

Hartmut Bossel
Universität Gesamthochschule Kassel
Kassel, Germany

Editorial, Sales, and Customer Service Office
US and Canada

A K Peters, Ltd.
289 Linden Street
Wellesley, MA 02181

Distributed in European markets by Vieweg & Sohn Verlagsgesellschaft GmbH, Braunschweig/Wiesbaden
Vieweg is a subsidiary company of the Bertelsmann Publishing Group International

Sales and Customer Service Office
Great Britain and other European countries

Verlag Vieweg
P.O. Box 5829
D-65048 Wiesbaden
Germany

Library of Congress Cataloging-in-Publication Data

Bossel, Hartmut
[Modellbildung und Simulation. English]
Modeling and simulation / Hartmut Bossel.
p. cm.
Includes bibliographical references and index.
ISBN 1-56881-033-4
1. Mathematical models. 2. Digital computer simulation.
I. Title.
TA342.B67 1994
003'.85--dc20 94-1847
CIP

ISBN 1-56881-033-4 (A K Peters)
ISBN 3-528-05419-0 (Verlag Vieweg)

Printed in the United States of America
98 97 96 95 94 10 9 8 7 6 5 4 3 2 1

Contents

Preface

It is the everlasting and unchanging rule of this world that everything is created by a series of causes and conditions and everything disappears by the same rule; everything changes, nothing remains constant.
The Teaching of Buddha, Bukkyo Dendo Kyokai, Tokyo 1966, p. 82

Our world is dynamic, but the human mind is ill-equipped for comprehending, assessing, and predicting dynamic processes. We often fail to understand and anticipate the dynamics of systems until it is too late—until previously available options have disappeared and we are trapped on a road to disaster. The dynamics of population growth, ozone depletion, carbon dioxide accumulation in the atmosphere, environmental deterioration, AIDS, civil wars, forest die-back, recession, unemployment, crime etc. have caught us largely unprepared, and have often led to totally inadequate responses.

It is possible to construct computer simulation models of dynamic processes and systems, and it is possible to use these models for gaining a better understanding of a system and its dynamics, and of the options available for coping with the ensuing problems. However, as in any other discipline, the process of modeling and simulation requires a certain amount of knowledge and skills: about systems and systems analysis, and about the approaches and tools available for modeling and simulation.

Modeling and Simulation is the English language version of a successful German textbook *Modellbildung und Simulation* (Vieweg Verlag, Braunschweig/Wiesbaden, 1992; 2nd ed. 1994). The English version has been considerably revised and improved compared to the German editions. The book evolved from courses on Modeling and Simulation taught to undergraduate and graduate students, managers and planners at Kassel University and at training seminars in Germany, China, and Malaysia. The book is meant for students of all disciplines—dynamic systems are found everywhere. A first course in differential equations, linear algebra and matrix analysis, or control systems, and minimal programming experience might be helpful for studying the subject, but it is not a prerequisite. The system-graphical method used for model development opens the subject to all with only a limited mathematical background.

Modeling and Simulation is a self-contained and complete guide to the methods and mathematical background of modeling and simulation of dynamic systems, covering basic system concepts, the modeling approach, computer simulation and its different uses, mathematical systems theory, example models and applications, and a unique collection of 50 elementary system models. A substantial amount of exercise material is provided, in particular in Chapter 6 SYSTEMS ZOO. The SIMPAS simulation software and the "Systems Zoo" are provided on the accompanying diskette.

Chapter 1 SYSTEMS AND MODELS introduces to the concepts of systems, models, and simulation.

Chapter 2 STRUCTURE focuses on the development of influence diagrams (causal loop diagrams) as the basis of systems analysis and demonstrates modeling and simulation by developing a small global model.

Chapter 3 SYSTEM STATE concentrates on the accurate description of system components and their functions using concepts of state space analysis and of dimensional analysis, and linking this information in a complete mathematical model. Several (mostly nonlinear) systems are introduced, and simulation models are developed for the rotating pendulum and a fishery operation.

Chapter 4 BEHAVIOR deals with the translation of simulation diagrams or model equations into running computer simulation models, using two different programming approaches: the SIMPAS simulator (provided on the diskette and based on TurboPascal) and the STELLA simulation software. The various simulation approaches to the mapping and understanding of behavior are demonstrated.

Chapter 5 CHOICE AND DESIGN introduces to the three fundamental tasks of systems analysis: path analysis, policy analysis, and system design. The three previously developed models (global model, pendulum, fishery) are applied to these tasks. The basic orientors of system behavior are derived as a framework for performance evaluation.

Chapter 6 SYSTEMS ZOO provides a unique collection of dynamic system models covering (mostly) elementary (nonlinear) processes of fundamental importance (examples: exponential and logistic growth, escalation, competition, dependence, predation, overshoot and collapse, population dynamics, production cycle, ecosystems, oscillators, chaos, etc.). Each model is fully documented; the systems zoo program with 50 models is also provided on the accompanying diskette.

Chapter 7 MATHEMATICAL SYSTEMS ANALYSIS provides a comprehensive survey of the mathematical background of dynamic systems analysis based on (vector) state space analysis.

The simulation software SIMPAS for DOS computers offers many interactive features, good (textual, tabular, graphic) documentation of models and simulation results, as well as additional features for the study of parameter sensitivities and global system behavior. The user can generate his/her own stand-alone simulation model with sophisticated features by inserting the model equations into SIMPAS and compiling under TurboPascal.

This book, the SIMPAS simulator, and the Systems Zoo have grown out of my research and teaching at Kassel University, the ASEAN Institute of Forest Management and the Forest Research Institute Malaysia (both Kuala Lumpur), the Sabah Forest Department (Sandakan, Borneo), and the Environment and Policy Institute of the East-West Center, Honolulu. I am grateful for the moral support provided by my friends of the International Network of Resource Information Centers (INRIC) during the gestation period of the project, in particular by Donella Meadows and Dennis Meadows. The first draft of the English edition was very thoroughly and critically reviewed by Donella Meadows, and her helpful comments have led to substantial improvements. I am deeply thankful to her, as well as to Ulla Marquardt for processing the text and mathematical formulae, to my wife Rika for providing moral and logistic support, and to our son Kendrik for drawing the system diagrams, and programming substantial improvements of the SIMPAS simulator.

November 1993

Hartmut Bossel

CHAPTER 1
SYSTEMS AND MODELS

1.0 Introduction

Models and simulations have always been an essential part of the human experience: as we get up in the morning, we crank up our mental model of the little world around us, run a few simulations in our mind on how we are going to deal with the problems and people we will meet during the day, try out different approaches, evaluate the likely outcomes, and start the day with a plan. It will not protect us from failures and surprises, but it will have prepared us to deal more effectively with whatever tasks await us.

In this book we shall deal with exactly the same approach—except that the mental model is translated into a computer model, and the simulations of alternative outcomes under different conditions are made on the computer. The main advantage is that the computer can track the multitude of implications of complex relationships and their dynamic consequences much more reliably than the human mind.

Models and simulations of many kinds are tools for dealing with reality: they are as old as humanity itself. Humans have always used mental models to better understand reality, to make plans, to consider different possibilities, to share their ideas with others, to try out changes and alternatives, to develop blueprints for realization of some ideas, or to convince themselves and others that certain ideas cannot be realized.

Even thousands of years ago, buildings, boats, and machines were first tested as small models before being constructed on a large scale. Children's games have always been simulations of the world of grown-ups—using models of people, animals, objects, and vehicles. The model worlds of mythology, legends, and religions, their characters, and the mental simulations they allow, have guided the behavior of generations in all cultures. Pictures and poems, novels and movies, party platforms and constitutions, models and simulations are also an important part of our world of experience. As scientists identify generalizable principles and processes, they construct models which are used to investigate and simulate new possibilities which lead to new technologies and opportunities.

Models range from miniaturized realistic representations of the original to technical drawings to functional diagrams. They may consist of stories, fables, and analogies or be expressed in mathematical formulae or computer programs in which case they also permit the simulation of dynamic behavior. As simulation games, they consist of a set of rules guiding behavior; gaming simulations teach one to cope with new and unusual situations.

With electronics came the possibility of representing complex dynamic systems—such as aircraft or vehicles—by equivalent systems of electronic components using physical analogies between electronic circuits and mechanical systems. The behavior of such systems could be simulated on the analog computer without even constructing and testing the real system. Simulators could then be employed for training operator personnel (for example, pilots and operators of nuclear plants) to operate complex technical equipment—even long before the first prototype went into operation.

The fact that computers can process quickly and precisely any mathematical or logical formulation in arbitrary combinations widened the applicability of modeling and simulation to anything—in any form—that can be formalized and made computable. This development led to new possibilities in almost all domains of human experience: for representing hitherto hardly accessible complex systems, for simulating their dynamics, and for understanding systems and dealing with them better than before.

Recent technological advances suggest that in the future computer simulations may include humans as acting systems to an even greater extent than in the realistic simulators used now for training pilots of advanced aircraft. Using electronic helmets which provide a three-dimensional view of 'virtual reality' and electronic gloves to interact with this simulated world, computer simulations can provide even physical and emotional experience in imaginary worlds which only exist in the bit states of a few computer chips. The future will show whether and in what domains of human experience this complete immersion in simulated worlds will contribute to human development; we shall not deal with this here.

Our task will be to present the procedures and methods for modeling and simulating the dynamic systems that exist all around us. We will be dealing with a common approach suitable for all disciplines: electrical and mechanical engineering, agriculture and forestry, ecology and economics, management science and regional planning. This is possible because the properties and behavior of dynamic systems are determined not by their physical structure and appearance but by their system structure and processes. In fact, when viewed and analyzed with the instruments of systems analysis, identical structures and behaviors are often recognized in superficially very different systems from widely differing fields. In addition to helping in the solution of particular problems, systems analysis, modeling, and simulation therefore also help us in gaining a much more general understanding of dynamic systems and their behavior and of the world around us.

In this chapter, a general overview over systems, models, and the purpose and approach of modeling and simulation will be given.

In Section 1.1, we discuss the reasons for model construction and simulation, the principal approaches to the task, and typical applications in different fields.

In Section 1.2, the possible spectrum of dynamic systems and their characteristics is reviewed, and the steps of the modeling, simulation, and analysis process are outlined in some detail.

In Section 1.3, the fundamental properties of systems, and elementary concepts of system processes and functions are introduced. These very basic concepts underlie all system studies.

In Section 1.4, the fundamental properties of models, their different types and respective advantages and disadvantages are discussed. The modeling process itself is subject to the strict principles of scientific investigation.

This chapter thus lays the groundwork for the concrete modeling, simulation, and systems analysis work in the remainder of the book.

1.1 Tasks of Model Construction and Simulation

1.1.1 Why modeling and simulation?

The lives of individuals, the future development of a region, and even the global future often depend on dynamic systems whose processes and behavior we do not understand sufficiently. The examples range from bridges, aircraft, vehicles, social processes, urban development, population explosion, wars, environmental pollution to global climate change.

Knowing what will happen or might happen under certain circumstances may therefore be a matter of life and death—not just a matter of curiosity. To predict how a dynamic system will respond under certain conditions is often—as we shall see—very difficult even for simple systems. In such cases we are challenged to generate reliable information about behavior with an acceptable effort. A model is sought that would be able to simulate behavior and to provide hints concerning necessary actions to avoid inadmissible or even dangerous developments.

The product we are interested in is therefore a reliable substitute description that can help one understand a real system and its expected behavior.

1.1.2 How can we simulate behavior?

Simulation of behavior can be achieved by two entirely different approaches. The first possibility is to arrive at a **description of behavior** from observations of one or several identical systems, observing how it behaves (output) under different conditions (input). Then one uses convenient mathematical relationships to relate input to output and imitate the behavior of the real system using these relationships which usually have nothing to do with the real processes in the system. The system is not described in all of its details and functions; it is treated as a "black box."

A second approach is to attempt an **explanation of behavior** by modeling the actual processes of the real system. In this case much has to be known about the system itself: Of what parts is it composed? How are they connected? How do they influence each other? Past observed behavior is only of secondary interest; the emphasis is on the description of structure and processes. Using this information, the system behavior can be simulated even for conditions not observed in the past. The system is treated as a "glass box," where all of its elements and processes are visible, or as an "opaque box," where at least the most essential elements and connections are discernible.

The representation of behavior by computer simulations has now superseded almost all other methods that formerly played a role (for example, hydraulic, electrical, or mechanical analogues). The reasons are obvious:

1. Independently of the type of system considered, computer simulations allow the use of a common methodology and generally applicable software programs.
2. The costs of model construction and simulation are usually only a fraction of the costs of real or analog physical models.
3. The time course of dynamic behavior can be significantly shortened or lengthened. Very fast processes of nature can be slowed down and studied in detail in the simulation; very slow processes can be sped up.
4. Dynamics that would lead to the destruction of the real system have no consequences on the computer model: the simulation program can be used

over and over. This allows in particular the detailed investigation of dangerous system developments.

5. There is no risk for the real system. Measurements that may interfere with the processes of the real system are not necessary.

1.1.3 Applications of dynamic simulation models

Simulation models are found to be useful in many fields, and new applications are discovered almost daily. Simulation models now play a very significant role in

1. scientific understanding
2. system development in technology
3. system management
4. development planning.

Scientific understanding: Modeling and computer simulation of a dynamic system may lead to a new understanding that would not follow directly from the original knowledge about the system. For example, a system may show oscillations, collapse, or chaos that would not be deducible from a knowledge of its elements and their individual interconnections alone. An example is the behavior of forests under pollution load. Models of this system show the possibility of sudden collapse even for a small constant stress and invariant system structure—something one could not guess by studying the components in isolation. Other systems may show chaotic (largely unpredictable) behavior that can only be understood by modeling the system and studying its simulated behavior under different conditions. Simulation also has particular significance for testing of scientific hypotheses: it is usually quite easy to formulate competing hypotheses as corresponding program statements, introduce them into a simulation model, and compare the simulation results with observations.

System development in technology: Traditionally, most applications of computer simulations have been in this area, mainly because the systems are usually precisely definable with regard to influence structure and parameters and can be conveniently described by mathematical models using the physical relationships in the system. Applications are: control technology and optimization; vibrations of buildings and machines; simulation of control and stability as a function of the structural, aerodynamic, or hydrodynamic loads on vehicles, aircraft, and ships; simulation of nuclear and chemical processes and of their control; simulation support for hydrological projects; applications of computer-aided design including simulation of motions, oscillatory behavior, or simulated interior views of a system by "walking" or "flying" through it (architecture).

Medical technology and pharmacology also use simulation; for example, for the control of the heart rhythm, the development of dialysis equipment, and the investigation of the decay dynamics in the different body organs of newly developed pharmaceutical compounds. In projects of this kind, computer simulation helps to find promising and safe solutions, to deal with threats and risks, and to eliminate most risks before the design is realized and marketed.

System management: Computer simulation is also used to better manage existing dynamic systems. In this case, parallel simulation of the system to be managed, using current real input data, allows anticipation and timely recognition of necessary control actions and their impacts. This approach is widely used in the chemical industry to control processes in chemical reactors, for example. Another example is agriculture where plant-available nitrogen and soil water change quickly as a result of the fast processes in the soil. Even the experts cannot easily intuit these changes. Computer simulations can help to apply nutrients and irrigation in an optimal manner. Using computer simulation of the growth of a pest population based on random sampling, appropriate countermeasures can be initiated at the right time, even eliminating the need for chemical pesticides. In forestry, silvicultural measures and their long-term consequences can be investigated and compared to develop economically favorable and ecologically sustainable planting, thinning, and harvesting policies. In industrial management, the simulation of production and warehousing policies plays an important role.

Development planning: In studies of long-term social development, the response of different social groups must be accounted for. The behavior of these actors can be included in simulations either by scenario assumptions, by simulation games, or even partially by simulation of behavior itself. Examples are applications in urban development planning, regional planning, and studies of alternative national development paths. The objective of these studies is often to understand the complete spectrum of behavioral possibilities in order to project likely behavior under given conditions, to discover possibilities for intervention and their impacts, and to determine in time chances for alternative developments.

For future global development this large-scale social use of modeling and simulation is of special significance. It will therefore be discussed in more detail.

1.1.4 Modeling and simulation for the study of development paths

The interacting dynamics of regional and global, ecological and social developments (population growth, economic development, resource use, pollution loads, climate change by greenhouse gases, deforestation, etc.) are today largely incomprehensible even for experts; their impacts on the environment and society remain uncertain. This uncertainty carries high actual and potential costs and risks. All possibilities for reducing these uncertainties by better assessment of future development prospects should be thoroughly investigated, especially if the cost of this information search only amounts to a fraction of the cost of failures.

We are not talking about exact forecasts of seemingly unavoidable developments. The development of society in its ecological environment is only partially predictable, at best. Some social developments, however, are determined by societal actors (individual persons, organizations, institutions) whose likely attitudes and actions can be assessed to some degree but will never be exactly predictable. Examples: birthrates, consumer reaction, and political and religious orientation. In other areas (weather and climate, animal and plant populations) development paths may diverge quickly and rapidly if the system is "chaotic." In both cases, predictability does not strictly apply, but the possible behavioral range can be traced out and possible development paths can be identified.

Under these conditions one can attempt to describe the possible development paths of the system and their characteristics and impacts taking into account possible bifurcations of development. With this information, one can attempt to identify possibilities for intervention and to determine their possible impacts. Since the decisions of individual actors and the possibly chaotic jumps in the behavior of individual subsystems remain unpredictable, the description of all possible development paths is not possible. The objective can only be to trace out the breadth of possible development and to identify the most probable "riverbeds" of development corresponding to certain parameter constellations ("scenarios").

Example: Development paths of energy supply and reduction potential of CO_2-emissions for different developments in efficient energy use and regenerative energy sources.

For future CO_2-dynamics and consequent climate changes, reliable assessment of the possible development paths of the energy supply is of great significance. The standard approach—analyzing past trends and extrapolating them into the future (descriptive modeling)—fails miserably because structural changes (changes in attitudes, technologies, environmental regulations, etc.) must be expected which have not been observed in past behavior. Fortunately, structurally valid simulation models (explanatory modeling) can be used to obtain relatively reliable information about possible long-term development since processes like

1. market penetration and saturation
2. product shifts and supply mix
3. introduction of new technologies
4. shifts in the relative shares of competing processes (modal split)
5. lowering of energy consumption per energy service unit

6. introduction of processes of efficient resource use
7. change in consumer awareness
8. development of energy service demand
9. pressures from environmental degradation
10. value change in society
11. sustainability assessments
12. international competition, etc.

can be directly included in systems analysis and model construction. This allows fairly realistic and reliable studies of the effects of certain policies such as carbon taxes or efficiency requirements.

Example: Development paths of resource supply and environmental pollution for different approaches in product design, recycling, re-use of materials, and waste disposal.

As in the energy sector, the impacts of different development options on the supply of material services, in particular on their (scenario-dependent) mutual interactions and impacts, cannot be investigated reliably without structurally valid models. These must be able to address questions such as: Which measures are particularly effective? Which are particularly consumer-friendly and easy to carry out? Which approach produces lower pollution loads? How does it affect the spectrum of compounds generated? Which approach can secure sustainability of supply at high environmental quality? Which responses must be expected from enterprises and unions? What will be the costs? Here also, a structural model provides an integral framework for comprehensive investigations, research, discussion, evaluation, and decision-preparation on the one hand and a tool for the simulation of different, scenario-dependent development paths on the other.

1.2 Dynamic Models and Model Development

1.2.1 Spectrum of dynamic systems and models

Dynamic systems may have (or lack) certain properties which require (or exclude) certain modeling approaches. We cannot deal with all possible approaches in detail and shall therefore focus on one large group of models which has come to be known as "system dynamics" models. It is the most appropriate approach for the types of problems mentioned previously. Our focus will therefore be on time-continuous non-linear deterministic systems with no spatial dependence of variables, i.e. on systems described by ordinary differential equations.

The spectrum of dynamic systems and models is best described using a list of pairs of terms. The terms in the first column are the ones that correspond most closely to the approach discussed in this book.

explanatory	—	descriptive
real parameter	—	parameter fitting
deterministic	—	stochastic
constant parameters	—	time-variant parameters
non-linear	—	linear
time-continuous	—	time-discrete
space-discrete	—	space-continuous
autonomous	—	exogenously driven
numerical	—	non-numerical

Explanatory—descriptive: This important distinction was already discussed above: the explanatory model reproduces the behavior of the real system by similar structure and elements while the descriptive model merely produces similar behavior; the structure and elements of this model may have nothing in common with the original. Our goal is not only to mimic behavior but also to understand the reasons for it. This requires an explanatory model. (These models are also called "process models," "mechanistic models," "real-structure models," or "structural models.")

Real parameter—parameter fitting: In real-structure (explanatory) models, the model structure should correspond closely to the essential (behavior-determining) structure of the real system. This requires using the parameters of the real system which can often be measured directly in the system or which may be available from other investigations. Where this is not the case, parameter fitting becomes necessary. The model parameters are then selected in such a way that the quantitative results of simulations agree with empirical observations of the behavior of the real system. When fitting is unavoidable, parameter fitting of structurally valid (explanatory) models is much preferred over fitting parameters to structurally invalid relationships, as in descriptive models (Richter/Söndgerath, 1990).

Deterministic—stochastic: Random changes of parameters, of influence relationships between system elements, and of influences from the environment are excluded in deterministic models. In stochastic models, such influences are explicitly considered; for example, by stating transition probabilities between system states or random fluctuations of environmental inputs, e.g. weather conditions. Stochastic models produce different results for each individual simulation run. A large number of simulations (e.g. Monte-Carlo simulation) can provide an overview of the statistical distribution of behavior, mean values, and scatter.

Often modeling has to describe the aggregated behavior of a large number of individuals or processes (animal and plant populations, production of a forest or a field as consequence of the photosynthesis of millions of leaves; the aggregated quantities pressure, temperature, and density in thermodynamics; gross national product; traffic accidents; etc.). The stochastics of the individual behaviors and processes can then be replaced by statistical averages describing an aggregate quantity. In this way, the (aggregated) behavior of the real system can often be satisfactorily approximated by a deterministic model. The aggregation actually removes the considerable uncertainty associated with individuals and produces more reliable (aggregated) results.

Constant parameters—time variant parameters: The parameters of a system are quantities which are either constant or functions of time only. (Other quantities are system variables). For a system with constant parameters, structure and relationships do not change with time, and identical behavior is obtained at a later point in time under identical initial conditions, environmental inputs, and elapsed time. The assumption of constant parameters is appropriate in particular for short-term simulations. It cannot be maintained when processes like aging play a role. Organisms are examples of time-variant systems: the behavior of an older person differs from that of a younger person or a child. Time-dependence of parameters can be introduced by simply introducing time-variant parameters.

Non-linear—linear: This mathematical distinction refers to the types of terms found in the formulations for the rates of change of the state variables (which are the central variables of a system, see Ch. 3). In linear systems, state variables can only appear in the first power in these (differential or difference) state equations. Linear systems can be treated by analytical methods while non-linear systems can usually only be dealt with by numerical simulation. Unfortunately, reality is almost always non-linear (while the vast majority of texts written on the subject deal exclusively with linear systems). This restricts the usefulness of linear formulations to approximations that are usually only valid in a very limited range. Moreover, non-linearity introduces a whole bag of often spectacular new phenomena (multiple points of equilibrium with different stability behaviors, different regions of attraction, bifurcations, limit cycles, chaos, catastrophe) that have no counterparts in linear systems. To be open to the requirements of real systems, we allow arbitrary (non-linear) formulations of the state equations.

Time-continuous—time-discrete: The systems of the world we experience are almost all continuous, i.e. they are defined and measurable at any instant. For example, the tree outside my window exists "continuously": I could touch it or take a picture of it at any arbitrary instant or time interval. The states of time-discrete systems, by contrast, are only defined and observable at certain discrete time intervals; for example: the pictures of a horse race projected by a movie projector every 1/20 of a second. Each picture is quite different and "discrete," and it is only our mind which produces an impression of continuous motion. For continuous systems, the methods of calculus can be used where the rates of change of a state variable can be formulated in terms of an infinitesimal change of the state in an infinitesimal time interval. Discrete systems, by contrast, require a calculus of finite differences of states at discrete time intervals. (See Sec. 3.2.2 for an example of each system. In Ch. 7, basic formulae are provided for both types of systems.) Since we wish to stay close to the properties of real systems, we use the time-continuous formulation in specifying models. Computer simulation actually requires a discretization which approximates the continuous result as the time step of computation is made very small.

Space-discrete—space-continuous: Real systems cannot be located at a single point; they spread over a certain region of space. In some cases this spatial distribution does not play any role in the dynamics of a system. For example, the pressure in a closed gas vessel is identical at each point of the vessel. Or, in considering the photoproduction of a leaf canopy, it may be admissible to compute the total production as a function of total leaf mass without considering the production of individual leaves as a function of location and orientation in the canopy.

In other cases the distribution of system quantities in space is of essential importance for the simulation. For example, in the simulation of air flows on aircraft wings or in the atmosphere, of stresses in complex load-bearing shells, or of groundwater flow, the complete spatially and temporally varying field of system variables must be considered. This amounts—mathematically—to a description by partial differential equations. Simulation can be made using the method of finite elements or the method of finite differences. It will be obvious that the computational complexity and the demands on computer time increase enormously with each additional dimension. In this book, we deal exclusively with systems which have no spatial gradients. This means that we only have to deal with ordinary differential equations with time as the independent variable.

Autonomous—exogenously-driven: Systems are embedded in an environment and will normally receive inputs from it to which they will have to respond. But this causes only part of their behavior. Another part is caused by their state variables affecting their own rates of change in feedback loops and hence generating system-characteristic eigendynamics. Systems not subject to exogenous inputs are called "autonomous." Strictly speaking, real systems cannot be completely autonomous in the long run; they have to replenish energy and other supplies eventually. However, system behavior is often dominated by its autonomous response modes and shows little correlation to exogenous inputs. One cannot understand the behavior of a system without knowing its autonomous response which is determined by the specific combination of system elements in the system structure. In simulations, we usually deal with systems with some exogenous inputs, but simulations will usually reveal the dominant role of the autonomous response, the eigendynamics of the system. The many examples in the systems zoo of this book (Ch. 6 and diskette) demonstrate this fact.

Numerical—non-numerical: The term "state," describing the current condition of the system, applies in a very broad sense. It is applicable to measurable and numerically quantifiable quantities (such as weight, volume, population number, etc.), but it also applies to qualitative attributes (such as red, hot, beautiful, etc.). The dynamics of a system may not be determined exclusively by measurable system quantities; they could also be a function of qualities and their changes (for example, traffic dynamics as a function of the yellow-red-green states of a traffic light). (In this case we deal with a system allowing only discrete states.)

Restricting modeling and simulation only to systems where all quantities can be expressed numerically would exclude large classes of dynamic systems that are of considerable practical interest (social systems, ecological systems, behavioral systems in general). It is also inadmissible on scientific grounds to exclude non-quantifiable variables that play an important role in the system (by omitting these variables as "unscientific"). For example, it will hardly be possible to develop a valid simulation of urban development without including such system quantities as "housing quality", "shopping attractiveness," etc. which are qualitative variables having a critical influence on urban development.

The inclusion of non-numerical state variables is mandated by the scientific requirement for completeness. In the past, the inclusion of such quantities in computer models was difficult since only numerical methods were available for computer simulation. Today it is possible with modern methods of computer-assisted knowledge processing and fuzzy systems analysis to include numerical as well as non-numerical components and relationships in the model formulation. These methods owe much to advances in artificial intelligence, fuzzy systems theory, and progress in knowledge processing and object-oriented programming.

1.2.2 Steps in the modeling and simulation process

The complete process of systems analysis from model development to simulation to behavioral analysis and system design proceeds in several stages:

1. development of the model concept
2. development of the simulation model
3. simulation of system behavior
4. policy analysis and system design
5. mathematical systems analysis.

We shall follow this sequence in this book. The connections between the different topics are outlined in Figure 1.1. The modeling task defines the model purpose; both determine the model concept. The model concept is developed into the simulation model which allows the simulation of system behavior. The simulation model is then used to study policy options or even to (re)design the system to meet given criteria. Mathematical systems analysis and the experience with similar systems help in developing and understanding the system and in improving its performance. The topics can be characterized by the concepts of "structure," "state," "behavior," "choice and design," "mathematical systems theory," and "generic systems" which also characterize the contents of the following chapters.

The models which will be developed and studied are characterized by the terms of the previous section: explanatory, real-parameter, deterministic, constant parameters, non-linear, time-continuous, autonomous, and numerical.

The steps of the modeling, simulation, and analysis process will now be outlined in some detail to provide the reader with some idea of what awaits him or her in the subsequent chapters. The steps apply to other types of models as well.

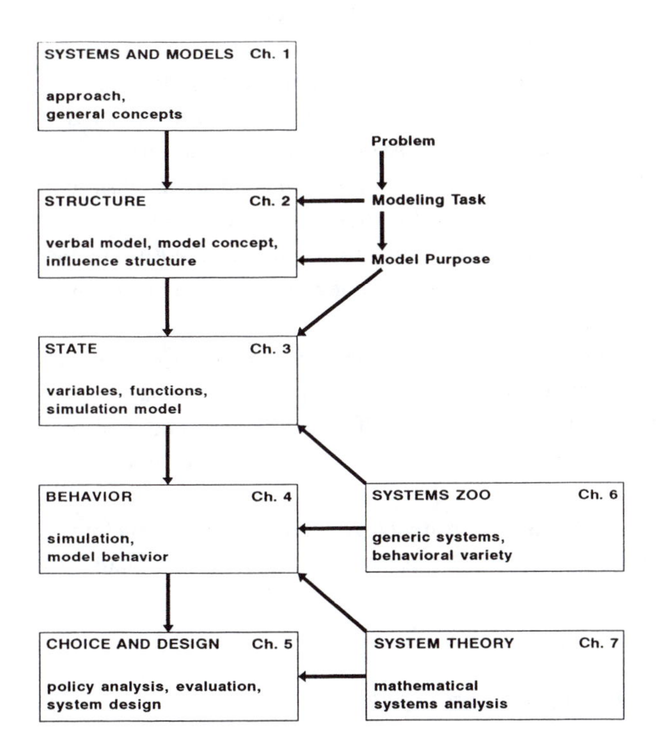

Fig. 1.1: The chapters of this book correspond to the different stages of the modeling and simulation process.

1.2.3 Developing the model concept

All models are sketches or "caricatures" of the real system. The model purpose determines the simplifications and aggregations used. The model purpose has to be clearly defined in response to the task to be solved.

The system definition requires a clear definition of system boundaries, i.e. of its borders with the system environment. These will depend on the model purpose. The expected influences from the system environment must be determined, and their points of influence on the system structure must be specified.

While descriptive modeling depends on an extensive collection of behavioral data, structural (explanatory) modeling concentrates on the definition and recognition of the behaviorally relevant system structure. In this process, the system analyst works closely with experts, scientists, and actual actors in, or operators of, the system in order to develop a model of the relevant system structure. This knowledge about structure and function of the system is formulated in everyday language; the initial model concept is a verbal model.

From this verbal model the influence relationships are extracted and combined in the influence structure, usually also in graphical form as an influence diagram. General system-theoretical knowledge may supplement this information.

This requires the following steps which will be discussed in more detail in Chapter 2. In this phase of the work, the emphasis is on **system structure**.

Definition of problem setting and model purpose: The modeling task must be clearly defined; it is the basis for the definition of the model purpose.

System demarcation and definition of system boundaries: With reference to the model purpose, one has to define what belongs to the system and what belongs to the system environment.

System concept and verbal model: The concept of the system is developed and is formulated as a verbal model in agreement with the system purpose.

Development of the influence structure: The system elements and their influence relationships must be isolated from the verbal model to develop the influence diagram.

Qualitative analysis of the influence structure: The influence structure, in particular its dominant relationships and feedback loops, allows a first qualitative analysis of system behavior.

1.2.4 Developing the simulation model

The influence structure contains only the qualitative description of influences (for example, "A influences B"). In order to obtain a model that can be used for simulation, all relationships must be specified in a way that allows computation (by algebraic or logical operations or functions). In this phase, expert knowledge is decisive. A formalized (mathematical and/or logical) model suitable for simulation evolves from these different steps.

The formalization of simulation models can be done in different ways using very different (general or special) programming languages. The formalization used should be appropriate to the problem setting, the model purpose, and the potential user. Since each modeling task has its own unique requirements, and each programming language has its specific advantages and limitations, a unified approach should be avoided. ("Model banks" are therefore not a good idea.) It is possible that in the future the share of object-oriented programming will increase since it offers a great variety of possibilities for model formulation (qualitative, numeric, logical, etc.) and also high flexibility in model formulation, model extension, data storage, and model documentation.

The following steps have to be performed; they are discussed in more detail in Chapter 3. In this phase of the work, the emphasis is on the description of **system states**.

Dimensional analysis: The elements identified in the influence diagram must be precisely specified in terms of their exact meaning and their units of measurement.

Determining the functional relationships: The functional relationships between the elements must be uniquely specified in terms of their functional dependence; dimensional analysis can be used as a tool for defining these relationships.

Quantification: Using the parameter values of the real system, the influence relationships are quantified.

Developing the simulation diagram: By maintaining the structure of the influence diagram and adding the functional relationships and parameter values, one obtains the simulation diagram as the basis for the simulation program.

Program statements and computable model: The program statements for the simulation model follow from the previously defined and quantified influence relationships. They can be read off directly from the simulation diagram. All influence relationships must be formalized in a computable manner. Default values must be defined for initial values, system parameters, and exogenous influences; they may be changed later in the simulation runs.

Validity test for the model structure: One has to test whether the structure of the real system is correctly represented in the model (and corresponds with the model purpose).

Development of alternative forms of representation: One should find out whether the simulation model could be made more transparent or comprehensible, without loss of validity, by permissible mathematical or structural transformations. In particular, one should check whether modularization is possible and permissible.

Attempting a compact representation: It is often possible to reduce the system structure to a simple elementary structure which simplifies the analysis and allows certain generalizations.

1.2.5 Simulation of system behavior

After programming, initial simulations, and validity tests of the response (behavioral validity), a model is available for routine simulations of alternative policies or development paths. While for historical investigations (for example, as part of the validity tests) the influences from the system environment are known, "scenarios" of expected exogenous inputs have to be assumed in order to compute future paths. These scenarios should be plausible, consistent, and complete. For large models, the development of consistent scenarios may require a significant effort. The spectrum of possible future exogenous influences on the system should be covered by a complete set of scenarios. These are then used in simulations to obtain an overview of the possible behavioral spectrum of the system.

It is an advantage of structural modeling that each step of model development produces new knowledge about the system and its processes, even if in the end there should not be any simulations. Quantification forces the analyst to develop clear concepts about the types and strengths of influences. The definition of consistent scenarios focuses on exogenous influences and their possible development. Each of these steps by itself contributes to better systems understanding.

For simulation of a system, the following steps, discussed in greater detail in Chapter 4, have to be undertaken. In this phase of the work, the emphasis is on **system behavior**.

Choice of simulation software: The formalized simulation model contains *all* model-specific information. Other program parts necessary for the simulation can therefore come from general purpose programs for dynamic simulation. The choice depends on the model type, the type of computer, the programming language used, and the personal preferences of the developer.

Programming the simulation: Depending on the simulation software used, the insertion of the model equations into the program is done as lines of program code using special program statements; as a description of the system elements (blocks) and their structural connections using the computer keyboard; or by constructing a simulation diagram on the screen using corresponding symbols and the mouse.

Selection of integration procedure: Dynamic models of the type discussed here are reducible to systems of ordinary, usually non-linear differential equations which have to be numerically integrated. There are several integration routines available for this purpose.

Run time parameters: The simulation computes the dynamic development as a function of time and therefore requires the definition of time points for the beginning and end of the simulation. The choice of the time step is important for the speed and precision of the simulation.

Initial values: The initial conditions of the state variables have to be set at the beginning of the simulation. These initial values must correspond to those of the real system under the conditions to be investigated. Normally, default values are provided that can be changed by the user in the simulation runs.

System parameters: It is the purpose of simulations to investigate the response of the model system to changes in the system parameters. These system parameters must be chosen before starting the simulation runs (or the default values will be used).

Exogenous influences: In addition, the response of the system to certain prescribed influences from the environment, to historically observed conditions, or to certain developments assumed for the future is of interest. These must also be specified before the start of the simulation. Default values will be provided to describe the most likely developments.

Scenarios: In more complex systems, many parameters and environmental inputs have to be investigated simultaneously. Since the number of possible combinations is large, the parameter sets influencing behavior should be aggregated in coherent, internally consistent, and plausible scenarios. This is of particular significance for the investigation of future perspectives, for technology assessment, and for risk analysis.

Presentation of results: Most simulation software offers several possibilities for presenting results, from a simple table of results to two- and three-dimensional graphics and animated presentations of the system dynamics. One should use effective illustrations that provide the user with a quick and reliable overview of the system dynamics.

State trajectories: Of particular significance is the presentation of the dynamics of the state variables, i.e. the trajectories in state space as a function of the chosen parameters and exogenous inputs. Comparison of the paths for different simulations produces clues about general system behavior (oscillations, points of equilibrium, collapse, chaos) and of the effect of individual parameters.

Sensitivity: The comparison of state trajectories as a function of the variation of sensitive parameters provides clues to the sensitivity of the model, to uncertainties in the formulation, and to necessary changes of critical parameters.

Validity testing: Since structural validity was previously tested during the development of the influence structure of the simulation model, the focus is now on other aspects of validity, i.e. behavioral validity, empirical validity, and application validity. One has to show that the simulated dynamics agree qualitatively and quantitatively with the observed or expected behavior, and that the model results and the knowledge gained correspond to the model purpose.

1.2.6 Performance evaluation, policy choice, and system design

The modeling and simulation task usually does not end with the investigation of behavior of the system. In most cases one would also like to find out how to influence or modify the system in order to guide it onto desired trajectories. There are three basic tasks: computing and evaluating possible development paths, finding "better" policies, and designing a system to show certain behavior in response to certain inputs.

Evaluation of system performance requires the definition of appropriate criteria—evaluation results always depend on the criteria set chosen. For large and complex systems, systematic evaluation is necessary to compare alternative development paths. Mapping of indicators of system state on quality criteria is required also when choosing among policies to manage a system. If the required system performance cannot be achieved by mere choice of parameters or appropriate management policy, system design (or redesign) may lead to the desired results. In this case, the system structure and its elements are chosen to provide the required output for the range of expected inputs.

Chapter 5 deals with the corresponding tasks. In this phase of the work, the emphasis is on **choice** of criteria and policies, and on **system design**.

Criteria for the evaluation of behavior: A precondition for system optimization or simply for an 'improvement' of system behavior is the definition of evaluation criteria. Occasionally it is possible to find a relatively simple criterion (like minimization of cost); but very often system solutions have to satisfy simultaneously a multitude of criteria. In complex decisions often one has to pay attention to criteria hierarchies where individual criteria contributions have to be mapped onto very fundamental criteria of system existence and system development (basic orientors).

Criteria for behavior may refer to the instantaneous system state (for example, constraints defining permissible states); others summarize performance over a time period by integration of certain criteria over time (for example, minimization of fuel consumption in spacecraft attitude control).

Policy choice and system design in search for better solutions: If the state development (both instantaneous and over a longer time period) is evaluated by appropriate criteria, this will provide clues for policy choices and even for system changes that would produce a better solution. Often, a search employing a large number of simulations will lead to acceptable results. This procedure is not very elegant, and one may still overlook far more favorable solutions. Numerical optimization procedures allow more systematic searches, but "optimization" is not appropriate for complex self-organizing systems whose performance cannot easily be evaluated by a simple optimization criterion and whose exogenous influences remain uncertain.

Stabilization of unstable systems by parameter and structural changes: This is an important topic, in particular for technical systems. Control theory has developed an extensive set of tools, in particular for linear systems. Basic concepts from control theory also apply in general systems theory.

1.2.7 Analysis of the model system

With the establishment of model validity and simulations in the parameter range of interest, the task of the system analyst is essentially completed. However, it is often possible and useful to go beyond these steps and attempt to obtain deeper insight by further analysis of the model system. While computer simulation has the advantage that it can deal with complex non-linear systems that are not open to mathematical analysis, mathematical analysis has the advantage that it may lead to solid proofs of certain system properties (e.g. stability) that can only be conjectured from simulations. Mathematical systems analysis becomes important in particular if a generic simulation model has been developed, i.e. a model that preserves its characteristic structure and properties when applied in other contexts. (An example is generic simulation models for tree growth which, after appropriate parameter adjustments, can be used for tundra trees as well as for the trees of the tropical rainforest.)

The points of departure of mathematical systems analysis are the state equations of the system, i.e. the ordinary differential equations for the rates of change of the states derived during the modeling process. From these state equations it is possible to obtain information concerning the points of equilibrium of the system, its attractors, its stability, and possible sudden behavioral changes.

In this area we find the following tasks which will be partially tackled in Chapters 2 through 6, and for which corresponding analytical approaches will be presented in Chapter 7. In this stage of the work, the emphasis is on **mathematical systems analysis**.

Obtaining the state equations: Although the state equations are contained in principle in the model formalization and the simulation statements, the necessary compact presentation for mathematical analysis usually requires some condensation and transformation.

Development of a generic model system: On closer examination, it is often discovered that beyond the specific simulation task, the model equations possess generic validity and can be applied to many related systems. For further mathematical analysis and the derivation of results of wide general applicability, the most general generic form of the state equations should be developed.

Equilibrium points: The system is in equilibrium if there is no net pressure for change, i.e. if the rates of change of each of the individual state variables add up to zero. This condition can be applied to find the equilibrium points of the system. They can be stable or unstable.

Finding further attractors: Higher dimensional non-linear dynamic systems may—in addition to their equilibrium points—possess certain state regions into which the system state is attracted (attractors). Finding the equilibrium points and attractors provides important information about the global system behavior.

Behavior at equilibrium points, and stability: The stability of an equilibrium point determines whether a system state in the neighborhood of this point will have the tendency to move away or to move toward an equilibrium. Information about stability is implicitly contained in the state equations.

Linearization at the equilibrium points: For non-linear systems, the stability analysis in the vicinity of equilibrium points becomes difficult if one has to work with the full non-linear system equations. If one assumes only small disturbances from the equilibrium point, the non-linear system equations may be linearized and can then be investigated using the tools of linear systems analysis.

Properties and behavior of linear systems: Linearization allows application of the tools of linear systems analysis (eigenvalues, behavioral modes, stability) for developing an understanding of the behavior of a non-linear system.

Behavioral change for parameter changes: Since non-linear systems may have more than one equilibrium point, they may switch to qualitatively different behavior (in particular, different stability behavior) if a critical parameter is changed. In dealing with dynamic systems, it is vitally important to know whether, and under what circumstances, such behavioral changes may occur.

1.2.8 Generic structures and the systems zoo

Again and again in model development for different applications, one runs into system structures that are generically identical and therefore show the same behavioral spectrum. A large and important group is the group of linear systems for which a refined apparatus of mathematical analysis exists. By contrast, non-linear systems do not show such generalizable properties. Even small non-linear systems differing only in "minor details" may show completely different behavior.

Certain generic system structures are encountered in many different sectors of reality. With their characteristic behavior, they determine the dynamics of the systems around us and shape dynamic developments in our world. Examples are: exponential growth or decay, logistic growth, predator-prey relationships, competition, addiction, resource exploitation and overshoot, disturbed equilibrium dynamics, dynamic cycles caused by delays, and many other phenomena. The systems analyst and modeler should be familiar with their characteristic structure and the resulting behavior.

In Chapter 6 a large number of relatively simple but fundamentally different systems, their structure, and their characteristic responses are described. The models, model equations, model parameters, and results of standard simulations are fully documented; they are also found as executable programs in the **systems zoo** on the accompanying diskette.

1.3 Fundamental Properties of Systems

So far we have been very pragmatic in our use of terms like "system" and "model" and other system terms. We wanted to get a start on the subject and had to start somewhere—i.e. at a general level of understanding of what these terms mean.

Their traditional scientific training has brought some people to believe that one should not begin to work without having a "clean" set of definitions. Some apply this to systems analysis also. I happen to believe that systems can only be understood by understanding systems, not by reciting definitions. This sentence is not meant to be quite as circular as it sounds, but it brings out the fact that learning about systems means learning from systems in an interactive feedback process. This is done in the following chapters.

Nevertheless, there is much that can be said about systems and models in general which does not require working with them hands-on. In the remainder of this chapter, I shall try to summarize what I find noteworthy. The information should provide valuable background knowledge for the material in the following chapters, but it is not essential for getting started in modeling and simulation. The reader who is eager to get into modeling can skip the remainder of the chapter for now and come back to it later when he/she has developed some basic understanding of systems and models and is ready for some more background information.

1.3.1 What is a system? System identity, integrity, and purpose

A system is a set of interrelated objects (elements, parts) that have certain general properties:

1. It fulfills a certain function, i.e. it can be defined by a **system purpose** recognizable by an observer.
2. It has a characteristic constellation of (essential) **system elements** and an (essential) **system structure** which determines its function, purpose, and identity.
3. It loses its **identity** if its integrity is destroyed. A system is therefore **not divisible**, i.e. the system purpose can no longer be fulfilled if one or several (essential) elements are removed.

Examples:

A chair is a system since it possesses a system purpose and a system structure (seat, backrest, legs with certain relationships between them), and the removal of certain elements (for example, of two legs) would lead to a destruction of system integrity, i.e. the original system purpose can no longer be fulfilled.

A pile of sand is not a system. While a system purpose can be defined (storing sand), the removal of even a very large fraction of the sand would not change its identity as a sand pile.

A drinking glass is not a system. It obviously has a purpose, and its identity as a drinking glass would be destroyed by cutting it in half, but (for the purposes of this analysis) the drinking glass consists of a single element without any relations.

The Strasbourg cathedral is a system since system purpose, elements, and relationships can be recognized, and the removal of certain elements and relationships would lead to a loss of its identity. Organisms, machines, organizations, and the interacting processes of the ecological environment are also systems.

Systems are therefore characterized by elements and an essential influence structure between them, which allows the fulfillment of certain functions defining system purpose and system identity. In modeling and simulation, as we understand it here, the principal task will be to find the essential elements and their functions and to establish the essential influence structure of the system.

Basic system concepts are shown in Figure 1.2. A system exists in a particular system environment, from which it is separated by the system boundary. It receives inputs from the environment, and it provides outputs to the environment. The system elements are connected by a characteristic system structure. Some of the structural connections may be parts of feedback loops.

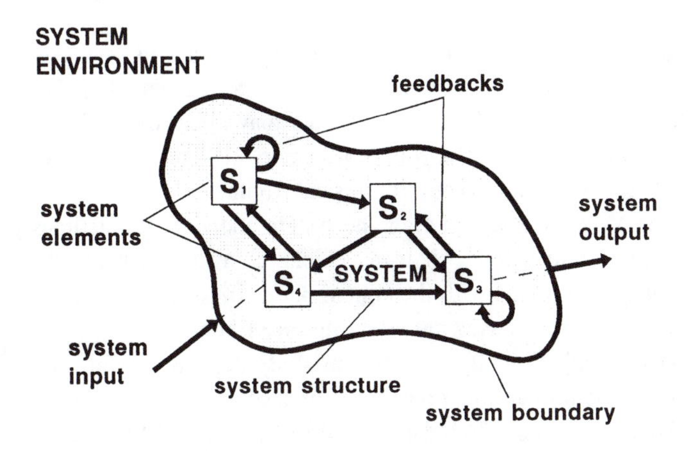

Fig. 1.2: System concepts: elements, structure, feedback, system boundary, system environment, system inputs, and outputs.

1.3.2 Dynamic systems, system behavior, and time period

On closer look, all systems are dynamic systems, even systems which seem rather static (chair, Strasbourg cathedral). They show aging over a long period of time; there are dynamic stresses and strains under load (use of a chair, wind pressure on the cathedral) that may play a role in certain investigations. However, we shall reserve the term "dynamic system" for systems that change their state and therefore show dynamic behavior over the time period of interest. We must not restrict ourselves to directly observable behavior. Relevant state variables are often not observable and are still important for the function of the system.

In practical applications, we are interested in statements about the behavior of the system, i.e. its changes over time. We already suspect that in many cases, knowledge of past system behavior is not sufficient for obtaining reliable statements about future system behavior. Also, in many cases, the internal state of the system itself has to be analyzed. For example, from the behavior of a vehicle (fuel consumption, velocity, emissions) it may not be obvious that the oil level is too low, the motor is overheating, and the car will suddenly stop with a broken piston. Description of a system therefore requires more than is observable from the outside. The essential variables that completely describe the actual system state are termed **state variables**. They play a decisive role in systems analysis, modeling, and simulation.

1.3.3 System boundary, environment, and inputs and outputs

Systems of many different kinds populate our world; each system has its own **system environment**. Elements in this environment may exert an exogenous influence on system development; conversely, system outputs may affect the environment in different ways. For a system study it is necessary to define a **system boundary** which clearly separates the system from its environment.

It is not always easy to know where to draw the system boundary. Sometimes the answer is simple; for example, for a chair, a vehicle, or a human being. In these cases, physical surfaces coincide with the system boundaries. In other cases, in particular for ecological or social systems, a definition of boundaries is more difficult. Since complexity and difficulty of a system study depend crucially on where the system boundary is drawn, this process requires considerable attention. Where for example should the system boundary of a forest be drawn? Should the change of soil water and ground water supply, air humidity, and rainfall caused by the forest be considered in a closed system presentation, or is it possible to consider rainfall as an exogenous influence independent of the forest itself? The criteria for the definition of a system boundary all attempt to find a system surface within which the system can behave in relative autonomy. Boundaries are most usefully drawn:

1. Where the coupling to the system environment is much weaker than the internal couplings in the system (for example, the skin of an organism).
2. Where the existing couplings to the environment are not functionally relevant. Example: to study the bodily functions of an ant, it can be studied as an individual in isolation; to study its social functions in the ant-hill, it must be considered as part of a larger system.
3. Where the inputs from the environment are not significantly affected by the system itself or by feedback of system output (in ecosystems for example: solar radiation, temperature, rainfall).

The definition of the system boundary may depend on the purpose of the system description. If, for example, the objective is to study the effect of a forest on the local climate, then the atmospheric processes such as the recycling of water by evaporation, condensation, and rainfall must be considered.

1.3.4 When is a system observable? Behavior and state

A system acts on its environment by its **behavior variables** (output variables). Only these can be observed in the environment. If behavior does not respond directly to influences from the environment, it must have been caused by changes in the system itself, i.e. changes of the system state with time, to which we refer as state changes. The behavioral variables possibly reflect only a part of the internal processes of the system; often not enough information about the system state is observable from the outside. The actual system state (and the corresponding rates of change of the system state) may therefore be only partially or not at all observable from system output to its environment. For the development of the system, however, the system state, i.e. the totality of its state variables (the state vector) is decisive—even if it should not be observable from the environment.

State variables are defined as those variables that determine completely the state of the system at any instant, including all other system variables derivable from the system state. They are independent of each other, i.e. a state variable cannot be obtained by an algebraic combination of other state variables, and each individual state variable is required for the complete description of the system.

Example: In a regional study, the number of people and the total area of wheat fields are state variables. They are independent of each other: the number of people cannot be computed from the wheat field area and vice-versa. Other system variables *can* be computed from a knowledge of these state variables; for example, the total annual consumption of wheat by the population and the annual production of wheat from these fields (using appropriate parameters, i.e. the normal per capita consumption and normal per hectare production).

State variables are often not uniquely definable; i.e. different quantities in the system can stand for a certain state variable. For example, a state variable is necessary to measure the amount of water in a bathtub. However, for the description of the system dynamics it is irrelevant whether the volume of water (in liters), the mass of the water (in kilograms), the depth of the water (in centimeters), or even the number of water molecules is used. Obviously, the choice of units for the state variable may have practical significance; for example, for purposes of measurement.

Of special importance is the fact that although the state variables are not uniquely determined, a given system can only be described by a certain unique number of state variables. This number is termed the **dimension of the system**. Since in a system model the rate of change of each state variable must be defined, this dimension is exactly equal to the number of differential or difference equations describing the rates of change of the system. If the dimension exceeds the number of state variables used, the system description is incomplete. If the number of state variables exceeds the dimension of the system, the description is redundant and over-determined. For example, a mechanical system composed of a mass and a spring needs two state variables for its description (to represent kinetic and potential energy), a description by only one state variable is not possible; a description by three state variables would also be incorrect.

1.3.5 State variables are memory variables

State variables make up the "memory" of the system. Typically they are "stocks" of energy, resources, money, or individuals that change over time. The new level is determined from the level at the previous point in time where the state was last determined and from the inflows and outflows during the time interval. The state variables therefore reflect the sum of all state changes over a long period of time; they contain the "history" of the system.

In a field, for example, the soil water content, the amount of plant-available nitrogen, and the biomass of the crop are state variables. The biomass of the crop is an observable behavioral quantity. The two others are (normally) not observable, but they are vital for the development of the total system (and the harvest) and can be measured or estimated.

In searching for state variables it turns out to be useful to imagine the system as being suddenly frozen. In this case, all processes, i.e. state changes, cease. Only the contents of reservoirs (memories, stocks, levels, i.e. candidates for state variables) would still be measurable. After a sudden thaw the system would continue its dynamic behavior exactly at the point where it had been frozen with the corresponding levels of the state variables as its new initial conditions. In later chapters we will come back to the definition of suitable state variables.

1.3.6 Elements and structure determine rates of change

There are two fundamental causes of state changes: firstly, influences from the system environment may lead to state changes, and secondly, processes in the system itself may cause state changes (Fig. 1.3). This makes clear that the influence structure of the system itself (which determines how exogenous and internal influences are processed) determines the state changes and therefore the state variables and the system behavior.

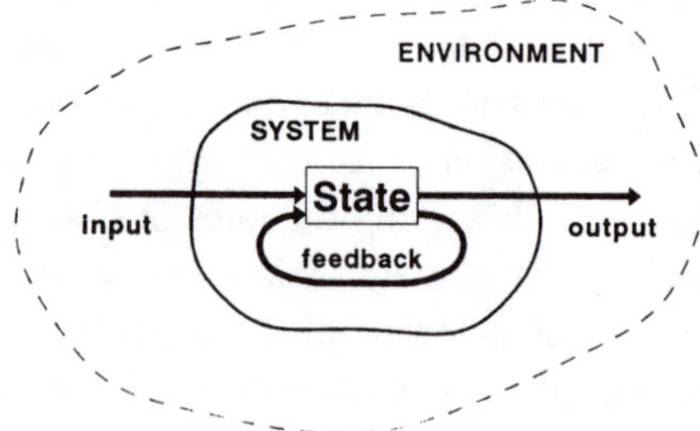

Fig. 1.3: System dynamics have two distinctly different causes: 1. inputs from the environment and 2. feedback effects from the system states.

In the normal definition of the system boundary, we assume that the inputs from the environment are completely independent of the system behavior, i.e. there is no feedback of behavior (of system outputs) on the environment producing changes there. The system behavior then results from the (system-independent) inputs from the system environment and from the feedback action within this system. Both groups of effects are transmitted and transformed by the system structure which must therefore be known for the description of behavior.

Note that identical or almost identical behavior can be produced by different system structures. This means that while system structure defines behavior, the opposite is not true: behavior does not uniquely define a corresponding system structure. We therefore cannot deduce system structure from observed behavior. On the other hand, if only an accurate description of behavior (not of the system) is required, a simpler system may be able to "mimic" the behavior of a complex system. (Much as a parrot mimics certain phrases of its owner while otherwise not having much in common with the person).

1.3.7 Feedback generates system dynamics

If a system consists of a single state variable without feedback, the state will change solely by inflows or outflows which are unaffected by system state. (Example: bathtub with fixed inflow and outflow.) But even in this simple case it is not possible to draw a conclusion about the current system state from observation of current inflows and outflows. For large instantaneous inflow the current water level may be low or high; for strong outflow the water level may be (still) high, etc. A correct description can only be obtained by integration (summation) of inflow and outflow over a certain time period; in addition, the initial value of the state variable has to be known for a certain point in time. This shows again that state variables cannot be computed directly (algebraically) from other known quantities in the system or its environment. This observation removes the legitimacy from many descriptive models.

Everything becomes even more complicated if the system contains feedback loops, i.e. if state variables can influence the rates of change of state variables. Even if the system structure should be simple, the response of such systems can rarely be assessed reliably even by experienced analysts. We therefore have to rely on mathematical analysis (not always possible) and simulation (always possible) to produce information about system behavior.

Take as an example the mutual coupling of two state variables: Variable A affects the rate of change of Variable B while Variable B affects the rate of change of Variable A. Such systems may oscillate with a fixed frequency, but few people would be able to conclude this from looking at the system structure. Systems like this are part of our everyday experience: a weight hanging from a spring is an example, where A is kinetic energy and B is potential energy. Another example comes from warehousing where A is inventory and B is order backlog. Other, relatively simple feedback couplings may lead to "deterministic chaos" where behavior can no longer be predicted with precision (only regions of attraction can be given where the system state will be found).

1.3.8 System behavior: eigendynamics and forced dynamics

Feedbacks in a system can therefore generate characteristic behavior which is determined by system structure and system elements and which cannot be directly related to exogenous influences on the system. This characteristic behavior is sometimes termed eigendynamics. The spring-mass system and the inventory system oscillate even without external excitation after an initial perturbation. Here again the descriptive approach fails since it would have to try to explain an oscillating motion by the system input, when there is *no* oscillating input.

Oscillations generated by the system itself, without external excitation, can appear unexpectedly and may have significant effects on system development. An example is the interplay between long-term investment and capital stock leading to the Kondratieff cycles which have periods of several decades and enormous consequences for national economies.

The response of systems with weak or strong eigendynamics to exogenous inputs may differ widely. The impacts may range from hardly noticeable change to resonance and excitation of self-destructive vibrations. In general, we can only say that exogenous inputs and eigendynamics determine the system behavior. The exact response is a consequence of the elements of the system, of their structural connections, and finally of the exogenous inputs as function of time (in particular, periodic excitation). These responses can be determined by computer simulation without subjecting the real system to time-consuming, costly, and possibly destructive experiments.

1.3.9 System and environment parameters determine response

By definition, exogenous inputs from the environment are not dependent on the dynamics of the system. To study system response, they must be prescribed as functions of time. In the system itself, other constant or time-variant parameters will have an influence. Examples of time-variant parameters are parameters that change with aging of system components. Examples for constant parameters are spring constants, damping parameters, lever lengths, bore and stroke of an engine, capacity limits, etc. The presence or absence of an influence relationship can also be interpreted as a parameter which may be time-dependent or event-dependent.

Parameters may have a decisive influence on system behavior, in particular if they weaken or strengthen important influence relationships. It is to be expected that a system may show different behavior if one or several critical parameters are changed. These critical parameters must be identified by sensitivity studies. On the one hand, sensitive parameters cause the system to react strongly to small perturbations of these parameters; on the other hand, it is just these parameters that can be used to influence the system behavior in desired ways.

1.3.10 Subsystems and modularity

The systems of our technical, social, and ecological environment are rarely simple and usually relatively complex. They almost always consist of relatively autonomous subsystems for which system boundaries can be drawn; they can then be studied in terms of their own behavior. Examples: the human organism consists of a multitude of very specialized organs which can be individually analyzed and for which system boundaries and exogenous inputs can be specified: stomach, intestine, heart, brain, etc. The same is true for technical equipment and machines; for example, a vehicle with its engine, gear-box, suspension, brake system, etc. An ecosystem also is composed of a large number of subsystems which carry out qualitatively very different functions: plants (producers), animals and decomposers (consumers), etc.

For systems studies of these complex systems, it makes sense to adopt this modularity, to define corresponding subsystem boundaries and to study the subsystems and their behavior in response to relevant inputs. If the structures of the subsystems are known and their behavior can be determined, the behavior of the complete system can be investigated and understood as the result of the interaction of the various subsystems.

This analysis of individual subsystems reduces system complexity significantly. In general, the analysis remains transparent and comprehensible. Different specialists can assist in analyzing the subsystems. Since this approach requires a clear definition of interfaces (and hence of the mutual influences between submodels) problem areas and critical parameters are more readily identified and changed, appropriate controls can be designed, and detrimental influences decoupled.

Analogous to the real system, the complete system study focuses on the coupling of the subsystems. If the subsystems are well-understood, it is often possible to find compact descriptions of the processes in the subsystems without representing all the detail of their internal processes. This means that the actual complexity of a subsystem must not necessarily be represented in the description of the total system.

Finally, modularization is almost mandatory for understanding complex systems. Dynamics of systems with more than half a dozen state variables can rarely be understood intuitively; reliable prediction is all but impossible. If, however, the subsystems (each containing only a few state variables) are well-understood, one can usually also understand how their interaction produces the behavior of the total system. This means that models and behavior of even large systems can still be transparent; we are not forced to rely blindly on the results of complex computations.

1.3.11 Superior systems and hierarchies

Modularity is particularly significant for system behavior if subsystems are part of a hierarchy, i.e. if superior and subordinate subsystems can be identified. The subsystems of complex systems are often organized in a "hierarchy of responsibility" to assure efficient functioning of the total system. If system states remain within some "normal" range, individual processes will be controlled by the responsible subsystems within the range of their own autonomy. However, if circumstances arise that move the system state out of its normal range, these transgressions must be handled by a superior unit. The superior subsystem then generates a suitable system response. Only if the range of responsibility is again exceeded at this level is an attempt made to solve the problem by actions of another unit even more superior in the hierarchy. Systems may have several levels of hierarchy. The hierarchy also operates in the opposite direction. If one of the superior units initiates a (global) behavioral change, it is the task of the subordinate units to find (locally) appropriate solutions.

Example: In normal operation of a heating system, the room thermostat will regulate the room temperature by turning on a valve which circulates hot water in the radiator. However, as heat is lost (for example, by low outside temperature or an open window), the water temperature in the boiler drops to the point where the preselected temperature setting can no longer be maintained. At this point, the boiler thermostat will start the fuel burner which will supply a temporary heat surplus. If, on the other hand, the heating system cannot supply the necessary heating power because the fuel tank is empty, the next higher system in the hierarchy (a human operator) will be responsible for either obtaining new heating fuel or firing the wood stove.

In systems like these, modularization ensures that complex control and decision functions remain transparent and comprehensible.

1.3.12 System development: control, adaptation, and evolution

Environmental influences partially determine the system behavior as we have seen. The magnitude of their effect on the behavior depends on the influence structure of the system. Sometimes systems can be controlled by controlling the inputs from their environment.

However, the feedbacks in the system itself are usually more important for system control and adaptation of behavior to environmental conditions. Feedback means that the system state influences itself. Behavior changing internal feedbacks are possible on several hierarchical levels in complex systems with different typical response characteristics and time constants (typical response times). These possibilities are also shown in Figure 1.4.

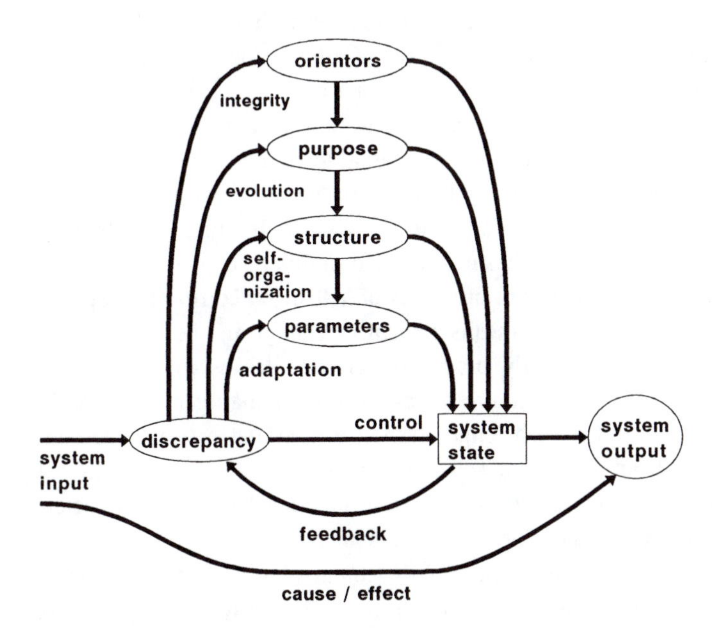

Fig. 1.4: System response can be caused by different processes with very different time constants: cause-effect, feedback, adaptation, self-organization, evolution, maintaining system integrity.

response time	**level**	**response**
immediate	process	cause-effect
short	feedback	control
medium	adaptation	parameter change
long	self-organization	structural change
very long	evolution	change of identity
always	basic orientors	maintaining integrity

The simplest type of system response is the **cause-effect relationship**. It occurs at once as in for example, the flow of an electric current after throwing a switch. It is the only type of system behavior which can legitimately be described by relating the output directly to the input. Unfortunately, it is often assumed that the same simple relationship is also applicable to other types of system response (such as the following), and this erroneous assumption often leads to fundamental mistakes.

On the next higher level we find responses which are generated by **feedback** in the system, involving at least one state variable or delay. Control processes belong to this category. The response time is short, and influence structure and system parameters remain invariant. An example is the thermostat.

On the next higher level we find processes of **adaptation**. In this case the system maintains its basic influence structure, but parameters are adjusted to adapt to the situation, possibly changing the response characteristics in the process. For example, a tree may adapt to the gradual lowering of the groundwater level by growing its roots to greater depth. This constitutes a parameter change (root length and root surface). The fundamental system structure of a tree, in particular, the function of the roots, has not changed in this case.

On the next higher level we find processes of **self-organization** in response to environmental challenges. This means structural change in the system. For example, a company that originally produced kerosene lamps may decide to produce electric light bulbs in the future in response to changing market conditions. Processes of this kind have a longer response time and can only be conducted by systems with an ability for self-organization. Organisms or technical systems rarely or never belong to this category; on the other hand, this characteristic is often found in social systems, organizations, or ecosystems.

A system may also change its identity in the course of an **evolutionary process**. This means that its functional characteristics, and hence its system purpose, change with time. Adaptations of this kind may take place as a result of reproduction and evolution of living organisms and also in the course of evolution of products (for example, the development from horse-drawn carriage to a modern sports car). It is characteristic of this process that the system change coincides with a possibly drastic shift in system identity (change of goal function and of system purpose). An evolutionary example is the development of flying animals (birds) from water-dwelling reptiles.

All of these system responses to challenges from the environment in essence constitute an attempt to maintain **system integrity** (possibly over many generations and over a long time period) even if it means changing system identity, i.e. system purpose. From this observation it is possible to deduce that a system must orient its development with respect to certain basic criteria to assure its long-term existence and development in an often hostile environment. This orientation may be implicit (forced upon the system) or explicit (actively pursued by the system). These basic criteria of behavioral orientation (basic orientors) can be defined by the concepts of existence, effectiveness, freedom of action, security, adaptivity, and regard for others. They are derived in Chapter 5 (Sec. 5.1) by considering the challenges from a system environment characterized by the general properties of an environment: scarce resources, variety, variability, change, and other systems.

Normally, we will only encounter the lower levels of the system response hierarchy in systems analyses and model development. However, it is important to be aware of the spectrum of possibilities since processes such as identity change for maintaining integrity may play a significant role, for example, in social systems and may therefore be important for studies of future developments.

It is particularly important to distinguish between processes that leave the influence structure of the system invariant and processes that change the structure. In control or adaptation processes (with continuous parameter changes) the influence structure is not changed; the behavioral repertoire of the system remains qualitatively the same. If the influence structure is changed, however, the behavioral potential of the system may change fundamentally. In the simplest case this may occur if a latent structural connection in the system, which previously did not play any role, is suddenly activated by given circumstances, or if, conversely, a structural relationship is deactivated or removed. (For example, if an important structural connection breaks, the system response may change completely.)

1.3.13 Actors in the environment: orientation and interaction

By "actors" we mean systems that do not respond to environmental influences by unconditional reflex but which have behavioral options and whose behavior is consciously or unconsciously guided by reference to certain criteria, goals, or principles. Examples are: individuals (consumers!), organizations, and states. In these cases one can infer likely behavior from analyzing the impacts of decision alternatives on satisfaction of the "basic orientors" (or other relevant decision criteria) of the actors. In the investigation of future development paths, this assessment procedure can be used to estimate likely behavioral tendencies of actors and to improve the certainty and validity of forecasts.

In the environment of a given system there are usually other systems with which it has to interact. This means that its behavior will influence other systems and affect their behavior while it is itself influenced by other systems and necessarily responding to them. Also, there are indirect influences from the actions of other systems on the environment which may then affect the system at hand. A classical example is a predator-prey system: the prey population is partly determined by the ecological carrying capacity of a region and its change (which also depends on its utilization by the prey population) while the predator population depends on the prey population and its change.

When systems interact, the total dynamics are determined by the interactions and are not simply the sum of the individual behaviors; for a correct behavioral description, the total system must be considered.

1.3.14 Unpredictability even for deterministic systems

Until recently, the common assumption for deterministic systems (whose behavior is independent of random events and only depends on system state and non-probabilistic influences from the system environment) was that from knowledge of the initial state and the exogenous influences in the intervening time period, any later system state could be computed. Further, it was generally assumed that after a small perturbation of the initial conditions, the system would converge to the same state trajectory as before.

This is true for the majority of deterministic systems, but we are now more aware of the fact that the state trajectories of many deterministic systems may rapidly diverge even for almost identical (initial) conditions and may then end up on totally different state trajectories. This destroys the assumption of predictability of these systems. For such "chaotic" systems, it is only possible to state regions of attraction where the system state might be found later—an exact prediction of the future system state is no longer possible. Examples of chaotic systems are found in studying insect and fish populations, weather patterns, turbulent flows, and flutter vibrations of airplane wings. Chaotic behavior often turns up in unexpected places. However, only a small fraction of systems is found to be chaotic. Moreover, even in these systems, chaos may only appear for a limited range of parameters, or it can be avoided by appropriate control measures.

Chaos is one reason why systems may be unpredictable. A second reason follows from the fact that conscious actors (individuals or organizations) may occasionally act contrary to "rational" principles in unexpected ways. A third cause of uncertainty are random events; for example, a hurricane, an earthquake, an accident, or the random distribution of seeds in a forest.

However, even if a system is impacted by an unpredictable event, its resulting behavior is not arbitrary. System behavior always has its limits and constraints (energy and resource constraints, possible behavioral ranges) that restrict the behavioral spectrum. This is particularly true for the behavior of human actors: much of their behavior is guided by cultural and social norms and values. The range of possible system behaviors can therefore usually be stated even if behavior itself cannot be exactly described.

1.4 Fundamental Properties of Models

1.4.1 Using models: advantages and disadvantages

The simplest method for obtaining predictions of behavior of a system is simply to observe its response under different conditions. This procedure is appropriate in many cases, for example, for chemical experiments or the observation of animals, but in other important areas it is inappropriate or even impossible to carry out. For example, it would take from decades to centuries to study the development of artificial mixed forest ecosystems by observing research plots; the flight dynamics of moonlanders could not be tested on earth; and large-scale experiments concerning the greenhouse gases in the atmosphere are simply not permissible. In cases like these, the only remaining approach is to work with models and simulations instead of experimenting with the real system.

The advantages of using models for obtaining information about behavior are numerous: there is no need to experiment with the original system; there is no threat to the system; results can be obtained quickly; the investigations can cover a much broader range than would be possible with the real system; alternative development paths can be studied and compared; and the costs of the investigations are relatively small, particularly if the model is a computer model which does not require any physical implementation.

The modeling approach also has its disadvantages: the model is obviously not the original system, and there is always an uncertainty whether the model does indeed describe the system behavior correctly in all of its aspects. Thorough validation of the model can remove much of this uncertainty.

1.4.2 Model as a representation of limited validity

A model is always a simplified representation of a particular domain of reality. It is valid, if at all, only for this particular domain and only for a specific purpose. For example, a road map is a model of the major roads which is completely sufficient for purpose of a driver's orientation; it is valid for this particular purpose. Otherwise, the map printed on a piece of paper has almost nothing in common with the geography of the country or the physical surface of the roads.

A model for the simulation of behavior must itself be able to generate dynamic behavior. Therefore it must have the general features of all dynamic systems: its elements must be linked by an influence structure, and it must be able to respond to inputs from the system environment. Often this model is nothing but a mathematical formula from which system behavior can be deduced in response to inputs (simulating exogenous influences).

The model is therefore not the original system. It can only reproduce a limited set of behaviors of the original. This set is determined by the model purpose and the model formulation. A well-functioning model may lead one into assuming that its behavior is representative of that of the real system in all aspects. One should always remember the limited scope of the model, and one should use caution in drawing conclusions about real system behavior from model results. In discussing the model or its results, one should avoid talking about the "system" and "system behavior" (unless it is clear that one means the model system).

1.4.3 Model purpose determines system representation

The original problem setting circumscribes a certain domain of application in which the model can be expected to supply valid answers. This means that the range of possible answers is limited. This range of answers determines the model purpose. Modeling efficiency requires limiting the application domain and the model purpose. A generally valid supermodel can only be constructed with enormous effort and would be inefficient for specific problem settings. Since the possibilities for error also increase as complexity increases, it is found that for answering specific questions the reliability and validity of a supermodel will usually be poor. Therefore it is usually not true that a bigger model is a better model—the best model is the simplest one that fulfills its specific purpose.

The model purpose is therefore the most important prerequisite of model development. The more precise the specification, the more accurate, precise, and compact can be the model formulation. The formulation of the model purpose therefore belongs at the beginning of model development; it requires some care.

The model purpose determines the type and extent of the model formulation in the same way as the problem statement determines the model purpose. This means that in order to deal with different model purposes, the same system must be represented by different models. Since a one-to-one mapping of the system to model is in general impossible (except for simplest cases), focusing on certain aspects required by the model purpose is necessary to develop an efficient and compact representation.

The influence of the model purpose on model construction becomes evident if one considers, for example, the possibilities of simulating a forest: very different models will result depending on whether the purpose is the description of a silvicultural management unit, a natural ecosystem, a system of ecological succession, the photosynthetic production in the course of a day, or the dynamics of forest development under light and nutrient competition.

Because of these multiple possibilities, it is always recommended to specify the model purpose precisely in writing at the beginning of a systems analysis and to remind oneself of the purpose during the process of model development. If this is not done, it is easy to get carried away by a fascinating modeling process which although leading to a beautiful model is in the end not able to provide answers to the original questions.

1.4.4 Alternatives: imitating behavior or modeling structure

There are two fundamentally different possibilities for simulating behavior: imitating behavior or representing the system structure in order to use it to simulate system behavior (Sec. 1.1.2). A third possibility of some practical relevance is a mixture of both approaches.

The first possibility consists of **imitating system behavior** by an arbitrary model that merely has to satisfy the requirement of producing the same behavior. In this case any construction that is able to imitate the behavior of the original is acceptable. This approach also means that the original system is understood as a "black box," i.e. its actual influence structure is of no interest. Since in this case only behavior has to be imitated, observations of past behavior must be available, but the data requirements are merely restricted to these observations.

The second possibility consists of **representing the essential structure** of the original system at least as far as it is necessary for the model purpose. In this case a model of the system, not a model of behavior, is developed. This means that the influence structure of the original system must be known and understood. The system is now recognized as a transparent "glass box." Accordingly, there are entirely different data requirements as opposed to the first case: observations of behavior are not required; instead the system structure with its real parameters must be known, inasmuch as it is relevant to the model purpose.

The third possibility is a **mixture** of both approaches that is often used in practical applications when influence structure and parameters can only be partially determined. In this case one tries to represent the influence structure of a system according to the best available knowledge in order to obtain at least qualitatively correct behavior (behavioral validity). The unknown model parameters are then fitted in such a way that the model behavior also agrees numerically with the observed behavior of the original (empirical validity). In this case the system is understood as a "grey box" or as "opaque." For this kind of model construction, fundamental knowledge about elements and influence relationships in the system must be available as well as behavioral observations.

1.4.5 Imitation of behavior

The **imitation of behavior** without further analysis of the system leads to the modeling of observed behavior as a function of time and certain exogenous inputs (**descriptive model**) (Fig. 1.5). If the response of the system shows a certain regularity and repeatability, one can draw conclusions about future behavior under the same conditions. For example, from observation of the hands of a cuckoo clock, one can draw a conclusion about their position after another 60 minutes without having any knowledge of the clock's internal mechanism. These "time series" observations could be formulated as a mathematical "model" (position as a function of time) and could then be used for "simulation" of "system behavior." This "model" will fail in its conclusions, however, if previously unobserved events affect the behavior of the clock (for example, if the pendulum is stopped or the driving weight reaches the floor).

Descriptive models are applicable only for conditions that are identical to the conditions applying when the observations were made. If one can assume that these conditions will change little or not at all, then the descriptive approach is permissible. However, applying a descriptive model to conditions differing significantly from the conditions of observation is inadmissible.

Forecasts of economic developments in many industrial countries are usually based on the descriptive approach. The economic model is continuously "fitted" (parametrized) anew with recent data and is then used to compute a future trend based on the trends of the recent past. The actual processes of the economic system are not represented in the model. This means that a valid forecast of responses to "novel" events is simply not possible.

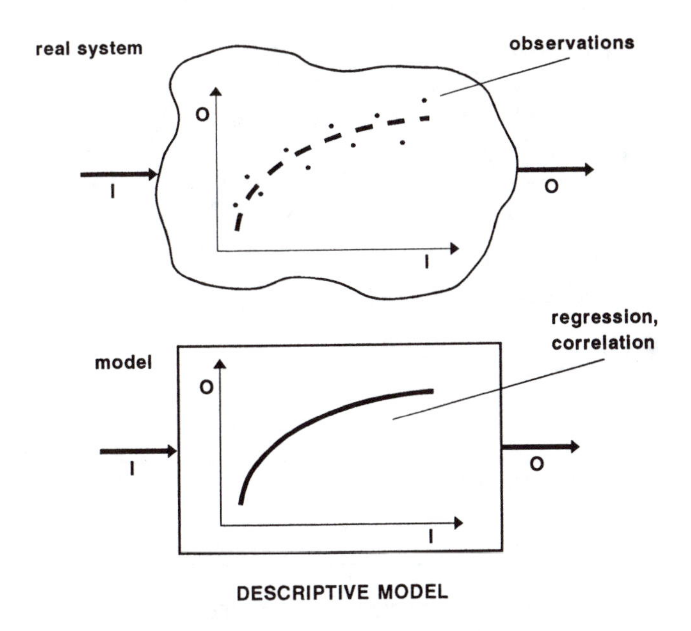

Fig. 1.5: Descriptive models imitate behavior of a system by mimicking the observed input-output relationship by a convenient mathematical relationship. It is usually unrelated to the actual system structure.

1.4.6 Explanation of behavior

The **representation of a system's structure** and of its components and their connections allows one to understand system behavior without any observations of behavior **(explanatory model)** (Fig. 1.6). For example, the time behavior of a cuckoo clock can be readily deduced from its mechanics, the length and weight of its pendulum, its gears, etc. This information can be used to answer many questions about the system: What are the dynamics of the pendulum? Where will the hands be after 60 minutes if the clock is started at the 12 o'clock position? Is the clock too fast or too slow? When and how often will the cuckoo call? In what intervals will the clock have to be rewound? How does the time behavior change if the pendulum weight is changed?

The answers can be given even if the clock has never run at all: a correct structural description allows conclusions concerning future behavior even under conditions never before experienced.

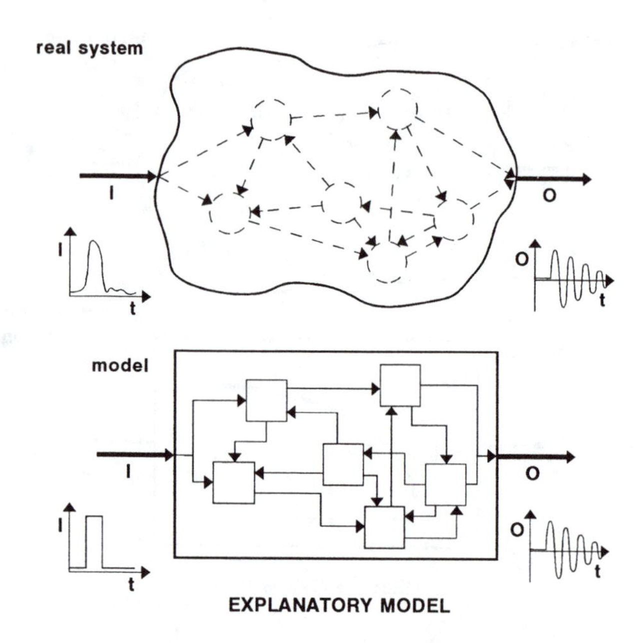

Fig. 1.6: Explanatory models represent the (essential) system structure of the original. They are therefore able to produce dynamics similar to those of the real system, even under circumstances that have not yet been observed in reality.

The results of structural analysis of a system can be expressed by a system of (ordinary) differential or difference equations with time as independent variable. This mathematical model can then be used for a multitude of parameter studies. If the model describes the behaviorally relevant structure correctly, it is "structurally valid." Since the behavior of the system follows from a correct presentation of the structure, the model can also be expected to be "behaviorally valid." If the real parameters (pendulum mass, gear ratios, etc.) have been correctly determined, it should also be "empirically valid," i.e. the simulation of system behavior using this model, starting with the same initial conditions and environmental inputs as the original, should show the same numerical results.

Because of its ability to explain behavior as a function of structural relationships, structural modeling is also referred to as "explanatory" modeling. Its validity is independent of historical observations of behavior. It can even be employed to simulate response to conditions that have not yet occurred, to investigate the whole range of possible behaviors and development paths, and to study and understand conditions and possibilities for system change.

Using the partial differential equations of physics describing the flows and thermodynamics of the atmosphere, weather forecasts, for example, are based on structural, explanatory modeling. Obviously, a weather forecast based on the trend obtained from data of last week's weather would not make much sense. A structurally valid weather model, however, can be fed with the full range of possible weather data in almost arbitrary combinations, in order to produce the likely weather conditions for the next few days. In addition, it can also be used for the simulation of new and extreme weather developments which might occur after a climate shift. The reason weather forecast models cannot make valid forecasts beyond a few days is not that they are structurally inaccurate; it is that weather is a chaotic and not a fully predictable process.

Flight simulators use a relatively detailed mathematical description of the motion of an aircraft in three-dimensional space as a function of aircraft attitude, velocity, and control inputs by the pilot. This mathematical model can be used for realistic simulations not only of normal take-offs and landings, but also of dangerous flight conditions to which one would not subject a real aircraft. In this way it is possible to test new aircraft in the simulator even before the first prototype has left the ground, and to train pilots with no risk and little cost.

1.4.7 Descriptive components in explanatory models

The distinction made here between descriptive models and explanatory models applies generally. For example, the historical development of energy consumption is often used to forecast future energy consumption without ever making an attempt to understand essential determinants of consumption. Foresters use "growth and yield tables," i.e. past observations of tree growth, to forecast future yields. Since they played no role decades ago, environmental pollution and climate change are not reflected in the tables which are becoming increasingly unreliable under the changing environmental conditions.

Explanatory modeling of energy consumption, on the other hand, has to consider the relationships and processes determining energy demand. Energy consumption of private cars, for example, is a function of population development, disposable income, time budget, settlement pattern, market saturation phenomena, efficiency improvements in vehicle engines, modal split of the means of transportation, energy mix and resource availability, environmental regulations, oil price changes, international innovation and competition, etc. For long-term forecasts of forest growth under rapidly changing environmental conditions (pollutants, soil acidification, CO_2-increase, changes of temperature, solar radiation, rainfall, etc.), a comprehensive description of the system structure for the growth-determining ecophysiological processes of nutrient cycling and energy flows in trees and forests becomes necessary: photosynthesis, transpiration, respiration, litter decomposition, mineralization, etc.

Despite the clear theoretical distinction between descriptive and explanatory models, pure explanatory models are seldom found. Explanatory models usually depend on descriptive submodels to describe individual relationships in an aggregated form. An example is the use of (measured) light sensitivity functions of leaf photosynthesis; another is the (measured) relationship between spring compression and spring force for a progressive (non-linear) spring. The full inner structure of the leaf or the molecular structure of the spring is not represented—it doesn't need to be as long as it is unlikely to change over the forecasting period.

Finally, in testing the validity of explanatory models, time series observations must be used to test whether the behavior derived from systems analysis agrees with observed behavior. Both the behavioral and the empirical validity of the model can be tested only in comparison with behavior of the real system.

1.4.8 Different modeling approaches, and different data needs

"Modeling" may describe very different processes, and "models" may differ in type, as we have seen. The differences between descriptive and explanatory models are often overlooked even by model developers. Not only does the process of model development differ, but at the end of the development process one also finds fundamentally different model formulations even for the same system although the results may be more or less identical for a given set of conditions.

For model developers of the descriptive tradition, the emphasis is on fitting large data sets from time series observations to (usually simple) mathematical relationships that are unrelated to the real influence structure of the system. The fitting parameters are estimated with high precision by statistical procedures since small differences have a decisive influence on the results. There is no attempt to understand the processes of the real system or to apply this understanding to the model formulation. Descriptive model development is characterized by extensive and time-consuming data acquisition and parameter estimation.

For the model developer of the explanatory school, the emphasis is on correct recognition of those processes that are decisive for system behavior. The system analyst has to investigate the structure and function of the system together with experts who are familiar with the system and its operation. Structure and function are then described in mathematical (differential) equations whose formulation is determined by the real system (and not by available algorithms for parameter estimation). The formulation can therefore be of arbitrary complexity.

In explanatory models it is important to avoid unnecessarily complex formulations unless they enhance model validity. The parameters in the model equations usually describe the parameters of real system processes that can be measured directly; the parameters are not obtained from the time behavior of the real system (as in descriptive models). The data requirements are therefore limited to data about influence relationships and parameters. In principle, all data can be obtained from the real system, either from knowledge about presence or absence of influence relationships, or from quantitative measurements of functional dependencies. For model construction, time series data of system behavior are not required, but they are necessary for model validation for which an additional independent data set would also be required in the descriptive model.

In the descriptive modeling approach, the data requirement consists of a multitude of quantitative data from observations of system behavior. In the explanatory modeling approach the data requirement is limited to (mostly) qualitative information about the system structure and the numerical values of (usually) a few real parameters. The correct identification of an important feedback link (qualitative information!) is much more relevant for understanding the system's dynamics than another costly time series of its behavior. This means that explanatory models can usually be constructed with a far smaller data acquisition effort. Although requiring comparatively little behavioral data, the validity of explanatory models is generally better since "structural validity" is one of the development goals. However, the effort to understand structure and function will be much greater than for descriptive models.

Structural modeling does not require representing in detail each relationship of the real system. On the contrary, behavioral validity only requires identification of the behaviorally relevant influence structure which often constitutes only a small subset of the total structure. In identifying this structure, one has to work with a good knowledge of the specific system and of systems in general. The result is a compact model representing the smallest possible simulation model still exhibiting the same behavioral spectrum as the original system.

Where large models are unavoidable, they should be constructed in a modular fashion by using compact submodels that are individually tested for validity before they are coupled to the other submodels to make up the complete model.

There is no alternative to structural modeling whenever feedbacks between state variables play a role in the system, complex influence relationships are present, non-linearities (for example, in saturation and limitation processes) determine the behavior, bifurcations of behavior may occur, actors orient their behavior with respect to the system state and their own basic orientors, or responses to novel conditions must be reliably determined. This is particularly true for systems in the techno-economic and ecological domain and their future development.

1.4.9 Understanding future dynamics requires system understanding

If future behavior in response to new challenges is to be assessed, descriptive imitations of historical behavior are of little help. Also, imitation of past behavior captures only a small part of the potential behavioral spectrum of a system.

If, on the other hand, a valid representation of structure and function has been developed in an explanatory model, one can expect reliable results even for conditions which did not apply in the past. The model will be able to respond "realistically" to new challenges, and it can be used with some confidence to investigate possible future development paths.

The distinction is not merely one of data fitting (in the descriptive model) versus structural representation (in the explanatory model). Above all, it reflects the fact that system structures can develop characteristic eigendynamics and a surprising behavioral variety. This variety is captured in the explanatory model, but it is not reproducible in the descriptive model.

If the spectrum of possible future developments is to be investigated, the behavior of the essential actors in the system (consumers, entrepreneurs, international competitors, etc.) must be represented. The inclusion of historically observed behavior to predict future behavior is usually inappropriate. The complete failure of traditional energy forecasts in the seventies should be a clear warning. However, one should not conclude from this that the behavior of actors under novel conditions is completely "open," arbitrary, and unpredictable. On the contrary, actors have to weigh their decisions with respect to their likely impacts. This forces attention on satisfying "basic orientors," as explained in Sections 1.3.12 and 5.1, and restricts the set of behavioral alternatives significantly. Likely developments can be more reliably identified.

1.4.10 Model validity: when can the model stand for the original?

As is true for scientific theories in general, model development also faces the problem that "correctness" of a model cannot be proven; a model cannot be "verified." The fact that a model delivers correct results in a specific setting (i.e. reproduces the behavior of the original) does not constitute proof that it will work correctly in all or even other circumstances. A model (or theory) can only be proven false by showing that reality and simulation differ. This requires applying it under different sets of critical conditions, in a conscious attempt to "falsify" it.

We therefore do not speak of the "correctness" of a model but only of its validity relative to the model purpose. This validity can be established by extensive falsification trials, but it is only true until evidence to the contrary appears. In order to show that the model system can represent the original system well enough for the model purpose, validity must be demonstrated with respect to four different aspects: behavioral validity, structural validity, empirical validity, and application validity.

Behavioral validity: Here one has to show that the model system produces (qualitatively) the same dynamic behavior as the original system under the same initial conditions and exogenous influences as the original system. For example, if the original shows damped oscillations under certain conditions, the model should also show them.

Structural validity: Here one has to show that the influence structure of the model corresponds (within the constraints of the model purpose) to the essential influence structure of the original. For example, the model must have the same number of (essential) state variables, and they must be connected by the same feedback structure as the original.

Empirical validity: Here one has to show that (within the constraints of the model purpose) the numerical or logical results of the model system correspond to the empirical results from the original system under the same conditions or that (if observations are not available) they are at least consistent and plausible. For example, the population development computed by a model must numerically agree with the observations. Behaviorally valid models need not be empirically valid but can be made so by fitting appropriate parameters. This approach may have to be used when the structure can be identified clearly but when crucial system parameters cannot be measured and must be inferred by fitting (for example, the capacity parameter of logistic growth).

Application validity: Here one has to show that the model and its simulation capabilities correspond to the model purpose and the requirements of the model user. For example, in a model developed for silvicultural management, all of the quantities relevant to the forester must appear (stemwood volume, diameter at breast height, height, basal area, etc.), and it must be possible to study their development as a function of thinning and logging and other silvicultural measures over several rotation cycles.

1.4.11 Scientific approach and model development

Model construction always means simplification, aggregation, omission, and abstraction. Modeling is therefore not possible without selections often involving difficult decisions, some of which can only be justified in retrospect when model results can be compared with reality. These decision processes can be largely formalized and systematized, but—as in any decision—they do require evaluations which can only be partially objective. Subjectivity is therefore unavoidable in modeling even if it should be based on objective knowledge as, for example, in the collective experience of a scientific discipline. The choices and simplifications in the model construction process must be substantiated by comprehensive validation tests and by corresponding falsification trials.

Critics of modeling often stress the subjectivity aspect, but modeling's scientific approach does not differ in any way from the scientific approach used and accepted elsewhere. In any scientific investigation, subjective choices must be made. Modeling has to meet the same requirements of the testability and reproducibility of assumptions, hypotheses, statements, and results. Completeness and precision in the use of facts is mandatory. Chains of conclusions have to be complete and correct. Comprehensive falsification tests have to be undertaken for validation. And finally, complete documentation has to be produced so that others can understand all assumptions and derivations and replicate all results.

If modeling and simulation are occasionally accused of being "unscientific," this is not because modeling and simulation are unscientific but because some modelers may have disregarded the principles of scientific work—as in other scientific fields. However, some criticism of modeling comes from the fact that the work has to be interdisciplinary, cutting across established disciplines and schools in order to represent a complex domain of reality. Disciplinary scientists then recognize only sections of their work; they may find that their complex and detailed knowledge has been much aggregated or even simplified, that influences were included which they thought could be neglected, that hypotheses were used which originated in other scientific schools and that, in general, system scientists seem to have a somewhat different view of the world with which they can only partially agree (for example, the emphasis on structural characteristics).

(Structural) modeling is an attempt to look through and beyond the outward appearance and physical shell of systems, at the processes which drive their dynamics and determine their behavior. Traditional science stresses observation and description while systems analysis and structural modeling stress the understanding of processes and their resultant dynamics. Systems analysis and modeling are not possible without the methods and results of traditional science, but it is also true that the traditional sciences stand to benefit much from modeling and simulation and the new insights they offer.

CHAPTER 2
STRUCTURE

2.0 Introduction

The first task in building a structural model of a system is to identify the important system elements and their interconnections. The product of this effort is the influence diagram, an initial qualitative model. Later, the simulation of the system will require a more exact specification and quantification of the components in the influence diagram. However, even this first qualitative picture of the system provides some useful insights.

In this chapter we will be concerned with the development of the influence diagram (or influence graph) and its step-by-step differentiation into a working simulation model.

Model development starts with a problem definition which leads to a more precise definition of the model purpose. From the model purpose follows the definition of the system boundary. The next step is the description of components and interconnections in everyday language, the verbal model. This model leads to the identification of the relevant system elements and of the relationships between them which are then represented pictorially in the influence diagram.

The influence diagram is the first rough sketch of the system structure. It shows how effects (impacts, influences) are passed on in the (model) system from one system element to the next. It is therefore tempting to use this sketch to deduce information about possible system behavior. This can be done by qualitative assessments, numerical investigations, logical deduction, or mathematical analysis of the influence diagram. For example, negative feedback loops in the influence diagram suggest damped, stable motion, whereas positive feedbacks suggest amplification of initial disturbances and instability. However, the reliability of such assessments using the influence diagram alone is limited; usually a more differentiated model description is required.

In the first part of this chapter (Sec. 2.1), we will deal with the development of the influence diagram using as an example a small "global model" describing the essential relationships between population development, level of material consumption, and environmental pollution. The basic rules for developing influence diagrams will be described and applied. Since we are dealing with qualitative relationships which can also be formally processed by computer, a brief introduction to the use of knowledge processing for impact analysis is provided.

In the second part of the chapter (Sec. 2.2), we continue the construction of the "global model"' by building on the system structure developed in Section 2.1, and differentiating it sufficiently to provide (qualitatively) correct descriptions of dynamic processes. In this model building process, the advantages of modularization become obvious since submodels for subsystems can be individually developed and tested before being coupled to each other. For the global model, we employ a dimensionless representation using relative quantities to obtain the fundamental dynamics of "global" development independently of concrete units of measurement. This approach is also meant to demonstrate that dynamic simulation models can be used to represent relationships that are only vaguely quantifiable or only quantifiable in relative quantities, as is often the case in the social sciences or in regional planning.

In the third part of the chapter (Sec. 2.3), we translate the model equations into a primitive TurboPascal program and run several simulations for different choices of the policy parameters. Model validity is discussed, and results are compared with those of a "real" world model.

The results of the chapter will be briefly summarized in Section 2.4. Additional information on methods of qualitative and dynamic analysis of influence diagrams by linear systems analysis can be found in Chapter 7.

2.1 Developing the Influence Diagram

We shall demonstrate the development of the influence diagram using the example of a "global model." This will provide an introduction to the approach of model building and simulation, and it will also show that in many cases, even using highly aggregated descriptions, systems analysis may be capable of describing basic behavioral trends of complex systems that cannot be deduced from other approaches.

2.1.1 A small global model: purpose, verbal model, and relationships

Beginning with the world model of Forrester (1970), numerous "global models" have been developed in an attempt to describe the dynamics of global development with reference to a few central variables. Although these models, by necessity, had to use a considerable number of simplifications to represent often very complex relationships, there can be no doubt that they can provide reliable descriptions of the behavioral trends of some variables of central importance (population, industrial development, environmental pollution) (Meadows et al., 1972; Meadows et al., 1992). For our purposes even these aggregated models are much too complex; we will here have to work with a much more compact description.

System description and model development are very much determined by the modeling purpose. Any systems study must begin with a clear statement of the reason and purpose of the undertaking. During the systems study, one should keep the model purpose in mind; otherwise the outcome of the exercise may be a superb model that is not able to provide answers to the original questions. We therefore begin with the definition of the purpose of the model.

Model purpose: *Using the smallest possible number of variables, the "global model" should be able to provide qualitatively correct information about the dynamic development of population, economic activity, and pollution. If there is a hint of unstable long-term behavior, the model should provide information about possibilities for long-term stabilization. Concrete prescriptions for action are not expected of the model.*

The starting point of any system study is a verbal description of the relevant facts in everyday language. This knowledge will usually not be sufficient to describe a system completely. Almost always, hypotheses and additional information from scientific studies, statistical data, diagrams, interviews, etc. will have to be added later to complement the verbal description of the system. But the verbal model is the starting point.

The **verbal model** for the "global model" could be formulated as follows:

"Worldwide we observe today an increasing stress on natural resources and the natural environment. The reason for this is a constant increase in population and economic activity and, as a consequence, the consumption of the different resources and the dumping of wastes of all kinds in the environment. An important determinant of this resource and environmental load is the consumption of resources and energy per capita. This consumption has the tendency to increase as the pollution stress increases (since resource exploitation becomes more

difficult and more measures of environmental protection are required). As consumption increases, the material standard of living also improves with a corresponding effect on population development. However, pollution and the diminishing natural resource base have feedback effects on the health and life expectancy of the population. Environmental pollution and the strain on the natural resource base lead to growing societal costs. As a result, an increase in societal action can be expected in order to deal with detrimental developments."

What **elements** must be included in a model to provide a reasonably valid description of the system? The number of elements should be kept to a minimum without, however, leaving out quantities that are important to the behavior.

In this verbal model, complex relationships of the real system are presented in a highly aggregated form. We have reduced complexity by limiting ourselves to processes that we consider to be the most relevant and by aggregating many state variables into one representative quantity. For example, the verbal model speaks of "population" and "pollution," not of specific nations or pollutants. This simplification is typical for human information processing which always relies on complexity reduction, pattern formation, and pattern recognition. This process is determined by previous experience and accepted knowledge; it is therefore partly subjective. There is therefore a real possibility that the verbal model is not an adequate description of reality. We should be aware of this, and further model development may reveal deficiencies in the first verbal formulation. To begin, however, we will use the general model elements from the verbal description.

The important elements in the verbal model are:

1. population
2. pollution (environmental and resource stress)
3. (material) consumption per capita
4. societal costs
5. societal action.

In setting up the **influence relationships** two points must be observed:

1. *Only direct influences* are considered.
2. Each influence relationship is *considered in isolation*, as if the remaining part of the system were "frozen" (*ceteris paribus* conditions).

For example, the verbal model says that material throughput (for example, energy required to produce one unit of food) increases if the state of the environment deteriorates (since food may have to be imported from less polluted regions or grown with more chemicals). At the same time, the growing societal costs connected with increasing pollution stress will lead to corresponding societal actions (for example, investments in smokestack scrubbers, energy efficiency, or recycling) that will subsequently reduce the material throughput. For the purposes of the influence analysis it would not be correct to abbreviate the influence chain to an indirect statement of "deterioration of environmental conditions leads to a reduction of material throughput." Rather, each direct influence has to be listed separately ("deterioration of the environment leads to societal action," "societal action leads to energy and materials conservation and reduced material throughput"). The indirect statement follows from chaining the direct influences.

The following direct influence relationships are contained in the text of the verbal model:

1. If population grows, pollution will grow.
2. If pollution grows, consumption per capita will increase.
3. If per capita consumption increases, then pollution will increase.
4. If the per capita consumption increases (leading to a corresponding improvement of material conditions), then the population will increase. Note: at high consumption levels, the opposite effect (population decrease) may take place.
5. If pollution increases, the population (growth) will be reduced.
6. If pollution increases, then societal costs also increase.
7. If societal costs increase, societal action will increase.
8. Societal action can reduce population (growth).
9. Societal action can reduce consumption (growth).

2.1.2 Logical deduction

The verbal model connects statements like these in a web of interconnected logic which can be used to draw conclusions. For example, from the statements above, one can draw the conclusion that at high pollution level, the societal action may reduce per capita consumption in order to reduce the pressure on the environment.

Even at this qualitative stage, a formalized model can be developed. This formal model can be processed by computer to generate logical conclusions implicitly contained in the verbal model using non-numerical knowledge processing (or "artificial intelligence"). This is of some interest if the verbal model contains a large amount of expert knowledge that can no longer be analyzed and processed correctly and consistently by mental processes alone. In order to use computer-aided knowledge processing, the information in the verbal model must be carefully formalized in a "knowledge base" such that the computer can link the statements in a logically correct manner to draw conclusions.

Statements such as those shown above can be represented, for example, in the knowledge processing DEDUC language (cf. Bossel et al., 1989). The basic elements of this language, which is based on the predicate calculus, are objects and object structures, premises, and rules (implications). These elements can be freely defined by the program user. They are connected by using a simple grammar containing a few key words (such as **is**, **if ... then**, **and**, **or**, **not**) which the program is able to recognize and process in a logically correct manner.

Some types of knowledge are best represented by **object structures**. For example,

apple, pear, orange **is** fruit.
wheat, rye, corn **is** grain.
grain, fruit, meat **is** food.
Rika, Derk, Karen, Kendrik **is** person.

This allows statements to be written on the most general level of objects (for example, one statement about "food" instead of many similar statements about all types of food).

Statements describing states of a system are called **premises**. For example,

available(apple).
hungry(Derk).

Such statements can be linked by **implications** or rules in order to produce new statements. For example, the implication

if hungry(person) **and** available(food) **then** eat(person,food).

is a very general statement of what happens when a person is hungry and food is available. If this general-purpose "knowledge base" is prompted by the two premises above, it would produce the **conclusion**

eat(Derk,apple).

To return to our global model, let us first define an **object structure** describing subsequent intervals of time (a so-called "time chain" which allows introducing simple dynamics).

yesterday, today, tomorrow, after-tomorrow **is** time.

Statements about initial conditions are formulated as **premises**, for example,

increase_population(today).

Other knowledge in the verbal model can be represented in the form of **rules** (implications):

If increase_population(time) **or** increase_consumption(time)
then increase_pollution(time).
If increase_pollution(time)
then increase_costs(time), increase_action(time).

```
If increase_action(time)
then decrease_pollution(+time).
```

The (+time) object in the last implication refers to the following time period. For example, if the current value of "time" is "today," then that of "+time" is "tomorrow," according to the time chain above.

Using this knowledge base, the DEDUC computer program generates the following logical **conclusions**:

```
increase_pollution(today).
increase_costs(today).
increase_action(today).
decrease_pollution(tomorrow).
```

This primitive example is only meant to show how computer-assisted knowledge processing operates; it only represents a small part of the verbal model. A useful and reliable non-numerical model usually requires a complex, coherent, and well-designed knowledge base.

Computer-assisted knowledge processing has an advantage if a large number of influences must be considered simultaneously in order to draw logically correct conclusions and if qualitative implications rather than precise numbers are sufficient for understanding (or are the only kind of understanding possible).

This is very often the case in impact assessments. Figure 2.1 shows part of the conclusions from an ecological impact assessment of cassava production in Thailand (from Bossel et al., 1989; p. 62-104). In this case, the knowledge base contains 86 rules about relationships, influences and impacts in the fields of agriculture, ecology, economy, and plant physiology as well as 20 object structures. This knowledge base represents very general knowledge in these fields and was not written specifically for Thailand or cassava production. It could be used (with minor alterations) also for potato production in Idaho, for example. The specialization to a particular problem setting is provided by premises describing the initial situation (about 40 premises in this application). Depending on the specification of the initial conditions, a knowledge base, once developed, can therefore be used for very different impact assessments.

In the particular application shown in Figure 2.1, the knowledge base was used to find out whether computer-based knowledge processing could have foretold the rapid growth of cassava production in Northeastern Thailand and its ecological, economic, and social impacts as a result of a legal loophole in the agricultural policy of the European Community. Although knowledge processing of this type cannot make precise quantitative assessments, the impact assessment correctly predicted the spectrum of impacts to be expected.

transportation(EC,NorthEast,present)
populationGrowth(Thailand,nearFuture)
land_scarce(Thailand,nearFuture)
export_possible(maniok,Thailand,present)
NOT **employment**(NorthEast,nearFuture)
poverty(NorthEast,nearFuture)
NOT **fertilization**(NorthEast,nearFuture)
wish_to_grow(maniok,NorthEast,nearFuture)
NOT **pest**(maniok,Thailand,present)
NOT **harvest_risk**(maniok,Thailand,present)
NOT **spraying**(maniok,NorthEast,nearFuture)
wish_to_export(maniok,NorthEast,EC,nearFuture)
possible_to_sell(maniok,NorthEast,EC,nearFuture)
favorabCondition(maniok,NorthEast,nearFuture)
production(maniok,NorthEast,nearFuture)
income(NorthEast,nearFuture)
NOT **water_retention**(NorthEast,nearFuture)
yield_decrease(maniok,NorthEast,nearFuture)
erosion(NorthEast,nearFuture)
flash_floods(NorthEast,nearFuture)
groundwater_drop(NorthEast,nearFuture)
income_drop(NorthEast,nearFuture)
soil_loss(NorthEast,nearFuture)
silting_of_lakes(NorthEast,nearFuture)
envirDestruction(NorthEast,nearFuture)
flooding(NorthEast,nearFuture)
wells_dry(NorthEast,nearFuture)
debt_burden(NorthEast,nearFuture)
pressureToClear(NorthEast,nearFuture)
nutrient_loss(NorthEast,nearFuture)
forest_clearing(NorthEast,nearFuture)
yield_loss(NorthEast,nearFuture)
interest_burden(NorthEast**,**near Future)
species_loss(NorthEast**,**near Future)
forest_loss(NorthEast**,**near Future)

Fig. 2.1: Excerpt from the conclusions of an ecological impact assessment of cassava production in Thailand using the knowledge processing program DEDUC. The conclusions describe impacts (for example: soil erosion, silting of lakes, debts, illegal land clearing) to be expected in the near future in the Northeast of Thailand (from Bossel et al., 1989).

Computer-assisted knowledge processing can be used to link qualitative information (knowledge or concepts about a system) in a formalized model and to process this information in a logically correct manner. Note that in this case the computer in each specific application constructs a specific "model" from the stored knowledge by linking appropriate statements corresponding to the given initial conditions (premises). Such a knowledge base therefore does not represent a single model but rather the bits and pieces to construct *ad hoc* models as needed. Moreover, the model construction itself is handled by the computer. The program user supplies the problem description and relevant questions, and the program generates all the answers that the knowledge base allows.

Computer-assisted knowledge processing will not be discussed further here. We now turn again to the construction of the influence diagram.

2.1.3 The influence diagram

The influence relationships derived from the verbal model show that the different system quantities influence each other in complex fashion. However, it is rather difficult to visualize all relationships in the system at once by merely considering the verbal model or the individual influence relationships contained in it. An influence diagram can provide the necessary overview. The system elements are represented by the "nodes" of the diagram and marked by their respective names; the influences between them are represented by arrows in the direction of the influence. If the arrows are provided with a plus or minus sign (symbolizing the sense of the influence) and a numerical value ("weight," symbolizing the strength of the influence), we refer to these influence diagrams as "weighted influence graphs."

The influence diagram captures the behaviorally relevant structure of the system. It is therefore the basis for any simulation model. Because of its importance for the success of model development, the influence diagram has to be developed with care and precision. The following rules should be observed:

1. **System elements are represented by the "nodes" of the influence diagram:**

First, the relevant system elements are placed on the diagram as points or "nodes," taking into account the mutual influences between nodes. Variables strongly connected by mutual influences should be drawn next to each other in order to avoid too much crossing-over of influence arrows.

Example: Figure 2.2a shows the location of the five elements "population," "pollution," "consumption," "costs," and "action" for the global model.

2. **Influences are represented by arrows connecting the nodes:**
Next, the relationships between the nodes are marked by arrows. An arrow pointing from A to B has the meaning: Element A influences Element B.

In this book we adhere strictly to this particular meaning in all diagrams, including the simulation diagrams shown later. This has the advantage that the structure of the influence diagram is identical to that of the simulation diagram; the latter can be copied from the former. It should be kept in mind, however, that in some kinds of system diagrams arrows may also have other meanings (e.g.: "Event B follows Event A," "B is sub-ordinate to A," or "Something flows from A to B"). In system dynamics diagrams using DYNAMO or STELLA notation, thick (or double) arrows represent flows while dashed (or thin) arrows represent influences—which is occasionally confusing for beginners. (For example, an outflow in STELLA is represented by a double arrow pointing out of the corresponding state variable box while the influence arrow would have to point into the box since the outflow affects the level.)

Example: In Figure 2.2b, the five elements shown in Figure 2.2a are now connected by (nine) arrows indicating the influences described by statements 1 through 9 of the verbal model (Sec. 2.1.1). For example, Statement No. 1 ("If the population grows, pollution will grow") requires an arrow from "population" to "pollution."

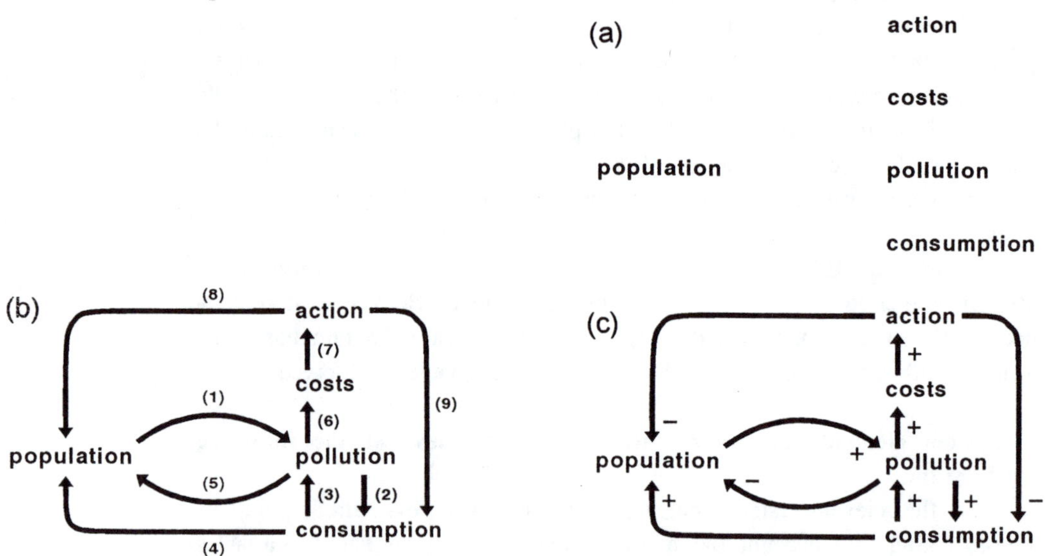

Fig. 2.2: Steps in developing the influence diagram for the simple "global model. (a) Arranging the elements. (b) Drawing the influence arrows. (The numbers refer to the verbal model in Sec. 2.1.1). (c) Establishing the sense of the influences.

3. A minus sign indicates influence in the opposite direction:

The verbal model usually provides information on whether Quantity B will increase or decrease as Quantity A increases. (For example, Statement No. 5 says "If pollution increases, population will be reduced.") The sense of the relationship is indicated in the influence graph using the plus or minus sign. The "+" sign indicates that the receiver quantity (B) will change in the same direction with a change in sender quantity (A) (i.e. as A increases, B increases; as A decreases, B decreases). A change in the opposite sense, however, is characterized by a "–" sign. (As A increases, B decreases; as A decreases, B increases.)

Sometimes only opposite influences are actually marked on the diagram. A missing sign implies an influence in the same direction.

Note that the plus and minus signs do not necessarily mean addition or subtraction. Influences in the same or opposite sense may also be caused by other mathematical formulations (for example, multiplication and division).

Example: Plus and minus signs for the global model relationships are shown in Figure 2.2c. There are three influences (Statements No. 5, 8, 9) which have an opposite effect and therefore carry a minus sign. All other influences are in the same direction. The plus sign need not be shown for these arrows.

4. The influence diagram must contain only direct influences:

In developing the influence diagram, care must be taken to avoid unintended double counting. This may happen if direct and indirect influences are confused.

For example, if it is known that a change in A will cause a change in B, while a change in B will cause a change in C, then the (correct) conclusion "A changes C" may tempt one to include also an influence arrow from A to C (wrong). This would only be permissible if there is an additional, separate, and direct influence of A on C in addition to the one that operates through B.

The influence diagram must contain **only** direct relationships; it then automatically also contains indirect influences. In entering a relationship one should always make sure that it is indeed a direct influence relationship and not an indirect one.

Example: In Figure 2.2c one may be tempted to draw an arrow with a minus sign from "pollution" to "consumption" to account for the fact that too high a pollution level would surely have a damping effect on the consumption level. However, on more careful analysis it becomes obvious that this effect results from the high "societal costs" of pollution leading to "societal action" which reduces the "consumption level." These elements and the corresponding direct influences (Statements No. 6, 7, 9) are already shown in the diagram; only this path is allowed.

5. When considering one influence, all other influences must be "frozen":

In developing the influence diagram, one is often tempted to simultaneously consider other concurrent influences. Not only is this unnecessary, it also leads to errors and is therefore not permissible. Each single influence must be analyzed under *ceteris paribus* conditions ("the other (influences) remain the same"). Viewing each influence in isolation without having to simultaneously consider other influences makes the modeling task much easier. The correct mutual interplay of all influences is found later as a result of the relationships and feedback links of the influence diagram.

Example: When considering what happens to "pollution" as "consumption" increases, typical reasoning often goes as follows: "As consumption increases, pollution will increase" (correct). "At the same time, we must assume that population will increase (for other reasons), causing still higher pollution. However, there will be societal pressure to reduce pollution. The net result of all of this will be a decrease (or increase?) of pollution." One must strictly focus on the one relationship and resist the temptation to consider simultaneously all influences which could play a role. It is the task of the simulation model to determine the net result of simultaneous influences from other components of the system.

6. Define clearly the conditions for which the influence diagram is to be valid:

In the time course of development of a system, new influences may appear (or old ones disappear), and the signs of influences may change. For different stages of development, different influence diagrams may therefore be necessary. It is therefore necessary to define uniquely the initial state for which the influence graph is to be valid. This definition must not be changed during the entire process of developing the influence diagram.

Example: Our small global model will have a somewhat different influence structure, depending on the conditions we have in mind when we develop it. If we consider a medieval rural society, the negative feedbacks on "population" development and "consumption" level caused by a societal response to pollution growth (Statements 6 - 9) would play no role and would not be included: in this case, the population and its pollution are too small to cause environmental concern. Also, demographic observations tell us that the sign of the relationship between "consumption" and "population" becomes negative as a population becomes more affluent. The influence structure and the signs shown apply approximately to a society with a low consumption level and a high pollution level. The diagram would have to be modified to better describe other applications.

7. An odd number of minus signs reverses the feedback sense in a loop:

The signs of the influences in an open or closed influence chain (feedback loop) can be used to determine the sense of the influence from the first to the last node. Minus signs on two arrows of an influence chain, for example, mean a double reversal of the original sense, i.e. a restoration of the original sense. The general rule is: "Minus times minus equals plus."

Example: One of the six (!) feedback loops in our small global model (population—pollution—cost—action—consumption—pollution—population) has a total of four plus signs and two minus signs. The effect of the two minus signs cancels; the overall feedback effect of this loop is in the direction of the original disturbance. By contrast, the sense of an initial disturbance reverses in the loop: population—pollution—population.

8. The feedback sense can be indicated by a plus or minus sign:

A feedback loop as a whole will be positive if the loop contains an even number of minus signs and negative if it contains an odd number of minus signs (Rule No. 7). The feedback sense can be shown in parentheses inside the loop.

The sign of a feedback loop has some practical significance: a negative feedback usually has a stabilizing effect; a positive loop a destabilizing effect.

In a feedback loop with a negative sign, the sense of an initial perturbation is reversed after passage through the loop. In many cases (not always!) this means a damping (i.e. stabilization) of the initial input (disturbance). By contrast, in a loop with a positive sign, the original sense of the perturbation is preserved. This will often (not always!) lead to an amplification (i.e. destabilization) of the original input. In a more careful analysis, the weights (= amplification factors) of the individual influences have to be considered (see Sec. 7.5.5 on pulse dynamics).

Example: The small global model has a total of six feedback loops, three of which are positive and may therefore contribute to unstable behavior, unless their effect is checked by the three negative loops. The overall effect cannot be decided by inspection of the influence graph alone.

The use of these rules is also shown in the example of Figure 2.3 which contains the most relevant influences for climate change. Among the four system quantities: global average temperature, area of arctic ice, albedo (reflected radiation), and heat absorption there are four influence relationships:

1. As temperature increases (decreases), the ice area decreases (increases) (reversal of direction, therefore minus sign).
2. As the ice area increases (decreases), the reflection of radiation (albedo) increases (decreases) (influence in the same direction, therefore plus sign).

3. As the albedo increases (decreases), the absorption of solar radiation decreases (increases) (reversal of direction, therefore minus sign).
4. As absorption of solar radiation increases (decreases), global average temperature increases (decreases) (same direction, therefore plus sign).

The four influences are connected in a feedback loop that has a positive overall sign. We therefore can expect from this loop an amplification of the original perturbation: if the average temperature is increased (decreased) for some reason, this trend will be amplified via the feedback in the direction of a further temperature increase (temperature decrease). Note that the influence diagram together with the signs on the influence arrows is valid for the case of global warming as well as global cooling (ice age).

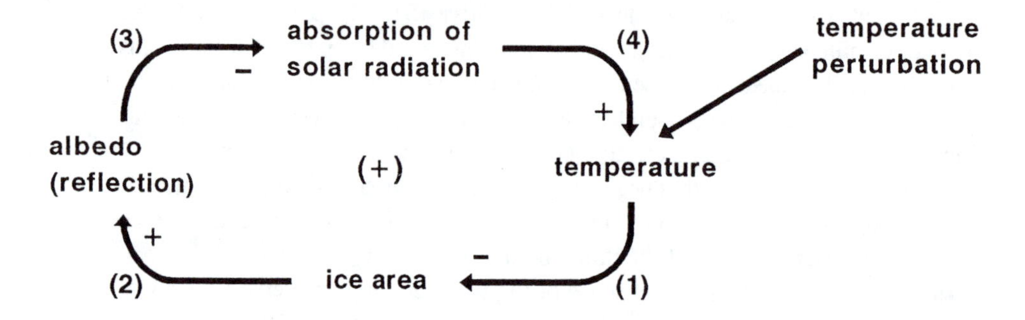

Fig. 2.3: Influence graph of the relationships between solar radiation, atmospheric warming, and ice area.

2.2 Differentiation of the Model Concept

2.2.1 Differentiation of the system components of the global model

Our goal is to develop a simulation model for the influence structure developed in Section 2.1 (Fig. 2.2). This model should provide a more or less correct description of the global dynamics resulting from the interplay of population, pollution, and consumption. Of particular interest are the influences on the overall development of explicit measures to control population and consumption.

A first naive approach might be to use the influence diagram directly in an attempt to compute the dynamics of the corresponding system. After all, each arrow describes the influence of one element on another while its sign (plus or minus) even defines the direction of the influence (same or opposing sense). All we have to do in principle is to begin with an initial state of the elements represented by the different nodes of the influence graph and then follow the propagation of influences from one element to the next as time goes on. Ideally, the system would initially be at rest at some reference state. A perturbation would then be added at one of the elements, and we would then compute how this disturbance "reverberates" in the system. If it eventually dies out, the system would be stable. However, if the initial perturbation is amplified in the feedback loops of the system, this would point to instability of the system.

The hope is that this approach would yield some reliable information about the dynamic behavior of the system. Obviously, in order to do a numerical calculation, we first have to quantify the influences in some way. Secondly, we have to be more specific about how one element influences another element.

Reality teaches us that system elements can react and exert influence on other elements in a myriad of different ways, most of them non-linear (meaning that the response of an element can be a very complex function of the input). This reminds us that there is usually no simple and quick approach to an accurate and reliable computation of system response. Exceptions are linear systems or systems studied under small perturbations from an equilibrium state. In this case, the system equations can be linearized, and the perturbed system reduces to a linear system (see Secs.7.1.11 - 7.1.14). If we stay within the assumption of small perturbations, the analysis of the influence graph might at least produce some indication of the dynamic stability or instability of the system.

This is indeed the case, although the approach cannot substitute for a more accurate system description as we will develop it later in this section. Two methods in particular have been used for investigating dynamic characteristics of influence graphs: pulse dynamics and rate dynamics. Pulse dynamics calculates the propagation and amplification of a perturbation pulse; it amounts to the computation of a discrete linear system. Rate dynamics uses the states of the nodes to calculate rates of change of neighboring nodes; it amounts to the computation of a continuous linear system. Both methods are described in Section 7.5.5.

Simulations of the pulse dynamics and the rate dynamics of the influence system of Figure 2.2 show that lack of negative feedback of sufficient strength, counterbalancing the effects of the positive feedback loops, will lead to unstable development. Stabilization only occurs in a relatively small range of the control parameter determining the strength of societal action dealing with population and consumption growth: if this action parameter is too weak, it cannot prevent unstable divergence; if it is too strong, the system shows unstable oscillations. These results only provide some qualitative hints about system behavior. In particular, we must remember that the computation of influence graph dynamics by these methods is based on the assumption of small perturbations from a reference condition. The results bear no relationship to the full dynamics of the real system—unless it happens to be linear. ("Linearity" means that changes at a node can only come about by additive combinations of changes at neighboring nodes. Example: $x_1 = a\, x_2 + b\, x_3$, where a and b are constants and x_i are the node variables.)

These first approximations of a dynamic simulation model using a direct translation of the influence diagram are far—too far—removed from the real processes; they cannot even provide an approximate description of the real dynamics. This becomes obvious, for example, by considering the link between "pollution" and "population" on the one hand, and "consumption" and "population" on the other; population losses would then only depend on pollution and population gains on consumption. However, population is definitely (and primarily) dependent on the population itself: the more people, the more births and deaths. We therefore have to reconsider and improve the model formulation.

If we drop the assumption of small perturbations and of linear (additive) coupling of perturbations at the nodes, we are forced to consider more accurate representations of the different system components and their functional dependencies. We will then find that the different system quantities have fundamentally different properties. In particular, we have to distinguish:

1. **Input quantities** such as system parameters or exogenous inputs from the system environment which are independent of the system development.
2. **State variables** (levels, stocks, storage variables) which describe the state of a system at each point in time. They cannot be replaced or expressed by other system variables. The state variables form the "coordinates" of the behavior space of a system.
3. **Intermediate (auxiliary) variables** which are directly computable from the momentary values of the state variables, the parameters of the system, and/or the exogenous inputs.

This differentiation between input quantities, state variables, and intermediate variables is of fundamental significance for system representation. In the mathematical system description, a differential or difference equation must be written for each of the state variables. For the intermediate variables we merely need algebraic equations (or logic relationships). The number of state variables therefore gives the dimension of the system which is equal to the number of the differential or difference equations describing it.

The beginner usually finds it difficult to distinguish between state variables and intermediate variables. The following concept often helps: state variables are those system variables that would be measurable or observable if the system were suddenly "frozen," i.e. if its dynamic development were suddenly interrupted. They would have to be recorded if the dynamics were to be continued at a later time as if the interruption had never taken place.

Of course, this can happen only in fairy tales: in the Grimm brothers' "Sleeping Beauty" everyone and everything in the palace fall asleep at the instant the fairy casts her spell. Even the arm of the cook trying to strike the kitchen boy is instantaneously arrested in mid-air. When the prince kisses the princess a hundred years later, every activity instantaneously continues where it stopped a century before—and the kitchen boy finally gets slapped in the face.

This "Sleeping Beauty" thought experiment shows that one also has to consider quantities which are not obvious and not directly observable. In this fairytale example, not only would the position of the hand of the cook preparing to strike the kitchen boy have to be recorded but also the kinetic energy stored in the hand at the instant when the magic spell was cast over the palace.

In our global model (Fig. 2.2) "population" is a state variable. Its value cannot be computed from the momentary values of the other system quantities. It would be measurable after freezing the system, and it would have to be known to continue system development without any breaks at a later point in time. Either the absolute number of people, or a relative population number defined with respect to a given reference number (for example, population in 1970), could be used as measures for this state variable.

"Pollution" (representing environmental and resource stresses) is also a state variable. As a corresponding state measure one could use: the concentration of certain pollutants in the environment, the amount of irreversibly wasted natural resources, the number or species which have become extinct, etc.

If we suddenly freeze our miniworld we will still be able to measure "population" and "pollution," but we would not be able to measure "consumption." However, we would still see all the capital equipment normally consuming the resources: machines, buildings, factories, cars. From their number and normal resource use we could easily compute the approximate per capita level of consumption, however. Hence we decide to introduce a state variable "capital" measuring the share of each person in the equipment and infrastructure causing the resource consumption by their presence and use. Again, this state variable can obviously not be computed from the momentary values of other system quantities.

In contrast to these state variables, the system element "costs" (Fig. 2.2) as a measure of the costs of pollution to society is a direct function of the momentary environmental pollution. Since it can be determined from this quantity, "costs" is definitely not a separate state variable but rather an intermediate variable. The same is true for the system element "action" as a measure of societal activities to counteract the costs of pollution. This quantity is again directly dependent on "costs." Therefore, "action" is also an intermediate quantity.

We have identified a total of three state variables. This means we will have to deal with three differential or difference equations for "population," "pollution," and "capital." These are still embedded in the same structure as before (Fig. 2.2). Identifying the state variables by boxes, we can redraw the system (Fig. 2.4).

In the following, we proceed in a modular fashion, by first developing submodels for the subsystems population, pollution, and capital, and testing these modules individually. Only after we have found a valid formulation for each of the system components will we return to complete the model by adding the influence relationships between population, pollution, and capital as shown in Figure 2.4.

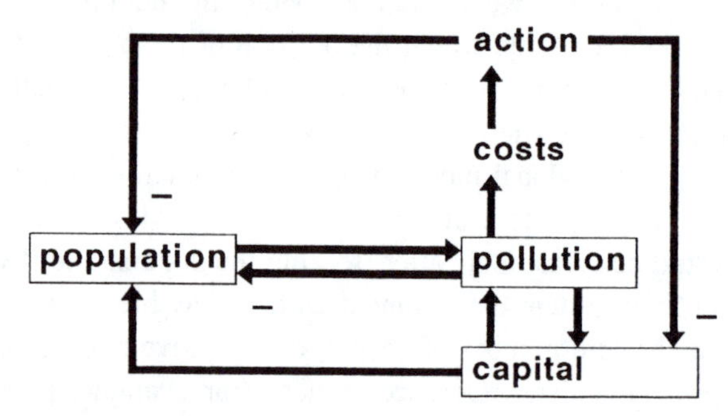

Fig. 2.4: The structure of the primitive global model as a starting point of a differentiated system analysis.

2.2.2 Submodel "population"

It is characteristic for the development of a population (population dynamics) that the annual number of births as well as the annual number of deaths depends on the population itself. Ignoring migration, we can formulate the verbal model:

1. The larger the population, the greater the number of births.
2. The larger the population, the greater the number of deaths.
3. The higher the birthrate, the greater the number of births.
4. The higher the deathrate, the greater the number of deaths.
5. The more births, the larger the population.
6. The more deaths, the smaller the population (minus sign!)

These relationships are shown in the influence diagram for population development (Fig. 2.5).

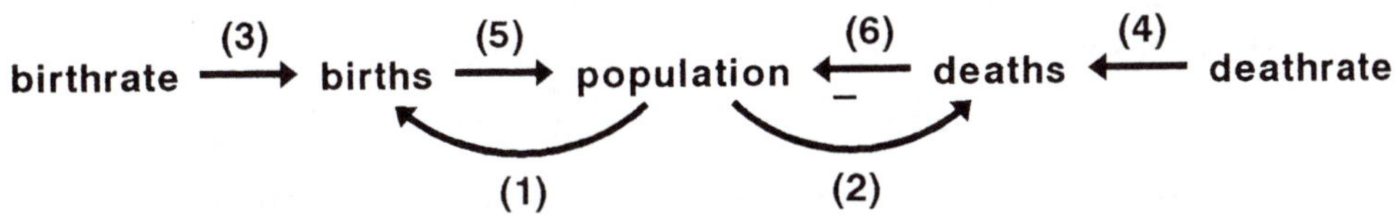

birthrate (3) → births (5) → population (6) ← – deaths (4) ← deathrate
(1) (2)

Fig. 2.5: Influence diagram of the submodel for population development.

On closer look, the annual number of births depends on the age-specific fertility of women and on the number of women in each ageclass. This is too much detail for our purpose. As a first approximation, the annual number of births can be expressed as a proportional dependence on population number:

births = birth_rate * population.

The birthrate is between one percent per year (most industrial countries) and four percent per year (many developing countries).

In a similar way—by neglecting age-dependent mortality—the annual number of deaths can be expressed as a proportional dependence on the population number:

deaths = death_rate * population.

The deathrate in all countries is around one percent. (In population equilibrium it would be equal to (1/life expectancy); for example, 1/80 = 1.25 percent per year.)

Using these two equations, the new population number can be computed after a one year interval from

new_population = old_population + (births – deaths) * 1 year.

More precisely, we should write

population(time) =
= population(time – time_step) + (births – deaths) * time_step

where "time" and "time_step" must be measured in years since we have previously defined "births" as "births per year" and "deaths" as "deaths per year." This provides the "recipe" for computing the population at time (time) from the population at the previous time (time – time_step) and the annual gains by births and losses by deaths. Note that we can repeat the same procedure at the next time step and again at the one following it. In this way we can continuously compute the population development. All we have to know is the initial value of the population at the time when the computation begins and the values for the "birth_rate" and "death_rate" which apply at each time step. (These values may actually change with time).

Actually this recipe is not restricted to population dynamics. It applies to all kinds of state variables. More generally, we can write

state(time) = state(time – time_step) + (state_change/time_unit) * time_step.

Apart from the initial condition for the state, the important thing is obviously to know the (state_change/time_unit), i.e. the rate of change of the state variable at each time interval. It usually changes continuously as a function of time and of state variables, intermediate variables, parameters, or exogenous inputs. Its momentary value is expressed by the infinitesimal change d(state) of the state variable in an infinitesimal time interval dt, or

state_change/time_unit = d(state)/dt = rate.

Since "rate" can be computed from current values of the system quantities at each point in time, the fundamental equation for the computation of state development can now be written as

state(time) = state(time – time_step) + rate(time – time_step) * time_step. (2.1)

Returning to population dynamics, we can now collect the essential equations for computing population development starting with an initial population value ($population_{init}$).

Rate of change of state variable:

d(population)/dt = population_rate = births – deaths (2.2)

where

births = birth_rate * population (2.3)
deaths = death_rate * population. (2.4)

Computation of population level:

population(time) =
= population(time–time_step) + population_rate*time_step. (2.5)

We have now developed four equations for "population_rate," "births," "deaths," and "population" which can be used to compute the population development for a given initial condition for "population." They represent different types of equations, and we therefore introduce corresponding distinctions in the simulation diagram that is developed from the influence diagram. Figure 2.6 summarizes the symbols and notation we shall use for simulation diagrams.

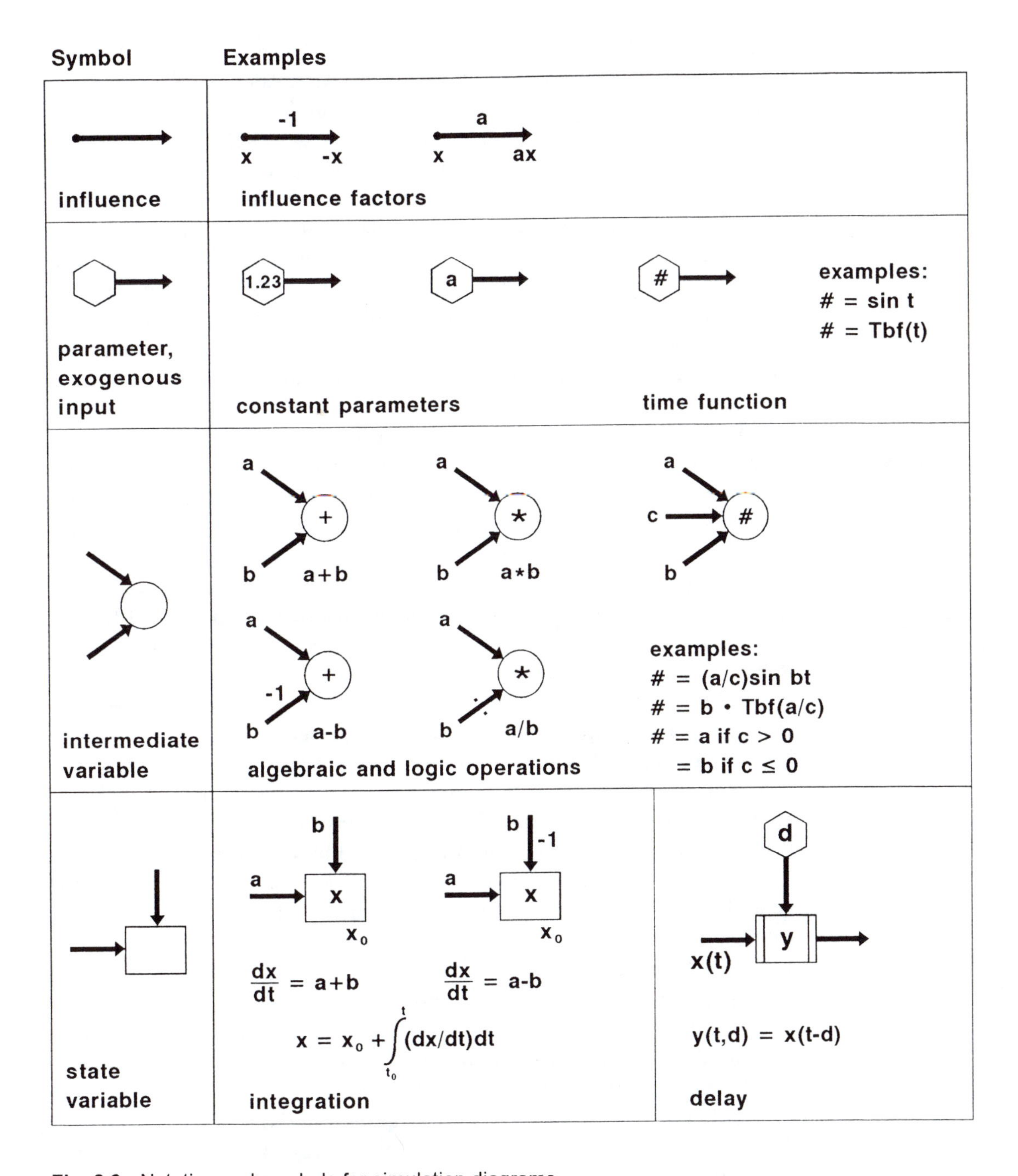

Fig. 2.6: Notation and symbols for simulation diagrams.

Equations 2.2 - 2.4 are algebraic. In algebraic equations one quantity is computed from algebraic combinations of other (simultaneously defined) quantities. For this type of relationship between system elements we introduce the **circle** as characteristic symbol. All influence arrows pointing into the circle have to be connected by the algebraic operation specified for this circle. Simple algebraic operations (addition and subtraction, multiplication and division) can be indicated by corresponding symbols (+, *) in the circle. More complex algebraic or logic relationships must be documented in the simulation diagram, if possible.

In order to compute Equations 2.3 and 2.4, numerical values for "birth_rate" and "death_rate" must be provided. Such numbers are called parameters. They are usually constant, but they may also be prescribed functions of time. In any case, they do not depend on any of the system variables, hence they cannot have any influence arrows pointing into them. For these input quantities, we use the **hexagon** as characteristic symbol.

Finally, the computation of the state variable "population" by Equation 2.5 is again of a different type. It is the numerical integration of a state variable, for which we use the **rectangle** as characteristic symbol. Such boxes also symbolize containers, the contents of which (also called "levels" or "stocks") gradually change in response to inflow and outflow rates.

Equation 2.1 represents the numerical integration method of Euler and Cauchy. We shall use it again and again in the following. With this general integration formula in the background (or another more exact procedure like the Runge-Kutta method, cf. 7.1.6), the specification of the rate of change: rate = d(state)/dt (i.e. of the differential equation; here Eq. 2.2) and an initial value of the state variable suffice to completely define the model.

Finally, we adopt signed weights of influence relationships: multiplicative **factors** and their signs are written directly next to the influence arrows. Where an arrow is drawn without a weight, this always means a weight of +1, i.e. no change in the original magnitude and sign. A weight of –1 means a reversal of sign.

Using these symbol definitions, we can now develop from the influence diagram (Fig. 2.5) the corresponding simulation diagram (Fig. 2.7). By comparing with Equations 2.3 - 2.5 we note the following general results:

1. Each block (hexagon, circle, or box) corresponds to a system equation or parameter specification and vice-versa.
2. Each box corresponds to (the integration of) a state variable or the corresponding differential equation for the state variable.
3. Each circle corresponds to an algebraic equation.
4. Each hexagon corresponds to an input quantity or fixed parameter.
5. Different rate components can be indicated by separate rate arrows pointing into a state "box"; it is not necessary to combine them first in one rate block.

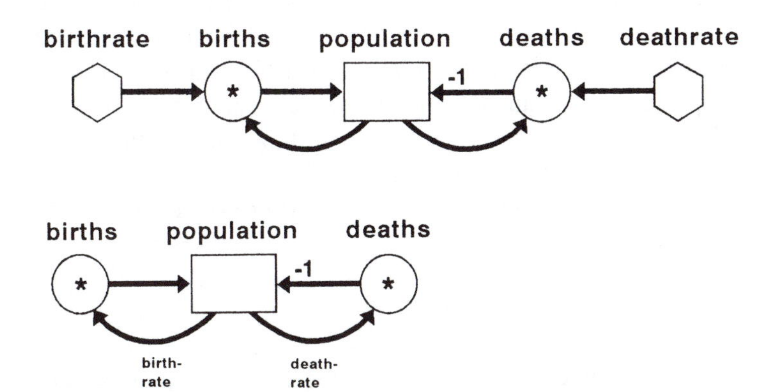

Fig. 2.7: Simulation diagram of the submodel for population development using two equivalent presentations: constant factors can be entered as multipliers ("weights") on the proper arrows.

There is obviously a direct correspondence between the model equations on the one hand and the simulation diagram on the other: the simulation diagram was developed using the equations, and the equations can be written down if the simulation diagram is available (and the algebraic relationship for each intermediate variable [circle] is specified). The simulation program can therefore be developed from either the model equations or the simulation diagram. This means that—if direct development of the simulation diagram from the influence diagram is used—a mathematical formulation is not required for developing a simulation program. This should be a matter of relief to many model developers.

We have now formulated the submodel for population development (Eq. 2.2 - 2.5) and represented it as a simulation diagram (Fig. 2.7). We now have to check whether this formulation produces an acceptable result and can be employed for the purposes of our global model.

Using Equations 2.2 - 2.5 we could now write a small computer program (a programmable pocket computer suffices) and compute results for different values of the parameters "birth_rate" and "death_rate," and the initial condition "population_{init}." However, in this case an analytical solution can be written down:

$$\text{population}_t = \text{population}_{init} \cdot e^{(\text{birthrate-deathrate})t}$$

This can be easily checked by differentiating population with respect to time: d(population)/dt and substituting the expression for population_t. We therefore obtain, depending on the sign of (birthrate-deathrate), exponential growth or exponential decay.

As noted before, our intention is to find the fundamental dynamics of the model system. Therefore, we can employ relative (dimensionless) state variables. We therefore select a reference state of "1" and consider the changes from this initial state. If the population eventually doubles, the state variable would then have the value "2." We also have to choose values for birthrate and deathrate. As noted above, the deathrate is around 0.01 in most countries (one percent per year), while a birthrate typical for most countries is around 0.03 (three percent per year):

$$
\begin{array}{lll}
\text{population}_{\text{init}} & = 1 & [-] \\
\text{birthrate} & = 0.03 & [1/\text{year}] \\
\text{deathrate} & = 0.01 & [1/\text{year}].
\end{array}
\tag{2.6}
$$

The rate of change of a state variable always has the dimension [unit of state variable / time unit]. Since in our case population is dimensionless, "births" and "deaths" have dimension [1/year], the same as "birth_rate" and "death_rate."

Before we integrate a submodel into a larger model, we should always test its behavior under reasonable assumptions to make sure that it functions as intended. In this case, there is not much to test: we will obtain exponential growth or decay, depending on the sign of "population_rate." Developments for a deathrate of 0.01 and different birthrates are shown in Figure 2.8. For a birthrate of 0.03, the population will double in about 35 years and increase by a factor of 7.5 in 100 years.

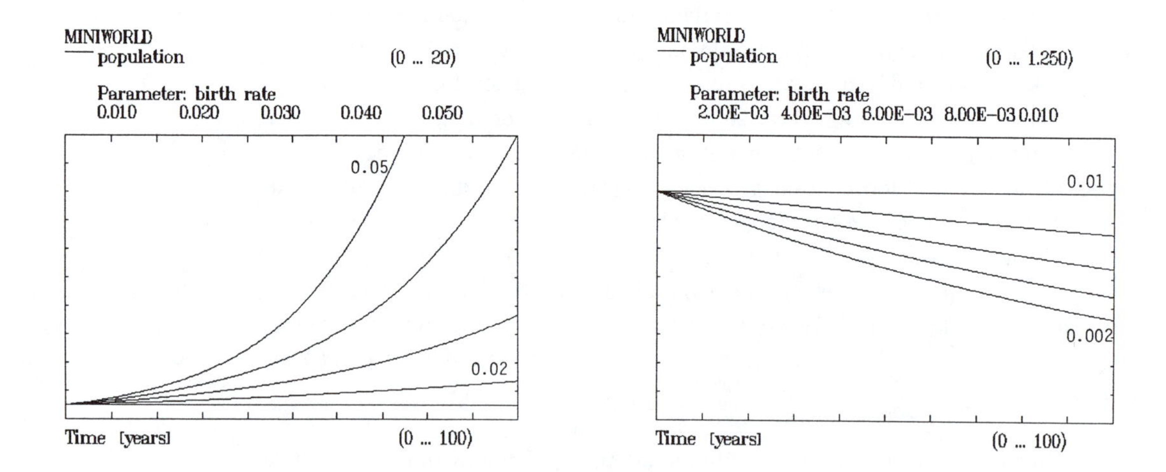

Fig. 2.8: Population development for positive and negative net growth rate = birthrate – deathrate. (death_rate = 0.01.)

2.2.3 Submodel "pollution"

Most environmental pollution can be broken down in the course of time by ecological processes in the soil, water, and atmosphere. Important exceptions are chemical compounds created by man with which organisms have no evolutionary experience and which are therefore not (or almost not) bio-degradable. Degradable material is broken down and absorbed at a certain rate—i.e. by a certain percentage per time unit—as long as the ecosystem has not been overloaded by it. As an exception, the decay rate of radioactive material cannot be influenced in any way and is therefore independent of environmental conditions.

Ecological absorption processes always have a capacity limitation that is given by the limits of the required ecological conditions (for example, nutrient, light, and water limitations). If overloaded, the absorption process can at most operate at this upper limit; very often, however, a collapse of the system, and a consequent significant deterioration of pollutant absorption, is possible. For example, overloading an ecosystem with pollutants may destroy the very microorganisms on which its pollution absorption capacity depends.

From this general knowledge, we can formulate our verbal model of the pollution sector:

1. As pollution input increases, the pollution level increases.
2. The larger the absorption rate, the greater the absorption of pollution.
3. The more pollution, the higher the absorption.
4. The more absorption of pollution, the lower the pollution level.
5. The better the environmental quality, the greater the absorption.
6. The higher the pollution level, the lower the environmental quality.
7. The lower the absorption threshold, the worse the environmental quality.

This information translates into the influence diagram shown in Figure 2.9. The seven relationships are transformed into seven arrows in the diagram.

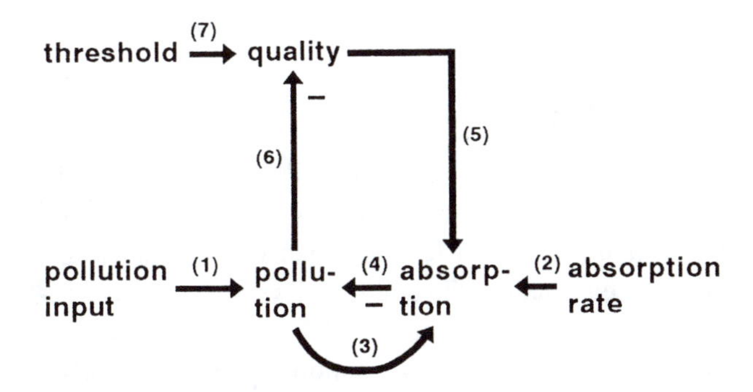

Fig. 2.9: Influence diagram of the submodel for pollution load.

In the submodel for environmental pollution, we therefore have to consider two response modes. If the environmental load is below a critical threshold value, the decay of pollution per time unit will be proportional to the current environmental load (the more pollution, the more absorption). However, if the pollution load exceeds the critical value, then only a (constant) amount corresponding to the capacity limit can decay per unit time. The ratio of pollution level to critical threshold level therefore determines which of the two response modes applies. We assume a continuous inflow of new pollution per time unit; this is an exogenous input to the submodel. (In the full global model, the pollution input will be a function of "population" and "capital," as indicated in Fig. 2.4.)

In both cases, the rate of change of the pollution level is given by

d(pollution)/dt = pollution_input – absorption. (2.7)

In order to model the threshold behavior we must first define a suitable measure for determining whether the critical level has been exceeded. We define an index of environmental quality by

quality = threshold / pollution. (2.8)

The variable "quality" is greater than unity if the pollution level is below the critical threshold. It is less than unity if the pollution level exceeds the critical threshold.

For the subcritical case where the pollution load has not reached the capacity limit, i. e. for the case (quality > 1) we have:

absorption = absorption_rate * pollution.

In the supercritical case, where the pollution load exceeds the capacity limit, i.e. for the case (quality ≤ 1), the absorption rate cannot exceed the value defined by the absorption threshold, it therefore remains at a constant level. If the pollution input rate continues to exceed this constant absorption rate, the pollution level will increase continuously.

absorption = absorption_rate * threshold
= absorption_rate * pollution * quality.

These two cases show different system structure, and will therefore cause different types of behavior. This difference becomes obvious by comparing the simulation diagrams in Figure 2.10 and their characteristic responses. The subcritical case corresponds to exponential adjustment of the pollution level to a level consistent with the constant "pollution_input" (see Model M103 in the systems zoo). The supercritical case leads to a constant increase in pollution level (Model M101 has the same structure).

The alternative model formulations can now be combined into a single submodel for pollution by introducing an appropriate switching function:

```
if quality > 1 then
        absorption = absorption_rate * pollution
else
        absorption = absorption_rate * pollution * quality.          (2.9)
```

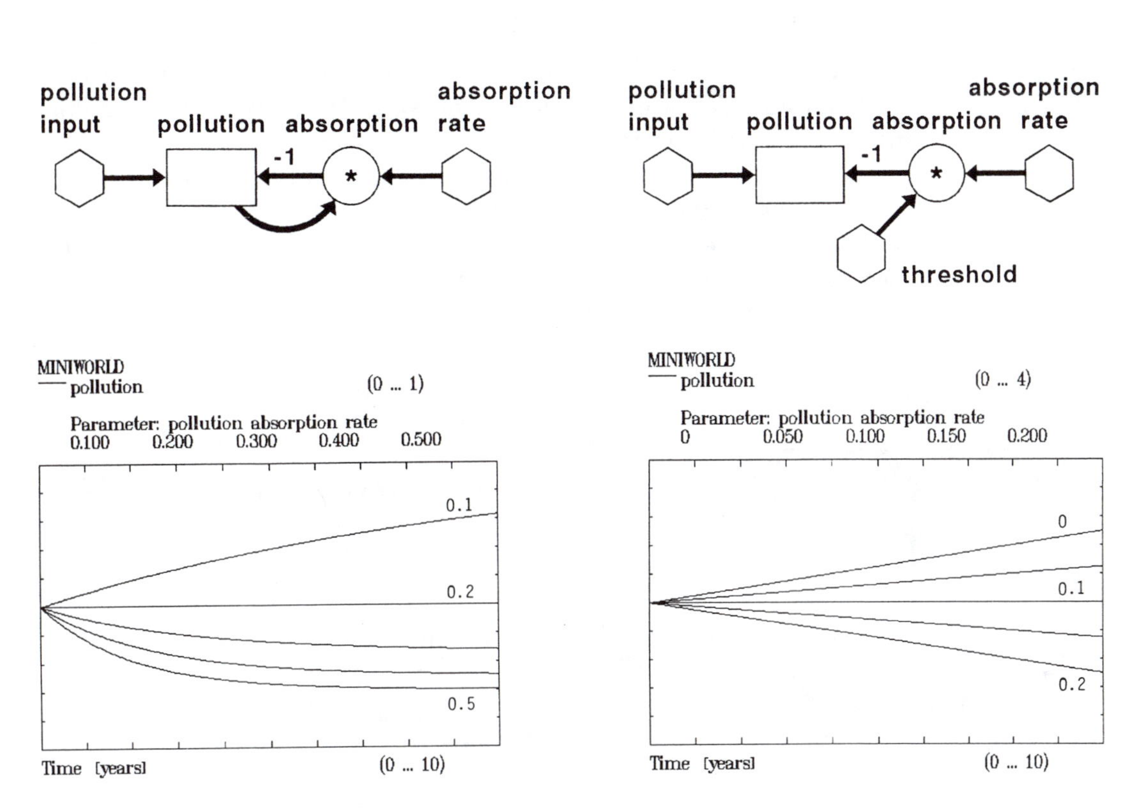

Fig. 2.10: Simulation diagram and behavior of the submodels for the pollution load (pollution_input = 0.1). Left: subcritical pollution load; right: supercritical load.

The simulation diagram corresponding to the Equations 2.7 - 2.9 for the pollution submodel is shown in Figure 2.11. Note the correspondence to the influence diagram (Fig. 2.9): the seven arrows representing the influence structure remain unchanged, while the six elements shown in the influence diagram are now replaced by six different "blocks"—three hexagons for parameters and exogenous inputs, two circles for the intermediate variables "quality" and "absorption," and one rectangle representing the state variable "pollution."

We can now quantify the parameters. Since we want to work with relative state variables, we assume for the absorption capacity limit: threshold = 1. For the absorption rate, the absorption threshold, and the initial value of pollution load, we select the following quantitative values:

$$\begin{aligned} pollution_{init} &= 1 \quad [-] \\ absorption\_rate &= 0.1 \quad [1/year] \\ threshold &= 1 \quad [-]. \end{aligned} \tag{2.10}$$

At this absorption rate, 1/10 of the pollution level is broken down per year; the time constant of environmental pollution is therefore 1/(0.1) = 10 years. (The time constant is the inverse of the specific change rate.)

In order to test this submodel, we subject it to various values of (constant) pollution input. Some response curves are shown in Figure 2.11. They clearly demonstrate the change in behavior as the system switches to a different absorption mode as it becomes overloaded. In the full "global model," the pollution input will vary as a function of population and consumption.

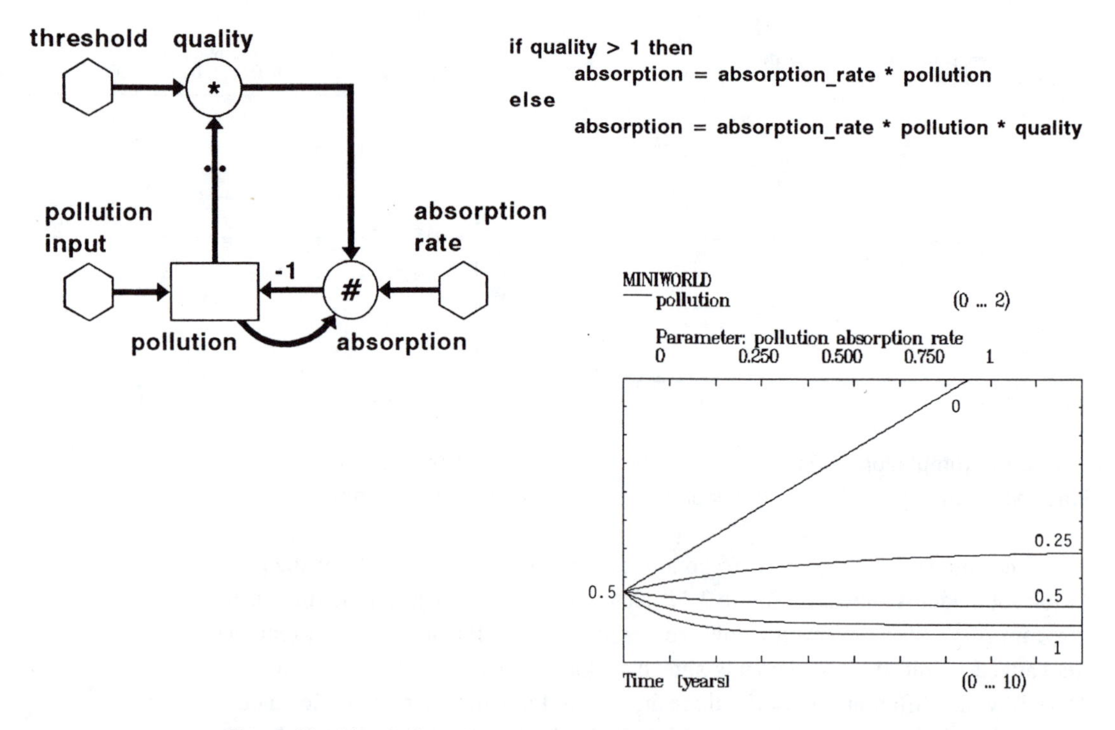

Fig. 2.11: Simulation diagram and response curves of the submodel for pollution with a structural switch for supercritical pollution load (pollution_input = 0.2).

2.2.4 Submodel "consumption"

As explained earlier, resource consumption is mainly a function of the capital stock (infrastructure, machines, vehicles, buildings) requiring energy and resource inputs for operation. Hence consumption can be computed by applying an appropriate multiplier to the capital stock. The development of the capital stock itself is mostly "autocatalytical," i.e. there exists a positive feedback between the capital stock and its growth rate: existing machines and infrastructure usually produce capital equipment in excess of necessary replacements—the total stock grows. In economic statistics this shows up in the (roughly exponential) growth of capital investments, gross national product, and ensuing consumption in many countries. By dividing by the population number, one obtains per capita figures. In our submodel, we consider capital stock per person as an indicator of per capita consumption level. I hasten to add that in more accurate models one would have to introduce widely differing (resource and energy) consumption intensities of capital stock: for example, a poorly insulated and a well-insulated house having the same capital cost and providing the same "energy service" can differ by a factor 10 in their energy consumption.

While per capita consumption has been observed to grow exponentially in many countries in the past, this certainly cannot be expected to continue: there is an upper limit to the amount of capital stock that uses up energy and resources for each of us. We must therefore expect eventual saturation of consumption. Available technology would even allow significant reductions while maintaining a high material standard of living. However, it is unlikely that the capital stock would be similarly reduced: more sophisticated technology usually requires greater capital investments. We will assume that the saturation level is a function of "consumption control": stricter control of consumption will produce saturation at a lower level. Correspondingly, the capital stock at saturation will be lower if "consumption control" is greater.

With these considerations, we can formulate a verbal model for the consumption submodel:

1. The more capital stock there is, the more (net) growth of capital stock.
2. The higher the capital growth rate, the more capital growth.
3. The more capital growth, the more capital stock.
4. The more capital stock, the more consumption.
5. The more consumption control, the smaller the growth space.
6. The more capital stock, the smaller the growth space.
7. The more growth space, the more capital growth.

The influence diagram corresponding to this verbal model is shown in Figure 2.12.

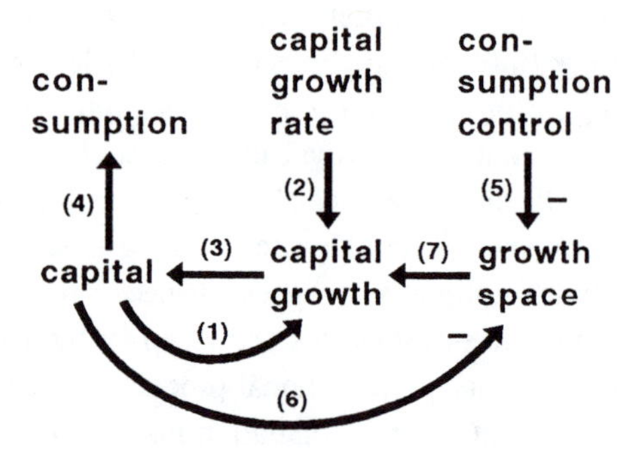

Fig. 2.12: Influence diagram of the consumption submodel.

If there is no consumption control, the relationships in the influence diagram suggest the following model formulation resulting in exponential growth:

capital_growth = capital_growth_rate * capital.

Saturation can be achieved by changing the formulation in such a way that as the capital stock approaches a saturation level, the capital rate of change "capital_growth" is reduced to zero. This can be achieved, for example, by the logistic formulation

capital_growth = capital_growth_rate * capital * growth_space (2.11)

where the "growth_space" becomes smaller as the capital stock increases:

growth_space = (1 – consumption_control * capital).

Here "consumption_control" determines at which level saturation will set in. If "consumption_control" = 0.1, for example, growth will stop as "capital" approaches the value 10. The rate of change of the state variable "capital" (its state equation) is given by

d(capital)/dt = capital_growth. (2.12)

Since we decided to use non-dimensional relative quantities, we can express the consumption level by

consumption = capital. (2.13)

(This means, for example, that a relative "capital" level of 150 percent would result in a relative "consumption" level also of 150 percent, compared to some reference index of 100 percent.)

The simulation diagram corresponding to the influence diagram Figure 2.12 and Equations 2.11 - 2.13 is shown in Figure 2.13. The initial value for the (relative) capital stock, and the (normal) growth rate of consumption are quantified as

$$\begin{aligned} &\text{capital}_{init} = 1 \quad [-] \\ &\text{capital\_growth\_rate} = 0.05 \quad [1/\text{year}]. \end{aligned} \tag{2.14}$$

Figure 2.13 also shows the typical logistical response curves of this system for different parameter values (see also Models M107 and M108 for logistic growth). The "consumption_control" parameter remains undefined for now; in the final simulation model the program user can change it and observe the resulting system response.

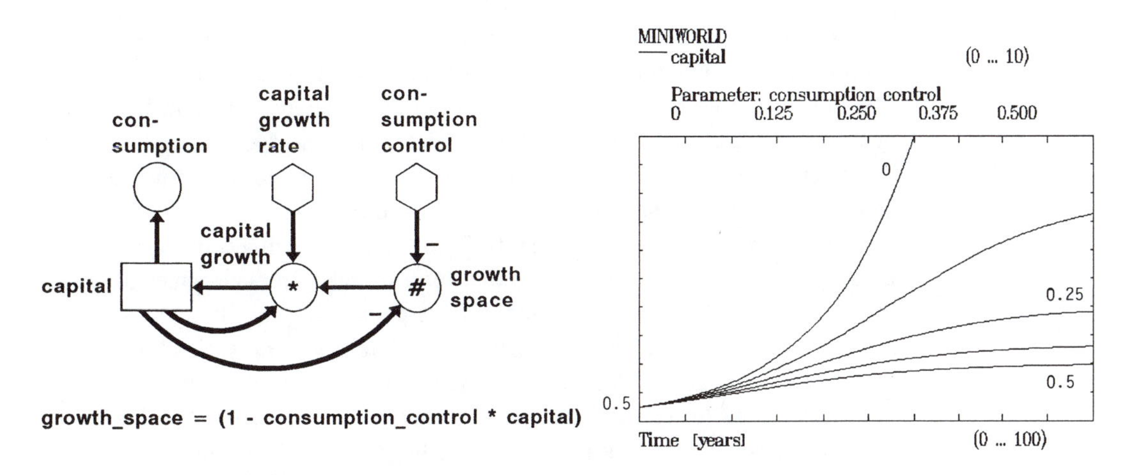

Fig. 2.13: Simulation diagram and response curves of the consumption submodel. The model produces logistic growth to a limit value (capital_growth_rate = 0.05).

2.2.5 Linking the submodels

We now intend to link the three submodels for population, pollution, and consumption as indicated in the original influence structure (Fig. 2.4). In doing so, we have to examine the specific formulation of each link, i.e. in particular: which variable of one submodel has to be linked to which variable of another submodel? In the following considerations we have to keep in mind that "population," "pollution," and "capital" (and "consumption") were defined as relative quantities whose value should have an order of magnitude of "1."

We first collect the three differential equations for the three isolated subsystems from the Equations 2.2 - 2.14:

```
d(population)/dt = births – deaths                                        (2.15)
      = birth_rate*population – death_rate*population

if (quality >1) then
d(pollution)/dt   = pollution_input – absorption_rate*pollution
else                                                                      (2.16)
d(pollution)/dt   = pollution_input – absorption_rate*threshold

d(capital)/dt     = capital_growth                                        (2.17)
      = capital_growth_rate*capital*(1 – consumption_control*capital).
```

We start by considering the additions that have to be made to the differential equation for "population" (Eq. 2.15). The influence diagram (Fig. 2.4) shows a first influence from the pollution submodel to the population submodel expressed by the arrows from "pollution" via "costs" and "action" to "population." This link was intended to represent population control in response to high environmental pollution. It therefore has to have an effect on births. As a measure of environmental quality, we can use the quality factor "quality" defined in Equation 2.8. Linking this to "births" would mean that for a decrease in environmental quality the number of births would be reduced correspondingly. The strength of this influence can be quantified by the "birth_control" representing societal action. We therefore should multiply "births" by the factor "(birth_control*quality)."

The second (direct) linkage from "pollution" to "population" in Figure 2.4 (with a minus sign) was meant to describe the effect pollution has on health by lowering the average life expectancy. The linkage is therefore from the "pollution" level to "deaths:" the higher the pollution load, the higher the (absolute) deathrate. Since the normalized value of "pollution" is of the order of magnitude of 1, we can use "pollution" directly (weight 1) as a factor for "deaths." We should therefore multiply "deaths" by the factor "pollution."

The third influence on "population" indicated in the influence graph (Fig. 2.4) comes from the "consumption" ("capital") submodel. It was intended to describe an increase in the number of (surviving) children per family as the material standard of living increases. Analogous to the coupling of "pollution" to "deaths" it makes sense here to couple "consumption" directly to "births" with a weight of 1. We therefore multiply "births" also by the factor "consumption." Note, however, that at higher levels of consumption, the opposite effect is usually observed: the rich have fewer children. (You may wish to modify this relationship in the model and observe its effect!)

With these additions we can now formulate the modified differential equation for population:

```
d(population)/dt =                                                    (2.18a)
      = birthrate * population * (birth_control*quality) * consumption
      – deathrate * population * pollution.
```

We now consider the additions that have to be made to the differential equation for "pollution" (Eq. 2.16). The coupling of "population" to "pollution" in the influence diagram (Fig. 2.4) can only be seen together with the coupling of "consumption" to "pollution." It describes the effect of the population and consumption levels on pollution generation. The pollution input depends on population and on the level of per capita consumption and is therefore proportional to the product (population * consumption). To determine the corresponding pollution input, we must introduce an appropriate "pollution_rate." The "pollution_input" in the submodel formulation Equation 2.16 must therefore be replaced by "pollution_rate*consumption*population." (We assume a pollution_rate = 0.02 for this link. This means, for example, that for population = 5 and consumption = 1, the pollution input rate corresponds to the maximum possible absorption rate of (absorption_rate * threshold) = 0.1.)

The modified differential equation for the pollution submodel therefore becomes:

```
if (quality >1) then
d(pollution)/dt = pollution_input – absorption_rate*pollution
else                                                                  (2.18b)
d(pollution)/dt = pollution_input – absorption_rate*threshold
```

where

```
pollution_input = pollution_rate*consumption*population.
```

We finally consider the additions to be made to the differential equation for the submodel for "consumption" (Eq. 2.17). The linkage from "pollution" to "consumption" (i.e. "capital") in the influence diagram (Fig. 2.4) was intended to describe the fact that specific consumption increases with an increase in pollution (by more resource intensive measures of environmental protection and resource exploitation, greater use of fertilizers and biocides, more difficult resource mining, etc.). This means that "pollution" influences "capital_growth." Here again, we can assume a simple proportional influence with weight 1. We therefore multiply "capital_growth" by the factor "pollution."

The remaining linkage from "pollution" via "costs" and "action" to "consumption" in Figure 2.4 is intended to limit the growth of the consumption level by appropriate societal action. In order to simulate this effect, it makes sense to couple "pollution" to "capital_growth" in such a way that it leads to an earlier saturation of the capital growth process. This means it has to affect the term "(1 – consumption_control*capital)" in Equation 2.17. The magnitude of the saturation effect can be adjusted be using the parameter "consumption_control" (to be specified by the user) as a factor. If there is no control, i.e. if "consumption_control" = 0, there is no saturation, and capital and consumption grow exponentially.

The modified differential equation for the capital submodel is therefore written as

```
d(capital)/dt = capital_growth_rate*capital*pollution*
    *(1 – consumption_control*capital*pollution).                (2.18c)
```

The solution of this set of differential equations (Eq. 2.18 a, b, c) representing the "global model" requires initial conditions for the states, for example:

$$
\begin{aligned}
\text{population}_{init} &= 1 \\
\text{pollution}_{init} &= 1 \\
\text{capital}_{init} &= 1
\end{aligned}
\qquad (2.18d)
$$

and specification of the parameters, for example:

```
threshold               = 1
absorption_rate         = 0.1
birth_rate              = 0.03
death_rate              = 0.01                                   (2.18e)
pollution_rate          = 0.02
capital_growth_rate     = 0.05
consumption_control     = 1
birth_control           = 1.
```

The complete model with its submodels and their linkages is shown in Figure 2.14a. We recognize the previously developed submodels in this diagram. If the links of the submodels are redrawn (Fig. 2.14b) and compared with the original influence structure (Fig. 2.4), we find agreement except for the multiplicative (non-linear) coupling of "population" and "consumption" with "pollution." In the discussion of non-linear couplings in Section 7.5.5 (cf. the linearization of the predator-prey model, Section 7.1.12) it is shown that (if linearization is admissible) a multiplicative coupling of two quantities can be resolved into two (additive) links by linearization. This then again leads to the influence diagram of Figure 2.4. The basic structures of the simple linear graph of Figure 2.4 and of the non-linear model of Figure 2.14 therefore correspond to each other.

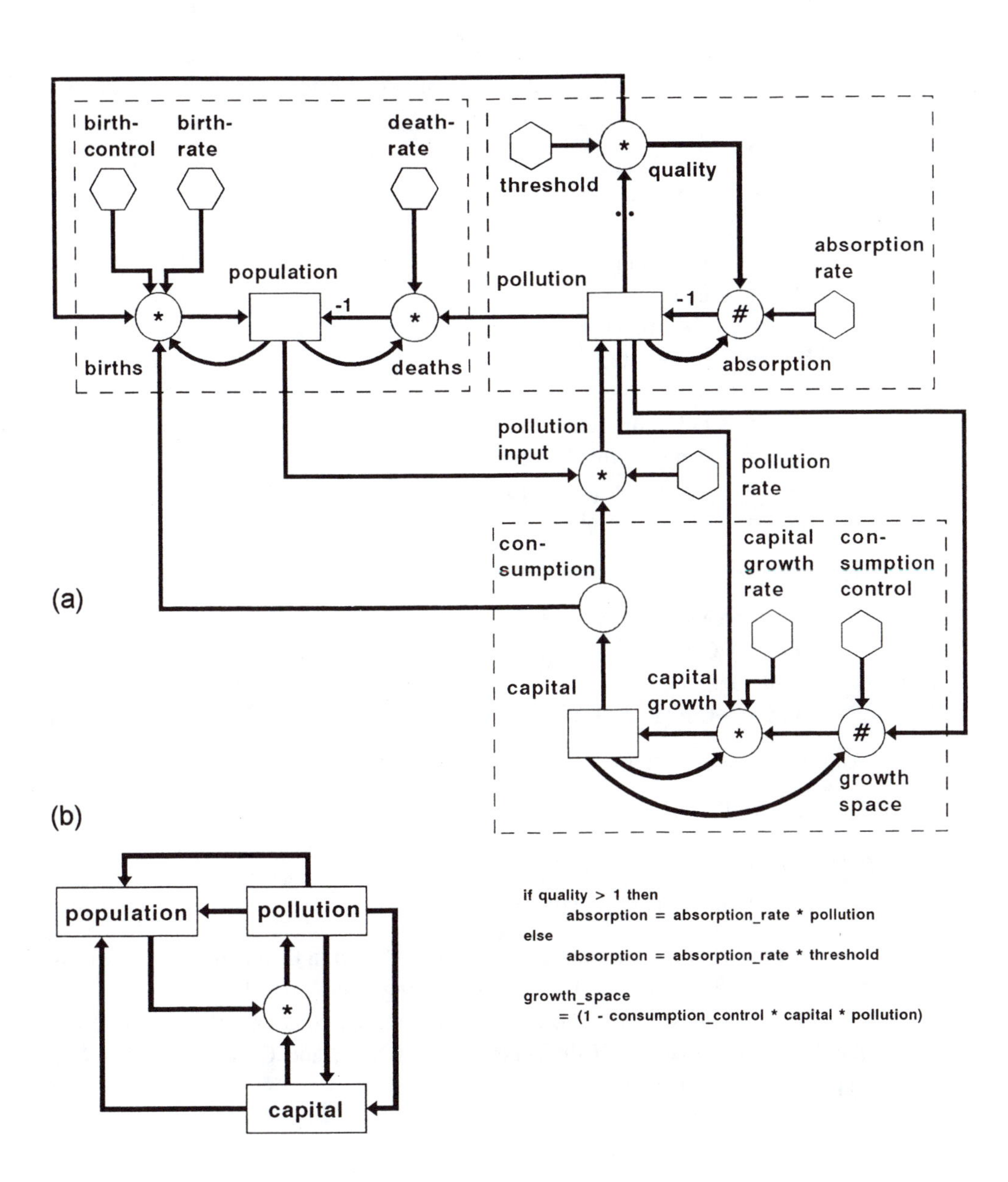

Fig. 2.14: The global model after coupling of the submodels for population, pollution, and consumption. Above (a): simulation diagram; below (b): coupling structure of the submodels.

A more compact form of the model equations (Eq. 2.18a-e) is often convenient, in particular if mathematical manipulations have to be performed. We therefore rename variables and parameters:

N	=	population
P	=	pollution
C	=	capital
P^*	=	threshold
Q	=	quality $= P^*/P$
a	=	absorption_rate
b	=	birth_rate
d	=	death_rate
e	=	pollution_rate
k	=	capital_growth_rate
f	=	consumption_control
g	=	birth_control.

The model equations can then be written as a compact set:

$$dN/dt = b \cdot N \cdot g \cdot (P^*/P) \cdot C - d \cdot N \cdot P \tag{2.19a}$$

$$dP/dt = e \cdot C \cdot N - a \cdot P^* \qquad (\text{for } P > P^*) \tag{2.19b}$$

$$dP/dt = e \cdot C \cdot N - a \cdot P \qquad (\text{for } P \leq P^*)$$

$$dC/dt = k \cdot C \cdot P\,(1 - C \cdot P \cdot f) \tag{2.19c}$$

with the initial conditions

$$N_0 = 1,\ \ P_0 = 1,\ \ C_0 = 1 \tag{2.19d}$$

and the parameters

$$a = 0.1,\ \ b = 0.03,\ \ k = 0.05,\ \ d = 0.01,\ \ e = 0.02. \tag{2.19e}$$

The birth control parameter g should be chosen in the vicinity of 1; the consumption control parameter f can be in the range from 0 to 10.

Note that most of the terms in the differential equations (Eq. 2.19a-e) are non-linear combinations of the state variables N, P, and C. We must therefore expect complex behavior.

2.3 Simulation and Model Behavior

2.3.1 Simulations with a simple simulation program

Using the mathematical model (Eq. 2.19a-e), we now want to develop a simulation program that will give us a first quick overview of the system behavior. For this we do not need any professional simulation software; we can directly program the simulation with a general purpose programming language.

A simulation program GLOBSIM.PAS written in TurboPascal is presented in Program 2.1 in the Program Listings Appendix. In the first part of the program, the control parameters g and f (birth control and consumption control) are requested from the user. In the subsequent statements, some parameters for the graphic output are defined. Next, the different system parameters and the initial conditions for the state variables are defined. With this information, the simulation can begin. The numerical integration of the three differential equations for population N, pollution P, and capital C is carried out in a "time loop" using time steps dt from the initial time t = 0 to the final time t = final. At each time step, the computed results for the state variables are also plotted on the screen. Pollution P is shown by a thin line, population N by a medium line, and consumption level C by a fat line. After completion of the integration loop, the chosen control parameters and the final values of the state values are also printed on the screen.

In the following runs, we fix the birth control parameter g at the value 1 and experiment with different values of the consumption control parameter f, i.e. the saturation level for consumption per capita.

For unlimited consumption growth (consumption control parameter f = 0), an explosive growth of consumption and pollution occurs after about three decades of relatively slow growth of N, P, and C. This leads to a rather sudden collapse of the population (Fig. 2.15a).

For very low values of the consumption control parameter (f < 0.1), this tendency of explosive growth of C and P with subsequent collapse of the population remains. After collapse, population remains at a value near zero until pollution has decayed to a level which allows a new growth episode. This leads to a strongly damped periodic behavior. Figure 2.15b shows a run over 500 years for f = 0.03. In this case the model oscillation has a period of about 120 years.

Less dramatic behavior appears only if the consumption control parameter f is increased to values above 0.1 (Fig. 2.16). These runs show a damped oscillating approach to an equilibrium state with constant values for the three state variables. We find the following equilibrium values:

control parameter	f	N	P	C
	0.1	1.558	3.119	3.208
	0.5	4.543	1.817	1.101
	1.0	7.211	1.442	0.693

The investigation shows that under equilibrium conditions a high consumption level corresponds to a high pollution and a low population level while a low consumption level allows a higher population level at a lower pollution level.

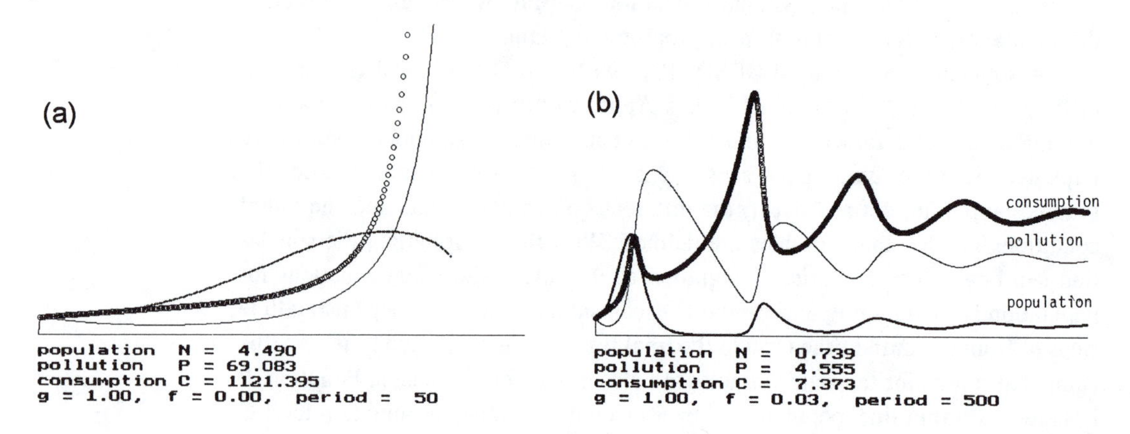

Fig. 2.15: Simulation results of the global model. Left (a): collapse for unlimited growth; right (b): strong oscillations if there is only weak damping of consumption growth. Population: thin, pollution: medium, consumption: fat line.

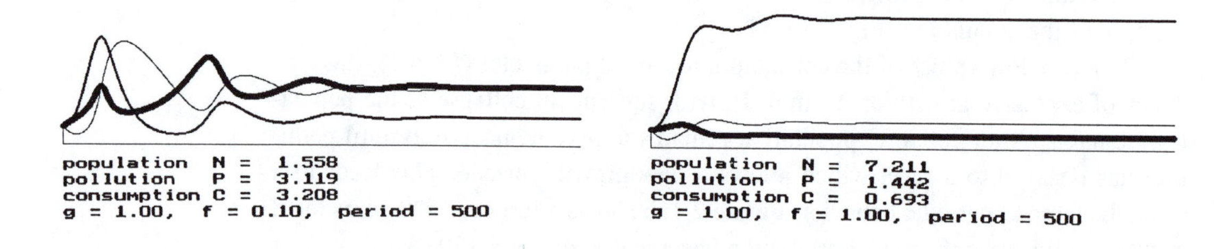

Fig. 2.16: Simulation results of the global model for stronger damping of the consumption growth. The equilibrium values are strongly dependent on the consumption control parameter f.

2.3.2 Validity of model formulation and simulation results

In Section 1.4.10 it was stated that we must judge the validity of a model with respect to behavioral validity, structural validity, empirical validity, and application validity. And of course, the discussion of validity only makes sense with respect to the **model purpose**. If, as in this case, we wanted to achieve "qualitatively correct information about the...dynamics...using the smallest possible number of variables" (Sec. 2.1.1), we must not complain if the model does not allow us to predict the exact population or number of cars in Tanzania in 2013.

On the whole, the GLOBSIM model produces plausible results: if material consumption is allowed to increase unchecked, pollution will increase correspondingly and will produce a collapse of population; population does not recover unless pollution has dropped to an "acceptable" level; the dynamics of population, pollution, and consumption are interdependent; per capita consumption soars as the population is decimated by the effects of pollution; there are oscillations, especially if there is little "damping" by consumption control; and the system shows stable equilibrium points which depend in a logical way on the consumption control parameter. We note, by the way, that the response of the complete global model is quite different from the response of any of the three submodels: the coupling of the submodels produced a system with new and different characteristic behavior.

However, the first impression of qualitative agreement should never lead one to believe the simulation results without further validity tests of the model and the results. This becomes particularly important if the model is to be used for decisionmaking.

As a first step, the mathematical derivations and the programming of the equations must be checked to make sure that they do not contain any errors. In subsequent steps, structural validity, behavioral validity, empirical validity, and application validity must be tested within the constraints of the given model purpose.

Structural validity was the main focus of our attention in the development of the submodels and their linkages. We introduced all the connections of elements which appeared to be necessary and permissible, such as the effects of pollution on births and deaths. With reference to the model purpose determined in Section 2.1.1 "to obtain qualitatively correct information using as few variables as possible" the formulation seems to be sufficient and structurally valid. Note that structural validity requires "correctness" of the influence **structure**, and "correct" identification of the system **elements** (state variables, parameters, intermediate variables). The precise numerical relationships are not at issue here.

Behavioral validity is related to structural validity since behavior is mainly a consequence of structure. However, behavioral validity also requires behavior comparable to that of the real system for a choice of realistic system parameters (Eq. 2.19e). There should be good agreement with respect to the following items:

1. speeds of processes; i.e. rates of change
2. oscillation periods
3. maximum and minimum values
4. phase shifts between state variables
5. equilibrium values
6. stability behavior
7. behavior under extreme conditions
8. plausibility.

With respect to these items, the results of GLOBSIM are not implausible, as we observed in the initial analysis of the time plots. Within the limits of the model purpose we conclude behavioral validity. Note that behavioral validity requires "correct"' reproduction of the main *qualitative features of system behavior*; again, the precise numerical values are not at issue here.

Empirical validity would require that the numerical values of the variables should agree closely with observations. For this "global model," empirical validity, i.e. quantitative agreement with reality, definitely does not apply. (It is therefore not permissible to draw concrete conclusions from the simulations!) In order to be empirically valid, the model would have to be formulated in real quantities using measurable counterparts of the real system. The parameters also, in particular the different relationships (e.g., the influence of pollution on the birthrate), would have had to be formulated with more precision and complexity.

Application validity requires that the model adequately serve the purpose for which it was developed. This has probably been achieved if we consider the model as a simple pedagogic tool for the demonstration of the dynamic effects of elementary relationships between the environment and human society.

We shall return to the modeling with real, dimensional quantities after we have dealt with some fundamental aspects of systems in the next chapter.

2.3.3 Comparison with a "real" world model

In 1972 a computer modeling team at the Massachusetts Institute of Technology (MIT) led by Dennis and Donella Meadows published the results, and later also the complete documentation of a rather complex world model World3, (Meadows et al., 1972; Meadows et al., 1974). This model was based on a simpler world model World2 developed by Jay Forrester, the "father" of the system dynamics modeling method which we also use (with some modifications) in this book (Forrester, 1970). The title of the Meadows' book *The Limits to Growth* became the slogan of the ecological debate of the seventies and eighties. This book appeared in 29 major languages and has affected the thinking of millions. In 1992 the Meadows and other members of the former MIT team updated World3 slightly using more recent data and used the model (World3-91) to look again at possible paths into the 21st century. They determined that we are already *Beyond the Limits* (the title of their book, Meadows et al., 1992); that there is a possibility of global collapse, but that there is also the possibility of a sustainable future.

World3 is a very complex model compared to the "global model" we developed in this chapter. World3 has 18 state variables, 60 parameters, 52 table functions, and some 200 complex equations for intermediate variables and rates. Obviously, this is still a very simplified description of the real world, but a huge amount of empirical data was used in this model to substantiate and formulate the many individual relationships and submodels making up the full model. The model captures the major processes driving capital investment and growth, population development, pollution emission and absorption, resource depletion, agricultural production, land development, and employment in industry and services in considerable detail. It links all these processes together in a complex dynamic system. The dynamics that the model produces are the result of the dynamic interactions of the many individual processes. If the individual processes have been correctly described, and if the complex relationships are faithfully represented in the model, we should expect reasonable agreement with developments in the real world. For the period 1972 to 1992 this is indeed the case, and scenario results beyond this period are plausible.

A system dynamics model like World3 (and the models in this book) is not developed as a forecasting tool, however. Its major purpose is to investigate the options (still) available, to clarify the spectrum of possibilities, and to help in the selection of options which might produce desired outcomes (more in Ch. 5).

It is interesting to compare the results of World3 with our "global model." Both models claim to capture the essential structure of global developments—their outcomes should therefore show similar behavior. Both models can be run on the SIMPAS simulation software we use in this book. Figure 2.17 shows the result of this comparison.

The three state variables, population, (per capita) consumption, and pollution, show similar time developments with respect to each other. Consumption and population peak at about the same time. Pollution continues to increase and has a maximum some 30 years after the population and consumption peaks. Note that "Miniworld" uses relative quantities—the numerical values of the two models are therefore different.

This much agreement between two models that differ so much in their complexity and detail is remarkable—it suggests the conclusion that the elementary system structure of both models has much in common. More agreement cannot be expected—as a quick look at the complex formulations and in particular the table functions of World3 would show.

This little exercise carries an important message: in order to understand the characteristic dynamics of a complex system, it may be worthwhile to attempt to isolate the "essential structure" of the system responsible for its typical behavior. This not only may lead to better understanding, it may also provide simpler model formulations. In some applications (e.g. forest growth models), simpler structures (with much faster computing time!) can be used to replace the "full" and much more complex models, often with only little loss of accuracy.

Chapter 6 of this book presents in the "systems zoo" a collection of such elementary structures, which are common in even the most complex models.

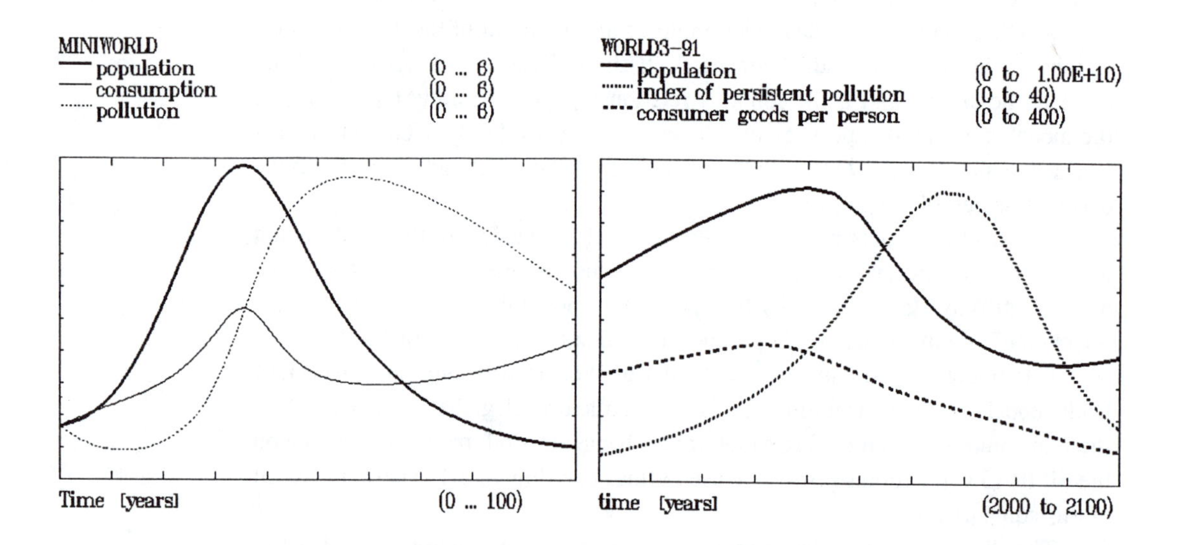

Fig. 2.17: Comparison of simulation runs of the "Miniworld" global model (GLOBSIM, left) and of the World3-91 global model (Scenario 2 in Meadows et al., 1992, right). The qualitative behavior of both models is similar. Note in particular the relative position of the peaks of the three curves .

2.4 Summary of Important Results

In this chapter we used a simple "global model" to demonstrate the steps of the modeling process. We started with the definition of the model purpose, wrote down the verbal model, used it to develop the influence diagram, identified the different elements of the model system, specified their functions and interconnections with other elements, developed the corresponding equations, wrote a simulation program, used it to compute several development scenarios, and assessed the validity of the model. These essential steps will have to be repeated in other modeling efforts, irrespective of their subject.

We record some important results:

1. The **representation of the system** in the influence diagram **depends on** the problem setting and the **model purpose**. This must be specified precisely in the beginning. It determines the choice of system boundary and the way the system is analyzed by the modeler.
2. The **influence diagram** focuses on the **influence structure** of the system, not on the exact function of its elements. In general, it is therefore not a complete and valid description of the real system. By itself, it cannot be used to simulate dynamic behavior except in rare exceptions (a linear system or linearization under the assumption of small perturbations).
3. Reliable and precise information about the **dynamic behavior** of a system can only be produced by a **simulation model** which correctly represents the properties of the **system elements** and of their (often non-linear) **interconnections**.
4. In dynamic systems we can distinguish **three different categories of system elements**, represented by different symbols:
   Input elements (parameters, initial values of the state variables, and exogenous inputs from the system environment). They are not influenced by the system development. Representation in simulation diagrams: **hexagon**.
   State variables (storage or memory variables of the system). They embody the history and development of the system. They are changed continuously by their rates of change (with dimension [state variable dimension / time unit]). Representation in simulation diagrams: **rectangle**. Arrows pointing into the rectangle represent rates of change of the state variable.
   Intermediate variables (converters, auxiliary variables). They are algebraic or logic functions of input elements, state variables, and/or other intermediate variables. Representation in simulation diagrams: **circle**.
5. The new value of a **state variable** is found by **numerical integration** in which the increment during a time step (the rate of change of the state variable times the time step) is added to the previous value of the state variable.

6. The characteristic feature of influence diagrams and corresponding simulation diagrams is their **feedback structure**. In a **positive feedback** loop, the sense of an original perturbation is not reversed. In each pass through the loop, it may be amplified, introducing a **destabilization** tendency. In a **negative feedback** loop, the sense of an original perturbation is reversed. In each pass through the loop, the previous perturbation is counteracted, introducing a **stabilization** tendency.
7. Complex simulation models can usually be broken down into **submodels** which can be **individually developed** and tested before they are linked to construct the full model. The behavior of the **full model** may be quite **different** from that of the individual submodels: the whole is not the sum of its parts but may be qualitatively different.
8. The **validity** of a model is defined by reference to the **modeling purpose**. It has four aspects: structural validity, behavioral validity, empirical validity, and application validity.
9. The characteristic behavior even of a complex system may be caused by an **elementary system structure** which may be generic, i.e. it may be found in many different systems which superficially bear no resemblance to each other but show **similar behavior**. Identification of the elementary system structure is a major step towards understanding a complex system.

CHAPTER 3
SYSTEM STATE

3.0 Introduction

Recognition of the behaviorally relevant system structure and its representation in the influence diagram constitute the first phase of model development. The influence diagram permits some qualitative conclusions about the system it represents, but these results are neither sufficient nor reliable for understanding exactly how a system may behave over time.

Modeling of dynamic systems therefore also requires a close look at the individual elements and their particular role and function in the system. Introducing their specifications into the influence diagram, we obtain the simulation diagram which now contains all the information necessary to program and compute the simulation.

In going through this process of model development in the previous chapter, we found that system elements can only be of three distinct types which we distinguished by corresponding symbols: input quantities (hexagons), state variables (rectangles), and intermediate variables (circles).

In this chapter, we will take a closer look at the modeling process and the description of dynamic systems. We will find that the notion of "state," and the description of dynamic systems by their "state equations," are useful concepts for developing the simulation model and for analyzing the system which it represents. The development of the state equations is therefore at the heart of the modeling process.

In Section 3.1 we review the properties of the three types of system elements in order to arrive at a general system diagram from which the central role of state variables becomes evident.

In Section 3.2 we focus in more detail on system state, state variables, and state computation. We will find that the essential dynamics of a (continuous) system are contained in its state equations—differential equations describing the rates of change of the state variables.

In Section 3.3 we show, for a few elementary systems, how their specific elements and structure—as expressed in the state equations—determine their behavior. Typical structures—often recognizable in complex simulation models—produce typical behaviors.

In Section 3.4 we are reminded of the need for dimensional consistency in all model equations. Dimensional analysis not only permits finding correct conversion factors where needed, it can also assist in the formulation of correct model statements: dimensional disagreement is an important indicator of errors in the model formulation.

In Section 3.5 some of these concepts are applied to the development of a simulation model for a physical system—a rotating pendulum, starting from the verbal model.

In Section 3.6 the modeling process is repeated to develop the simulation model for an ecological-economic system—a fishery operation.

In both of these cases, the end product of the modeling process is the set of model equations or, equivalently, the simulation diagram. Either can be used to program the simulation. This will be done in the next chapter.

In Section 3.7 we show how by condensation and conversion to non-dimensional variables, model equations can be rendered into a form that may have much wider applicability and reveal structural similarity and kinship to other systems. This may lead to "generic" state equations which describe whole classes of systems.

In Section 3.8 the major results of the chapter are summarized.

In this book we use different kinds of diagrams to depict systems. We will use the following terminology:

Influence diagrams show only influences between system elements; they do not show functional differences between elements.

Simulation diagrams contain the influence structure and precise specifications for all elements (either in graphical form or as formulae); the simulation program can be written down by reference to these diagrams.

Structural diagrams, block diagrams, or system diagrams contain the system structure and distinguish between system elements but do not contain all model information required for simulation. (The diagrams used in the STELLA simulation language are examples.) Model equations have to be supplied separately.

3.1 System Elements and Basic System Structure

Before continuing with the development of additional simulation models in the following sections, we shall now discuss in greater detail (with the experience from the development of the global model) the following:

1. the types of system elements (inputs, state variables, intermediate variables), and
2. the fundamental structure of their couplings in a system.

This leads us to a basic system structure that is common to all dynamic systems and is therefore also the basis of mathematical analysis of systems using the "state space" approach. The graphic symbols for system elements in the simulation diagram will often be referred to as "blocks."

3.1.1 Input elements

Input elements, i.e. exogenous inputs from the environment, system parameters, and initial conditions, are characterized by the fact that they remain independent of the development of the system. We therefore characterize them by a special symbol (hexagon, see Fig. 2.6). This means that they cannot have any system variable as an input, i.e. there can be no connecting arrow from any system element to the input element. Inputs can change *only* as function of time or stay constant. A possible dependence on time is shown in the simulation diagram by specifying a corresponding time function. We therefore have the following rule for the simulation diagram: exogenous quantities and system parameters (hexagons) cannot have any input arrows. The explicit functions of time used depend on the concrete application and can be of any form.

If a time dependence is given as a data series (for example, weather data), then the use of table functions is advisable. In table functions numerical values for the dependent variable (output) are given for a sequence of values of the independent variable (input) in the form of a table of numbers. Example: for January through December (months number 1 through 12), the average monthly temperatures are listed as a table (Fig. 3.1). To supply output corresponding to input that is not directly listed in the table, a table function interpolates between neighboring table function values. Linear interpolation is used most often: this amounts to approximating the (unknown) function by a straight line between neighboring points and reading off the output value corresponding to the input value (Fig. 3.1).

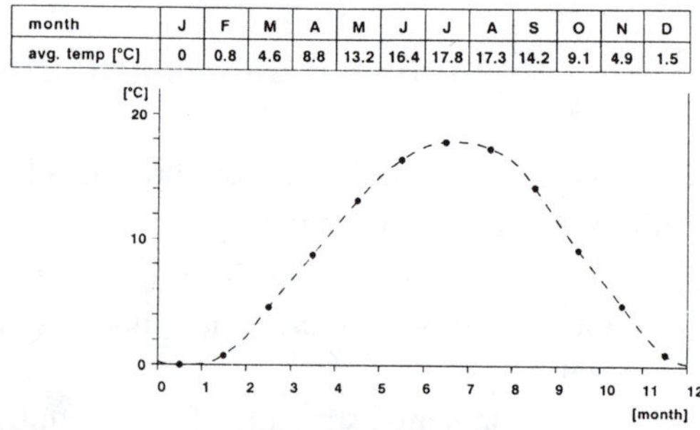

month	J	F	M	A	M	J	J	A	S	O	N	D
avg. temp [°C]	0	0.8	4.6	8.8	13.2	16.4	17.8	17.3	14.2	9.1	4.9	1.5

Fig. 3.1: Table functions are used to express empirical relationships for which no mathematical expressions exist.

If a simple time function is used, its specification together with the required parameters can be provided near the corresponding hexagonal blocks. For example, if a system is excited by a sinusoidal forcing function of amplitude a and frequency w, this could be indicated by a hexagon with the specification "a sin (w · time)."

If the function can be expressed by a (complex) mathematical expression, we characterize the block by its name and add the expression to the simulation diagram. For example, the mean daily temperature in Central Europe can be approximated by a sine function of the seasonal time [year], with mean annual temperature T_{avg} and temperature amplitude T_{amp}:

$$T_{day}(time) = T_{avg} + (T_{amp}/2) \cdot \sin(2\pi(time - 1/12) - (\pi/2)) .$$

This equation is too long to be written next to the block "T_{day}," and it is therefore listed at some other convenient place in the simulation diagram.

Occasionally, "time" itself has to be used in the computations. We characterize it by a hexagonal block with the designation TIME.

If it is necessary to compute an exogenous system input from a complex mathematical expression, a separate "submodel" with its corresponding simulation diagram can be developed. This submodel can then be shown in the simulation diagram as a single input block.

3.1.2 State variables

State variables are always storage (or memory) variables. In the system diagram they are therefore characterized by a rectangular box ("reservoir," "level," "stock") for which, however, a computational procedure must be specified. Two types of storage quantities are of importance: integrators and delays.

The **integrator** is the most common way of representing system states. It is characterized in the simulation diagram by a rectangle (Fig. 2.6). The task of the integrator is to determine the present state from a given initial condition at a previous time and the subsequent history of the rates of change of the state variable (the inflows and outflows of a level). A simple example of the function of an integrator is a reservoir with adjustable inflow and outflow (for example, a bathtub). As we have already seen in Section 2.2.2 (Eq. 2.1), this example readily provides a "recipe" for the computation of the current value of the state variable by numerical integration; i.e. adding to the previous value of the state variable the inflows and outflows of the system in the time interval between "previous time" and "current time." These inflows and outflows depend on the length of the time interval considered while the valve setting may actually remain completely unchanged. In order to eliminate the influence of the time interval, the inflows and outflows are expressed as flows per time unit and are called "rates," for example, "inflow rate = inflow of 1 liter per second." Multiplication by the time interval then produces the correct incremental change of the state variable during this interval.

The arrows pointing into the rectangles designating integrators therefore represent rates of change of the corresponding state variables. The dimension of a rate is always equal to the dimension of the state divided by the time unit:

dim (rate) = dim (state) / dim (time_unit).

Occasionally the term "rate" is used for a "normalized" or "specific" rate, i. e. rates of change of a relative (non-dimensional) variable per unit time. The dimension of such a normalized rate is always:

dim (normalized rate) = dim (normalized state) / dim (time_unit)
= dim (time_unit^{-1}).

The normalized rate is therefore identical to the inverse of the time constant of that particular process. Dimensions will be given in square brackets [...].

Examples:

State variable:

population [people]

Normalized state variable:

relative_population [-] = population/reference_population [people/people]

Rate:

births [people/time_unit]

Normalized rate:

(specific) birth_rate [time_unit^{-1}]
= births/reference_population [(people/time_unit)/people].

Delays also have to store (earlier) system states and hold them for later use. Their properties are therefore similar to those of state variables. For this reason we again use the rectangular symbol, now adding additional vertical bars (Fig. 2.6). Note that the dimensions of the input and output of a delay are identical (which is not the case for integrators!).

In a production system we find production and delivery delays: products ordered today can only be shipped to the customer in six weeks, for example. People entering an escalator as a certain function of time will leave it again in the same order, according to exactly the same function of time shifted by the amount of time the escalator takes to move one person from start to finish. Such a delay, where the time function entering the delay is only time-shifted and remains otherwise unchanged, is called a **transport delay**.

The task of the transport delay (holding delay) is to hold the value of a variable for a given delay period T. A signal received at time t is held unchanged for the period T and then passed on at a later time (t + T). Put differently, the transport delay output at time t corresponds to the input it received at time (t – T). This means that all inputs received in the meantime have to be remembered to produce the correct output later. (This may mean a large number of storage variables in the simulation.) The holding delay is important, for example, for the representation of transport delays (assembly line, rail transport, trucking, etc.) where the object does not undergo any changes during the transportation process.

Of practical importance is also another type of delay, the **exponential or smoothing delay**. For example, a business firm may daily receive a randomly changing stream of orders for its product. For obvious reasons, the firm would try to maintain a fairly constant stream of production, for example, a daily output of one-tenth of the total number of orders on the books. This obviously introduces a delay, but it also means a smoothing of the output function (deliveries) compared to the input function (orders).

The process taking place in a smoothing delay (where inflows accumulate as they come while outflows are a function of the amount accumulated) can be represented by an integrator with a negative feedback (drain). This gives rise to a solution determined by an exponential function of time; we therefore also refer to it as exponential delay. The delay time corresponds to the inverse of the feedback factor: a weak feedback coupling (a small drain) results in a relatively long delay. The exponential delay differs fundamentally from the transport delay: the delay time of an exponential delay is only an average delay time; in actuality the exponential delay passes information about the new state right away. A significant computational advantage of the exponential delay is the fact that (in contrast to the transport delay) it can be represented by a very few state variables—usually one or three (see Sec. 4.1.4 and Models M 103, M 201, and M 301 in Ch. 6).

3.1.3 Intermediate variables

Variables that are not input or state variables can be computed from current values of input and/or state variables at any time by algebraic or logic operations. We therefore call them "intermediate variables," "auxiliary variables," or "converters." They are characterized by circles in the simulation diagram (Fig. 2.6).

For the functional blocks symbolizing **algebraic or logic operations** the following rules apply:

1. Each block must have at least one input (arrow pointing into the circle). It can affect any number of other elements (any number of arrows may lead out of the circle).
2. It must be shown which functions are to be performed with these input quantities. Simple algebraic functions can be indicated in the circle; others must be specified separately on the simulation diagram (see Fig. 2.6).
3. Each block represents exactly one output variable.

The most common blocks are those of the elementary algebraic operations (Fig. 2.6): adders (subtraction is indicated by a factor of "–1" at the corresponding input arrow) and multipliers (division is indicated by the division symbol ":" at the corresponding input arrow). As required, other algebraic operations or functions (including table functions) are indicated in or at the block symbol (circle or ellipse) (Fig. 2.6). A circle may also represent a complete subprocess; however, an individual block symbol must be used for each output quantity of that process.

Rates of change for state variables as well as other output quantities usually have to be computed as algebraic functions of input variables, state variables, and time. Occasionally, relationships have to be provided as table functions, or logic statements have to be used. A rate of change should be represented by its own block (circle) although it will often be identical to a particular state variable or input quantity. However, even if no separate rate block is used, the rate is always clearly defined by the quantity represented by the arrow entering the state variable. (In the systems zoo diagrams in Chapter 6, separate rate blocks are occasionally not shown in order to keep the diagrams compact.)

In realistic models of complex systems, the full graphical system description using these symbols may lead to very complex algebraic influence chains and nets which may be difficult to grasp and understand. These algebraic computation nets must not contain any algebraic loops, i.e. closed loops containing only circle blocks are not permitted. The reason is obvious: algebraic (or logic) operations instantaneously deliver an output corresponding to a given input. In a (closed) algebraic loop, this would mean that the input of a block is instantaneously determined by its own output (returned by way of the loop) which is physically impossible. Feedback loops must therefore always contain at least one state variable (or delay) to break the instant return of signals. There are no further restrictions on intermediate variable formulations: the choice of elements and linkages is solely determined by the relationships found in the real system. This implies the unrestricted use of non-linear relationships where necessary.

In a mathematical formulation of a simulation model, one would attempt to reduce complex networks of algebraic relationships to the most compact expressions possible by using all permissible transformations and operations. In modeling, this standard mathematical procedure has severe disadvantages. The computational and memory requirements for the computer model are reduced only insignificantly by this approach, but it makes the system structure and the processes taking place in it much more difficult to understand. It is therefore usually advisable to retain in the simulation diagram all elements and relationships identified in the verbal model and in the influence diagram, even if their aggregation into more "elegant" condensed mathematical expressions is possible.

One may want to deviate from this general rule if: 1. in a very complex system it may actually increase clarity by aggregating subprocesses; 2. the same subprocess appears repeatedly in the system and can therefore be described by its own subsystem block; or 3. a subprocess consists of relatively trivial and well-known steps whose complete presentation in the system diagram would not contribute to system understanding. In these cases the subprocess should be represented by its own system block (circle) with the corresponding inputs and output connections. This block is then specified in detail by its equations, or a separate simulation diagram, and computed in a corresponding subprogram. For each required output quantity, an individual system block must be defined.

3.1.4 Elementary block diagram of a dynamic system

We can deduce the essential structure of a dynamic system from the influence relationships permitted among the three types of system elements (input quantities, state variables, intermediate variables), where rates are to be represented as intermediate variables:

1. Input quantities (hexagons) cannot be influenced by other system elements.
2. The inputs of state variables (rectangles) can only be rates (intermediate variables, circles).
3. Intermediate variables (circles) can be functions of inputs (hexagons) and state variables (rectangles). (The influence of other intermediate variables can be formally eliminated by substitution.)
4. States (rectangles) can only affect intermediate variables (circles) directly.
5. Not all intermediate variables are rates (with arrows pointing into states).

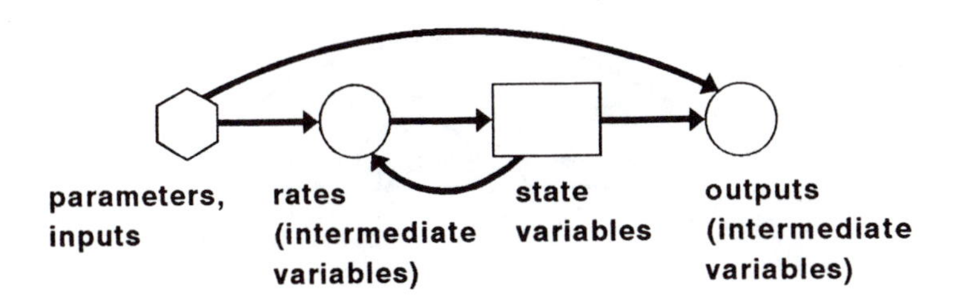

Fig. 3.2: Elementary block diagram of a dynamic system. This structure applies to all (continuous) dynamic systems. Blocks designate vectors, arrows represent matrices of connections.

If we now try to draw a diagram using these rules, we obtain the elementary block diagram of a dynamic system shown in Figure 3.2. This diagram is very fundamental and completely general—we have not imposed any restriction on the system. On first sight, it may look too simple to represent any of the complex systems in the world around us. However, if we let each of the symbols stand for a whole set (a "vector") of elements of the same type (but perhaps very different functions), and regard each arrow not as a single wire between two elements but as a whole bundle of wires between two sets of elements (a matrix of functional connections, where individual wires connect one element from the "sender" set to one element of the "receiver" set), then we realize that this simple structural diagram can indeed represent an almost limitless number of different systems. This "geometric" view of a system can be rephrased mathematically in terms of the "state equations" of a dynamic system which will be discussed later.

Example: To illustrate the one-dimensional case, let us assume we have to compute the food consumption of a population which changes dynamically in response to time-dependent birth- and deathrates. We have the relationships:

1. Total food consumption is population times per capita food consumption.
2. Population increases by births.
3. Population decreases by deaths.
4. Births (per year) are given by population times birthrate.
5. Deaths (per year) are given by population times deathrate.

In this example, "per capita food consumption" is a parameter, "birthrate" and "deathrate" are inputs (exogenous functions of time), "population" is a state variable, "births" and "deaths" are the two components of the rate of change of "population," and "food consumption" is an output variable. With this information, we can draw the simulation diagram in Figure 3.3. Its structure is identical to the general system structure in Figure 3.2.

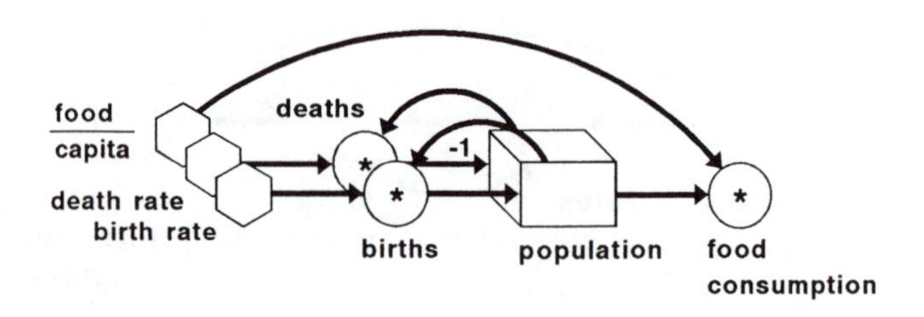

Fig. 3.3: Simulation diagram for computing the dynamics of food consumption. Note that this one-dimensional system has the same structure as the general system diagram in Fig. 3.2.

In the general system diagram (Fig. 3.2), the central role of the state variables becomes obvious: the history of the system is "remembered" by the current state which determines its own rate of change and thereby its future development and the output of the system. The state variables are therefore the central variables of a dynamic system: if we know how they change, other variables of the system can also be computed. But since we know their initial values and the inputs of the system, the problem is reduced to determining the differential equations for the rates of change. The algebraic expressions defining the rates of change are called the "state functions" **f.** The algebraic expressions combining state variables and input quantities to represent system output are called the "output functions" **g**. Both sets of equations together are referred to as "state equations" of the system.

Focusing on system structure allows us to recognize similarity or even identity in systems which may bear no resemblance in their physical appearance. For example, the completely different systems of Figure 3.4 (a mass-spring-damper system and an electrical circuit) have identical system structures and can be described by identical simulation diagrams and identical equations.

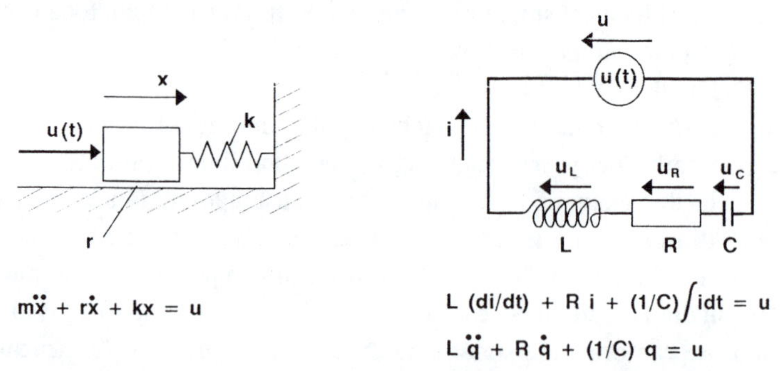

Fig. 3.4: Physically different systems may have identical system structure.

3.2 System State and State Computation

3.2.1 System state and state variables

For modeling and simulation the choice of state variables and the determination of the time-dependent system state are of central importance. We therefore have to deal with these concepts in somewhat greater detail.

A system that is completely independent of earlier system states, i.e. a system which is without memory or inertia, can be shifted instantaneously to a new state (within its physical limits) by a corresponding input **u**(t). Such systems are rare in reality. We find them as transformers of inputs (for example, a loudspeaker which should ideally respond instantaneously to high-frequency oscillations without any memory of the sounds it just produced in the previous split-second; a mechanical typewriter that has no memory of the letters which were just typed on it; an electrical transformer; a gear box which reduces the fast rotation of an engine crankshaft to a slow rotation of the rear wheel.) All of these memory-free systems have no state variables (at least in their idealizations); they reduce to a simple algebraic transformation of input into output. They are therefore categorized as **input-determined systems** or **memory-free systems.**

Most systems, however, have memories of some sort and some inertia since in general the current state of the system will not allow an instantaneous change to an arbitrary different state.

Examples:

1. A full reservoir cannot be half empty at the next instant.
2. A room cannot be instantaneously heated.
3. A vehicle at rest cannot move at 100 km/h at the next instant.

In these and similar cases the present system state, together with the exogenous inputs, obviously determines the new state. This also means that the history of the system has determined the present state. We refer to systems characterized by memory as **state-determined systems**.

This now poses an obvious question: what has to be known in these systems about the previous state in order to determine the new system state? Obviously, not all system quantities will be needed since, as discussed earlier, some of them can be computed by algebraic relationships from other system quantities. The system quantities from which all others can be calculated are the state variables.

State variables are the smallest set of endogenous time-varying variables that allow—for specified inputs and a given model purpose—the complete description of the system state. The **order** or **dimension** n of a system is the number of its state variables.

In general, a state variable is not unique. Equivalent representations of the same system using different state variables are generally possible. However, if different but equivalent representations are used, the number of the state variables always remains the same.

Example: In order to describe the time-dependent change of the amount of water in a bathtub one of the following state variables could be chosen:

1. the depth H of the water in the tub,
2. its volume V, or
3. its mass M.

Correspondingly, the inflows and outflows could be specified as water depth per time, volume per time, or mass per time. If one of the variables (depth, volume, mass) is known, the other two equivalent variables can be simultaneously computed (applying corresponding conversion factors). From the water depth follow water volume and mass, from the water volume follow water depth and mass, and from the water mass follow the volume and the water depth. The system has *one* level, therefore *one* state variable; its dimension therefore is "one," i.e. it is a system of first order.

The **system state** is the smallest set of current information about the system whose knowledge, together with knowledge about the time-dependent input functions and system parameters, allows the determination of future system states (in a deterministic system). Example: If you know how much water is in the tub and what the inflow and outflow rates will be in the next minute, then you know how much water will be in the tub after one minute.

The **state vector** is the vector of the state variables in state space. As the system state changes, the tip of the state vector moves to another point of state space (see Fig. 3.5).

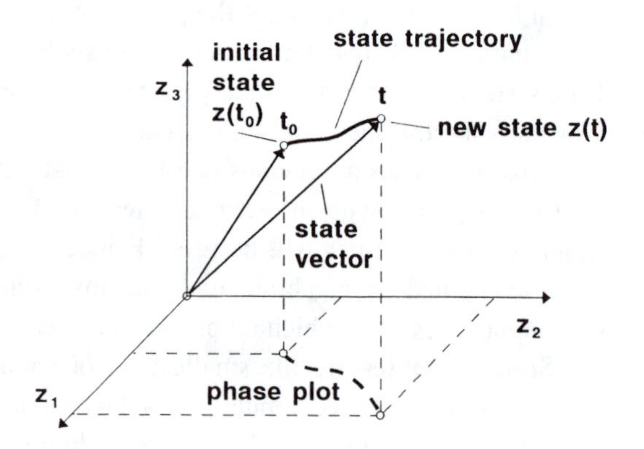

Fig. 3.5: State vectors and state transition in state space.

The **state space** is the space spanned by the n components (the n individual state variables) of the state vector **z**. The current system state is represented by a point in this coordinate space (= tip of the state vector) (see Fig. 3.5).The state space in two state coordinates is usually called the "**phase plane**" (state plane). In state space, time t appears as a parameter.

Example: A certain ecological system is represented by three state variables: vegetation biomass B, deer D, and wolves W. These three variables span a three-dimensional state space with the three coordinates: biomass, deer, and wolves. Any combination of biomass, deer, and wolves can be drawn as a point in this three-dimensional space (B, D, W). This point represents a particular state of the system. The state vector is the arrow from the origin of state space (0, 0, 0) to the current state (B, D, W). As the state changes, it moves to another point of this state space. The state vector changes its length and direction accordingly. The state trajectory is the path of the state in state space during this change, or the path of the tip of the state vector.

In the modeling of systems, the determination of the relevant state variables is often difficult on account of the non-uniqueness of the state variables and the often complex relationships in the system. Possible candidates for state variables are all reservoirs, stocks, or memories of the system (for example: populations, energy, matter, information) as well as all delay or inertia elements (for example: transport delays, transport belts, queues, etc.). In selecting state variables, one has to make sure that the set does not include storage quantities that can be directly determined from other state variables. (In the bathtub example, water volume and water height cannot be used as separate state variables.)

Clever choice of the state variables can lead to particularly simple system descriptions. In linear systems, for example, a general system description can be transformed to a much more elegant canonical description with decoupled subsystems of first order by use of a transformation involving the eigenvector matrix (modal matrix). This is discussed in Section 7.3.

Since the system state $\mathbf{z}(t_o)$, defined by the state variables at time t_o together with the input vector $\mathbf{u}(t_o, t)$ over the time interval (t_o, t), allows determination of the new system state at time t, this also applies for any number of subsequent time intervals. This implies for a deterministic system that the state vector $\mathbf{z}$ (or the n state variables z_i, respectively) can be uniquely determined at each point in time $t > t_o$ if

1. the initial states $z_i(t_o)$ are known for each state variable z_i, and
2. the input vector $\mathbf{u}(t_o, t)$ during the interval (t_o, t) is given.

In practical terms this means that in order to compute the dynamic behavior of a system, the contents of all **storage elements** and/or the states of all **delays** must be known at a certain point in time in addition to the input vector $\mathbf{u}\ (t_o, t)$

over the time interval in question. Thus one can decide what the state variables of a system must be by searching for those system quantities that must be recorded in order to allow a continuation of a system process after an interruption ("Sleeping Beauty" problem).

Examples:

1. In a production system, the following initial conditions must be recorded: the stocks of the stored material and information, the current loading of transmission belts, the velocities and loading of machines, etc.
2. In a forest ecosystem, one would have to record as initial conditions: nutrient pool in the soil, present biomass stocks, organic litter mass, etc.
3. In a national economy, one would need initial data for: population, capital stock, savings, stocks of supplies, agricultural area, etc.

In considering dynamic systems, we have to distinguish between continuous systems and discrete systems. The states of continuous systems are defined at any point in time; the states of discrete systems are only available at discrete points in time. The system equations, analytical solutions, and stability conditions of both system types differ from each other (see Ch. 7). In this book we deal almost exclusively with continuous systems.

3.2.2 Computing the system state

Viewing a system as a "black box" from the outside (Fig. 3.6), only two types of quantities are discernible: those which act as input quantities from the environment into the system (**inputs** u_i), and those which can be observed as behavior outside of the system (**output quantities** v_j). The different input and output quantities can be summarized by an input vector **u** and an output vector **v**. The system therefore can be viewed as a transformer converting inputs **u** to behavior **v**. The inputs as well as the behavior are functions of time, hence **u**(t) and **v**(t).

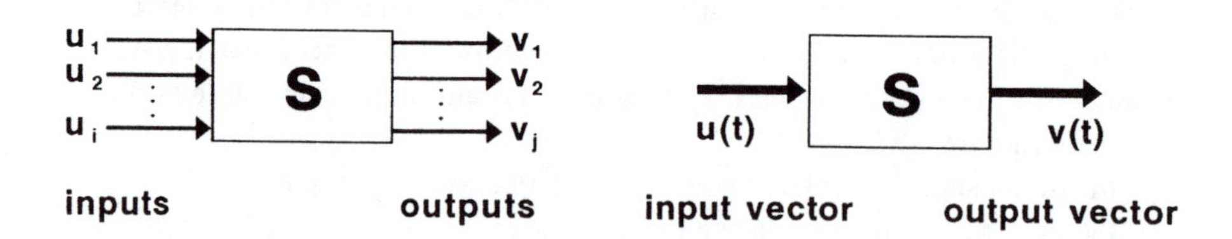

Fig. 3.6: System as "black box" with inputs **u** and outputs **v** (behavior). Right hand side: vector presentation of the system and its inputs.

In system dynamic models, we are not content with the mere description of the behavior of the "black box" as a function of its inputs. Even if a huge number of observations of behavior (output) as a function of input were available, we could only reliably use this information if the system were memory-free and not state-determined. If state variables are involved, we must expect that the changing system state also changes the response, and hence a careful modeling of the system's elements and structure would be required. For example, my old mechanical typewriter, an input-determined system, will always print an "x" if I press the "x"-key. For it, an input-output description makes sense. The reaction of my notebook computer, however, is state-determined since it depends on the state of its batteries: it will not react if they are empty. In order to model its response, I should therefore take this state variable into account.

In describing state-determined systems—which constitute the majority of systems around us—the intent is therefore to determine the behaviorally relevant elements and relationships in the system itself and to represent the real system structure as well as possible and necessary. We have seen how a system's elements and structure determine behavior. We therefore have to find out about the contents of the "black box" and the processes taking place in it.

In this approach to the analysis of a system and its behavior, the **state variables** z_i play a central role, as will have become obvious by now. These variables can also be collected in a vector $\mathbf{z}$ which changes with time: $\mathbf{z}(t)$.

In order to determine the state development over time we have to assume that the state $\mathbf{z}(t)$ results from the inputs $\mathbf{u}(t)$ as well as—by way of the feedbacks—the state itself. This relationship may depend on time if, for example, certain parameters are functions of time. In general, we therefore have to write:

$$\mathbf{z}(t) = \mathbf{F}\big(\mathbf{z}(t), \mathbf{u}(t), t\big).$$

This formulation requires the satisfaction of a simultaneous condition for $\mathbf{z}$, $\mathbf{u}$, and t, i.e. the state at time t is a function not only of the input signal but also of the instantaneous feedback of the as yet undetermined state. In principle, this task can be dealt with by iteration, but a real system does not usually iterate due to finite transmission velocities, system inertias, etc. The new state at time $t + \Delta t$ is rather a function of the conditions at the preceding point in time t (the small time step between the two points in time is denoted by Δt). From this follows the state equation:

$$\mathbf{z}(t + \Delta t) = \mathbf{F}\big(\mathbf{z}(t), \mathbf{u}(t), t\big).$$

In this formulation, the new state can be computed immediately. In order to do this, the preceding state must have been stored and must be available for the computation: the system description requires a memory (storage) for each state variable. Note that there is structural agreement with the state variables of real systems: these are also always storage quantities (levels, stocks, reservoirs)! In addition, this formulation corresponds to the processes taking place in discrete dynamic systems and in particular in numerical calculation on digital computers.

In a continuous system, $\mathbf{z}$ and $\mathbf{u}$ are constantly available, not only at discrete points in time with interval Δt. If $\mathbf{F}$ is continuous and differentiable, the state at time $t + \Delta t$ can also be approximated by:

$$\mathbf{z}(t + \Delta t) = \mathbf{z}(t) + (d\mathbf{F}/dt) \cdot \Delta t$$

or

$$\mathbf{z}(t + \Delta t) - \mathbf{z}(t) = (d\mathbf{F}/dt) \cdot \Delta t = \mathbf{f} \cdot \Delta t$$

where $\mathbf{f}$ is now defined as $(d\mathbf{F}/dt)$.

Dividing by Δt and applying the limit process $\Delta t \to dt \to 0$ produces the **state equation**:

$$d\mathbf{z}/dt = \mathbf{f}\big(\mathbf{z}(t), \mathbf{u}(t), t\big). \qquad (3.1)$$

Whatever we observe as the behavior of the system (behavior vector $\mathbf{v}(t)$) will be in large part a function of the state variables $\mathbf{z}(t)$ of the system. In addition, however, part of the output will be a direct function of the input signals $\mathbf{u}(t)$ which may have been passed through the system and transformed in some way. The relative contribution to the behavior vector from these two components can be expected to change with time in the general case (for example, by aging or wear of components). The general form of the **output equation** is therefore:

$$\mathbf{v}(t) = \mathbf{g}\big(\mathbf{z}(t), \mathbf{u}(t), t\big). \qquad (3.2)$$

Using these two relations we can now draw again the general block diagram for arbitrary dynamic systems which we derived earlier from considering the three types of system elements and their possible relationships (Fig. 3.2). This block diagram describes all types of dynamic systems, linear as well as non-linear, with either constant or time-dependent parameters (Fig. 3.7). Even arbitrarily complex models of system dynamics correspond to this general structure.

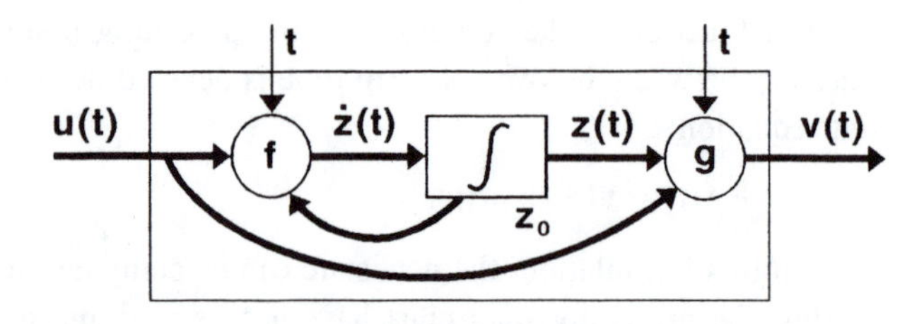

Fig. 3.7: General block diagram for dynamic systems (vector quantities).

Let us now look again at the computational procedure first for an integrator and then for the dynamic system as a whole. (The numerical integration procedure was already explained briefly in Sec. 2.2.2.) To visualize the process, consider computing the changing water level in a reservoir (e.g. a bathtub). Let us assume initially that the setting of the inflow and outflow valves can only be adjusted at discrete points in time with a time interval Δt. Then the new system state at time t follows from the old system state at time $(t - \Delta t)$ and the net flow during the time from $(t - \Delta t)$ to t, which likewise follows from the valve settings at time $(t - \Delta t)$ (i.e. the inflow rate and the outflow rate) and the time step Δt:

new state = old state + rate of change · time step

$$z(t) = z(t - \Delta t) + (dz/dt) \cdot \Delta t \qquad (3.3)$$

where the rate (dz/dt) is measured at time $(t - \Delta t)$. For continuous systems, where the rates of change can vary at any instant, the limit process $\Delta t \rightarrow dt \rightarrow 0$ results in the new state as a result of an integration over time with z_0 as initial value:

$$z(t) = z_0 + \int_0^t (dz/dt)\, dt. \qquad (3.4)$$

This equation shows once again that the rates (dz/dt) must always have dimension (dim (state variable) / dim (time unit)). Note that there is justification for calling rates "flows." They describe "flows" of the state variable into or out of the state reservoir. Often, as in the bathtub example, the rates are physical flows.

The instantaneous rate of change (dz/dt) follows from the system-specific state function $\mathbf{f}$. It is determined separately and constitutes the input of the integrator. In the simulation diagram, all arrows pointing into the integrator block are therefore understood as rates of change. The initial values of the integrator are provided separately (as numbers next to the states or as input quantity (hexagon)).

Simulation requires numerical integration. Several well-tested numerical procedures are available which differ with respect to their numerical precision and stability and their speed of computation.

For the majority of applications in system dynamics the simplest procedure, **Euler-Cauchy integration**, is quite adequate. It corresponds to Equation 3.3. In this case it is assumed that the rates of change remain invariant during the computational step (the time period from $(t - \Delta t)$ to t). This causes a small numerical error since the rates of change may in reality change in the meantime. The error decreases as the step size of the computation is reduced. However, if the step size becomes too small, this will lead to long computation times and truncation errors which may again cause errors in the results. It is therefore necessary to find a meaningful compromise in choosing the computation step. As a rule of thumb, the step width should be about 1/20 of the smallest time constant in the system or 1/100 of the smallest period of oscillation in the system.

The computational recipe for the integrator (state variable z) is therefore:

$$z_t = z_{t-\Delta t} + (dz/dt)_{t-\Delta t} \cdot \Delta t$$

or as program statement

STATE := STATE + RATE * DT (3.5)

Where higher precision at efficient computation speeds is required, the Euler-Cauchy method is not adequate. Much better results can be obtained with more sophisticated integration methods. Probably the most reliable workhorse of numerical integration is the **Runge-Kutta integration** of fourth order. The method is described in Section 7.1.6. It uses a weighted average of the rates at both ends and at the midpoint of the integration interval Δt. Either one of these methods—Euler-Cauchy or Runge-Kutta—can be selected in the SIMPAS simulation software and the systems zoo models which are found on the diskette.

Occasionally, one has to simulate systems in which simultaneously very fast processes (small time constant) and very slow processes (large time constant) play a role (for example, in tree growth, the stomata regulation is in minutes, the biomass increment in years). These so-called "stiff" systems may cause severe problems in numerical integration. If the Euler-Cauchy procedure is used, one encounters very long computation times and computational errors. Another approach is to use more efficient and precise integration procedures which are able to adjust the step width for minimum error and optimal computation time. It is also possible to separate the system into "fast" and "slow" subsystems and to compute the fast processes using a small step width. Finally, in some cases one might develop aggregated response functions for the fast processes from separate simulations which are then inserted into the model in place of the original processes. It may also be worthwhile to critically review the modeling attempt in the light of the model purpose—perhaps the problems can be avoided without loss of relevant information by focusing on either the slow or the fast processes.

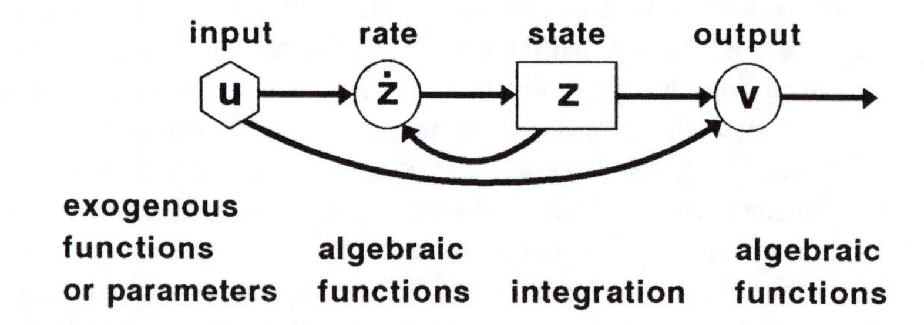

Fig. 3.8: General simulation diagram for dynamic systems (vector quantities).

We can now summarize the **computational steps** which will have to be carried out in computer simulations of dynamic systems (see Fig. 3.8):

1. Specification of all **initial conditions** of the state variables $\mathbf{z}_0 = [z_1(t_0), z_2(t_0), \ldots, z_n(t_0)]$ and of all **constant parameters**.

   For each time step $t, t+\Delta t, t+2\cdot\Delta t, \ldots, t+n\cdot\Delta t$ *of the simulation interval (time loop):*

2. Determination of the **current input variables** $\mathbf{u}(t)$ (influences from the system environment).
3. Determination of possible **time-dependent parameters**.
4. Computation of the **rates of change** of the state variables:

   $$d\mathbf{z}/dt = \mathbf{f}\big(\mathbf{z}(t), \mathbf{u}(t), t\big).$$

5. Numerical **integration** of rates of change to determine the state variables:

   $$\mathbf{z}(t) = \mathbf{z}_0 + {}_0\!\int^t (d\mathbf{z}/dt)\, dt.$$

6. Computation of the **output variables** (behavioral variables) $\mathbf{v}(t)$:

   $$\mathbf{v}(t) = \mathbf{g}\big(\mathbf{z}(t), \mathbf{u}(t), t\big).$$

While the steps must be carried out in this order, the computational sequence of equations plays a role only in the computation of the algebraic equations determining the rates of change (Step 4) since in this case quantities have to be computed simultaneously. The algebraic equations must therefore be written down in the direction of the influence flows, i.e. from the exogenous quantities (hexagons) and the state variables (rectangles) in the direction of the arrows across other intermediate quantities (circles) back to the state variables.

We note that the characteristic properties of the system, which determine its development, are *all* contained in the (generally non-linear) vector function **f**, i.e. in the rates of change of the state variables. The output function **g**, by contrast, has no effect on the system development: it only represents internal processes externally.

The decisive role of the state variables becomes again obvious from Figure 3.7 and the corresponding system equations (Eq. 3.1, 3.2). Given the time-dependent parameters, the time-dependent environmental input functions **u**(t), and the initial conditions of the state variables **z**, the further development of a (deterministic) system can be computed: no additional information is necessary. In particular, all **f** and **g** follow from these quantities. Conversely, in the case of an interruption (for example, a system breakdown or an interruption of the simulation) it is necessary to store only the state variables.

In this discussion no distinction was made between real dynamic systems and corresponding dynamic simulation models. The analysis was valid for both.

Before we look at some distinct elementary systems and their characteristic behavior, let us demonstrate the difference between a continuous and a discrete system, and the computation of their states, by looking at two examples.

Example: Continuous system

Figure 3.9 shows a water reservoir with an adjustable inflow and an adjustable outflow. The system has the state equation:

$$dV/dt = r_{in} - r_{out}$$

or

$$V(t) = V_o + {}_o\!\int^t (r_{in} - r_{out})\, dt.$$

Here V_o is the given initial value of the reservoir and r_{in} and r_{out} are inflow and outflow rates (volume per unit time) which are in general time-dependent. If the inflow and outflow rates remain constant during the time from 0 to t, the integration immediately gives the result

$$V(t) = V_o + (r_{in} - r_{out})\, t ,$$

i.e. the volume increases or decreases linearly with time and proportionally to the difference between inflow and outflow rate. Figure 3.9 shows the volume change for the case where the inflow rate is greater than the outflow rate. The state equation can also be represented by the block diagram in Figure 3.9. The water reservoir appears as an integrator with an initial value V_o, the input $dV/dt = r_{in} - r_{out}$, and the output V(t). (Don't confuse inflow and outflow with input and output, inflow and outflow are rates of change of the state variable while the output in this case is the state variable itself.)

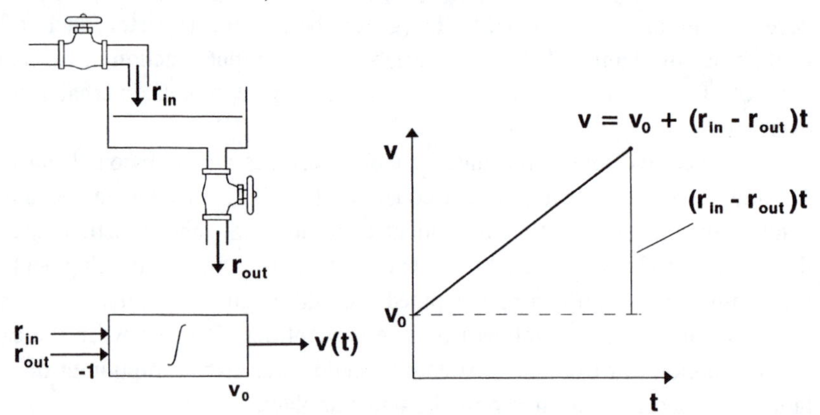

Fig. 3.9: Continuous one-level system with constant inflow and outflow.

Example: Discrete system

Now consider a savings account with an annual computation of the interest as a function of the current level of the account. The new account level follows from:

$$V(k+1) = V(k) + r \cdot V(k) \cdot T$$
$$= (1 + rT) \cdot V_k.$$

Here V is the current account, r the interest rate per year, and T the time period after which the computation takes place (here: one year). Starting with an account of 1 in year 0, the account level in future years can be computed using this state equation:

$$V_1 = (1 + rT) \cdot V_0$$
$$V_2 = (1 + rT)^2 \cdot V_0 = (1+rT) \cdot V_1 = (1+rT) \cdot (1+rT) \cdot V_0$$
$$\ldots$$
$$V_k = (1 + rT)^k \cdot V_0.$$

The solution is defined only for the discrete points in time, 0, 1, 2, etc., since it is assumed that interest accrues only once a year. The development is shown in Figure 3.10. The solution is a geometric series; the savings account will therefore have geometric growth.

The simulation diagram for this example is also shown in Figure 3.10. Note that in this case the increase is not constant and is not exogenously given as in the previous example but changes as a function of the level of the account and must be computed anew at each point in time.

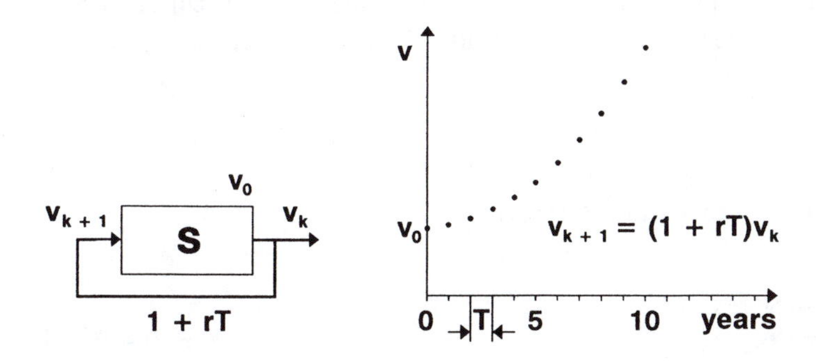

Fig. 3.10: Discrete one-level system with an (annual) increase which is proportional to the current level.

3.3 Some Elementary Systems and Their Behavior

With the block symbols and graphic conventions introduced in Figure 2.6 it is possible to construct dynamic system models of any complexity containing any subprocesses. However, most dynamic systems owe their behavior to only a very few elementary system structures. It is therefore appropriate to deal first with some simple elementary systems and their behavior in order to be able to recognize them in larger systems and to better understand how their behavioral features result from their system structures. Some of the elementary systems discussed in the following we have already encountered in the "global model" (exponential growth, logistic growth, exponential delay, oscillating system). The systems discussed in the following can all be studied in more detail with models in the systems zoo (Ch. 6 and diskette).

3.3.1 Memory-free system

The basic structure of a memory-free system is shown in Figure 3.11: the simulation diagram can only contain blocks for input quantities (hexagons) and intermediate variables (circles) which together correspond to the behavioral equation (output equation)

$$\mathbf{v}(t) = \mathbf{g}(\mathbf{u}, t).$$

In a memory-free system the input quantities are merely transformed by algebraic or logical relationships to the output quantities; the instantaneous values of the output variable $\mathbf{v}(t)$ follow from these relations at any time. Since there are no state variables, the history of the system has no influence on the current behavior which follows directly and instantaneously from the current environmental input $\mathbf{u}(t)$.

Memory-free system (transformer)

Fig. 3.11: Memory-free system: simulation diagram and corresponding equation.

The example in Figure 3.11 shows the (non-linear) computation of an output variable v from a time-dependent sinusoidal oscillation as input variable u:

$$v(t) = a \sin^2(wt).$$

A more practical example is shown in the photoproduction Model M110 in Chapter 6 and the systems zoo. It describes photoproduction as a function of solar position and radiation. In this case the photoactive solar radiation s is computed as a function of seasonal time, of day time, and of geographic latitude. This is a mere transformation—there are no state variables involved, and there is therefore no effect of the past history of the system.

3.3.2 Exponential growth and decay

Input-determined systems can change only if the input changes. State-determined systems can change even if there is no input since the state may have an effect on itself. The simplest structure to achieve this is that for exponential growth or decay shown in Figure 3.12 (first order system) which shows a feedback loop from the state back to itself (self-loop). Since the input to a state variable is a rate of change, this loop means that the state determines its own rate of change. This basic structure can be found in many processes; it may lead to both exponential growth or exponential decay depending on the sign of the feedback connection.

For example, assume a rabbit population increases by ten percent per month. This means that the number of additional rabbits at the end of the month can be computed from the current number of rabbits (by multiplying by 0.1): i.e. the rate of change is a function of the state. Similarly, in radioactive decay, the amount of material which decays in a given time period (rate) is always a given percentage of the material present (state): again, the state affects the rate and therefore its own dynamic development.

The state equation for this process is therefore:

$$dx/dt = r \cdot x.$$

The dynamics depend completely on the sign of the (relative) rate r: if it is negative, the state decreases exponentially; if it is positive, it increases exponentially. This system can be studied in more detail using the Model M 102 EXPONENTIAL GROWTH AND DECAY in the systems zoo. Figure 3.12 shows some representative results for different growth or decay rates $a = r$.

In a population, the number of births and deaths are proportional to the size of the population x and the specific birthrates b and deathrates d. Instead of accounting for gains and losses separately, the net relative growth rate can be obtained as the difference of the positive and the negative growth rate contributions: $r = b - d$. Depending on the sign of this difference, the system will show exponential growth or decay. We have already encountered this formulation of exponential growth in the "population" submodel of our small global model.

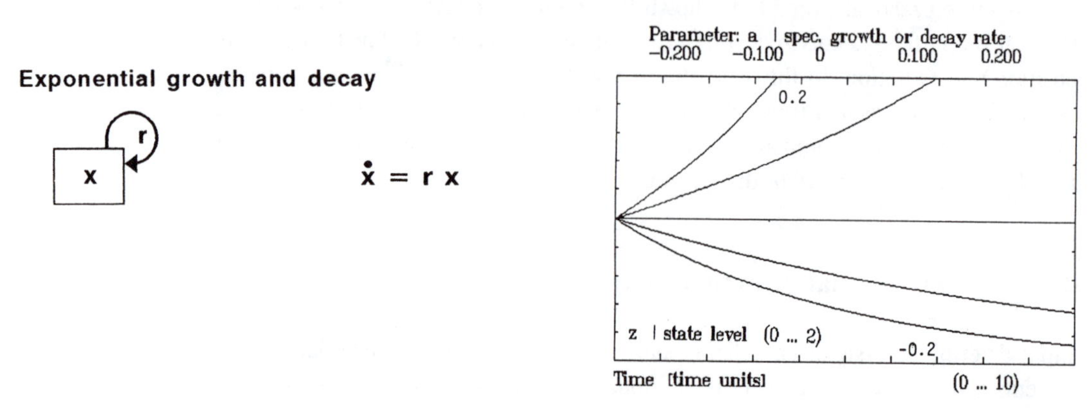

Fig. 3.12: Exponential growth and decay: simulation diagram and some results.

3.3.3 Logistic growth

In physical processes, growth cannot continue indefinitely. It will eventually approach a physical limit. Some processes are only stopped when they "hit the ceiling;" in others, the limit makes itself felt long before it has been reached. For example, crowding effects and the competition for scarce food slow down the growth of populations as they approach their limits and finally reduce growth to zero. This type of growth is called "logistic" (sigmoidal) growth.

The basic system structure for logistic growth is shown in Figure 3.13 (first order system). In this system the growth rate depends on the remaining capacity for growth. As long as the state variable (for example, population) is much smaller than the carrying capacity of the environment, growth is hardly restricted, and the population increases exponentially. As the population approaches the maximum possible population k, the remaining free capacity for growth $(k - x)$ approaches zero. Since the exponential growth term $(r \cdot x)$ is multiplied by this difference, the growth rate of the population approaches zero as the population reaches its capacity limit. This simple non-linear system, which (with certain modifications) is found in many ecological, economical, social, physical, and technical processes, shows a typical S-shaped (logistic) development.

$$dx/dt = r \cdot x \cdot (1 - x/k).$$

We have encountered this basic system in the "capital" submodel of our small global model. This system can be studied in greater detail using Models M 107 LOGISTIC GROWTH FOR CONSTANT HARVEST and M 108 LOGISTIC GROWTH WITH POPULATION-DEPENDENT HARVEST in the systems zoo. Figure 3.13 shows typical growth curves as function of normal growth rate $a = r$.

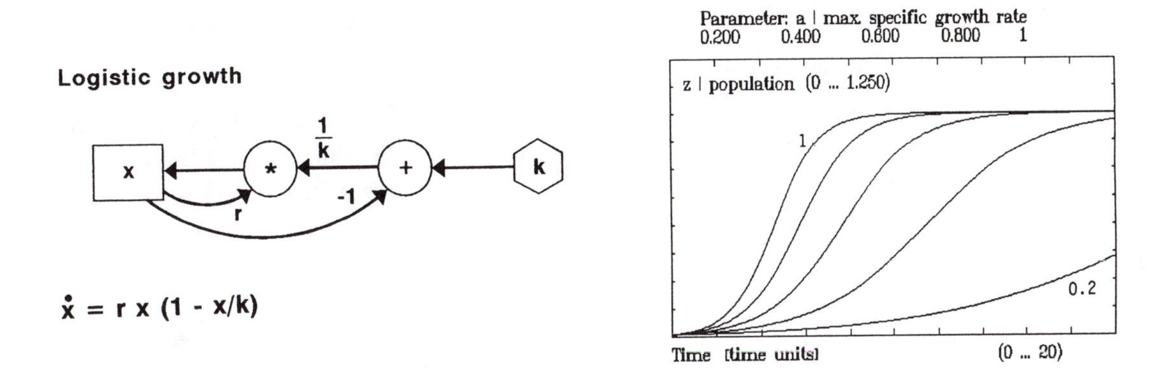

Fig. 3.13: Logistic growth: simulation diagram and representative results.

3.3.4 Exponential delay (exponential leak)

Another common process in which the state now depends on both the state itself and an input is shown in Figure 3.14 (first order system). In this system, there is a constant inflow that is independent of the state of the system. However, the outflow is determined by the state of the system: at a high level the rate of outflow will be higher than at low level.

The dynamics of the system can be understood by visualizing a bathtub with a leaking plug: if the water level (the state variable) is low, the leakage rate $(r \cdot x)$ will be small; if the water level is high, it will be larger. If for an initially empty bathtub the inflow u is set to a constant value $u > r \cdot x$, the leak rate of the plug will increase as the water level rises until the leakage losses are exactly equal to the inflow. As this point is reached, the water level will remain constant: the system has reached a state of flow equilibrium.

This process provides a description of the accumulation of pollution, for example, and we have already used it for this purpose in the "pollution" submodel of our small global model: the pollution input is independent of the pollution state while pollution absorption depends on the level of pollution present. The pollution level will stabilize where the absorption is equal to the (constant) pollution input.

The same process of adjustment takes place if the inflow rate u(t) is a function of time. In this elementary structure, the state x(t) will adjust to the input u(t) after a certain delay. The delay is longer if the specific leak rate r is smaller. Or, put differently: such a system can react faster to a changing input if it loses its history more quickly, i.e. if its leak rate r is large. A system having this structure therefore functions as a delay since its current state represents a recent part of its history. The time constant $T = 1/r$ is used as a measure for the delay.

$$dx/dt = u(t) - r \cdot x$$

This system can be studied in more detail using the Model M 103 EXPONENTIAL DELAY in the systems zoo (in particular by using different test functions u(t)). Figure 3.14 shows some typical response curves for different values of the decay rate $a = r$.

Exponential delay (exponential leak)

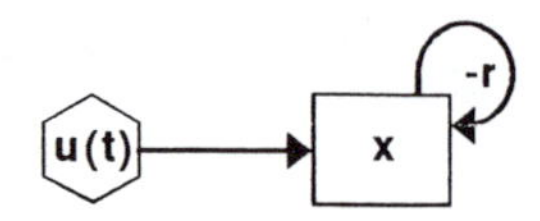

$$\dot{x} = u(t) - rx$$

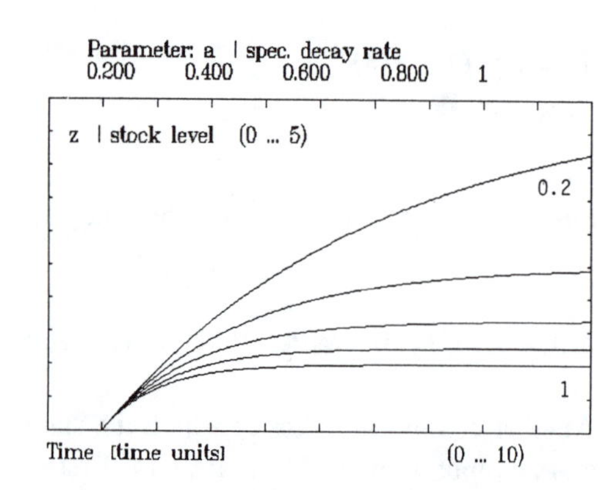

Fig. 3.14: Exponential delay (exponential leak): simulation diagram and representative results.

3.3.5 Linear oscillator

Pull down a weight suspended on a spring and release it: it will bounce upward until the tension in the spring is released, then it will fall down again, stretch the spring until the downward motion is broken, and then move upwards again. The motion will repeat several times: the system oscillates. Many systems can oscillate, but continuous systems can only do so if there are at least two state variables connected in a common feedback loop. In this case, one state variable would represent the kinetic energy of the weight, the other the potential energy in the spring.

The elementary structure of the linear oscillator is shown in Figure 3.15 (second order system). It has a feedback loop running across two state variables: the first state variable determines the rate of change of a second state variable, and this again determines the rate of change of the first state variable. For example, for the mass suspended on a spring, the spring displacement (state variable corresponding to potential energy) determines the spring force which determines the acceleration of the mass and therefore its velocity (state variable corresponding to kinetic energy), and this determines the location of the mass and therefore the spring displacement. The feedback loop is negative, and the system is likely to oscillate. (In the spring-mass system, an initial displacement causes a restoring force in the opposite direction.)

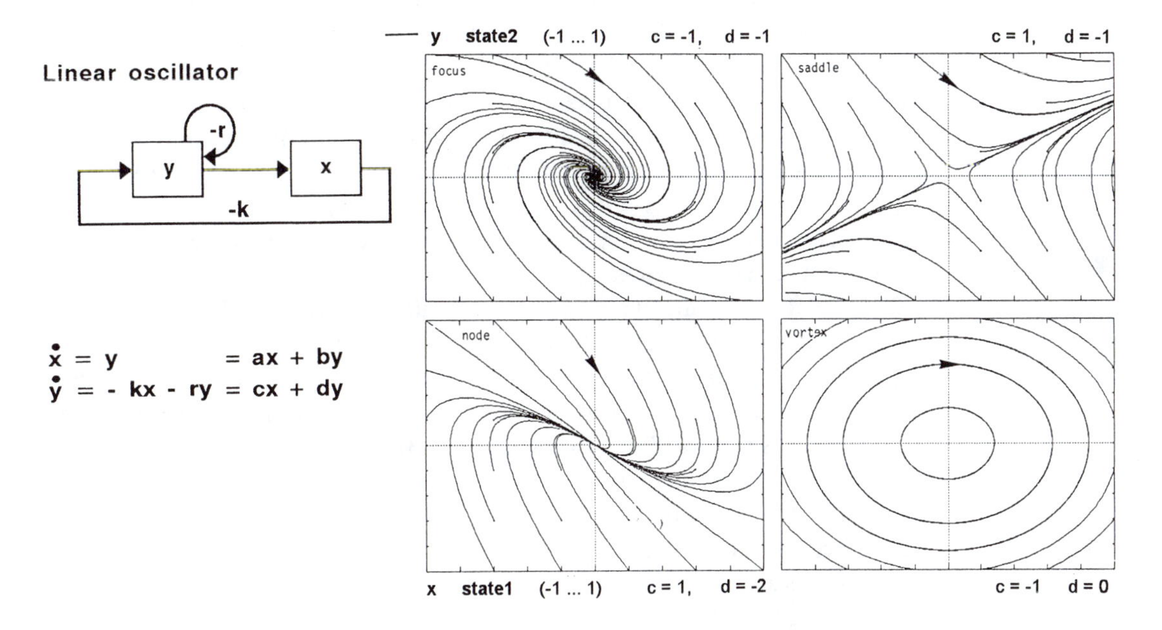

Fig. 3.15: Linear oscillator: simulation diagram and representative results.

In the spring-mass system the rate of change of displacement (dx/dt) is equal to the velocity y of the mass while the rate of change of the velocity (dy/dt), the acceleration of the mass, is equal to the acceleration caused by the restoring force of the spring (–k·x, opposing the motion), where k is the spring constant. The system corresponds to a harmonic oscillator with the coupling strength k and the state equations:

$$dx/dt = y$$
$$dy/dt = -k \cdot x.$$

In this form, the system has no damping, and it will therefore continue in undamped "harmonic" oscillation forever. In reality, systems lose energy by friction, and the motion is damped. Assuming a friction force proportional to the velocity of the mass, we must modify the system as in Figure 3.15: the damping force provides an additional acceleration ($-r \cdot y$) opposing the motion (Fig. 3.4). If such an (exponential) damping (proportional to one of the state variables) is introduced (see "exponential leak" in Sec. 3.3.4), the oscillations will be damped and will disappear after some time. The state equations of the system are then

$$dx/dt = y$$
$$dy/dt = - k \cdot x - r \cdot y.$$

Depending on magnitude and sign of the coupling k and damping r, the system shows very different modes of behavior. Some examples are shown in Figure 3.15. If one studies the behavior as a function of k and r (for example, using the systems zoo program M 203 LINEAR OSCILLATOR; note: system parameters $a = 0$, $b = 1$, $c = -k$, $d = -r$), one observes the following (Fig. 7.6):

1. Stable behavior is only possible if $r < 0$ and $k < 0$ (i.e. $d > 0$, $c > 0$).
2. Oscillations will occur if $k < (r^2/4)$ (i.e. $c > (d^2/4)$.

In the case of stable behavior, the system will return to a point of equilibrium ($x = 0$, $y = 0$) after a perturbation.

The distinct differences between the different modes of behavior become much more obvious if the solutions (the state trajectories) are plotted in state space (or "phase plane" in two dimensions). In the phase plane (x, y) we obtain characteristic response maps as functions of $c = -k$ and $d = -r$ which can be categorized as follows (Fig. 3.15 and Ch. 7.3.13):

1. saddle (always unstable)
2. node (stable or unstable)
3. focus (stable or unstable)
4. sink (stable)
5. source (unstable)
6. vortex (marginally stable)

These classifications refer to the typical shape of the trajectories in the neighborhood of the equilibrium point. In a stable focus, for example, all trajectories spiral into the focus and end at the equilibrium point which is at the center of the focus.

These behavioral patterns are of fundamental importance for the study of systems, since in the neighborhood of equilibrium points, non-linear systems can usually be approximated by their linearizations to study the local behavior (see Sec. 7.1.11 - 7.1.14). For autonomous linear systems (such as the present one), these equilibrium points are always at $x = 0, y = 0$. For linear systemss, the stability conditions are independent of the initial conditions.

This system can be studied in more detail using the Model M 203 LINEAR OSCILLATOR in the systems zoo. In this model the eigenvalues, as well as the response mode type and its stability, are shown for each parameter choice.

In general, oscillations of continuous systems are likely if negative feedback loops pass through at least two state variables. This is the case also for non-linear systems (predator-prey systems are an example).

3.3.6 Bistable oscillator

Small changes in system structure may introduce major qualitative changes in system behavior. Very often such structural changes are not readily apparent in the physical appearance of the system. In particular, if non-linearities are introduced, the system behavior may become very surprising indeed. Systems are only linear if their state variables appear in the first power and in additive combinations of the state equations; all other formulations are non-linear.

A certain non-linear modification of the linear damped oscillator (Fig. 3.16) produces an oscillator having not one but three equilibrium points, two of which are stable (second order system; Duffing system). This bistable oscillator has the property that the system trajectories end at one of the two stable equilibrium points, and the point at which the system comes to rest is determined by the initial conditions. The state equations are:

$$dx/dt = y$$
$$dy/dt = -x \cdot (x^2 - 1) - r \cdot y = x - x^3 - r \cdot y .$$

From the second equation it becomes clear that the coupling from x to y will become negative if $|x| > 1$: at some distance from the origin, the system will behave like the linear oscillator in Section 3.3.5. However, close to the origin, the feedback sign of the coupling from x to y reverses. The state trajectories (x, y) for this system now show a rather strange behavior (Fig. 3.16 and M 220, Ch. 6): the system has two stable equilibrium points (focus at $x = \pm 1, y = 0$) and one unstable equilibrium point (saddle at $x = 0, y = 0$). If the state amplitude is large enough and the damping small, the system oscillates around the three equilibrium points. As the state trajectory approaches the equilibrium point, it is finally captured by one of the stable equilibrium points and comes to rest there. However, it depends on the initial conditions at which of the two stable equilibrium points the system will come to rest. This system can be studied using Model M 220 BISTABLE OSCILLATOR in the systems zoo.

In contrast to linear systems, non-linear systems may therefore have several equilibrium points with different stability behaviors; the equilibrium state often depends on the initial conditions. Finally, there may be, in addition to equilibrium points, limit cycles and equilibrium surfaces of higher order (limit cycles around limit cycles: tori) on which the system can move (see, for example, Model M 219 LIMIT CYCLE OSCILLATOR in the systems zoo).

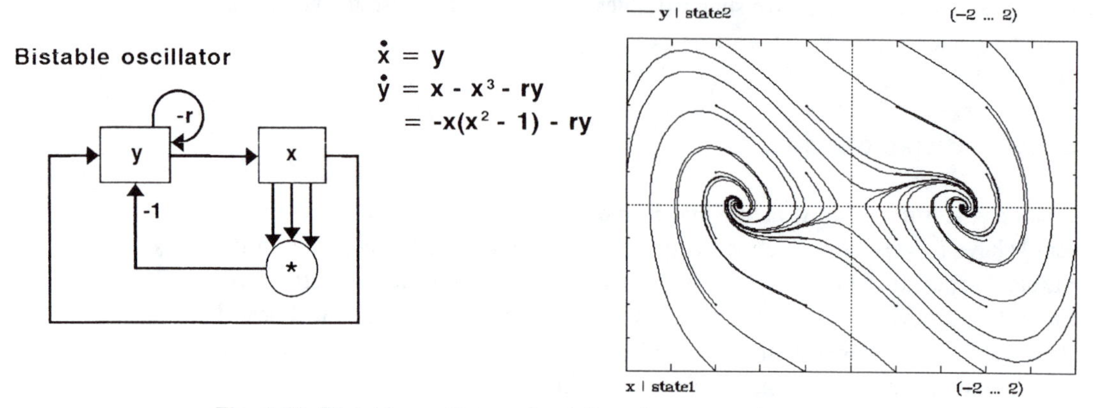

Fig. 3.16: Bistable oscillator: simulation diagram and some results. (r = 1).

3.3.7 Chaotic bistable oscillator

Except for the fact that it comes to rest at one of two equilibrium points, the motion of the bistable oscillator (Sec. 3.3.6) is quite similar in its regularity to that of the linear oscillator (Sec. 3.3.5). However, things change quite dramatically if we start shaking this system with a constant frequency: the system response becomes "chaotic." The oscillation now takes on a very irregular appearance (Fig. 3.17).

The elementary structure of the chaotic bistable oscillator is shown in Figure 3.17 (third order system; periodically excited Duffing system). The system is identical to the bistable oscillator (Sec. 3.3.6) except for the forcing by a time-dependent cosine function. In order to formulate it as an autonomous system (without exogenous forcing), time is introduced as a new state variable z, and we therefore use the three state equations:

$$dx/dt = y$$
$$dy/dt = x - x^3 - r \cdot y + q \cdot \cos(\omega z)$$
$$dz/dt = 1,$$

where z is identical to time t, since,

$$z = {}_0\int^t (dz/dt)\, dt = {}_0\int^t (1)\, dt = t$$

In certain parameter ranges the system now produces chaotic behavior (as in Fig. 3.17), which, as a function of time, is difficult to interpret. However, if we plot the same motion in state space, we find that the state trajectory moves in a certain region of attraction (here: a "figure eight," see Fig. 3.17), the "chaotic attractor" of the system. After orbiting one of the equilibrium points for a while, the system state diverges and orbits the other equilibrium point until it again diverges in a seemingly unpredictable fashion to the previous basin of attraction. Despite the deterministic state equations it is no longer possible to make an exact prediction of the state development. Continuous systems can only be chaotic if they have at least three state variables.

This system can be studied in more detail using the Model M 221 CHAOTIC BISTABLE OSCILLATOR in the systems zoo. Other chaotic systems in the systems zoo are M 309 ROESSLER ATTRACTOR, M 310 LORENZ SYSTEM, and M 311 COUPLED DYNAMOS.

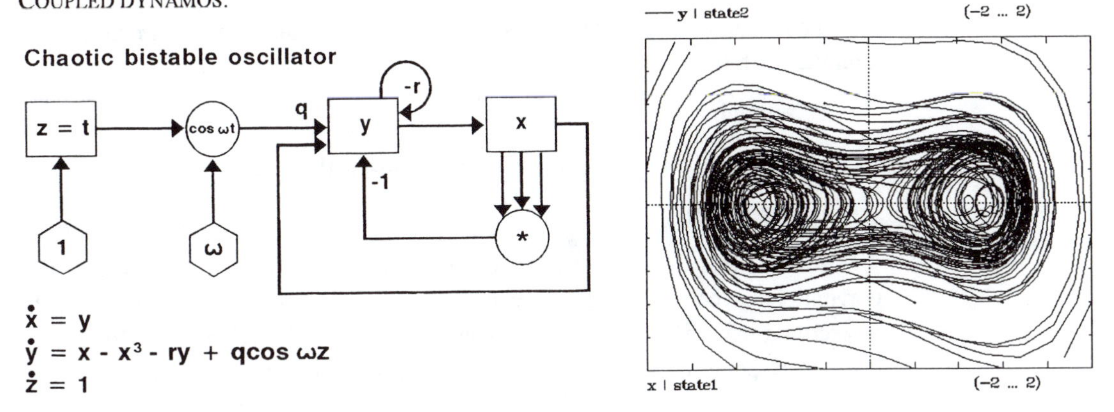

Fig. 3.17: Chaotic bistable oscillator: simulation diagram and some results ($r = 0.25$, $\omega = 1$, $q = 0.3$).

3.4 Dimensional Analysis

Mathematical equations involving dimensional quantities are correct only if the operations indicated on both sides of the equation agree not only in terms of the value of the quantities but also in terms of their units of measurement (dimension). In model development, dimensional agreement must be strictly enforced. The condition of dimensional conformity can be used to obtain important information for the model formulation. In particular, we can use it to:

1. check the validity of a model equation,
2. determine correct conversion factors, and
3. help us formulate model equations.

3.4.1 Checking dimensional validity

In dimensional analysis, we complement the terms of an equation by the unit of measurement (dimension) of each variable in brackets [unit]. In checking the equation, we must make sure that 1. the mathematical expression is legitimate, and that 2. units on both sides of the equation agree after performing the mathematical operations indicated by the mathematical expression. If there is no agreement, we must check for two possibilities: 1. the expression may be correct except for a conversion factor, or 2. the expression may be completely illegitimate.

Assume we wish to compute the total mechanical energy of a car rolling downhill. From elementary physics, we know that its total energy E_t is equal to the sum of potential energy E_p and kinetic energy E_k :

$$E_t = E_p + E_k .$$

The potential energy of a mass m lifted from an altitude zero to an altitude h on the earth (gravitational constant g) is given by the formula:

$$E_p = m \cdot g \cdot h.$$

For the kinetic energy of a mass m moving with velocity v we have:

$$E_k = m \cdot (v^2/2).$$

Assume that the mass M of the car is given in metric tons [t], the earth's gravitational acceleration g in [m/sec^2 = m/s^2], the height difference H (with respect to a reference altitude) in kilometers [km], and the velocity V [in km/h]. If the corresponding numerical values are simply introduced into the formula:

$$E_t = M \cdot g \cdot H + (1/2)\, M \cdot V^2,$$

this equation appears to be correct, but the dimensional test will show that the application of the formula in this form is not permissible:

$$E_t\,[?] = M\,[t] \cdot g\,[m/s^2] \cdot H\,[km] + (1/2) \cdot M\,[t] \cdot V^2\,[(km/h)^2].$$

We would be trying to add apples [t · m/s^2 · km] to oranges [t · (km/h)^2]. Obviously, several initial conversions have to be performed before the numerical values for M, g, H, and V can be inserted into the equation. However, the equation is correct in principle since all terms have dimension [energy] = [mass · acceleration · distance] or [mass · velocity ^{2}]. In fact, we could use arbitrary units of energy on both sides of the equation as long as we make sure that the units on both sides of the equation are identical.

In the expression for potential energy $E_p = m \cdot g \cdot h$, the units for m, g, and h on the right hand side can be arbitrary units of mass, acceleration, and length, respectively, as long as energy on the left side is expressed in corresponding units:

E_p [energy $unit_1$] = m [mass $unit_1$] · g [acceleration $unit_1$] · h [length $unit_1$];

for example,

[energy $unit_1$] = [kg] [m/s^2] [m] = [kg · $(m/s)^2$].

In the expression for kinetic energy $E_k = m\,(v^2/2)$, the energy unit on the left hand side follows from the mass and velocity units of the right hand side by

E_k [energy $unit_2$] = m [mass $unit_2$] · (1/2) · v^2 [(velocity $unit_2)^2$];

for example,

[energy $unit_2$] = [kg] [$(m/s)^2$] = [kg · $(m/s)^2$].

In order to add the potential and kinetic energy contributions, we must have

[energy $unit_1$] = [energy $unit_2$].

This does not necessarily mean that the same units of measurement (here for energy) have to be used throughout a simulation model. This is often not practical since original data have to be used which were obtained in specific measurement units. For example, the energy flows of photosynthesis in a forest are measured by plant physiologists in terms of milligram CO_2-exchange per square decimeter leaf area per hour [$mgCO_2 \cdot dm^{-2} \cdot h^{-1}$] while forest planners are interested in corresponding values in terms of tons dry matter per hectare per year [$t_{OTS} \cdot ha^{-1} \cdot year^{-1}$]. In this case it is legitimate and practical to compute leaf photosynthesis on an hourly basis using the first unit and to convert the result later to its annual and per hectare dry matter value using the second unit and the correct conversion factor.

The necessary conversions must be carried out very carefully as significant errors may easily be introduced. Conversion errors are not obvious since the equation appears to be "correct" upon superficial inspection.

3.4.2 Finding correct conversion factors

The search for the correct conversion factor starts with the choice of the proper units. For our example, we start by prescribing the energy unit which is to be used to express the potential and kinetic energies. We choose the Joule [J]. In agreement with the international system of units (SI system), using the Newton [N] as a (derived) unit of force, we have:

1 [J] = 1 [Nm] = 1 [kg (m/s^2)]·[m] = 1 [kg (m^2/s^2)].

We also have: 1 [t] = 1000 [kg] and 1 [km] = 1000 [m]. The gravitational constant g is a system constant which is expressed in the SI system by

$g = 9.81$ [m/s^2].

We now obtain for the potential energy [J] in the previous example as a function of the quantities M [t] and H [km]:

$$E_p\,[J] = M\,[t] \cdot 1000\,[kg/t] \cdot 9.81\,[m/s^2] \cdot H\,[km] \cdot 1000\,[m/km]$$
$$= (9.81 \cdot 10^6 \cdot M \cdot H)\,[kg\,m^2/s^2]$$
$$= (9.81 \cdot 10^6 \cdot M \cdot H)\,[J].$$

We now find the same dimensions on both sides of the equation. The equation is valid only if (as assumed here) M is expressed in [t] and H in [km].

For the conversion of the expression for kinetic energy we have to remember that one hour corresponds to 60·60 = 3600 seconds:

$$E_k\,[J] = (1/2) \cdot M\,[t] \cdot 1000\,[kg/t] \cdot V^2\,[(km/h)^2] \cdot (1000\,[m/km])^2 \cdot (3600\,[s/h])^{-2}$$
$$= (0.5 \cdot 10^3 \cdot 10^6 \cdot 3.6^{-2} \cdot 10^{-6} \cdot M \cdot V^2)\,[kg\,m^2/s^2]$$
$$= (38.58 \cdot M \cdot V^2)\,[J].$$

Again, this formula is valid only if M is specified in [t] and V in [km/h].

Since they are expressed in the same units [J], we can now compute the total energy [J] as the sum of the contributions of potential energy [J] and kinetic energy [J]. For the computation using M [t], H [km], and V [km/h], we now have the relationship:

$$E_t\,[J] = (9.81 \cdot 10^6\,M \cdot H) + (38.58 \cdot M \cdot V^2)$$

or

$$E_t\,[J] = a \cdot g \cdot M \cdot H + b \cdot (1/2)\,M \cdot V^2 \qquad \text{(i)}$$

with conversion factors $a = 10^6$ and $b = 77.16$. Use of the formula

$$E_t\,[J] = (m \cdot g \cdot h) + (0.5\,m \cdot v^2)$$

would only be permissible if all quantities were expressed in consistent SI quantities from the beginning, i.e. if m had been specified in [kg], g as 9.81 [m/s^2], h in [m], and v in [m/s]. We then obtain:

$$E_t[J] = (9.81\,[m/s^2] \cdot m\,[kg] \cdot h\,[m]) + (0.5\,m\,[kg] \cdot v^2\,[(m/s)^2]$$
$$= (9.81\,m \cdot h + 0.5\,m \cdot v^2)\,[kg\,m^2/s^2]. \qquad \text{(ii)}$$

Using the definition of the unit of force Newton [N], we can write

$$1\,[J] = 1\,[N\,m] = 1\,[(kg\,m/s^2) \cdot m] = 1\,[kg\,m^2/s^2]$$

to verify that we do indeed have the same unit [J] on both sides of the equations. We can also verify that Equations (i) and (ii) produce the same numerical result by substituting values for M (or m), H (or h), and V (or v).

3.4.3 Using dimensional analysis in formulating model equations

In our model development up to now we have only been concerned with the functional coupling of the system quantities without checking for dimensional conformity of the equations. In fact, we simply assumed that all quantities were given in compatible and consistent units (for example, SI units). This is generally not the case, and a careful checking of dimensional conformity is therefore mandatory in model construction.

But dimensional analysis can also help us in formulating model relationships for which we have no initial formula. From the influence diagram we know which variables and parameters (inputs) are to be used to compute a certain element (output); we also know in which units the inputs and the output are measured. Dimensional analysis then allows only certain combinations of the input variables to obtain the correct units for the output. This provides valuable information for formulating the correct relationship.

Example: A verbal model for the kinetic energy of a mass would contain the following statements for example:

"The larger the mass, the greater the kinetic energy."
"The larger the velocity, the larger the kinetic energy."

A first naive model formulation could therefore be

$$E_k = A \cdot m \cdot v$$

where A is an as yet undetermined factor. The dimensional test

$$E_k\ [J] = E_k\ [kg\ m^2/s^2] = A\ [?] \cdot m\ [kg] \cdot v\ [m/s]$$

immediately shows that A should have the dimension [m/s] in order to satisfy the dimensional equation. This is a suggestion to use v in the second power

$$E_k\ [J] = a\ [-] \cdot m\ [kg] \cdot v^2\ [(m/s)^2].$$

In this formulation a dimensionless coefficient a remains which has to be found from other considerations or investigations (for example, from the time integral of the impulse mv, $a = 1/2$).

This approach to the formulation of dimensionally correct model equations can be used in any arbitrary problem setting. It applies to "hard" mathematical models developed in the natural sciences using internationally defined measurement units as well as to "soft" models which may have to be developed in a social science context using *ad hoc* qualitative quantities. In all cases it is true that the dimensions—whatever they are—on both sides of an algebraic expression have to be identical. In both cases this requirement can also be used to find the mathematical form of an algebraic dependence or the missing dimension of a quantity in an equation.

Even if the model should only contain dimensionless quantities, the requirement of dimensional consistency also applies to the "dimensionless dimensions" of the quantities: they must agree on both sides of a (dimensionless) equation.

Example: If we define a relative (dimensionless) population number (for example, relative population = population/population$_{1970}$) and a relative (dimensionless) car ownership per capita [(cars/population)/(cars$_{1970}$/population $_{1970}$)], then the relative (dimensionless) number of cars obtained from the product of both numbers must still have the correct (dimensionless) "dimension" [cars/cars$_{1970}$]:

(relative number of cars) =
= (relative population number)·(relative car ownership per capita);
[cars/cars$_{1970}$] =
=[population/population$_{1970}$]·[(cars/population)/(cars$_{1970}$/population$_{1970}$)].

This condition in particular applies to the rates of change of the state variables: these must have the dimension of the state variable per time unit, even if a relative (dimensionless) time unit is being used.

We shall now apply influence analysis and dimensional analysis to develop two simulation models. The first, for a rotating pendulum, uses elementary relationships from physics; the second, for a fishery operation, employs ecological and economic relationships.

3.5 Model Development for the Rotating Pendulum

3.5.1 Problem statement, model purpose, and verbal model

A pendulum may be a simple mechanism, but it exhibits phenomena that are peculiar to non-linear systems. An understanding of these phenomena is useful also for understanding the behavior of other systems. It has stable and unstable points of equilibrium (lower and upper dead centers), it changes the characteristics of its motion with time (rotating or swinging motion), and its motion strongly depends on the initial conditions (position and angular velocity). In this section, we shall develop the model equations and the simulation diagram. The behavior of this system will be studied in detail in Chapter 4. We assume that the pendulum is free to rotate and that the pendulum mass is supported on a stiff rod (Fig. 3.18).

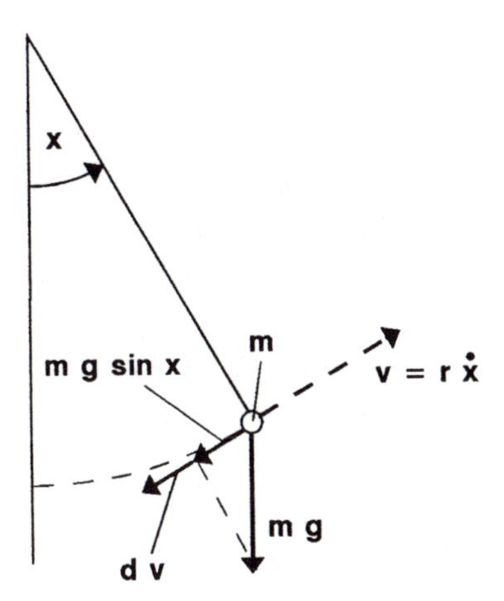

Fig. 3.18: Forces and parameters of the pendulum system.

If the initial velocity is high, the pendulum will initially rotate until its angular velocity has decreased to a point where it can no longer pass over the upper dead center. The pendulum will reverse its motion and swing around the lower dead center until it comes to rest at this point. Obviously, it is not easy to describe this motion: we are dealing with two qualitatively different types of motion (rotation in the same direction and swinging motion with alternating direction of motion). During the rotation and the swinging motion the velocity changes constantly. The direction of the driving force (the gravitational force) changes continuously with respect to the direction of motion. The pendulum motion is slowed down by air resistance. This resistance depends on the shape and cross-sectional area of rod and pendulum and on the viscosity of the air, which again depends on temperature and air density. The friction in the bearing also retards the rotational motion; it is determined by the quality of the bearing and the viscosity of the lubricant. If the mass is large with respect to the cross-section of the rod, the rod will be constantly stretched and compressed, thereby changing the radius of the pendulum (slightly).

Even the simplest system can show a number of complications on closer look. The first task of model building is therefore to determine which of the processes are indeed behaviorally relevant and must be included in the description, and which processes have little influence and can be neglected. Very often this decision cannot be substantiated *a priori*, and it may be necessary to revise it later, when more is known about the system. It is therefore necessary to start with hypotheses that later have to be validated or disproved by the results of the investigation. The choice of simplifications and hypotheses depends to a great deal on the model purpose. If the objective were the derivation of the exact equations for a pendulum chronometer, one would have to work with much more precision and detail than for the derivation of the equations of an idealized and simplified system.

Model purpose: *The goal of the study is to develop the equations of motion—the mathematical model—for the motion of an idealized rotating pendulum. The motion should be correctly described throughout the entire range of possible motion of the pendulum.*

For the rotating pendulum we use the hypothesis that the motion of an idealized pendulum will basically agree with that of a real pendulum of the same mass and the same radius. We therefore make the following assumptions:

1. The mass is concentrated at a point at the end of the rod.
2. The rod itself is completely stiff, weightless, and without air resistance.
3. The bearing is frictionless.
4. The velocity of pendulum motion is small enough to ensure laminar flow (low Reynolds number; air drag then increases linearly with velocity).

These simplifying assumptions are used in the **verbal model** which is the basis for the further model development. The idealized system to be described is shown in Figure 3.18. From well-known facts and our reflections about the system we develop the following list of statements (the verbal model):

1. Because of the stiff support, the motion is restricted to a circle of radius r.
2. Because of the stiff support, the components of motion (velocity, acceleration) can only be tangential to the circle; all components of motion in radial direction are equal to zero.
3. The position on the circular trajectory is uniquely determined by the angle of rotation measured from the lower dead center (x in radians).
4. The angular velocity on the circular path is the momentary rate of change of the angular position (the time derivative dx/dt).
5. The velocity of the pendulum mass m on its trajectory is equal to the angular velocity times the radius: $v = (dx/dt) \cdot r$
6. An air drag force acts on the pendulum mass, opposing the motion.
7. The air drag is proportional to the velocity of the mass.
8. The air drag is proportional to the cinematic viscosity of the fluid in which the pendulum moves.
9. The effects of viscosity, pendulum cross-section, and drag coefficient of the pendulum mass can be combined into a "damping constant" d such that the drag force is $d \cdot v = d \cdot r \cdot dx/dt$.
10. The gravitation force $m \cdot g$ (proportional to pendulum mass m and gravitational acceleration g) always acts on the pendulum mass in the direction of the earth's center.

11. The only component of this restoring force relevant for the motion is the component in the direction of the motion trajectory (tangential to the circle), i.e. $m \cdot g \cdot \sin x$ (Fig. 3.18).
12. Two forces therefore act on the pendulum mass: a. the damping force of the air resistance, and b. the restoring force of gravitation.
13. The acceleration acting on the pendulum mass is a result of these two forces.
14. The acceleration is proportional to the accelerating force.
15. The acceleration is inversely proportional to pendulum mass.
16. The angular acceleration is proportional to the acceleration of the pendulum mass but inversely proportional to radius. (For the same acceleration and a doubling of the radius, we obtain the same increase of velocity on the trajectory per time unit but half the increase of angular velocity.)
17. Positive angular acceleration causes increasing angular velocity.
18. Positive angular velocity leads to an increase in angle.

In this verbal model we now have collected all information about the system that appears to be important for the further model development. Ideally, the verbal model would contain all necessary and sufficient information, but normally it will also contain redundant information. Also, it may turn out that further information has to be added in the course of model development.

3.5.2 Developing the influence diagram for the rotating pendulum

The verbal model is the basis for the influence diagram representing the system elements and their influence structure. It is advisable to first develop a list of the system elements playing a role in the verbal model. We find the following:

radius	angular acceleration
mass	velocity
gravitational acceleration	trajectory component of gravitation
damping constant	damping force
angle	acceleration force
angular velocity	mass acceleration

The elements radius, mass, gravitational acceleration, and damping constant are constant parameters. The others are continuously changing system variables.

We begin with an arbitrary element and use the verbal model to determine which other elements have an influence on it. Continuing this for all system elements, and applying in particular Rules 4 and 5 for developing the influence diagram (Sec. 2.1.3), we find the following influence relationships:

1. The pendulum angle changes if the angular velocity is not equal to zero.
2. The angular velocity changes if there is a non-zero angular acceleration.
3. The angular acceleration depends on the acceleration of the pendulum mass.
4. An increase in radius means a decrease in the angular acceleration.
5. The acceleration depends on the mass of the pendulum: for an identical acceleration force, a larger mass means smaller acceleration.
6. The larger the accelerating force, the larger the acceleration.
7. The damping force retards the pendulum motion (negative acceleration).
8. The damping force increases with the velocity of the pendulum mass.
9. The damping force is higher if the damping constant (air viscosity) is larger.
10. The velocity of the pendulum increases with its angular velocity.
11. For the same angular velocity, larger radius means larger velocity.
12. Another contribution to the acceleration force is provided by the trajectory component of the gravitational force (that part of the gravitational force acting in the direction of the pendulum trajectory).
13. The trajectory component of the gravitational force is higher if the gravitational acceleration is larger.
14. The trajectory component of the gravitational force is higher for larger mass.
15. The trajectory component of the gravitational force depends on how much the direction of motion coincides with the direction of gravity, i.e. on sin x.
16. As the angle increases (from 0 to $\pi/2$), the direction of gravity agrees more and more with the direction of motion (sin x from 0 to 1).

These "atomic" influence relationships can now be drawn in the influence diagram (Fig. 3.19). Antagonistic influences are indicated by a minus sign.

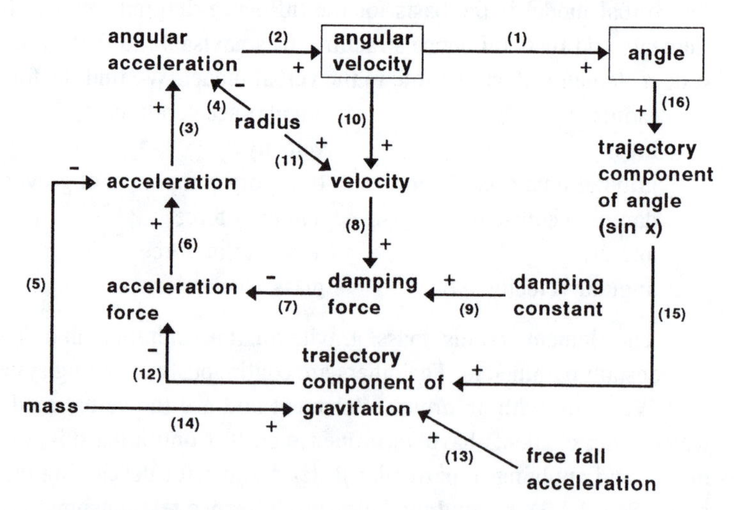

Fig. 3.19: Influence graph of the system elements of pendulum dynamics.

3.5.3 Quantities, dimensions, and relationships for the rotating pendulum

In the next step, we use the influence diagram to derive the mathematical expressions for the relationships between the system quantities. For each node, i.e. each system element, an individual relationship has to be derived. We start by introducing abbreviations, determining the units of measurement, and writing down the numerical values of parameters which are already known. Basic units of measurement are: m = meter, kg = kilogram, s = second, N = Newton, rad = radian.

r	radius	[m]	parameter to be chosen
m	mass	[kg]	parameter to be chosen
d	damping constant	[?]*	parameter to be chosen
g	gravitational acceleration	[m/s²]	g = 9.81 (physical constant)
x	angle	[1]	angle in radians [rad]:
x'	angular velocity	[1/s]	1 rad = 180/π = 57.3 degrees
x"	angular acceleration	[1/s²]	
v	velocity	[m/s]	
F_b	acceleration force	[N = kg m/s²]	
F_g	trajectory component of gravitation force	[N]	
F_d	damping force	[N]	
b	acceleration	[m/s²]	
sin x	projection on trajectory	[1]	

(* The dimension of d is here left unspecified; it will be determined in the dimensional analysis in the following [N/(m/s)]).

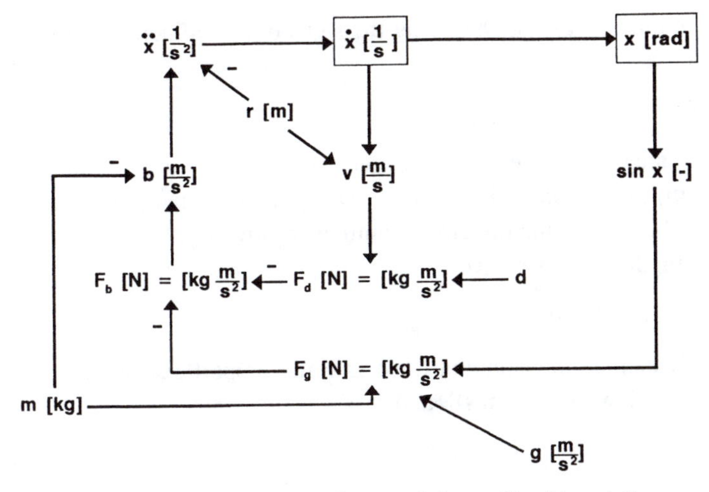

Fig. 3.20: Influence graph for the pendulum with abbreviations and dimensions.

We now redraw the influence structure of Figure 3.19, introducing at each node the corresponding abbreviations for the system quantities and their dimensions (Fig. 3.20).

From this diagram, we now read off at each node the influence relationship together with the dimensions of the quantities involved. We use the symbol "&" as an initially unspecified algebraic operator between the quantities. The correct algebraic operation will be found using dimensional analysis.

For the angular acceleration x" we have:

$$x'' \, [1/s^2] \Leftarrow r \, [m] \; \& \; b \, [m/s^2].$$

The equation can be satisfied only if

$$x'' = b \, / \, r.$$

This formulation also produces the antagonistic effect of r on x'' required by Relation 4 in Section 3.5.2.

For the acceleration b we have

$$b \, [m/s^2] \Leftarrow m \, [kg] \; \& \; F_b \, [kg \; m/s^2]$$

from which follows

$$b = F_b \, / \, m.$$

Again, this formulation assures the antagonistic influence of m, as required by Relation 5.

The acceleration force F_b is given by

$$F_b \, [N] \Leftarrow F_d \, [N] \; \& \; F_g \, [N].$$

Obviously, this can only be a sum. Observing the signs in the influence diagram, we must write

$$F_b = - F_d - F_g.$$

(Angular velocity, angular acceleration and acceleration force are counted as positive in the positive direction of the angle x. Damping force and gravitational component oppose the motion and therefore require a negative sign).

For the damping force F_d we have

$$F_d \, [N] \Leftarrow d \, [?] \; \& \; v \, [m/s].$$

Since we previously defined the damping force as being proportional to velocity, the dimension of d follows as $[N/(m/s)]$ and

$$F_d = d \cdot v.$$

The trajectory component of the gravitation force F_g follows from the relationship

$$F_g \, [N = kg \, m/s^2] \Leftarrow \sin x \, [1] \; \& \; g \, [m/s^2] \; \& \; m \, [kg].$$

This is satisfied by simple multiplication and therefore

$$F_g = \sin x \cdot g \cdot m.$$

For the velocity v we obtain

$$v \, [m/s] \Leftarrow x' \, [1/s] \; \& \; r \, [m];$$

therefore

$$v = x' \cdot r.$$

There are two remaining relationships in Figure 3.20, for angular velocity x'

$$x' \, [1/s] \Leftarrow x'' \, [1/s^2]$$

and for angle x

$$x \, [1] \Leftarrow x' \, [1/s].$$

Obviously, there is no algebraic relation which would lead to an equation with identical dimensions on both sides. The operation must be of a different kind; the difference in the time dimension provides a hint. The computational procedure

$$x' \, [1/s] = \int x'' \, [1/s^2] \, dt \, [s]$$

and

$$x \, [1] = \int x' \, [1/s] \, dt \, [s],$$

i.e. integration over time, again produces dimensional agreement. The two variables are state variables and therefore have to be determined from integrations over time (here: initial time $t = 0$) using given initial values x'_0 and x_0:

$$x' = x'_0 + {}_0\!\int^t x'' \, dt$$

and

$$x = x_0 + {}_0\!\int^t x' \, dt.$$

For all system elements we have now developed mathematical formulations that are in dimensional agreement. We now collect these model equations and use them to draw the simulation diagram.

3.5.4 Model equations and simulation diagram for the rotating pendulum

1. Parameters

gravitat. acceleration	=	g	$= 9.81$	$[m/s^2]$	
mass	=	m	$= 1$	[kg]	(default value)
radius	=	r	$= 1$	[m]	(default value)
damping constant	=	d	$= ?$	[N/(m/s)]	(scenario parameter)

2. Initial values of the state variables

angular velocity	=	x'_0	$= ?$	[rad/s]	(initial angular velocity)
angle	=	x_0	$= ?$	[rad]	(initial angle in radian)

3. Algebraic variables

velocity	=	v	=	$r \cdot x'$
traj. comp. gravitation	=	F_g	=	$m \cdot g \cdot \sin x$
damping force	=	F_d	=	$d \cdot v$
acceleration force	=	F_b	=	$-F_d - F_g$
acceleration	=	b	=	F_b / m
angular acceleration	=	x''	=	b / r

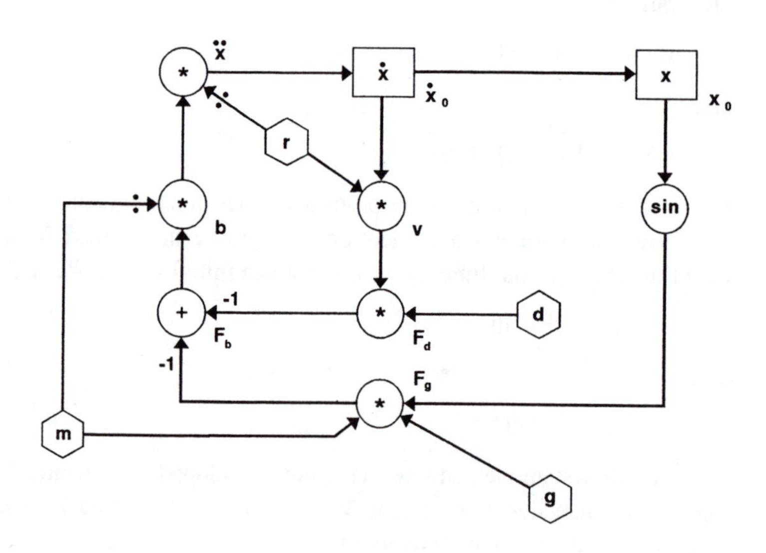

Fig. 3.21: Simulation diagram for the unconstrained dynamics of the rotating pendulum.

4. State variables and state equations

The state variables follow from the initial values and the time integral of the rates of change:

angular velocity $= x' = x'_0 + {}_0\int^t x'' \, dt$
angle $= x = x_0 + {}_0\int^t x' \, dt.$

The specification of the differential equations describing the rates of change (state equations) is therefore sufficient. Renaming angle $x_1 = x$ and angular velocity $x_2 = x' = dx/dt$ we obtain the differential equations for the rates:

angular velocity $x' = dx_1/dt = x'_1 = x_2$
angular acceleration $x'' = dx_2/dt = x'_2 = b/r.$

Using these model equations, we now develop the simulation diagram (Fig. 3.21) on the basis of the influence diagram (Fig. 3.20). Each specification of a parameter or equation for a system variable is represented by one "block," where the blocks differ in type and function. The block symbols are explained in Figure 2.6.

This diagram could be somewhat simplified if (corresponding to the conventions in Fig. 2.6) multiplicative parameters would be simply written as weights next to the influence arrows. This, however, has the disadvantage that the parameters are not as obvious (compared to the hexagonal symbol). Parameters are the "buttons" that can be used to adjust system behavior. Entering parameters as weights in the simulation diagram should therefore be restricted to those quantities which cannot, or are not meant to be changed (for example, the gravitational acceleration).

We shall learn in Chapter 4 how to construct a working simulation model from the model equations or the simulation diagram.

3.6 Model Development for a Fishing System

3.6.1 System description, model purpose, verbal model, and influence diagram

We begin with a **system description**:

A large lake is being used for fishing. Without fishing, the fish population would grow to its capacity limit. This capacity limit is constant and is determined by the nutrient supply in the lake and the corresponding growth of algae, phytoplankton, and zooplankton as a source of food for the fish. The reproduction processes of the fish population limit its maximum growth rate.

The average fish catch depends on the number of boats and the size of the fish population. Under favorable conditions, a certain amount of fish can be caught by each boat per year. Maintenance and operation of the boats cause certain expenses. The fishermen attempt to maximize their net income; they therefore invest part of it in the purchase of new boats. The purchase of new boats partially replaces older boats, but it may also increase the total boat number, thus increasing fish catch and profit. If no net income can be attained—either because of small catches and/or low prices for fish—no new boats will be purchased, and boat number will decrease as older boats are put out of commission.

An initial assessment (from the point of view of the fishermen) shows that the fish catch and hence the profit will be low if either too few or too many boats are used for fishing. It can be expected that there will be an optimum for the number of boats that will depend on the economic conditions (fish price and costs of boat operation and maintenance) and on the ecological conditions (number of fish that can be sustainably harvested). It would be important for the fishermen to know these conditions in order to control the fishing operation in such a way that 1. there will be neither an ecological collapse (of the fish population) nor an economic collapse (of the fishing operation), and that 2. a sustainable management of fishing on the lake can be achieved under optimal economic conditions. Since we are dealing here with complex interwoven ecological and economic processes, the development of a model for computer simulation under different assumed conditions and the search for an optimal solution is in order.

Model purpose: *By taking account of the population dynamics of the fish under harvest conditions as well as the economic conditions of the fishing operations, the dynamic processes of the fishery system should be represented in such a way that the results can be used as a decision aid. In particular, the model should allow an overview of the possible behavior modes of the system (for example, ecological and economic collapse).*

Keeping this model purpose in mind, we can now collect information for the **verbal model**. This information will come partly from the system description and partly from additional sources or considerations in order to close some knowledge gaps in the influence paths.

For the submodel of the **fish population** we develop Relationships 1 through 9 in the following list; they describe how the fish population changes as a function of natural growth conditions and "predation" by fishermen. For the submodel of the **fishing fleet** we define Relationships 10 through 27; they describe how the boat number changes in response to earnings from fishing.

Using this information the influence diagram can now be developed (Fig. 3.22). Obviously, the fish population and the number of boats are state variables; they are therefore marked by boxes.

1. The net growth of the fish population ("fish growth") depends on the current population size.
2. Fish growth is higher if the maximum growth rate of the fish population is higher.
3. Fish growth is reduced as fish density approaches a critical limit.
4. The fish population is increased by fish growth.
5. The fish population is reduced by the fish catches (annual rate).
6. Fish density increases as the fish population increases.
7. If the maximum fish capacity of the area is lower, the relative density will be higher for the same fish population.
8. The maximum fish capacity depends on the area of the lake.
9. The maximum fish capacity depends on the fish carrying capacity of the lake. If there is no fish harvest, the fish population can grow up to its capacity limit which is given by ecological conditions. If the population approaches this limit, the annual net growth will eventually be reduced to zero.
10. The number of boats increases by the annual purchases of new boats.
11. The number of boats decreases by the number of (old) boats put out of commission annually.
12. The maximum possible annual fish catch of the fleet (catch potential) is determined by the number of active boats.
13. The catch potential depends on boat performance (maximum annual catch rate per boat).
14. The actual annual fish catch increases with the catch potential.
15. The fish catch increases with fish density.
16. The annual catch proceeds increase with the fish catch.
17. Catch proceeds are proportional to the price of fish.
18. Net income is higher if catch proceeds are higher.
19. Net income is lower if fleet operating costs are higher.
20. Fleet operating costs are higher if operating costs per boat are higher.
21. Fleet operating costs increase with the number of boats.
22. If net income is higher, more capital can be invested in new boats.
23. If the fraction of the net income reserved for investment in new boats is higher, more capital is available for new boats.
24. If more investment capital is available, more boats can be purchased.
25. If the cost of new boats is higher, fewer boats can be purchased.
26. The number of boats decommissioned each year depends on the normal decommissioning rate (depreciation) which is inversely proportional to the average boat lifetime.
27. The number of boats decommissioned each year depends on the current number of boats.

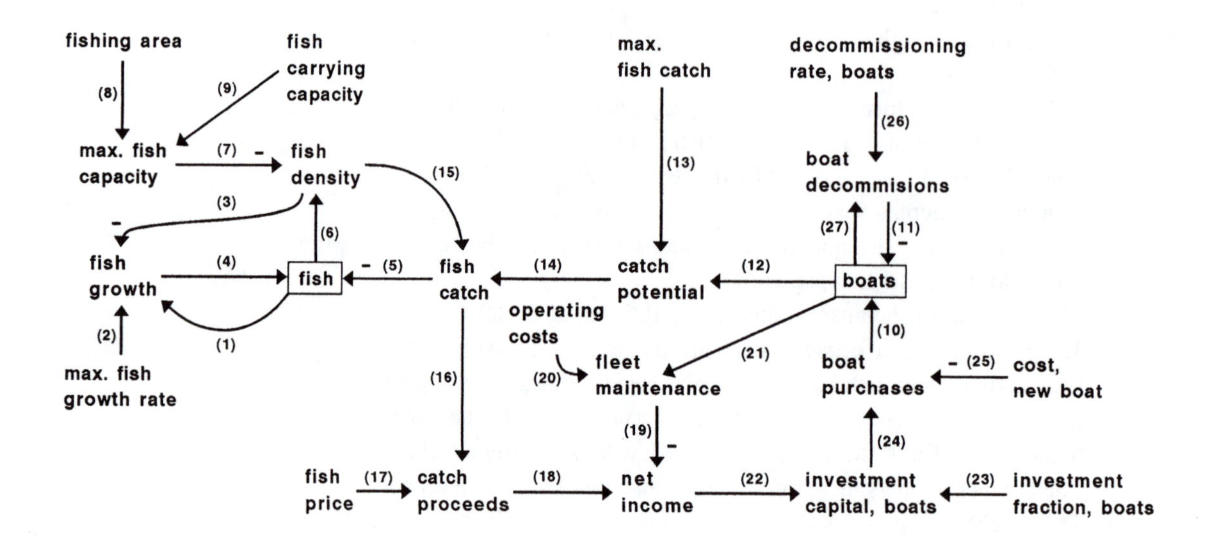

Fig. 3.22: Influence diagram of the system elements for fishery dynamics.

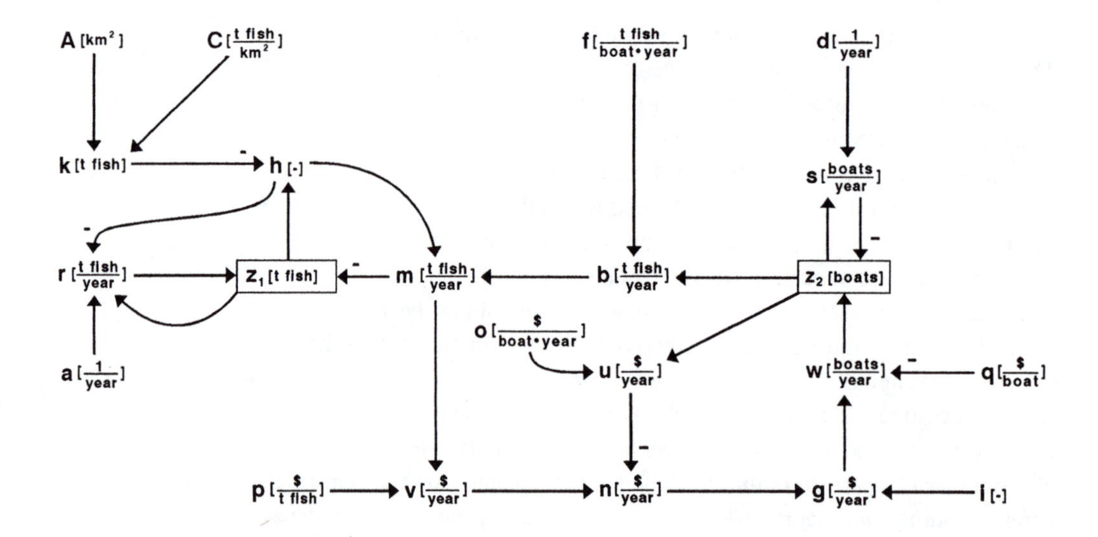

Fig. 3.23: Influence diagram using the abbreviations and dimensions of the system elements of the fishery dynamics system.

3.6.2 Quantities, dimensions, and relationships of fishery dynamics

By referring to the influence diagram, we now make a list of the system quantities, their dimensions, and the mathematical abbreviations we want to use for further analysis. In selecting abbreviations, one should be guided by practical considerations. We shall use here single letter abbreviations in order to make the mathematical manipulations in this chapter somewhat more transparent. Where this is not necessary, it is advisable to use longer self-explanatory names. This makes it a lot easier to understand the statements in the simulation program.

Example: In the following, the operating and maintenance costs (per boat) will be denoted by "o." To improve the readability of the program, we will later use "operating costs" (Ch. 4, 5).

Let the system **parameters** be specified as follows (yr = year):

A	fishing area	1000	[km^2]
C	fish carrying capacity	100	[t fish/km^2]
k	max. fish capacity	$C \cdot A$	[t fish]
a	max. fish growth rate	1	[1/yr]
f	max. fish catch	100	[t fish/(boat·yr)]
o	operating costs	50000	[$/(boat·yr)]
q	cost, new boat	100000	[$/boat]
d	decommissioning rate, boats	1/15	[1/yr]
1/d	boat lifetime	15	[1/yr]
i	investment fraction, boats	1/2	[-]
p	fish price	1000	[$/t fish]

The other system quantities represent **variables** which change during the simulation. Initial values have to be specified for the state variables:

z_1	fish (initial value)	5000	[t fish]
z_2	boats (initial value)	100	[boats]

We now redraw the influence structure of Figure 3.22, inserting at the nodes the abbreviations of the system quantities and their dimensions (Fig. 3.23). By paying attention to antagonistic influences (minus sign) and the dimensions of the influence relationships at each node we now obtain the following relationships:

h	fish density	$h = z_1 / k$	[-]
r	fish growth	$r = a \cdot z_1 \cdot (1 - h)$	[t fish/yr]
b	catch potential	$b = f \cdot z_2$	[t fish/yr]
m	fish catch	$m = b \cdot h$	[t fish/yr]
v	catch proceeds	$v = p \cdot m$	[$/yr]

u	fleet maintenance	$u = o \cdot z_2$	[$/yr]
n	net income	$n = v - u$	[$/yr]
g	investment capital, boats	$g = i \cdot n$	[$/yr]
w	boat purchases	$w = g / q$	[boat/yr]
s	boat decommission	$s = d \cdot z_2$	[boat/yr]

3.6.3 Model equations and simulation diagram for fishery dynamics

We now collect all data and equations for the simulation model:

1. Parameters

A	=	100	[km^2]	fishing area
C	=	100	[t fish/km^2]	fish carrying capacity
k	=	$C \cdot A$	[t fish]	max. fish capacity
a	=	1	[1/yr]	max. fish growth rate
f	=	100	[t fish/(boat·yr)]	max. fish catch
o	=	50000	[$/(boat·yr)]	operating costs
q	=	100000	[$/boat]	cost, new boat
d	=	1/15	[1/yr]	decommissioning rate, boats
i	=	1/2	[-]	investment fraction, boats
p	=	1000	[$/t fish]	fish price

(Some of these values may be changed in scenario investigations.)

2. Initial values of the state variables

z_{10}	=	5000	[t fish]	fish
z_{20}	=	100	[boats]	boats

(to be changed in scenario investigations)

3. Algebraic intermediate variables

h	=	z_1 / k	fish density
r	=	$a \cdot z_1 \cdot (1 - h)$	fish growth
b	=	$f \cdot z_2$	catch potential
m	=	$b \cdot h$	fish catch
v	=	$p \cdot m$	catch proceeds
u	=	$o \cdot z_2$	fleet maintenance
n	=	$v - u$	net income
g	=	$i \cdot n$	investment capital, boats
w	=	g / q	boat purchases
s	=	$d \cdot z_2$	boat decommission

4. State equations

$dz_1/dt = z'_1 = r - m$ rate of change of fish
$dz_2/dt = z'_2 = w - s$ rate of change of boats

Using these relationships we can now draw the simulation diagram on the basis of the structure of the influence diagram (Fig. 3.24). The simulation diagram contains all information necessary for programming the simulation model. The parameter values and initial values are found in the previous list.

We could now write the simulation program and run simulations. Before we do this in the next chapter, we shall have a look at the model equations for the pendulum and the fishery system to discover more about their general properties.

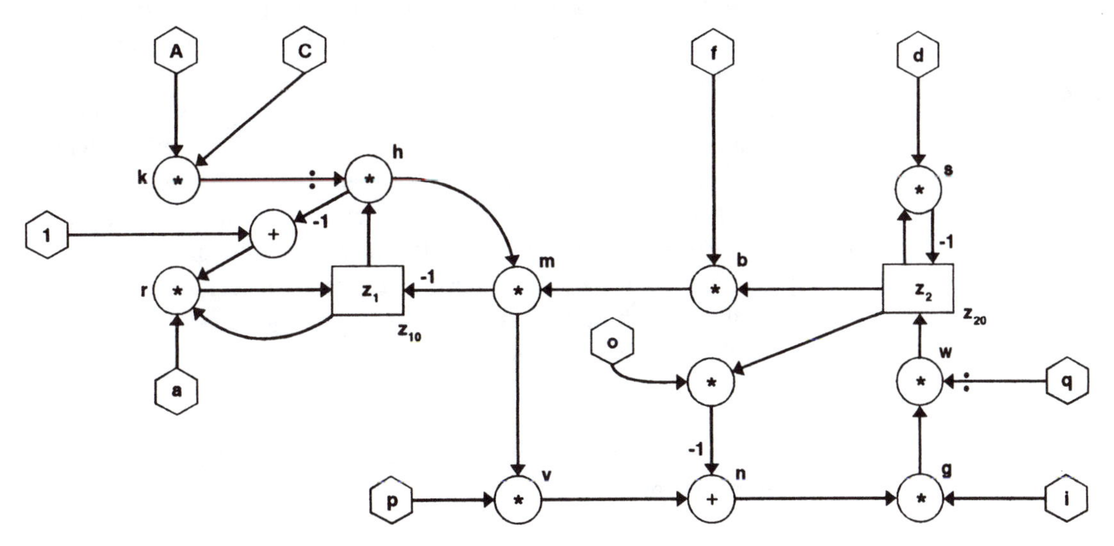

Fig. 3.24: Simulation diagram for the computation of fishery dynamics.

3.7 Non-dimensional Model Equations and System Similarity

In developing the equations for the simulation models, we have written one equation for each of the elements (or "blocks" in the simulation diagram). In the investigation of real systems, it is usually necessary to keep the model formulation close to variables and parameters of the real system in order to better identify and understand the influences of variables and parameters and to facilitate changes in the model structure which may be necessary in the course of model development.

The mathematician would normally attempt to combine the algebraic equations of the model, to eliminate intermediate variables by substitution and to obtain a compact system of differential equations describing the complete model. This procedure makes sense for mathematical analysis (if it is possible at all) and in particular for discovering the system's elementary influence structure. However, mathematical condensation usually complicates working with a model and understanding its behavior, and it should be avoided in routine model development.

Once a simulation model has been developed, condensation to its essential structure may provide some interesting insights. In particular, it may turn out that its structure is generic, i.e. that it belongs to a whole class of systems. Such similar systems show similar behavior, and by proper transformation of variables one may in fact use the solutions for one system to describe the behavior of another having identical generic system structure but totally different physical appearance or time constants.

It should therefore come as no surprise that in fields as widely apart as physics and psychology, economy and ecology, engineering and ethics, biology and banking, we encounter systems which, when viewed through the special glasses of the system analyst, turn out to be identical. What is learned in one discipline can also be applied to similar systems in another discipline.

The steps for deriving a more compact and general model formulation are:

1. condensation by substitution, and
2. deriving dimensionless equations.

3.7.1 Derivation of generic state equations for the rotating pendulum

By successive substitution of the relationships for the rotating pendulum (Sec. 3.5.4), we can obtain a very compact system description:

$$\begin{aligned} F_b &= -F_d - F_g \\ &= -m\, g \sin x - v\, d \\ &= -m\, g \sin x - r\, d\, x'. \end{aligned}$$

Also because $b = F_b/m$ and $x'' = b/r$

$$F_b = m\, b = m\, x''\, r$$

giving

$$m\, x''\, r + m\, g \sin x + r\, d\, x' = 0$$

or

$$x'' + (d/m)\, x' + (g/r) \sin x = 0. \tag{3.6}$$

This is the differential equation of second order for the rotating pendulum. It is non-linear on account of the term sin x. For a complete specification of the system, the initial values x'_0 and x_0 have to be specified.

Using the symbols of Figure 2.6 we can draw the simulation diagram for this mathematical model (Fig. 3.25a). We begin with the state variable x; its only input is the rate of change x'. This variable follows from integration of the rate of change x". By solving the differential equation just derived for x", we obtain an algebraic expression for the second derivative:

$$x'' = -(d/m)\, x' - (g/r) \sin x.$$

We therefore have to introduce corresponding feedbacks from x' to x" and from x to x" in the diagram. The first feedback is linear in x', with weight (–d/m). In the second feedback we first have to determine the non-linear term sin x before weighting the result by (-g/r).

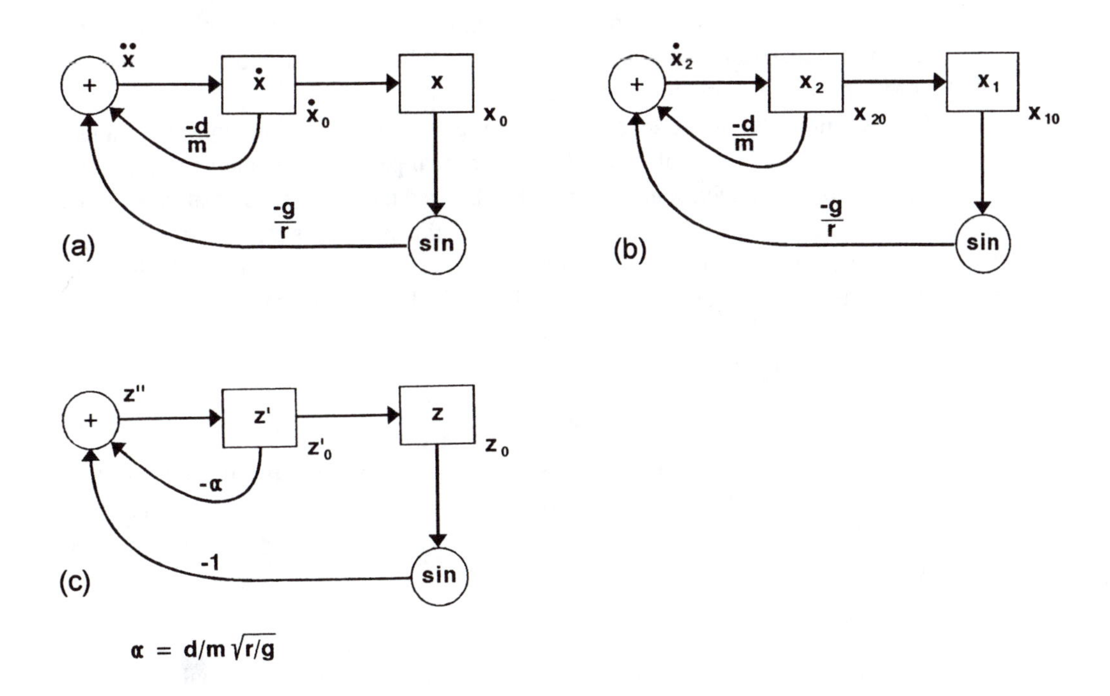

Fig. 3.25: a: Compact representation of the pendulum system, b: elimination of the second derivative by introduction of new state variables, and c: generic model with a single similarity parameter .

Obviously, x' in addition to x is also a state variable. We therefore rename these two variables

$$x_1 = x$$
$$x_2 = x'_1$$

and introduce them into the simulation diagram Figure 3.25b. Reading off the state equations for both boxes, we obtain *two* differential equations of *first order*:

$$x'_1 = x_2 \tag{3.7a}$$
$$x'_2 = -(d/m)\, x_2 - (g/r) \sin x_1 \tag{3.7b}$$

with the initial conditions $x_{10} = x_0$ and $x_{20} = x'_0$. The general procedure for converting a differential or difference equation of n-th order to a set of first order differential or difference equations is explained in Sections 7.1.8 and 7.1.9.

This set of state equations has four parameters (d, m, g, r). In order to study the full range of behaviors of this system, it seems that we have no choice but to vary them systematically, to produce an immense number of simulations covering the whole spectrum of possible parameter combinations and to then try to understand the system from this mass of data.

Fortunately, this is not so. If we rewrite the equations using non-dimensional variables, we will normally find that the original parameters combine to a smaller number of non-dimensional parameters, thus reducing the number of cases to be investigated significantly. The resulting non-dimensional state equations are generic, i.e. they describe a broad range of similar systems. For the same choice of non-dimensional parameters, the solutions are identical and can be easily recalculated in terms of the correct dimensional variables.

In the original pendulum equation

$$x'' + (d/m)\, x' + (g/r) \sin x = 0, \tag{3.6}$$

we first introduce non-dimensional state variables. In this case, the angle x is already dimensionless [radians], but the time dimension enters in the derivatives x' and x''. Rename $z = x$. We introduce dimensionless time τ by referring the real time to some reference time:

$$\tau = t / T.$$

The time derivatives can now be rewritten in terms of non-dimensional time

$$x' = dx/dt = (1/T)\; dz/d\tau = (1/T)\; z'$$
$$x'' = d^2x/dt^2 = (1/T^2)\; d^2z/d\tau^2 = (1/T^2)\; z''.$$

Replacing x' and x'' in Equation 3.6 yields the differential equation:

$$z'' + (d/m)\, T\, z' + (g/r)\, T^2 \sin z = 0.$$

Now introduce a reference time T. Note that $T = (r/g)^{1/2}$ is a suitable candidate since it will lead to cancellation of the coefficient of the third term; the equation now becomes:

$$z'' + (d/m)\,(r/g)^{1/2}\,z' + \sin z = 0. \tag{3.8}$$

Actually, this reference time is the inverse of the natural frequency of the undamped system for small angle x, found by solving

$$x'' + (g/r)\,x = 0$$

which has angular frequency $2\,\pi\,f = \omega_0 = (g/r)^{1/2}$. The differential equation (Eq. 3.8) can be written in the **generic form**

$$z'' + \alpha\,z' + \sin z = 0 \tag{3.9}$$

where we now have reduced parameter dependence to a single (non-dimensional) pendulum parameter (similarity parameter)

$$\alpha = (d/m)\,(r/g)^{1/2}.$$

The simulation diagram shown in Figure 3.25c is therefore the most general representation of the pendulum system.

In the generic form, all pendulums having the same α will have identical solutions in terms of z, z', and z" and non-dimensional time τ. For example, the same equation (Eq. 3.9; with the same α) can be used to compute the behavior of a pendulum of radius 100 m and mass 10 kg, of radius 0.01m and mass 0.1 kg, or of radius 1 m and mass 1 kg. Note, however, that the time solutions must be stretched by the time factor $T = (r/g)^{1/2}$. In real variables, the solution is recomputed by using

$$\begin{aligned} t &= \tau\,T \\ x(t) &= z(\tau\,T) \\ x'(t) &= (1/T)\,z'(\tau\,T) \\ x''(t) &= (1/T^2)\,z''(\tau\,T). \end{aligned}$$

The introduction of non-dimensional variables and non-dimensional time is a sure way to reduce the mathematical formulation of a system to its essentials, to reduce the number of parameters to the minimum, and to identify the generic structure of the system. Simulations can then deal with the generic system, and a simple scaling of the general results can provide solutions for any number of special cases.

We have now encountered the same simulation model in several equivalent descriptions which can all be transformed from one into the other and which will produce identical results:

1. detailed simulation diagram
2. set of simulation equations (specification of parameters and exogenous quantities, algebraic equations, and differential equations)
3. one differential equation of n-th order
4. n differential equations of first order
5. compact simulation diagram
6. generic state equations.

We will return to the simulation of the behavior of the rotating pendulum using these formulations in the next chapter.

3.7.2 Condensing the fishery model to the generic predator-prey system

The differential equations for the two state variables z_1 and z_2 are given in Section 3.6.3; they can also be copied from the simulation diagram (Fig. 3.24). For the fish population z_1, we obtain by successive substitution of the algebraic expressions describing its rate of change the differential equation

$$dz_1/dt = a \cdot z_1 \cdot (1 - z_1/k) - f \cdot (z_1/k) \cdot z_2$$

and similarly, for the number of boats

$$dz_2/dt = (p \cdot f \cdot (z_1/k) \cdot z_2 - o \cdot z_2)\ i\ /\ q - d \cdot z_2.$$

Introducing

$$c = (p \cdot i)/q$$
$$e = (o \cdot i)/q + d\ ,$$

the state equations of the model system can be rewritten as

$$z_1' = a \cdot z_1 \cdot (1 - z_1/k) - f \cdot z_2 \cdot z_1/k \tag{3.10a}$$
$$z_2' = c \cdot f \cdot z_2 \cdot z_1/k - e \cdot z_2. \tag{3.10b}$$

This system has exactly the structure of the classical predator-prey system of Lotka and Volterra with a logistic saturation at the capacity limit for the prey (see Model M 207 PREDATOR-PREY SYSTEM WITH CAPACITY LIMIT in the systems zoo and in Ch. 6):

$$x' = A \cdot x \cdot (1 - x) - B \cdot x \cdot y \tag{3.11a}$$
$$y' = C \cdot x \cdot y - D \cdot y\ . \tag{3.11b}$$

The first part of the first equation corresponds to logistic growth of the prey. The second part corresponds to the losses by predation; it depends on the predator population as well as on the prey population. The first part of the second equation shows the corresponding gain for the predator, the second part the (biomass or energy) losses by respiration.

If we now draw the simulation diagram for this compact form of the fishery model (Fig. 3.26), the generic structure becomes even more evident. On the left hand side of the state variable z_1 we find the structure of logistic growth (see Fig. 3.13). State variable z_2 has the self-loop typical of exponential decay. The two state variables are coupled by multiplication (non-linear!) representing predation (loss for the prey population, gain for the predator population).

In the next chapter we will deal with the simulation of this system and the analysis of its behavior. At this point we can already guess that over some range of its parameters this can be an oscillating system since a feedback loop connects both of the state variables.

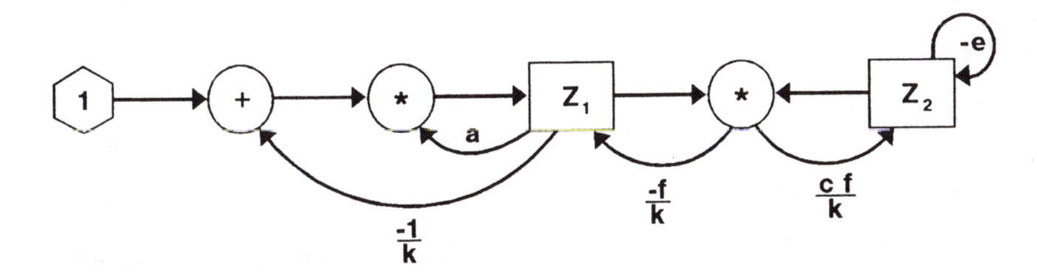

Fig. 3.26: Compact representation of the fishery system. The structure is identical to that of the predator-prey system with a capacity limitation of the prey population.

3.7.3 State equations using normalized state variables

Systems may show generically identical structures and still produce numerically different results. For example, population development for different countries when plotted in relative quantities (such as current population divided by initial population in 1950) may follow the same time function, even though one country had a population of five million in 1950, the other one of 500 million.

In order to better compare the behavior of such systems, it is advisable to use normalized state variables. A state variable is normalized with respect to some reference value. As a result the normalized state variable is dimensionless and will be of order of magnitude "1." A similar procedure can be applied to the independent variable time (see the following section).

When using normalization, it should be realized that it will usually lead to changes in the coefficients of the differential equations. One therefore has to use care in replacing the original dimensional state equations by new dimensionless normalized quantities in order to obtain the correct coefficients in the new state equations. The general procedure is as follows:

The original state variables z_i are normalized by corresponding reference states k_i (having the same dimension!) (for example, equilibrium states or the states at a certain reference point in time) in order to obtain the new dimensionless normalized state variables x_i:

$$x_i = z_i/k_i$$
or
$$z_i = k_i \cdot x_i .$$

These expressions are used to replace the z_i in the original state equations.

$$dz_i/dt = k_i\, dx_i/dt \;=\; f_i(z_1, z_2, \ldots z_i, \mathbf{u}, t) \tag{3.12}$$

which leads to

$$k_i\, dx_i/dt = f_i(k_1\, z_1, k_2\, z_2, \ldots k_i\, z_i \ldots, \mathbf{u}, t)$$
or
$$dx_i/dt = (1/k_i) \cdot f_i(k_1\, z_1, k_2\, z_2, \ldots k_i\, z_i \ldots, \mathbf{u}, t). \tag{3.13}$$

Example: For the fishery model it would be reasonable to use its capacity limit k to normalize the fish population and to use a "normal number of boats" j to normalize the boat number z_2. The normalized fish population x and the normalized boat number y are then defined by

$$x = z_1/k$$
$$y = z_2/j$$
or
$$z_1 = k \cdot x$$
$$z_2 = j \cdot y.$$

If these definitions are introduced into the state equations (Eq. 3.10a,b), the following normalized state equations are obtained:

$$dx/dt = x' = a \cdot x \cdot (1-x) - (j/k)\, f \cdot x \cdot y \tag{3.14a}$$
$$dy/dt = y' = c \cdot f \cdot x \cdot y - e \cdot y. \tag{3.14b}$$

Note that the generic structure has been preserved but that the coefficients of the terms have changed. Substituting

$$A = a \qquad B = (j/k) \cdot f$$
$$C = c \cdot f \qquad D = e,$$

we again obtain the generic state equations (Eq. 3.11a,b) of the predator-prey system. Both state variables x and y are now relative (non-dimensional) quantities of order 1. The four coefficients A, B, C, and D all have dimension [1/yr]. In this form, it is much easier to compare the behavior of systems whose original dimensional state variables have very different orders of magnitude. Note that the time scale has not been changed. A direct comparison of time developments is therefore possible.

3.7.4 Dimensionless state equations, normalized states, and normalized time

Sometimes systems have widely different time constants but are still generically identical. For example, the logistic growth of a bacterial colony may be generically identical to the expansion of the rabbit population in Australia—one reaching the carrying capacity of a Petri dish in hours, the other taking decades to populate a continent. In order to facilitate comparison of the time response of such systems with very different time dynamics, it is advisable to also normalize the time by introduction of relative (dimensionless) time, in addition to using relative state variables as explained in the previous section:

$$\tau = t/T.$$

Here T is a characteristic time (for example, the length of a normal period of oscillation). This leads to

$$t = T \cdot \tau$$
$$dt = T \cdot d\tau.$$

If we now replace dt in the state equations (Eq. 3.13) we obtain the dimensionless state equations

$$dx_i/d\tau = (T/k_i) \cdot f_i(k_1\, z_1, k_2\, z_2, \dots k_i\, z_i \dots, \mathbf{u}, t).$$

The normalized state variables x_i as well as the normalized time are now of order of magnitude "1."

Example: Introduction of normalized time $\tau = t/T$ into the normalized fishery model (Eq. 3.14) leads to the non-dimensional state equations

$$dx/d\tau = T \cdot a \cdot x \cdot (1-x) - T \cdot (j/k)\, f \cdot x \cdot y \qquad (3.15a)$$
$$dy/d\tau = T \cdot c \cdot f \cdot x \cdot y - T \cdot e \cdot y. \qquad (3.15b)$$

Again the generic structure has been preserved; only the coefficients of the terms have changed. The characteristic time T should always be expressed in terms of a parameter, or a combination of parameters, having the dimension of the time unit and being of particular significance for the system's dynamics. In this case, all of the coefficients (a, (j/k) f, c f, e) have dimension [1/year], and their inverse could therefore be used as characteristic time. However, the most meaningful measure is certainly the inverse of the normal net growth rate a of the fish population since this parameter determines the regeneration of the fish population and hence the dynamics of the system. We therefore choose the characteristic time

$$T = (1/a).$$

Introducing this into the state equations (Eq. 3.15a,b) yields the non-dimensional state equations

$$d x / d\tau = x \cdot (1-x) - (j/k) \cdot (f/a) \cdot x\, y \qquad (3.16a)$$
$$d y / d\tau = (c/a) \cdot f \cdot x \cdot y - (e/a) \cdot y \qquad (3.16b)$$

and use of the definitions

$$A = 1 \qquad B = (j/k) \cdot (f/a)$$
$$C = (c/a) \cdot f \qquad D = e/a$$

again leads to the generic state equations (Eq. 3.11a,b) of the predator-prey system. Note that in this form the logistic growth of the prey population, assuming absence of a predator, is completely parameter-independent: $dx/d\tau = x \cdot (1-x)$, except for the scaling of the time axis by $\tau = at$.

The use of normalized and dimensionless state equations is of great importance for the analysis and the comparison of system structure and system behavior. Without it, the comparison of generically related or identical systems having widely different time constants and absolute magnitudes of the state variables would not be possible. In the model systems of the systems zoo in Chapter 6, normalized state equations are used almost exclusively. From the numerical results for the normalized state equations x_i, the case-specific dimensional state variables z_i can be obtained by multiplying by the reference states k_i:

$$z_i = k_i \cdot x_i.$$

The dimensional time t follows from dimensionless time τ after multiplication by the reference time T:

$$t = T \cdot \tau .$$

The normalized state equations can be converted to the dimensional state equations (dz_i/dt) by replacing the dimensionless quantities

$$x_i = z_i / k_i$$
$$\tau = t/T$$

to obtain the corresponding dimensional state variables z_i and time t.

3.8 Summary of Important Results

While Chapter 2 focused on the influence structure of systems, the present chapter dealt with the more exact description of systems. This requires that the different nature of the system elements and of the processes taking place between them is explicitly recognized and modeled. The correct description of the influences in the system requires dimensional consistency of the model relations. The final products of the model development are mathematical relations or an equivalent simulation diagram which can both be used as the basis for programming a simulation model. Our investigations concentrated on continuous deterministic dynamic systems.

The most important results of this chapter are:

1. All dynamic systems with inputs $\mathbf{u}(t)$ and outputs $\mathbf{v}(t)$ can be described by a **state equation** and an **output equation:**

   $$\mathbf{z}(t) = \mathbf{F}(\mathbf{z}, \mathbf{u}, t)$$
   $$\mathbf{v}(t) = \mathbf{g}(\mathbf{z}, \mathbf{u}, t).$$

2. For **continuous systems** (which are defined at any point in time) the state equation can be written as a differential equation for the state rates of change $d\mathbf{z}/dt$. The **system equations** are then given by

   $$d\mathbf{z}/dt = \mathbf{f}(\mathbf{z}, \mathbf{u}, t)$$
   $$\mathbf{v}(t) = \mathbf{g}(\mathbf{z}, \mathbf{u}, t).$$

   The corresponding **general system diagram** is shown in Figure 3.7. The derivative with respect to time is often indicated in this book by a prime: $z' = dz/dt$.

3. In a continuous dynamic system the **state function** $\mathbf{f}$ and the **behavior function** $\mathbf{g}$ can be determined at any instant by **algebraic** (and/or logic) operations.

4. In a continuous dynamic system the **system state** $\mathbf{z}(t)$ must be determined **by integration** of the rate of change $d\mathbf{z}/dt$ over time. In the simplest case, this is done using the numerical integration of Euler and Cauchy over the time step t:

   $$\mathbf{z}_{new} = \mathbf{z}_{old} + (d\mathbf{z}/dt)_{old} \cdot \Delta t\ .$$

5. The **behavior of dynamic systems** depends to a significant extent on the number of state variables and the types of feedbacks between them. Systems with two or more state variables are able to oscillate; systems with three or more state variables may be chaotic.

6. All state equations (**f**) and all output equations (**g**) must be **dimensionally consistent**: both sides of the equation must have the same dimension. This requirement can assist in the correct formulation of model relationships.
7. If the model equations are normalized by introduction of reference quantities, one obtains normalized **dimensionless state equations**. In this form, systems of widely different time constants and magnitudes of the state variables can be better compared and analyzed.
8. Systems which appear very different in both structure and behavior may actually have identical **generic structure**. Normalization of state variables and time and the introduction of similarity parameters then also produces identical behavior in the new coordinates.
9. The **validity** of a simulation model must be tested with respect to four criteria: a. structural validity, b. behavioral validity, c. empirical validity, and d. application validity.
10. Wherever possible, simulation models should be developed in terms of **modules.** The validity of each module must be tested individually before it is coupled to other modules.

We now summarize some additional observations on state variables and systems which are useful for the assessment of deterministic systems. The remarks mostly concern continuous systems; possible differences for discrete systems are noted.

11. Storage or memory variables are always state variables.
12. From the states $\mathbf{z}(t)$ and the input vector $\mathbf{u}(t)$ all other system quantities at time t including the rates of change $\mathbf{z}'(t)$ of the states can be determined.
13. The initial states (reservoirs, stocks, levels) must be known for the computation of the subsequent system development. The state variables contain the "memory" of the system.
14. In general (exception: chaotic systems), the initial state of a stable non-autonomous system hardly has any influence on the system behavior after a sufficiently long time period since the effects of the input functions will dominate the system behavior.
15. The state functions $\mathbf{f} = d\mathbf{z}/dt$ describe the rates of change of the state levels as function of the current state variables $\mathbf{z}$, the input vector $\mathbf{u}$ and possibly the time t. The state functions are algebraic expressions which may be linear or non-linear in the state variables.
16. The new state level (of a continuous system) follows by time integration of the rate of change $\mathbf{z}'$. The state computation therefore is composed of an algebraic part (computation of $\mathbf{z}' = \mathbf{f}(\mathbf{z}, \mathbf{u}, t)$) and an integration over time.

17. Rates of change can be understood as inflows and outflows of the state quantity per unit of time. They always have the dimension [state variable/time unit].
18. The state level increases as long as the sum of the inflows is greater than the sum of the outflows. For constant inflow the state level may increase if the outflow rate decreases.
19. Even for high rates of change the state level cannot change instantaneously and drastically: state variables therefore act as buffers (delays, inertia). This results in a certain decoupling from the system parts dominated by other state variables and from sudden perturbations.
20. The state rate of change dz/dt is often dependent on the state $z(t)$ itself (feedback); i.e. the inflow and outflow rates of a state reservoir may depend on its current level.
21. Negative feedback (in continuous systems) tends to cause an approach to an equilibrium state and stabilization. While this is often desirable, a strong stabilization tendency may prevent or impede necessary changes in the system, on the other hand.
22. Positive feedbacks (in continuous systems) lead to amplification and often to exponential growth. Because of their "autocatalytic" effect, they are of great importance for the initiation of change as well as for growth processes, but they must eventually be compensated by negative feedback in order to prevent explosive, destructive growth.
23. The dynamic development of a state variable is often determined by negative as well as positive feedback. The resulting behavior depends on the relative dominance of one of the feedbacks, and that dominance may change over time, leading to complex behavior.
24. In general, a rate of change controlled by a negative feedback decreases as the state approaches the equilibrium level.
25. Feedbacks are often non-linearly dependent on the system state. These nonlinearities may change the relative dominance of feedbacks in the course of the system development; this may cause fundamental behavioral changes of the system.
26. If the feedback signal is delayed (for example, by an intermediate state variable or a delay), the state change will be determined by a delayed, no-longer-current state value. This may result in an overshoot of the equilibrium level. This again is corrected with some delay. Delayed feedback may therefore lead to oscillations.
27. In a continuous system, oscillations may occur if the system contains at least one state variable and a delay, or two state variables.

28. By contrast, a discrete system may have oscillations even if it contains only one state variable. In discrete systems, negative feedback may cause a sign reversal in the next time interval, followed by further sign reversal in future intervals.
29. Non-linear continuous systems with three or more state variables may be chaotic.
30. Identical system structures will produce identical behavior even if the systems are completely different in terms of their physical elements.

CHAPTER 4
BEHAVIOR

4.0 Introduction

In the previous chapters we focused on model development. We programmed a simple simulation only for the small global model. We now have to turn to the task of turning the mathematical models we developed into simulation models we can study under a wide range of parameters to determine their dynamic behavior.

It is possible to develop a complete individual simulation program for each model using a general-purpose programming language where the program would be able to deal with all necessary tasks of input, computation, evaluation, and presentation of results. However, it is obvious from the programming example for the world model in Chapter 2 (Prog. 2.1, GLOBSIM.PAS) that such an approach would be very inefficient: the program lines specifying the model amount to only a few percent of the total programming effort. The major part of the program is required for computation and data handling tasks that are not specific to the particular model and would be identical for other models as well.

Programming new simulation software for each individual application therefore does not make sense. For this reason, we have deferred the simulation of the rotating pendulum and the fishery dynamics until now and will now take up this task with suitable simulation software.

The use of generally applicable simulation programs is possible and sensible for two reasons; no matter what the content of the simulation model:

1. the necessary model specifications are similar, and
2. the necessary processing tasks are similar.

We saw in Chapter 3 that the necessary model specifications can always be developed in terms of three categories, independently of the content of the model;

in the simulation diagram these categories were characterized by different symbols:

1. **input quantities** (hexagons) representing parameters and exogenous quantities which are either constant or time-variant
2. **state variables** (rectangles) and their initial values
3. **intermediate quantities** (circles), i.e. algebraic (also logical) relationships including the equations for the rates of change of the state variables.

A generally applicable simulation software must be able to deal with model information in these three categories. It can then be used for arbitrary simulation tasks (of the class "dynamic systems"). The programming effort is then reduced to providing the specific model equations.

Similar conditions apply also to processing the model; independently of the specific subject of the model, the processing tasks are always the same:

1. inputting and changing parameters and initial values
2. calculation of model equations for each point in simulated time, updating of time, and recalculation
3. output of simulation results in numerical form (tables)
4. output of simulation results in graphical form (as time plots or in state space)
5. configuration of output formats by the user
6. model changes.

In addition to these basic required processing tasks, there are additional tasks that the simulation software should also facilitate:

7. multiple simulations to study parameter sensitivity
8. multiple simulations for "global analysis" of the model behavior in the relevant state space.

Today, a broad selection of simulation methods and simulation software is available. It ranges from developing one's own simulation program in a general programming language to drawing the simulation diagram on the screen using the mouse and the programmed query of all necessary model information by the computer. We can distinguish the following approaches:

1. **Programming a complete simulation program in a general purpose language:** For each simulation task, the model developer writes an individual program using a general purpose programming language such as Fortran, Basic, Pascal, or C. The advantage of this approach is that he/she may configure the program and its possibilities depending on his/her own individual requirements. However, this approach has the severe disadvantage that the significant investment of writing a complex program pays off only if it is necessary to fulfill specific tasks that cannot be handled by available simulation programs.

2. **Programming only the model equations in a general purpose language:** In this case, the model developer only has to write the program part describing the model using a general purpose programming language. Available software is used to provide the complete simulation environment; the user does not have to know or understand its internal workings. Examples are DYSYS and DYSAS for Basic (Bossel, 1987/89) and SIMPAS for TurboPascal which we will use here. The advantage of this approach is that the user can work in a programming language which is familiar to him/her and that he/she can use all the possibilities available in this language to formulate the model, including the use of specific functions and procedures written by the user. The disadvantage of this approach is that some programming knowledge is required.
3. **Programming the model in a special simulation programming language:** In this case, the user has to formulate the model in one of many special programming languages that have been developed specifically for the requirements of simulation. Examples are the well-known simulation languages CSMP and DYNAMO. The advantage of this approach is the usually simple programming of the simulation model. A disadvantage is that one has to learn a special simulation language which, in comparison to general purpose programming languages, usually only offers a limited number of programming possibilities and functions.
4. **Model development using computer-assisted design:** The information required for model development is a. of structural nature (information concerning influence connections), b. of qualitative nature (information concerning the functional type of system elements), and c. of quantitative nature (parameter values, functions). This suggests developing simulation software by employing the most user-friendly means available to request the necessary information from the user and to construct from this an executable simulation model. This means, for example, graphical input of the model structure by the methods of computer-assisted design, and a guided interactive query to determine as yet undefined model relationships and parameters. This approach is used in the STELLA simulation environment which will be discussed later in this chapter. This approach has the advantage that the user is relieved of all programming tasks. It has the disadvantage that the model formulation is restricted by the possibilities of the program (which, however, may be very extensive).

Of the four approaches, the individual programming (Approach 1) and the use of specific simulation programming languages (Approach 3) will lose importance in the future. They will therefore not be discussed here. By contrast, writing the model equations in a standard language for use with compatible simulation software (Approach 2) is of some interest because it is quite flexible and portable and has only modest hardware and software requirements. The use of user-friendly, graphically-interactive model development procedures (Approach 4) will become increasingly important in the future.

In this chapter we therefore deal with Approaches 2 and 4. In Section 4.1 the SIMPAS simulation environment (Approach 2) based on TurboPascal is introduced. In Section 4.2 we use SIMPAS to develop a simulation model of the dynamics of the rotating pendulum. Using this example, the processing and evaluation possibilities of SIMPAS will be demonstrated. In Section 4.3 we again use SIMPAS to develop the simulation model of fishery dynamics and to run corresponding simulations. We also deal here more extensively with finding the points of equilibrium of the system, i.e. those values of the state variables where the system is naturally at rest. In Section 4.4 the interactive graphical model environment STELLA (Approach 4) is described and applied to develop and run a simulation model of fishery dynamics. In Section 4.5 the most important results of this chapter are summarized.

Note: All SIMPAS models in this book are also found in the systems zoo (Prog. SIMZOO.EXE) on the accompanying diskette. To call up this program, type SIMZOO (and hit the RETURN key) from the DOS level. You can then go through all of the steps discussed below for using a SIMPAS model. Developing your own model with SIMPAS requires TurboPascal, however, as explained in the following section.

4.1 Simulator for a Standard Programming Language: SIMPAS

4.1.1 Ways of using SIMPAS models

The simulation program SIMPAS (**Sim**ulation of **P**rocesses **a**nd **S**ystems) has been written in TurboPascal for IBM-DOS computers. The simulation model itself is formulated as a TurboPascal unit and appended to SIMPAS. Both parts are compiled together yielding a self-contained executable (*.EXE) program that will run on all DOS-computers and common monitors. For further use, TurboPascal is not required. With SIMPAS, simulation models can be easily provided with a multitude of interactive processing possibilities ranging from interactive parameter query to simulation, global analysis, sensitivity studies, graphical presentation of results, and printed documentation.

Simulation models are developed as self-contained model units MODEL. PAS and then linked to the SIMPAS program parts SIMPAS.PAS, BASE.TPU, and SIMUL.TPU under TurboPascal to obtain a stand-alone executable program SIMPAS.EXE. There are therefore two different ways of using SIMPAS models:

1. Simulations under DOS (without TurboPascal) using a previously compiled executable SIMPAS simulation model (MODEL_ID.EXE). This allows complex simulations without specific software support even on simple DOS-computers. The model itself can then not be changed but extensive possibilities for change can be provided in the interactive parameter query.
2. Development of a simulation model by programming the corresponding model unit in TurboPascal (MODEL.PAS) and linking it to the other SIMPAS units (BASE.TPU and SIMUL.TPU) to obtain an executable simulation program (SIMPAS.EXE). This work has to be done in TurboPascal. The model unit can be altered as required using the TurboPascal editor.

Before discussing the development of SIMPAS model units for the rotating pendulum and fishery dynamics, the use of a compiled SIMPAS model (SIMPAS.EXE) will be briefly explained.

4.1.2 Use of compiled SIMPAS simulation programs

Simulation programs (MODEL_ID.EXE) already compiled under TurboPascal can be started directly from DOS by calling MODEL_ID. (MODEL_ID is used here as a place-holder for a particular model name; for example, PREDATOR). Preceding this, GRAPHICS must have been called from DOS in order to enable the screen graphics. In addition (but not for SIMZOO), the specific graphics interface for the monitor used must be available in the same directory (for the most common VGA monitors this would be EGAVGA.BGI). (The systems zoo already contains the graphics interfaces of common monitors.) The directory must also contain TRIP.CHR defining the TRIP font.

Example (on diskette): SIMZOO.EXE. The program is started by entering "SIMZOO" (under DOS). (SIMZOO contains 50 compiled SIMPAS models with English texts, see Ch. 6.)

SIMPAS is more or less self-explanatory. The program will produce a standard simulation run using the predefined default values of the model if the user presses the RETURN (ENTER) key only. Specific choices can be made at a number of points by moving the selection bar with the arrow keys and pressing RETURN. Where a parameter value is expected, the default value shown will be used upon entering RETURN, unless a new value has been entered.

A printer can be used to print parameters, tables, graphics, or other documentation. The correct printer setting must be chosen initially. If a pin printer is used, hardcopy is printed by simultaneously pressing CONTROL and P. Special printing software can also be used, such as HPSCREEN (for LaserJet and DeskJet printers) or CAPTURE (with WORD); these normally use the <Shift> and <PrtScr> keys. Simulation runs can be terminated early by pressing the ESCAPE key (ESC).

The different interactive modes of using SIMPAS for simulations will be presented later, when we discuss the simulations of the rotating pendulum and fishery dynamics (Sec. 4.2, 4.3).

4.1.3 Developing a SIMPAS model unit

A SIMPAS simulation model unit is developed as a TurboPascal unit according to the general scheme shown in Program 4.1 (Program Listings Appendix). All features available in TurboPascal can be used in the unit MODEL, including the use of additional functions, procedures, or units written by the user. In addition, the test functions Pulse, Step, and Ramp can be used as well as the predefined general purpose table function Tbf, and the delay functions Delay1 and Delay3. The event function Event can be used to define specific events. These functions and their use are explained in Section 4.1.4. Note that the punctuation in the following program statements is significant; in particular, TurboPascal statements must end with a semicolon (;).

The unit MODEL *must* have the format shown in Program 4.1; one must strictly adhere to the notation shown (here and in the model examples)! In developing a model it is best to follow the examples on the diskette (*.MOD) or refer to the models for the rotating pendulum and the fishery dynamics discussed in more detail below. To make matters more concrete, the model unit PREDATOR.MOD for a simple predator-prey model is shown in Program 4.2 (in the Program Listings Appendix).

The SIMPAS model unit (Prog. 4.1, 4.2) always starts with the unit identification UNIT MODEL. The next section INTERFACE lists required units (Base, Crt, Graph; Base is a SIMPAS unit while Crt and Graph are TurboPascal units) and lists the procedures (InitialInfo, ModelEqs, Summary) which are to be accessed from the other SIMPAS units (Base, Simul).

The next section of the model unit contains the IMPLEMENTATION. All variables and parameters used in the model unit are listed and defined by type following the "var" heading. This is followed by the three procedures which make up the model proper: InitialInfo, ModelEqs, and Summary.

In procedure **InitialInfo**, the text strings for the various labels and units that are to be used on program output and diagrams are defined; default values for initial values of the state variables, for parameters, and for the run time specifications are specified; and table functions are defined.

In procedure **ModelEqs**, we find the equations required for computing the time behavior of the model. At the beginning of the procedure, the values of the model parameters (ParamAnswer, ScenaAnswer), as provided by the user in the interactive query, are introduced into the simulation. This is only done once at the beginning of the simulation. The remaining part of this procedure lists equations which must be computed at each step of the simulation. First, the new values of the state variables (State[i]), as they follow from the numerical integration, are introduced. Then, all algebraic model relations are calculated in the correct order (of the influence arrows in the simulation diagram). Finally, the rates (Rate[i]) are defined. They are then used by the SIMPAS program to carry out the numerical integration, determine the new state values, and repeat the computations in the procedure ModelEqs for the next time step.

After completing the simulation, it is possible to use the computed results for user-defined output which can have any form: documentation, tables, graphs, or animation. The corresponding program statements must be provided in the procedure **Summary**.

The sample program PREDATOR.MOD in Program 4.2 follows the outline given in Program 4.1. Let us use this model to demonstrate the procedure of constructing a SIMPAS simulation program. The model unit PREDATOR.MOD can be found on the diskette. This program is a TurboPascal source program. Generally, the model unit can be written using either the TurboPascal text editor or any other text processing program such as Word, WordPerfect, or even Edlin. Once it has been written, the model should be saved as <MODEL_ID.MOD>, for example: PREDATOR.MOD.

For constructing the SIMPAS simulation program using TurboPascal, the model must be available under the name MODEL.PAS. It is therefore necessary to first copy the original model to a file having this name:

```
copy <MODEL_ID.MOD> MODEL.PAS
```

In our application:

```
copy PREDATOR.MOD MODEL.PAS
```

The name MODEL.PAS is essential for SIMPAS! Note that SIMPAS always works with the current file MODEL.PAS. If you forget to copy the current <MODEL_ID.MOD> to MODEL.PAS you should not be surprised to see another, recently used model on the screen.

For further work the following programs are required:

1. General: GRAPHICS.COM (comes with DOS)
2. TurboPascal: TURBO.EXE, TURBO.TPL, GRAPH.TPU
3. Suitable graphic interfaces: CGA.BGI, EGAVGA.BGI, HERC.BGI, etc. (these are routines which come with TurboPascal)
4. Letter font: TRIP.CHR (comes with TurboPascal)
5. SIMPAS units: SIMPAS.PAS, BASE.TPU, SIMUL.TPU (on the diskette)
6. Simulation model: MODEL.PAS.

You should copy these programs into a common directory SIMPAS which you then use for your SIMPAS modeling work. (The TurboPascal routines do not have to be copied into this directory if you define a corresponding path.)

Note that the SIMPAS TPUs (BASE.TPU, SIMUL.TPU) must have been compiled under the TurboPascal version used for setting up the simulation model (e.g. 5.0, 5.5, 6.0, 7.0). The correct TPUs are on the diskette. For example, if you are using TurboPascal 5.5, you must copy BASE.TPU and SIMUL.TPU from directory SIMPAS55 on the diskette.

TurboPascal with the main program SIMPAS.PAS is called from DOS by

```
TURBO   SIMPAS
```

The screen now shows the brief SIMPAS source program under the TurboPascal menu:

```
Program SimPaS;
uses simul;
begin
     simulate;
end.
```

Select the RUN option from the menu. TurboPascal should now compile the complete simulation program and start a simulation. (Note: Change the "destination" to "disk" under the menu heading COMPILE.)

(Practice this procedure using the program PREDATOR.MOD and the appropriate SIMPAS units on the diskette.)

4.1.4 Using special functions in SIMPAS

Using SIMPAS means being able to use all the programming features of TurboPascal, especially its procedures, functions, and units, including those that the program user has written. Anything that can be programmed, or is available in TurboPascal, can also be used in SIMPAS. However, there are some functions and procedures which are frequently required in simulation models and which are therefore made available in the SIMPAS units BASE.TPU and SIMUL.TPU:

1. Table function Tbf
2. Delay function Delay1 and Delay3
3. Pulse function Pulse
4. Step function Step
5. Ramp function Ramp
6. Discrete events Event
7. Euler-Cauchy and Runge-Kutta integration.

The features and the use of these SIMPAS functions will now be explained. We then develop a simulation program FCTNDEMO which demonstrates the use of these functions and their implementation in a model unit.

Table function Tbf

Many relationships between two variables cannot be easily formulated as mathematical expressions. This is especially true of empirical relationships; for example, a data series of noon temperatures for every second day of the month of April. Such data are usually provided as tables that are composed of data pairs (for example, day 20: 13 degrees Celsius; day 22: 17 degrees; day 24: 11 degrees, etc.). The table function is a way of supplying this information to the program (see also Sec. 3.1.1). Moreover, it also provides function values at intermediate points which are not listed in the table by using linear interpolation between neighboring points. For example, if the program needs a value for the noon temperature of day 21, it would compute it as the value halfway between 13 and 17, i.e. 13 + (17–13)/2 = 13 + 2 = 15 degrees.

SIMPAS table functions require a certain format. Input to the function, output of the function, and data pairs are defined in the procedure InitialInfo as follows (see also the example in PREDATOR.MOD):

```
TableFunction[<No. of Tbf>] :=
'<Name of output y> : <Name of input x>' + '/x1,y1/x2,y2/.../xn,yn//';
```

The $/x_i,y_i/$ represent data pairs of the table function. Note the apostrophes at the beginning and end of the text strings; they are required to define the strings correctly. <No. of Tbf> is the identification number for the table function; different functions must have a different number. <Name of input x> and <Name of output y> can be some descriptive labels.

In the program, the table function **Tbf** is called by using

```
Tbf(<No. of Tbf>,<Name of input quantity in program>);
```

<Name of input quantity in program> must be the exact label of the input variable, as it is used in the program.

Example:

```
TableFunction [1]: = 'outputY : inputX ' + '/0,0/0.7, 2.1/.../4, 0.5//';
Y := Tbf(1, X);
```

Note the following in using the table function:

1. Each table function used in a simulation model must have its own identification <No. of Tbf>.
2. The data pairs can be defined at arbitrary intervals of the input quantity (here X).
3. The data pairs must be ordered in the positive X-direction (as in a normal table).
4. The table function uses linear interpolation between two data pairs.
5. If input values X of the table function are below or above the range defined in the table, the output of the table function will correspond to the first or last table entry, respectively.

Delay functions Delay1 and Delay3

As the name implies, delays have the task of holding a variable and making it available again after a prescribed delay time. In a transport delay, the exact shape of the time function is preserved, and the entire function is simply shifted in time. Since all intermediate values have to be "remembered," this function requires storing (and shifting) a large number of values. In many applications it suffices to use the memory capabilities of the "exponential leak" (exponential delay) discussed in Section 3.3.4, whose output is a smoothed and delayed representation of recent input.

The SIMPAS delay functions are represented by exponential delays of first and third order whose outputs produce the same time integral (i.e. identical but delayed "impact") as the input variable. They are called in the program by using **Delay1** or **Delay3** (cf. Model M 216 STOCKS, SALES AND ORDERS in the systems zoo):

```
Delay3(<No. of delay>,<delay time>,<Name of input quantity in program>)
```

Example:

```
X := Delay3(1,DelT,U);
```

Internally, the exponential delay of first order of a signal u is computed as a state variable with a negative feedback of strength (1/T), where T is the delay time (time constant) of the system:

$$dz_0/dt = u - (1/T)\, z_0$$
$$\text{Delay1} = (1/T)\, z_0$$

with the initial value

$$z_{00} = u_0 T.$$

Similarly, three state variables are used to represent the exponential delay of third order:

$$\begin{aligned} dz_1/dt &= u - (3/T)\, z_1 \\ dz_2/dt &= (3/T)\,(z_1 - z_2) \\ dz_3/dt &= (3/T)\,(z_2 - z_3) \\ \text{Delay3} &= (3/T)\, z_3. \end{aligned}$$

The initial values are computed by the program from

$$\begin{aligned} z_{10} &= u_0\, T/3 \\ z_{20} &= u_0\, T/3 \\ z_{30} &= u_0\, T/3\,. \end{aligned}$$

Test functions Pulse, Step, Ramp, and Sin

In investigations of system behavior, in particular the response to external inputs, certain standard input functions of time are often used: the pulse function, the step function, the ramp function, and the sine function (Fig. 4.1).

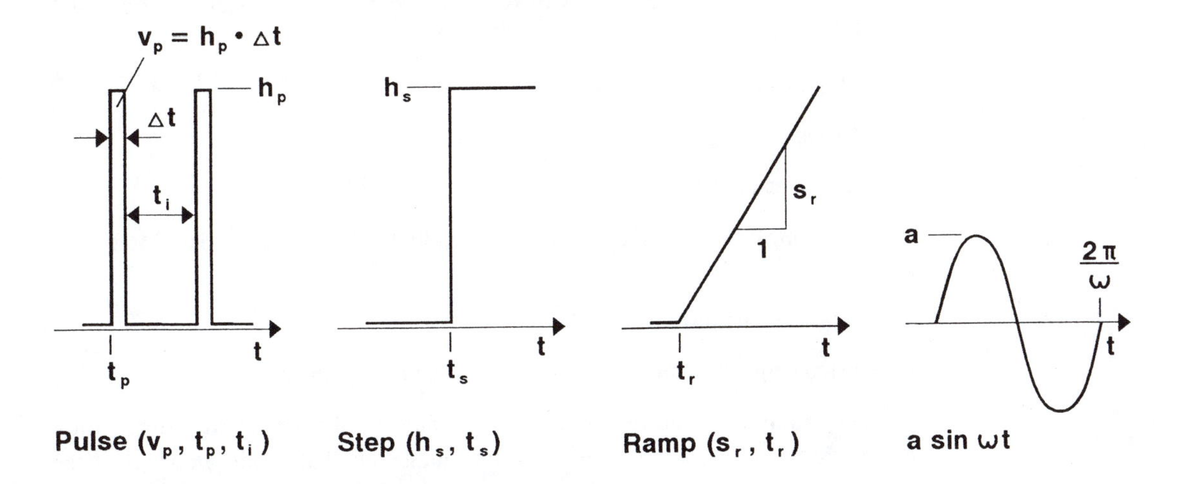

Fig. 4.1: Test functions: pulse, step, ramp, and sine. The time integral of a pulse is the step function, the time integral of a step is the ramp function.

The **pulse function** represents a sudden sharp "blow" to the system which happens in a very short (infinitesimal) time period. After this pulse, the input returns to its previous value (often zero). The "intensity" of the blow is measured by the "pulse area" which mathematically is the finite area of a pulse of infinite "height" and infinitesimal "width."

The pulse function **Pulse** is called up in a SIMPAS model (in Procedure ModelEqs) by

```
Pulse (<pulse area>, <pulse begin>, <pulse interval>)
```

Example:

```
U := Pulse (1, 10, 5);
```

This particular example defines a unit pulse (pulse area = 1), beginning at time $t = 10$ and being repeated again and again every five time units. Note that the pulse height depends on the timestep of computation:

pulse_height = pulse_area / timestep

In the example, the pulse height would be 100 for timestep = 0.01.

The arguments of the pulse function can also be defined as (real) variables:

```
U := Pulse (vp, tp, ti);
```

The **step function** represents a sudden increase of an input by a given amount. After this step, the input remains at this value. Note that the unit step function is the time integral of the unit pulse function.

The step function **Step** is called up in the SIMPAS model (in Procedure ModelEqs) by

```
Step (<Step height>, <Step time>)
```

Example:

```
U := Step (1, 10);
```

This function defines a unit step of magnitude "1" at time $t = 10$. More generally:

```
U := Step (hs, ts);
```

where step height hs and step time ts must be real quantities.

The **ramp function** represents a linear increase of an input with time with a given "slope" of the ramp (change per unit time). Note that a unit ramp is the time integral of the unit step function.

The ramp function **Ramp** is defined in a SIMPAS model (in ModelEqs) by

```
Ramp (<ramp slope>, <ramp time>);
```

Example:

```
U := Ramp (0.5, 5);
```

defines a ramp of slope 0.5 beginning at time $t = 5$. More generally:

```
U := Ramp (sr, tr);
```

where ramp slope sr and ramp time tr must be real quantities.

Sine functions (or cosine functions) of a given frequency are used to represent periodic input to a system. For test functions, single sine (or cosine) functions are commonly used but it is also possible to represent arbitrary periodic functions by a Fourier series of sine and/or cosine functions (see Sec. 7.4.3).

SIMPAS utilizes for the sine function the available TurboPascal function **Sin**(X: real), where X must be expressed in radians.

Example:

```
U := Sin (0.5 * time);
```

Note that the period of a sine oscillation sin (c·time) is given by the condition $c \cdot time = 2\pi$, i.e. period $= 2\pi/c$. In the example, the period is found from (0.5 period) $= 2\pi$, or: period $= 4\pi$. This sine function could therefore be equivalently written as

```
U := Sin (2 π * time/period);
```

By adding different test functions, it is possible to generate a great variety of complex test signals. Where this is not sufficient, a corresponding time-variant **Table Function** can be used as test function. A previously defined Table Function No. 3 is called up as a function of time by

```
U := Tbf(3, time);
```

Events

In simulation models, one often has to deal with the possibility of certain specific and time-limited events taking place which then shape the further development of the system. For example, a forest may be cut after 50 years, or it is hit by a storm (a random event) which destroys a certain percentage of the trees.

SIMPAS provides Boolean variables **Event**[i] that are automatically set to "false" at the beginning of a simulation. In the course of the simulation it is then possible to set the Event [i] to "true" (and later again to "false") to initiate whatever consequences follow from the event.

Example: Representing the impact of windfall in a forest:

```
if storm then Event [1]  := true;
if Event [1] then
     begin
          stemloss := 0.3 * stemwood;
          Event [1] := false;
     end;
```

The next to last program line assures that the (storm) event leads to windfall only once.

Numerical integration

SIMPAS offers two methods of numerical integration: the Euler-Cauchy method and the Runge-Kutta method of fourth order. The Euler-Cauchy method (explained in Sec. 3.2.2) is the simplest possible approach but it is often not accurate enough (especially if oscillations are involved). The Runge-Kutta method (explained in Sec. 7.1.6) provides much better accuracy but is often too slow for quick assessments of system behavior.

The Euler-Cauchy procedure of numerical integration uses the current state value and the current rate of change to compute a new state value:

$$z_{new} = z_{old} + (dz/dt)_{old} \cdot \Delta t$$

or

$$State[i]_{new} = State[i]_{old} + Rate[i]_{old} * TimeStep.$$

This procedure is sufficiently fast and precise for the models of this book—provided that the computation step Δt (TimeStep) has been chosen small enough. Often one has to compromise between the demands of computational precision and short computing time. The step width (TimeStep) must be chosen in such a way that it allows at least 20 to 100 computation steps over a period of significant change in the state variables (for example, one half period of a sine oscillation).

In applying this method, one always has to check for a possible dependence of the computation results on step width by varying the step width over a meaningful range. The step width must be chosen small enough to ensure that the observed computational error is insignificant.

For precise analyses and for so-called "stiff" systems (having very different time constants of their characteristic processes), the Euler-Cauchy integration is not suitable. In most practical applications, the fourth order Runge-Kutta method produces reliable results. Other procedures are also available (see for example, Press et al.: 1988, p. 547-577, 777-792).

Using the SIMPAS result file

The procedure SummaryEqs in the model unit can be used to process and present additional output as programmed by the user: graphs, tables, aggregations, and animation.

For programs of this kind written by the user it is important to know where the values stored during the simulation can be found. SIMPAS stores results according to the following scheme:

Result[1,i]	=	(simulation)time
Result[2,i]	=	state variable State[1]
Result[3,i]	=	state variable State[2]
Result[n+1,i]	=	state variable State[n]
Result[N+1,i]	=	last state variable State[N]
Result[N+1+1,i]	=	Outvariable[1]
Result[N+1+2,i]	=	Outvariable[2]
Result[N+1+m,i]	=	Outvariable[m]
Result[N+1+M,i]	=	last Outvariable[M].

Here i = running index, 1, 2, ... i, ..., N_{max}. Time START corresponds to i=1; time FINAL to i = N_{max}. For the SIMPAS version on diskette, N_{max} = 251.

4.1.5 Using SIMPAS functions: a sample program

The use of the pulse function Pulse, the step function Step, the ramp function Ramp, the table function Tbf, the sine function Sin, and the delay functions Delay1 and Delay3 will be demonstrated using an example program for use with SIMPAS. We use the test functions to define the input signals X which are then delayed by the exponential delays of first and third order. We define signals with identical positive and negative contributions leading to a time integral of zero. By integrating the delayed signals over time, we check that this condition also applies to the delayed signals. Most signals begin at time t = 1.

The **pulse signal** is composed of two opposite pulse functions:

```
X := Pulse (1, 1, 0) + Pulse (–1, 2, 0);
```

The **step signal** is constructed from three step functions:

```
X := Step (1, 1) – Step (2, 2) + Step (1, 3);
```

The **ramp signal** is constructed by adding four ramp functions:

```
X := Ramp (1, 1) – Ramp (2, 1.5) + Ramp (2, 2.5) – Ramp (1, 3);
```

We use the **table function** to define the following signal:

```
Table Function [1] :=
 'X : time' + '/0, 0/1, 0/1.2, 1/1,5, 0.2/2, 0/2.5, –0.2/2.8, –1/3, 0/4, 0//'
```

The **sine signal** is intended to produce a complete oscillation in the time period $1 < t < 3$:

```
X := sin (π·(Time – 1)).
```

In order to be able to select any of these test functions when running the model, we include a corresponding question in the interactive query:

```
ParamQuestion [1] := '1-Pulse, 2-Step, 3-Ramp, 4-Sin, 5-Tbf : (5) [-]';
```

The corresponding answer is introduced into the model definition (procedure Model Eqs) where it then controls (by an "if ... then" statement) the selection of the input signal:

```
Select := round (ParamAnswer [1]);
```

The delays of first and third order of signal X with delay time of DelT are computed from

```
Y1 := Delay1(1,DelT,X);
Y3 := Delay3(1,DelT,X);
```

These delays change the signal X; the time integrals of Y1 and Y3, however, should be equal to the time integrals of the undelayed function X. In order to test this, we integrate the input signal and both of the delayed signals:

```
Rate[1] := X;
Rate[2] := Y1;
Rate[3] := Y3;
```

SIMPAS automatically produces all state variables as outputs; hence we here have as outputs the state variables State[1] = IntX, State[2] = IntY1 and State[3] = IntY3 which are the time integrals of Rate[1], Rate[2], and Rate[3]. The corresponding identifications, initial values, and dimensions must be provided in the model definition (in procedure InitialInfo):

```
StateVariable[1] := 'Integral of X, undelayed: (0) [-]';
StateVariable[2] := 'Integral of X with Delay1: (0) [-]';
StateVariable[3] := 'Integral of X with Delay3: (0) [-]';
```

In addition, the values of X, Y1, and Y3 are to be available as output (defined in procedure InitialInfo):

```
OutVarText[1] := 'Signal: [-]';
OutVarText[2] := 'Signal with Delay1: [-]';
OutVarText[3] := 'Signal with Delay3: [-]';
```

Variables corresponding to these texts are defined in procedure ModelEqs:

```
OutVariable[1] := X;
OutVariable[2] := Y1;
OutVariable[3] := Y3;
```

To observe the effect of time delays, the delay time DelT with a default value of 1 is used as scenario parameter in the interactive query in procedure InitialInfo:

```
ScenaQuestion[1] := 'delay time: (1) [-]';
```

In the model part (procedure ModelEqs) we must then insert:

```
DelT := ScenaAnswer[1];
```

For the run time definitions we define default values (in procedure InitialInfo):

```
Start      := 0;
Final      := 10;
TimeStep := 0.01;
```

The corresponding program FCTNDEMO is shown in Program 4.3 (in the Program Listing Appendix). Simulation results are presented in Figure 4.2. They show a strong distortion of the input signal. The "effect" of the original signal is preserved even after the delay, as the integration results show.

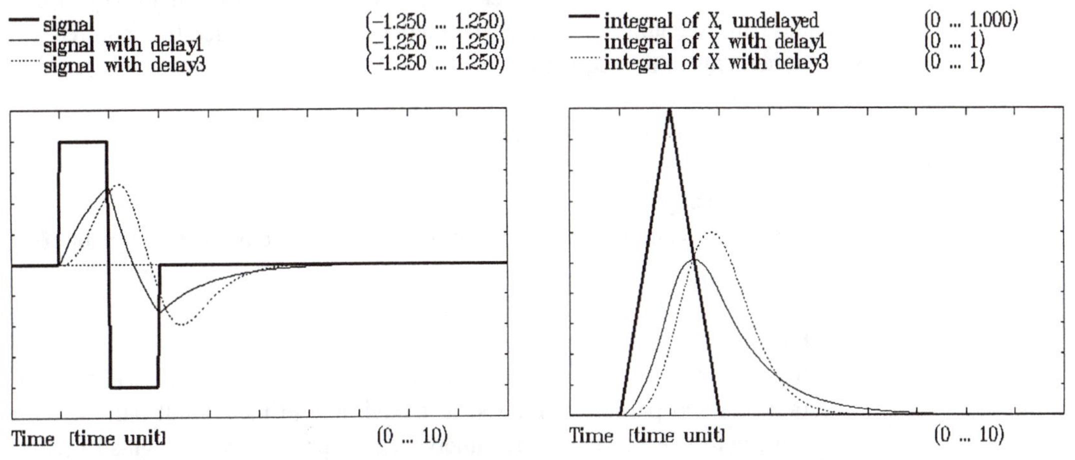

Fig. 4.2: Demonstration of SIMPAS-functions using the model unit FCTNDEMO. A rectangular wave is delayed (Delay1, Delay3); these signals are integrated.

4.2 Simulation of Rotating Pendulum Dynamics Using SIMPAS

4.2.1 Constructing the SIMPAS model from the simulation diagram

The model development starts with the description of the model contained in either the simulation diagram or the corresponding model equations and having the components :

1. parameters and exogenous quantities (hexagons)
2. state variables (rectangles) and their initial values
3. intermediate variables (circles) which can be determined from parameters, exogenous quantities, state variables, and other intermediate variables either by algebraic or logical statements
4. rates of change (circles) of the state variables following from 1., 2. and 3. which determine the system state at the next point in time.

For the rotating pendulum we obtained from the considerations in Chapter 3 (Fig. 3.21 and equations in Sec. 3.5.4) the necessary relationships for the simulation model. To make the program more readable, we introduce here names for the variables and parameters in plain English.

The easiest way to develop a new model unit is to use a similar available model unit as a template and to replace the old model formulations by those of the new model using the TurboPascal text editor or another text processing program. It is of course not necessary to develop the model unit line for line in sequence. In the following, we jump around in the model unit to introduce the required program statements in the sequence:

1. parameters
2. state variables and initial values
3. intermediate variables
4. rates of change
5. additional quantities, run time information, and additional model information.

1. Parameters

Since we wish to investigate the dynamics of the system for a broad spectrum of parameters and parameter ranges, it is advisable to include all parameters of potential interest in the interactive query. It is expedient to distinguish between relatively fixed system parameters (ParamQuestion) and more variable scenario parameters (ScenaQuestion). Parameters which are not to be changed under any circumstances (here: the earth's gravitational acceleration) are defined directly in the program. For the rotating pendulum model, we choose to define pendulum mass and pendulum radius as system parameters for the purposes of the interactive query, and damping as a scenario parameter. This leads to the following program statements in the procedure InitialInfo (default values in parentheses; dimensions in brackets):

```
ParamQuestion[1] := 'mass: (1) [kg]';
ParamQuestion[2] := 'radius: (1) [m]';
ScenaQuestion[1] := 'damping: (1) [N/(m/s)]';
```

The strings in '...' contain the question to be presented on the screen, the default value, and the dimension of the corresponding quantity. In order to introduce the corresponding answers into the model equations, they have to be available at the beginning of the simulation run. We therefore introduce into the procedure ModelEqs:

```
if Time=Start then
begin
      mass        := ParamAnswer[1];
      radius      := ParamAnswer[2];
      damping     := ScenaAnswer[1];
      gravitation := 9.81;
end; .
```

The gravitation constant is also defined at this point.

2. State variables and initial values

The state variables, their labels for screen output, the initial values, and the dimensions must be specified in the procedure InitialInfo:

```
StateVariable[1]    := 'angle: (0) [rad]';
StateVariable[2]    := 'angular velocity: (10) [1/s]';
```

The corresponding identifications used in the program have to be introduced in the model equations (procedure ModelEqs) (note: the time derivative of State[i] is Rate[i]):

```
angle               := State[1];
angular_velocity    := State[2];
```

3. Intermediate variables

The computation of intermediate variables from parameters and state variables and other intermediate variables is placed in the procedure ModelEqs. We insert the following equations:

```
velocity              := radius * angular_velocity;
gravitational_force   := mass * gravitation * sin(angle);
damping_force         := damping * velocity;
acceleration_force    := – gravitation_force – damping _force;
acceleration          := acceleration_force / mass;
angular_acceleration:= acceleration / radius;
```

Note that it must be possible to compute each variable from previously defined quantities. This determines the sequence of the equations.

4. **Rates of change**

The state equations are introduced in the procedure ModelEqs:

```
Rate[1]   := angular_velocity;
Rate[2]   := angular_acceleration;
```

The right hand sides follow from the previously determined intermediate variables.

5. **Additional quantities, run time information, and additional model information**

In order to achieve a more realistic description of the movement of the pendulum, a picture of the pendulum position in space as a function of time is of some interest. We therefore add in the procedure ModelEqs the computation of the horizontal and vertical position:

```
horizontal := radius * sin(angle);
vertical   := radius * (1 – cos(angle));
```

These quantities are defined as output variables:

```
OutVariable[1] := vertical ;
OutVariable[2] := horizontal;
```

Corresponding output texts have to be included in the procedure InitialInfo:

```
OutVarText[1] := 'vertical position: [m]';
OutVarText[2] := 'horizontal position: [m]';
```

The starting time, the finishing time, and the time step of the simulation must be specified in the procedure InitialInfo:

```
Start     := 0;
Final     := 10;
TimeStep  := 0.02;
```

The step width (TimeStep) depends on the integration method used (here defined for Euler-Cauchy), and on the dynamics of the model (cf. rule of thumb in Sec. 4.1.4 *Numerical integration*). The current value of simulation time is available for programming as a variable 'Time'. (It is not used in this model but it was used in FCTNDEMO.MOD.)

For the presentation of results, one has to specify in addition in the procedure InitialInfo a model title, a brief description, the time unit used, and the name of the model author and the date:

```
Title        :=  'ROTATING PENDULUM';
Description  :=  'Non-linear damped oscillation of pendulum'
                 + 'in circular motion';
TimeUnit     :=  'seconds';
Author       :=  'H.Bossel: Modeling and simulation 910612';
```

The brief description can have up to 255 letters (approximately four lines). Note that long text can be split up into several strings that are concatenated by the "+" sign. Finally, TurboPascal needs (under 'var') a complete listing of all quantities defined in the model with their type declarations. This list may also be written in the form of a table, giving dimensions and explanations (in commentary braces):

```
var
    angle, angular_velocity, angular_acceleration,
    mass, radius, gravitation, damping, velocity,
    gravitation_force, damping_force, acceleration_force,
    acceleration, horizontal, vertical: real;
```

This defines the simulation model for the rotating pendulum. The corresponding TurboPascal unit PENDULUM.MOD is fully listed in Program 4.4 (in the Program Listing Appendix).

4.2.2 Constructing the executable simulation program

Before continuing to work with the model unit as unit MODEL.PAS, it is advisable to save it under a descriptive name. We characterize all SIMPAS models with the file extension *.MOD. The model for the rotating pendulum is therefore saved by using (from DOS)

```
copy  MODEL.PAS  PENDULUM.MOD
```

We now call up TurboPascal with SIMPAS:

```
TURBO  SIMPAS
```

The TurboPascal screen now appears with the brief SIMPAS main program. In the TurboPascal menu we first select the option "Compile" and change the "Destination" to "Disk" (with RETURN). We then select the "Run" option. The simulation program will now be compiled and run by the computer—provided it is free of errors. If MODEL.PAS contains programming errors, they have to be eliminated in the usual way using the TurboPascal debugging procedures. The executable simulation program is now available as SIMPAS.EXE in the current directory on the floppy or hard disk. It should be renamed PENDULUM.EXE before further use:

```
rename  SIMPAS.EXE  PENDULUM.EXE
```

This simulation program can now be used independently of TurboPascal. In order to use it on another computer, it is only necessary to load in addition the corresponding graphics interface (for example, EGAVGA.BGI) and the letter fonts (TRIP.CHR) from the TurboPascal library. The following runs are all made with PENDULUM.EXE.

4.2.3 Standard run and interactive model use

To start the simulation program, enter the name of the model (PENDULUM) from the DOS level. In the upper left corner of the screen you will see the question

```
SimPaS in black-and-white (0) or color (1)?
```

The default setting is for a color monitor. In that case, you only have to press RETURN. Enter "0" and RETURN if you have monochrome monitor, or if you want to produce graphic output which will be reproduced in black and white. In that case curves will be distinguished by their line pattern.

Next, the SIMPAS title screen appears for a brief period, followed by the title screen for the model (here ROTATING PENDULUM). On the next screen we find the brief description of the model, a reference to this book, and some hints for using SIMPAS:

```
use RETURN for normal continuation
use ESCAPE for exit from simulation
use CTRL P to print screen display with pin printers.
```

Note as a general rule that using RETURN will always lead to the most "normal" or "frequent" mode of using the program. If a choice of parameters is involved, the default values which are always listed on the screen, will be activated upon pressing RETURN. Parameter values should therefore be entered only if they differ from the values listed on the screen.

At the bottom of this screen, you are also asked to:

Indicate PRINTER connection:
0 – none, 1 – 9-pin printer, 2 – 24-pin printer, 3 – Laser/DeskJet.

If you do not plan to use a printer (default setting), press RETURN (or "0" and RETURN). Choices "1" and "2" work for most standard 9-pin and 24-pin printers (in particular, the Epson FX and LQ series). If you press "2" and RETURN, you can on the next screen choose the graphics resolution, the vertical, and the horizontal enlargement of the graphics. Responding with RETURN to these questions will define hard copies of normal resolution and size (default settings). With pin printers, you can print the simulation parameters, tables of results, two- and three-dimensional graphics, and the table functions by pressing simultaneously CTRL and P after these outputs have appeared on the screen.

It is also possible to use other programs for producing hard copy output from screen displays, like HPSCREEN (for Hewlett-Packard LaserJet and DeskJet printers) or CAPTURE (with WORD). In that case, the <PrtScr> key is used instead of CTRL P. Choice of "3" in the previous question will produce some relevant information on the screen.

After pressing RETURN, the SIMULATION PARAMETERS screen appears. The first line identifies the model and gives the current date and time for the documentation of the simulation run. All system and simulation parameters that can be changed by the model user are listed together with their units of measurement and the current setting (normally the default value). If the list of parameters is too long to be shown on the screen in its entirety, not all parameters will be shown simultaneously but all of them can be accessed by scrolling the list using the arrow keys. (A hint will appear on the screen if this is the case.)

The simulation parameters are shown in the sequence:

1. system parameters
2. scenario parameters
3. initial values of the state variables
4. run time parameters.

For the PENDULUM model, the following list appears on the screen:

mass	[kg]	1.000
radius	[m]	1.000
damping	[N/(m/s)]	1.000
angle	[rad]	0
angular velocity	[rad/s]	0.000
Begin of simulation	[seconds]	0
End of simulation	[seconds]	10.000
Time step of simulation	[seconds]	0.020

If a printer is connected and has been initialized at the beginning of the run, a hardcopy of this parameter list can be produced by simultaneously pressing the CTRL and the P keys (or using <Shift> and <PrtScr>; see above).

All of the parameters listed can be changed at this point. The procedure will be explained below. For now, we accept this set of default parameters by pressing RETURN.

The next screen offers a:

MODE SELECTION:
Simulation using Euler-Cauchy
Simulation using Runge-Kutta
Parameter Sensitivity
Global Response
Table Functions
Quit.

Here the marker bar tags the "Simulation using Runge-Kutta" which is the default choice. Choice of "Simulation using Euler-Cauchy" would produce a quicker but less accurate simulation. The other functions will be explained in more detail later. "Parameter Sensitivity" permits investigations of the sensitivity of the model behavior with respect to changes of (system or scenario) parameters and initial values. "Global Response" allows the investigation of the global model behavior as a function of the initial conditions of the state variables. "Table Functions" shows the table functions contained in the model in the form of diagrams and tables. (PENDULUM does not contain any table functions but you can view the table function when running the FCTNDEMO model, for example.)

Pressing RETURN while the marker bar is on "Simulation using Runge-Kutta" starts the Runge-Kutta simulation. On the screen we now see a message informing us that the simulation is in progress and asking us to wait. At the same time, the rectangular bar at the bottom of the screen fills with 'happy faces,' indicating the progress of the simulation.

When the simulation is finished, the next screen offers the choice:

Continue OUTPUT PRESENTATION?
Yes, more results for this run
No, new simulation or quit.

The "Yes" option is marked as the default choice and pressing RETURN therefore leads to the presentation of the simulation results. The next screen OUTPUT SELECTION AND SCALING allows both the choice of output variables and of the scales to be used in the graphical output. On this screen, the variables are always listed in the order:

time
(all) state variables
preselected additional output variables.

Up to five of the variables (if available) can be chosen for simultaneous output. In the present case, the screen shows the list:

time	[seconds]	0	10.000
angle	[rad]	0	7.669
angular velocity	[rad/s]	-3.128	10.000
vertical position	[m]	5.46E-12	2.000
horizontal position	[m]	-0.997	0.990

Together with the names of the output variables and their units of measurement are listed the maximum and minimum values computed during the simulation which can be used later to assist in the scaling process. If the user merely presses RETURN, then the first five quantities are automatically selected. If less than five quantities are of interest, the user has to choose them individually.

Pressing RETURN now leads to the next screen:

OUTPUT PRESENTATION:
Table
Two-dimensional Graph
Three-dimensional Graph.

The second option is marked by the tagging bar; pressing RETURN therefore produces a two-dimensional graph of the results showing angle, angular velocity, vertical position, and horizontal position as functions of time (Fig. 4.3).

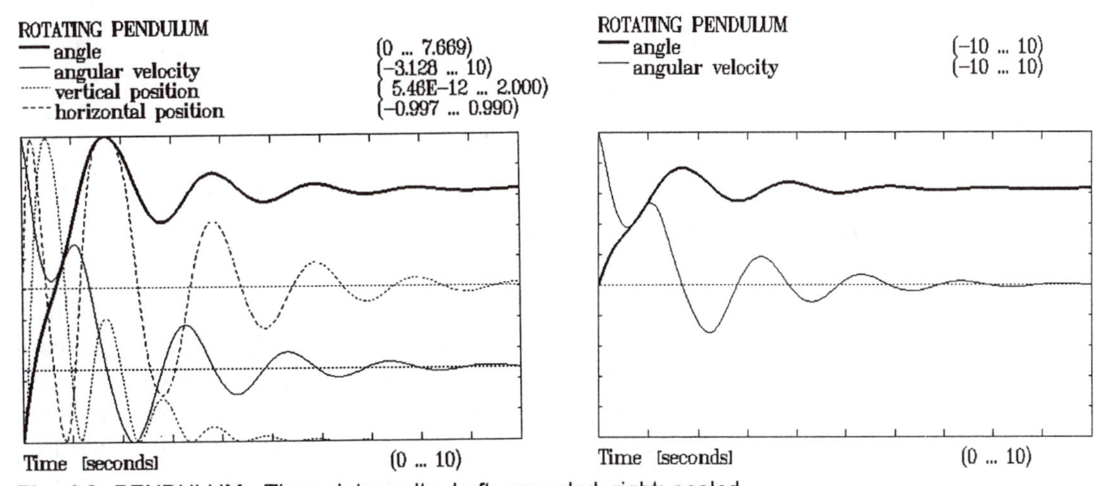

Fig. 4.3: PENDULUM: Time plot results. Left: unscaled, right: scaled.

Each of the curves fills the complete width of the diagram between minimum and maximum values given in the table above. The zero line is shown for the angular velocity and the horizontal position, both of which have positive and negative values. Because of the odd scales and the number of variables, this plot is not very convenient to use. Minimum/maximum scaling is usually not practical as it makes the comparing of results difficult. Below, we return to the selection and scaling of variables; let us first see how to produce output in the form of a table.

Pressing RETURN brings us back to the screen "Continue OUTPUT PRESENTATION?," and pressing RETURN again produces once more the OUTPUT SELECTION AND SCALING screen with the list of variables. Pressing RETURN again brings us to the selection of OUTPUT PRESENTATION. In order to produce tabular output, we move the tagging bar from "Two-dimensional Graph" to "Table" by using the "up" arrow and press RETURN.

The next screen shows a table with simulation results for 1. time (first column), 2. angle, 3. angular velocity, 4. vertical position, and 5. horizontal position (Fig. 4.4). The table contains the model name, the names and dimensions of the variables shown, and (normally) 11 rows of results which cover the selected time interval in ten equal time steps.

ROTATING PENDULUM

Time [seconds]	angle [rad]	angular veloc ity [rad/s]	vertical posi tion [m]	horizontal po sition [m]
0	0	10.000	5.46E-12	0
1.000	5.409	5.310	0.359	-0.767
2.000	7.233	-2.507	0.418	0.813
3.000	5.646	1.215	0.196	-0.595
4.000	6.683	-0.621	0.079	0.390
5.000	6.036	0.326	0.030	-0.245
6.000	6.435	-0.172	0.012	0.151
7.000	6.190	0.090	4.32E-03	-0.093
8.000	6.340	-0.046	1.61E-03	0.057
9.000	6.249	0.022	5.96E-04	-0.035
10.000	6.304	-0.010	2.20E-04	0.021

Fig. 4.4: PENDULUM: Tabular presentation of results for selected variables.

SIMPAS stores the results for 250+1 time steps for each simulation. All of these results are potentially available for output. They can be accessed by specifying a corresponding (small) time interval in the scaling process. If we leave again the tabular presentation by pressing RETURN, and select "Yes, ..." in response to "Continue OUTPUT PRESENTATION," we arrive again at OUTPUT SELECTION AND SCALING. The screen again shows the list of the available variables together with their minimum and maximum values. At the lower right hand corner of the screen we see two small boxes marked "OK" and "change." By using the "down" arrow, we move the tag marker to "change" and press RETURN. The cursor now blinks under the diamond marking the "Time."

In order to study the time interval from 5.0 to 5.1 in more detail, for example, we now move the cursor (and the marker) to the right using the "right" arrow which tags the minimum value for "time." We change the "0" to "5.0" using the keyboard. Using the "right" arrow once more, we mark the maximum and change it to "5.1." To finish the selection, we press the ESCAPE (ESC) key. The cursor jumps back to the "OK" field, and another RETURN allows us to select the "Table" again by using the "up" arrow. The table now contains only four rows of results for each of the output variables (for time = 5.0, 5.04, 5.08, and 5.12) since the interval for recording results was 10/250 = 0.04, and more results are not available in this interval.

Let us now use the scaling and selection options to produce custom-made graphical output. Using RETURN, we leave the tabular display and select "Yes, more results for this run" under the question "Continue OUTPUT PRESENTATION?" We use the "down" arrow and the RETURN key again to access the list of variables, and rescale the time interval again from 0 to 10. With the cursor back in the first column (under the diamond), we can move it and the tag marker by using the "down" and "up" keys.

The marks in the first column indicate which of the variables have been selected for plotting:

1. the diamond (♦) marks the (one) variable that is plotted on the horizontal axis,
2. the asterisk (*) marks variables (up to four) that are to be plotted on the vertical axis.

All variables can be plotted on either of the axes; it is merely required to mark them accordingly. The plotting markers are set or removed by pressing RETURN (with the cursor in the first column). An asterisk is converted to a diamond by using the "left" arrow key.

Before we select certain variables for plotting, let us first scale all of them to convenient scales. Moving the cursor first to the variable name, and then to its minimum and maximum values, we can rescale them as required. We then choose the following plot ranges: for the angle (-10 to 10), for the angular velocity (-10 to 10), for the vertical position (0 to 2), and for the horizontal position (-1 to 1). (This selection leads to a common zero line for the different plots which makes it easier to compare results.)

We first wish to produce a plot of "angle" and "angular velocity" as functions of time. This means that "vertical position" and "horizontal position," which now also carry an asterisk, must be deselected. We remove the asterisk in front of these variables by using the arrow keys and the RETURN key. Pressing the ESC key moves the cursor to the "OK" field. We acknowledge the selection (by RETURN), and select "Two-dimensional Graph" in the OUTPUT PRESENTATION selection. RETURN produces the time plot for the two variables "angle" and "angular velocity" shown in Figure 4.3b.

This plot now shows clearly that under the given initial conditions (high angular velocity in the lower dead center) the pendulum rotates once and then comes to rest after a damped swinging motion. The angle at which the pendulum comes to rest can be read off the scaling on the right hand edge of the plot (about 6.3). This corresponds to one full circle ($2\pi = 6.28$). The result is also confirmed by the table (Fig. 4.4). Note that the pendulum has not come to rest at time $t = 10$.

Another view of this process is presented by the phase plot (Fig.4.5) which we can obtain by changing the angle range to (0 to 10), by deselecting "time," and by selecting the variables "angle" and "angular velocity" where we mark "angular velocity" by a diamond for plotting on the horizontal axis. Choosing the "Two-dimensional Graph" under OUTPUT PRESENTATION, we now see a spiralling motion of the two state variables towards an equilibrium point with zero velocity and angle 2π . In this diagram, the angle is plotted as a function of the angular velocity, producing the time-dependent state trajectory with time as an implicit variable. After a full circle the motion is "caught" by the lower dead center at 2π ; the spiral motion in this diagram actually represents the swinging motion of the pendulum around the lower dead center (positive and negative angular velocity).

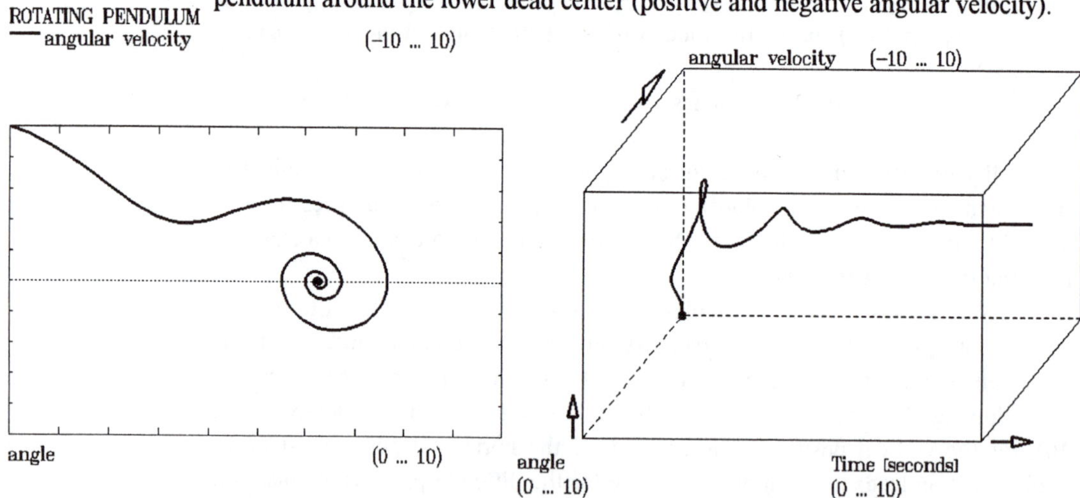

Fig. 4.5 (left): PENDULUM: Phase plot of angle and angular velocity.

Fig. 4.6 (right): PENDULUM: Three-dimensional view of the state (angle and angular velocity) as function of time.

This motion becomes somewhat more vivid in the three-dimensional presentation. By again using RETURN, we go back to OUTPUT SELECTION AND SCALING and mark "time" by a diamond, and "angle" and "angular velocity" by an asterisk. Choice of the "Three-dimensional Graph" produces the "corkscrew" in Figure 4.6. Again, the damped motion around the equlibrium point is obvious. Note that Figure 4.5 represents a two-dimensional view of Figure 4.6, looking from the right in the direction of the negative time axis. To aid in the visualization of the results, time slices appear in this presentation after pressing RETURN, and a small circle marks the point where the trajectory "pierces" each time slice.

A more realistic presentation of the pendulum motion is obtained by selecting under OUTPUT SELECTION AND SCALING the variables "time," "vertical position," and "horizontal position," picking "time" as the horizontal axis by marking it with a diamond and then representing these quantities in the three-dimensional display (Fig. 4.7). It can now be clearly seen that the pendulum begins at the lower dead center with a high angular velocity, moves through one full circle, is not able to reach the upper dead center again, and starts swinging back and forth in a strongly damped motion which finally ends at the rest position at the lower dead center. The time slices help in spatially orienting the trajectory.

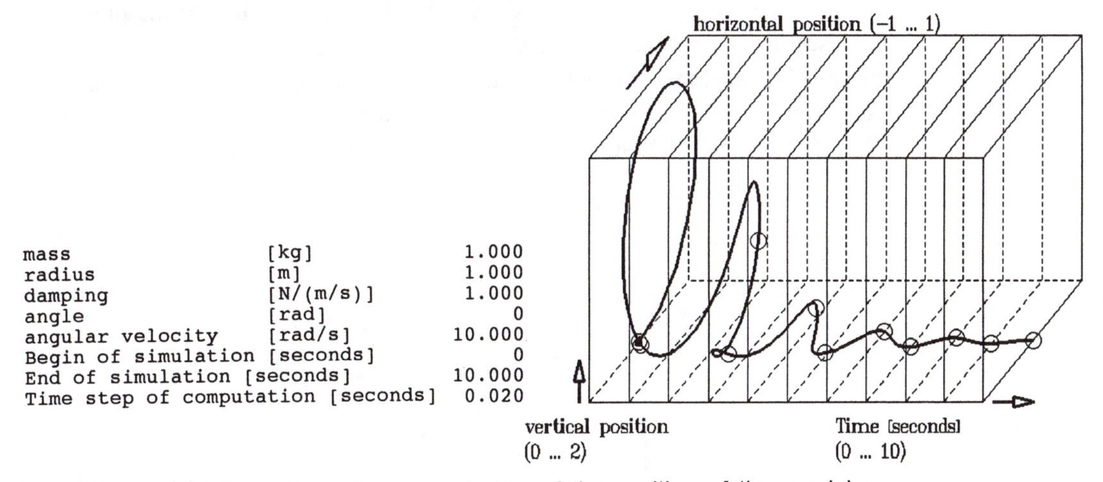

Fig. 4.7: PENDULUM: Three-dimensional view of the position of the pendulum mass as a function of time.

4.2.4 Changing parameters

If we want to change parameters of the system to study its behavior under different conditions, we have to change the default values in the SIMULATION PA-

RAMETER list which appears before the start of the simulation. To demonstrate the procedure, we leave the three-dimensional display by using RETURN. This brings us to the question "Continue OUTPUT PRESENTATION?" This time, we move the marker bar to "No, new simulation or quit," press RETURN, choose "Continue" in the CONTINUATION screen, and are back at the SIMULATION PARAMETER display.

We can now change the parameters "mass," "radius," and "damping," the initial values of the two state variables "angle" and "angular velocity," starting and finishing time of the simulation, and the time step of the computation. Let us investigate the effect of higher damping on the pendulum motion. To initiate the change process, move the marker from "OK" to "change" by using the "down" arrow. Then press RETURN to move the cursor to the first asterisk on the parameter list. Then move the marker up or down by using the "up" and "down" arrows to reach the parameter which is to be changed. We move the marker to "damping," use the "right" arrow button to select the numerical value (1.000), and change it to "2." By using the ESC key, we finish the selection, and acknowledge it by RETURN. This brings us to MODE SELECTION, where we select "Simulation using Runge-Kutta." After completion of the simulation, we use RETURN to take us to the OUTPUT SELECTION AND SCALING screen. Note that the previously used scaling and selection of variables has been preserved. This feature allows us to quickly compare the results of different runs, without having to reselect and rescale variables. We leave the settings unchanged, use RETURN to move to OUTPUT PRESENTATION, and select the "Three-dimensional Graph."

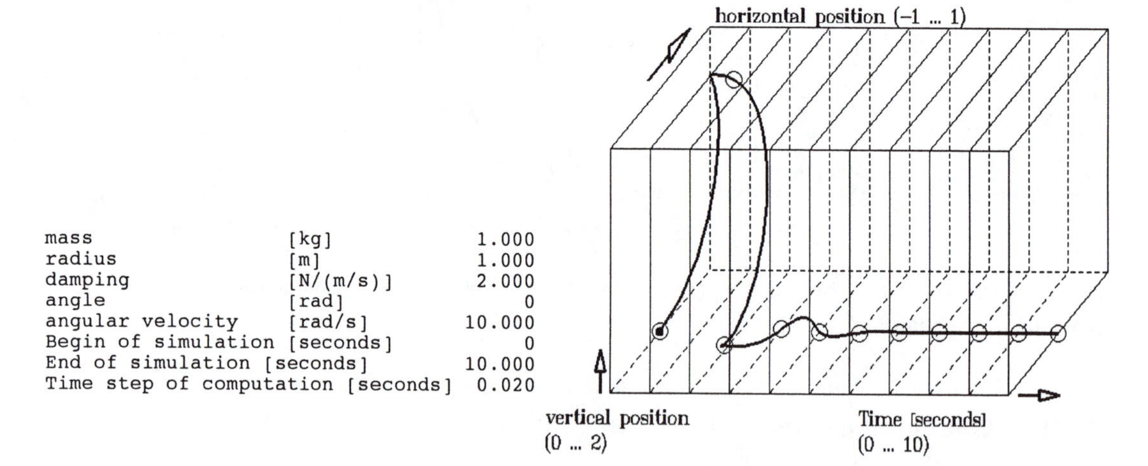

Fig. 4.8: PENDULUM: Three-dimensional view of the position of the pendulum mass as a function of time, for stronger damping.

Figure 4.8 shows the three-dimensional presentation of the pendulum motion for this case of stronger damping which should be compared with Figure 4.7. We note that in this case the damping prevents the motion from reaching the upper dead center. The pendulum falls back and quickly comes to a standstill at the lower dead center. In order to produce a time plot for angle and angular velocity, we go back to OUTPUT SELECTION AND SCALING, remove the asterisks from "vertical position" and "horizontal position," mark "angle" and "angular velocity" instead, and change the range of "angle" to (–5 to 5). The two-dimensional plot of these variables as function of time is shown in Figure 4.9. The equilibrium angle is now = 0; the strongly damped swinging motion ends at the lower dead center without completing a full circle. The strongly damped state spiral can also be clearly seen in the two-dimensional phase plot (as in Fig. 4.5).

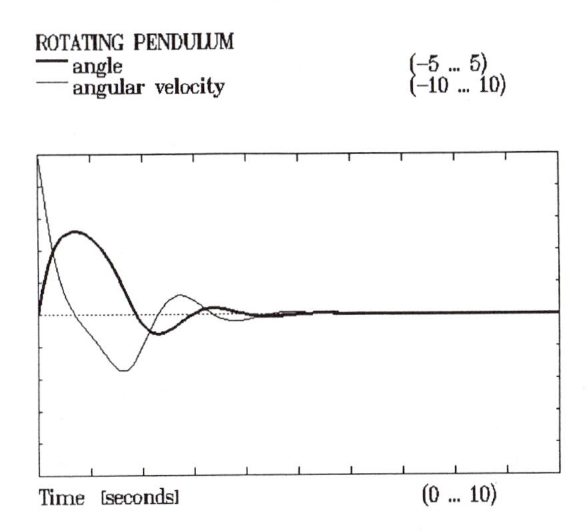

Fig. 4.9: PENDULUM: Time plot of pendulum motion for higher damping (d=2).

When parameters are changed, the simulation results often change significantly, and the scaling of the variables used in previous runs is no longer appropriate. To find appropriate minima and maxima for rescaling, the minimum and maximum values of the current simulation results can be used as a guide. The scaling range can be reset to these minimum and maximum values in the OUTPUT SELECTION AND SCALING screen.

To demonstrate this feature, we return to this screen and move the cursor from the "OK" to the "change" field using the "down" arrow key. At the bottom of the screen, a message appears briefly:

"for min/max scaling, enter "0" and <return> now!"

If we now enter "0" (zero) and press RETURN, all the minimum and maximum values for the output variables change. By selecting the "Two-dimensional Graph," we can verify that the plots are rescaled to the new minima and maxima. The new minima and maxima can be used to redefine the plotting ranges.

4.2.5 Parameter sensitivity

The initial phase of working with a new model is usually one of exploration in which the response is studied over a wide parameter range, often for a number of distinct parameter combinations ("scenarios"). If a large number of runs is necessary, the repetition of the parameter selection process to produce many different plots becomes tedious. SIMPAS provides two possibilities for presenting the results of several simulation runs as functions of selected parameter values in one diagram:

1. The effects of changes of **system and scenario parameters** can be analyzed using the option "Parameter Sensitivity."
2. The influence of **initial conditions** can be studied using the option "Global Analysis."

In the case of these multiple simulations, the total run time may become a problem. While the "Parameter Sensitivity" option requires five simulation runs, the "Global Analysis" option may require as many as $(21 \cdot 21) = 441$ runs. An efficient choice of integration method and of run time parameters, in particular of the simulation time step, is required.

The starting time for the simulations is usually given: for example, $t = 0$. The finishing time also is often known or can be determined by a few trial simulations. In the present example, we find for the chosen default values that the motion has almost come to rest after ten seconds—running the simulation for a longer time period would not make much sense. As before, the computation time step has to be determined in such a way that it will neither cause accumulating truncation errors (if the time step is too small) nor crude errors of extrapolation for the rates (if the time step is too large) (see Sec. 4.1.4 *Numerical Integration*). Fortunately, in most cases the numerical integration is not very sensitive to changes in the time step in a rather wide intermediate range.

In the case of the rotating pendulum the accuracy of the results of the Euler-Cauchy integration hardly improves as the step width is reduced to less than 0.05 (for the default parameter values). On the other hand, the computing time increases considerably if the step width is reduced to below 0.01. As a compromise, we therefore select time step = 0.02 for this application. The dependence of the results on step width can be analyzed by changing the time step in the range from 0.1 to 0.005, and comparing the tabulated results for $t = 10$, for example.

The accuracy of the Runge-Kutta method for the same time step is greater but the computing time is increased by approximately a factor of four. In exploratory investigations such as parameter sensitivity or global analysis, high accuracy is usually not required, and the use of the Euler-Cauchy method is adequate. One or the other method can be activated by using it for a simulation run just before selecting the "Parameter Sensitivity" or "Global Analysis" option.

For the following exercises with the model, all parameters have to be reset to their default values. The easiest way to achieve this is by starting the model PENDULUM again from DOS.

The influence of system and scenario parameters, as well as initial conditions, on the behavior of the model can be investigated by choosing "Parameter Sensitivity" under MODE SELECTION. If we do this immediately after starting the model, we are reminded to "Run one simulation first!" (to properly initialize SIMPAS). At this point we select "Simulation using Euler-Cauchy" to speed up the multiple simulations somewhat, at some loss in accuracy. After this initial simulation, we return immediately to the MODE SELECTION screen by choosing "No, new simulation or quit" under OUTPUT PRESENTATION. We then choose "Parameter Sensitivity" under MODE SELECTION.

The first PARAMETER SENSITIVITY screen presents a list of the available parameters, their measurement units, and preselected minimum and maximum values in the order:

1. system parameters
2. scenario parametes
3. initial conditions.

By a blinking "PARAMETER," the user is prompted to select one of the parameters by marking it with an asterisk (*), and to choose the minimum and the maximum values of the parameter range. SIMPAS will divide this range into five equal intervals. SIMPAS preselects a range from 0.5 to 2.5 times the default value of the parameter but the user is free to change this range.

In the present case, we have the list of parameters

mass	[kg]	0.500	2.500
radius	[m]	0.500	2.500
damping	[N/(m/s)]	0.500	2.500
angle	[rad]	0	0
angular velocity	[rad/s]	5.000	25.000

We now wish to study the influence of damping on the model behavior. In order to change the location of the selection asterisk and the parameter range, we move the cursor from the "OK" field to the "change" field using the "down" arrow as in the other screens and pressing RETURN. The cursor is now under the asterisk in the parameter field. We remove the asterisk by pressing RETURN, use the "down" arrow to mark "damping," use RETURN to set the asterisk at this position, and change the range from (0.5 to 2.5) to a minimum of "0" and a maximum of "2." This will allow us to investigate the range from undamped to strongly damped motion. When entering this selection by pressing ESC, a message appears briefly:

"Can this lead to DIVISION BY ZERO?"

The user should now check whether the value "0" is permissible for the selected parameter. In particular, one should check whether the parameter appears as a factor in the denominator of a model equation. (This would cause the simulation to crash.) If there is no danger in this regard, the user can continue without taking any action. Otherwise, a different parameter range (excluding "0") can be chosen by repeating the selection process. To finish the selection process, press the ESC key and acknowledge the selection by RETURN.

In the second PARAMETER SENSITIVITY screen, the list of available output variables appears together with their units of measurement, and the minimum and maximum values for the plotting range. The variables are shown in the order

1. time
2. state variables
3. additional output variables.

Note that the minimum and maximum values are identical to those currently found in the OUTPUT SELECTION AND SCALING screen. The user is free to change the values in the parameter sensitivity analysis, however.

By a blinking "OUTPUT VARIABLES," the user is prompted to select two of the output variables for graphical output. As before, the variable for the horizontal axis is marked by a diamond while the variable for the vertical axis is selected by using the asterisk.

For the parameter sensitivity analysis, we change the minimum and maximum values in the list of output variables:

time	[seconds]	0	10
angle	[rad]	-5	15
angular velocity	[rad/s]	-10	10
vertical position	[m]	0	2
horizontal position	[m]	-1	1

We pick "angle" as the horizontal variable, marking it by a diamond and "angular velocity" as the vertical variable, marking it by an asterisk. We finish by pressing ESC and RETURN.

On the screen now appear the results of five successive simulations (Fig. 4.10). The parameter range selected for the damping has now been divided into five identical intervals. The parameters of the five curves are noted above the plot.

Parameter: damping
0 0.500 1 1.500 2

The phase plot (Fig. 4.10) shows different response modes as a function of the parameter choice: the undamped motion (upper curve for d = 0) is a continuous rotating motion which will not come to rest. The angular velocity changes (approximately) sinusoidally during the rotation. For light damping (d = 0.5) the pendulum goes through two rotations before eventually coming to rest at the lower dead center (4π). For larger damping (d = 1 and 1.5), the pendulum only rotates once before coming to rest at the lower dead center (2π). For a somewhat larger damping (d = 2), the pendulum is not able to pass over the upper dead center and swings back to its resting point at the lower dead center (angle = 0). If finally the damping is very high (d = 5 and higher), the pendulum does not swing at all; it merely moves towards the lower dead center without changing its direction of motion before coming to rest. A different presentation of the same motion is produced by selecting time for the horizontal axis and the angle for the vertical axis (Fig. 4.11).

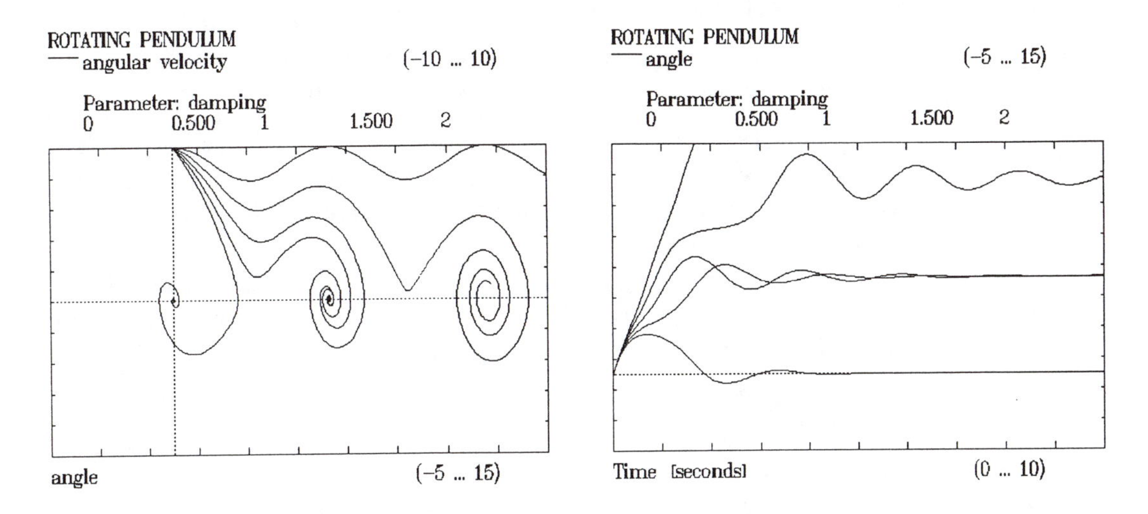

Fig. 4.10 (left): PENDULUM: Investigation of parameter sensitivity with respect to the damping parameter; phase plot.
Fig. 4.11 (right): PENDULUM: Investigation of parameter sensitivity with respect to the damping parameter; time plot.

4.2.6 Global response analysis

Non-linear systems often show a dependence of their qualitative behavior on initial conditions, such as different regions of attraction with different stability behavior: if started under one set of initial conditions, the system may eventually settle at a stable equilibrium point; if started under different conditions, it may be unstable. With SIMPAS, investigation of the dependence of the results on the initial values of the state variables is possible using the option "Global Response" under MODE SELECTION.

Choice of "Global Response" produces a first screen CHOICE OF STATE VARIABLES AND INITIAL VALUE RANGE. It lists all state variables, their units of measurement, and the minimum and maximum values of the state range to be investigated (the same settings as currently used in OUTPUT SELECTION AND SCALING). The graphical output of "Global Response" is in the form of phase plots, i.e. two state variables can be picked for plotting on the horizontal and vertical axis. "Global Response" is therefore most useful if the model has only two states—in this case, the complete response appears in one plot. If the system has more than two state variables, the phase plot may become difficult to interpret as it only shows a two-dimensional projection of a multi-dimensional system.

In our present example, we change the minimum/maximum range of the state variables to read as follows:

angle	[rad]	−10.000	10.000
angular velocity	[rad/s]	−10.000	10.000

When selecting this range which includes the zero value, the message "Can this lead to DIVISION BY ZERO?" briefly appears on the screen. We can ignore the warnings in good conscience since both state variables do not appear in denominators in the model.

We mark "angle" by a diamond for display on the horizontal axis and "angular velocity" by an asterisk for display on the vertical axis. Pressing ESC and RETURN, a selection menu appears on the screen:

```
GRID of initial values:
 5* 5 points
10*10 points
20*20 points.
```

SIMPAS will lay a corresponding grid over the range of state variables defined previously by minimum and maximum values. It will then use the state values corresponding to the grid points as initial values of the individual simulations. The corresponding state trajectories will be traced on the phase plot.

The choice of "5*5 points" means a computation of $(5+1)\cdot(5+1) = 36$ simulations, choice of "10*10" means 121 simulations, and choice of "20*20" means 441 simulations. We select "5*5," which is the default value (marked by the tagging bar), by pressing RETURN. The next screen now shows the selection

LENGTH of simulation:
normal simulation
1/10 of normal length
1/50 of normal length.

This choice concerns the length of the individual simulations. Choice of "normal simulation" retains the default setting for the simulation period. Choice of "1/10 of normal length" reduces this period to one tenth, the choice of "1/50 of normal length" to one fiftieth. This allows one to obtain the directions of the state trajectories in a reasonable time even for a very dense grid (e.g. 20*20).

In this case, we initially select the "normal simulation" (default setting) by using RETURN. The solution trajectories for each of the 36 initial value combinations are now traced on the screen (Fig. 4.12). We observe an obvious dependence on the initial values: for high initial angular velocities, the pendulum will undergo several rotations before switching to swinging motion. The point of equilibrium will therefore move to a multiple of 2π. If the angular velocity is initially low, the pendulum will remain in its original region of attraction and swing back and forth in damped motion—which corresponds to spiraling motion in the phase plot.

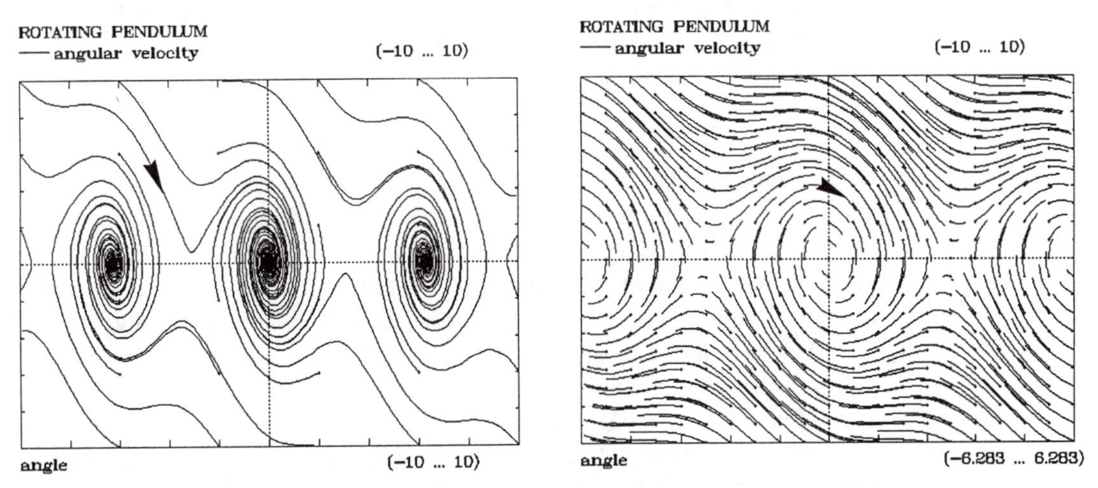

Fig. 4.12 (left): PENDULUM: Investigation of global behavior as a function of different initial conditions of the two state variables (angle and angular velocity).

Fig. 4.13 (right): PENDULUM: State trajectories with lower dead centers (stable; middle, and at left and right border); upper dead centers (unstable; in between).

In the phase plot the lower dead centers (at $2\ \pi n$, $n = 0, \pm 1, \pm 2...$) appear as stable focuses. They attract the motion in their neighborhood. By contrast, the upper dead centers (at $2\ \pi(n+1)$, $n = 0, \pm,1, \pm 2...$) represent saddles. As the motion approaches the saddles, it slows down. If the magnitude of the angular velocity at the upper dead centers is greater than zero, the motion will pass over the dead center. If it becomes zero before reaching the dead center, the motion reverses, and the pendulum begins to swing back and forth. The upper dead center is unstable: an infinitesimal disturbance at this equilibrium point will cause motion towards the lower dead center.

In order to investigate the solution field in more detail, we now return (via CONTINUATION, SIMULATION PARAMETERS and MODE SELECTION) to "Global Response" and select in CHOICE OF STATE VARIABLES AND INITIAL VALUE RANGE the range for the angle from (-2π to 2π), i.e. from (-6.283 to 6.283). In addition, we choose a finer grid (20*20) and the shortest simulation time (1/50). The lower dead center (of order zero) is now in the middle of the diagram, the lower dead centers of first order exactly at the right and the left edges of the diagram, and the upper dead centers (of first order) in the middle between the two (Fig. 4.13). If we now follow the direction of the individual 'flags,' we can construct the state trajectories for arbitrary initial values. If, for example, the motion starts in the upper left hand corner of the phase diagram (i.e. the lower dead center with $x = -2\pi$, $x' = 10$), it will initially slow down and approach the upper dead center, pass over it, and finally end at the lower dead center after several swinging motions (x represents the angle, x' the angular velocity).

The length of the flags represents the velocity of motion in state space, i.e. the total rate of change dr given by:

$$(dr)^2 = (dx)^2 + (dx')^2.$$

In Figure 4.13 it is clearly seen that this rate disappears at the upper and lower dead centers (at $x = 0, \pm\pi$, $\pm 2\pi$): these points are equilibrium points. They obviously differ with respect to the motion in their neighborhood: we have already identified focuses and saddles. In order to investigate the motion in the neighborhood of the equilibrium points, we now focus on their immediate vicinity.

We now choose again "Global Response" and specify the corresponding ranges of the state variables. For the lower dead center we select the range for the angle (from -0.02 to 0.02) and for the angular velocity (from -0.05 to 0.05). We again select a 20*20 grid density and the briefest simulation time (1/50). We obtain the solution of Figure 4.14. This shows very clearly the focus character of this equilibrium point: all flags point towards this equilibrium point. The total rate of change (flag length) diminishes in proportion to the distance from the lower dead center; at this point it disappears entirely.

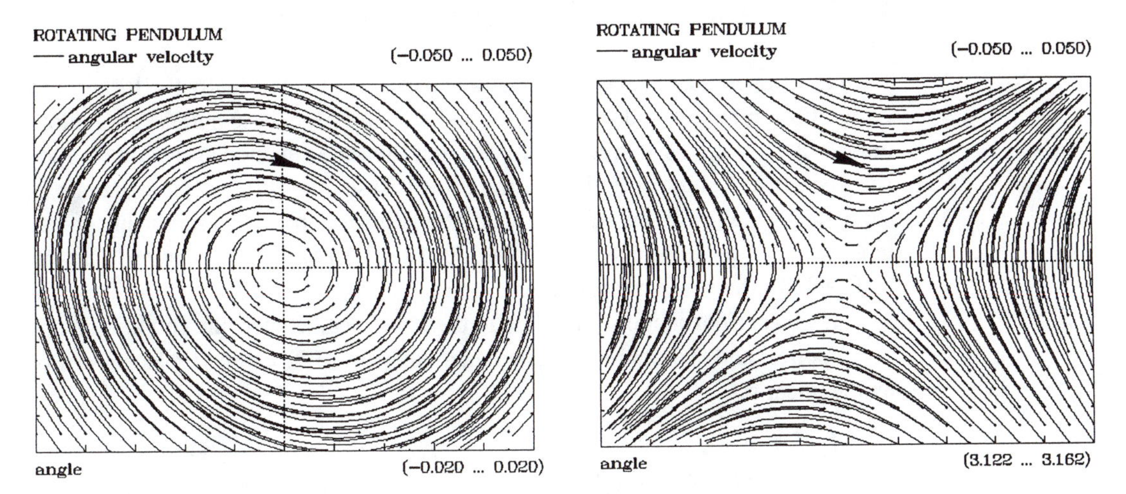

Fig. 4.14: PENDULUM: State trajectories in the neighborhood of equilibrium points. Left: lower dead center (focus); right: upper dead center (saddle).

The next upper dead center is at $\pi = 3.14159$. We select the range for the angle (from 3.12159 to 3.16159), and the range for the angular velocity again (from −0.05 to 0.05). For the same grid as in Figure 4.14 we now obtain the state trajectories at the upper dead center in Figure 4.15. It now can be clearly seen that the total rate of change again disappears at the dead center, and the motion initially approaches the dead center but then changes direction and moves away from it. This motion of the state trajectory is characteristic for a saddle.

The equilibrium points we found here by the "phase portrait" composed of state trajectories for many different simulations with different initial conditions can also be determined analytically. At equilibrium points, the rates of change vanish. Denoting the vector of state variables by **z**, the condition for equilibrium is

$$d\mathbf{z}/dt = \mathbf{0}.$$

The state equations for the pendulum are (Eq. 3.7)

$$dx/dt = y$$
$$dy/dt = -(g/r)\sin x - (d/m)\,y.$$

The condition for the equilibrium points is therefore

$$0 = y$$
$$0 = -(g/r)\sin x - (d/m)\,y.$$

This is obviously possible only if $y = 0$ (the angular velocity vanishes), and if $\sin x = 0$. Sin x vanishes for $x = 0, \pm\pi, \pm 2\pi, \ldots, \pm n\pi$, $n = 1, 2, 3, \ldots$.

4.2.7 Linearization of the equations of motion

The most important information about a dynamic system is usually the location of its equilibrium points and the behavior and stability near these points. This information can usually be obtained by linearizing the state equations near the equilibrium points and employing the resulting linear system of differential equations to study the dynamics in the neighborhood of equilibrium points using the analytical tools of linear systems analysis (see Sec. 7.1.11 - 7.1.14). We will demonstrate the approach for the nonlinear pendulum.

If we consider motion very close to the lower dead center, the condition $\sin x \cong x$ applies, and the nonlinear state equations (Eqs. 3.7) can be replaced locally by the linear set

$$dx/dt = y \qquad \text{lower dead center}$$
$$dy/dt = -(g/r)\,x - (d/m)\,y$$

At the upper dead center, we redefine x as the small angular deviation from the vertical position. Still counting counterclockwise angular motion x as positive, the projection $\sin x$ now points to the left (in the opposite direction as at the lower dead center). The sign of this contribution therefore has to be reversed, and we can therefore replace the nonlinear state equations (Eqs. 3.7) locally by the linear set

$$dx/dt = y \qquad \text{upper dead center}$$
$$dy/dt = -(g/r)\,(-x) - (d/m)\,y$$

These equations can also be derived more formally by using the Jacobian matrix (Sec. 7.1.14).

We can now use these two sets of linear differential equations to investigate the state trajectories in the neighborhood of the two sets of equilibrium points. Model M 203 LINEAR OSCILLATOR OF SECOND ORDER of the systems zoo has exactly the same generic form

$$dx/dt = A\,x + B\,y$$
$$dy/dt = C\,x + D\,y$$

and we can therefore use it to compute the behavior of the pendulum near the two equilibrium points. All we have to do is to substitute the appropriate parameters as before: $A = 0$; $B = 1$; $C = \pm(g/r) = \pm 9.81$; $D = -(d/m) = -1$. Plotting the results of the global analysis on the same scales as for the nonlinear simulations in Figure 4.14, we obtain the trajectories of Figure 4.15. Note that the two sets of figures are nearly identical. We conclude that the two linear sets of equations can indeed substitute for the full nonlinear system in the neighborhood of the equilibrium points. In particular, we can use them to investigate the stability characteristics.

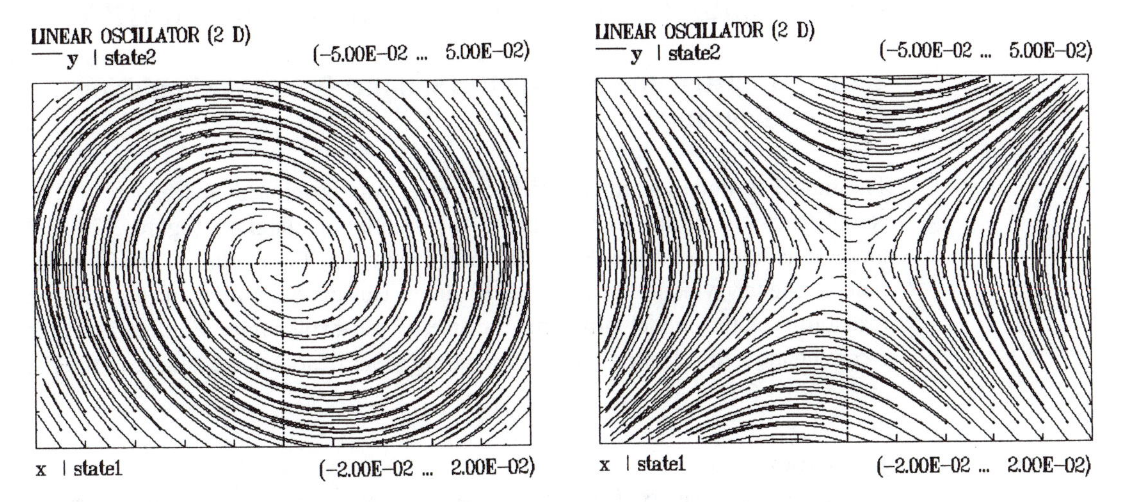

Fig. 4.15: State trajectories in the neighborhood of equilibrium points, computed from the linearized pendulum equations. Left: lower dead center (focus); right: upper dead center (saddle; the angle is now counted from this equlibrium point).

4.2.8 Summary of the observations with the rotating pendulum model

Using the model of the rotating pendulum, we have demonstrated the different application modes of SIMPAS. A number of observations concerning the behavior of a dynamic system were made which will be pursued in greater detail later:

1. The qualitative behavior of a dynamic system depends on the choice of parameters and initial values (for the non-linear system).
2. A (non-linear) system can have several equilibrium points where its "drive," i.e. all rates of change, disappear. Once this state has been reached, there will be no more motion in the system (in case it remains unperturbed).
3. Equilibrium points may be stable or unstable. A small perturbation from a stable equilibrium point will lead to motion back to the equilibrium point. For an unstable equilibrium point, even an infinitesimal perturbation will cause accelerated motion away from this equilibrium point.
4. Depending on the choice of parameters and initial values, the same system may show qualitatively different modes of behavior. The pendulum may perform rotating as well as swinging and aperiodically damped motions.
5. The motion is strongly determined by the damping of the system. Small damping leads to oscillations; high damping will produce aperiodic motion.

4.3 Simulation of Fishery Dynamics with SIMPAS

4.3.1 Constructing the SIMPAS model from the simulation diagram

The model formulation for the fishery model was developed in Chapter 3. The complete specifications were given in the simulation diagram Figure 3.24 and model equations in Section 3.6.3. In order to simplify the mathematical formulation, one-letter symbols were employed there. However, in order to improve the readability of the model program, the use of self-explanatory English terms is recommended. In selecting the corresponding identifications, we adhere to the list of Section 3.6.2.

1. Parameters

We distinguish fixed parameters, system parameters, and scenario parameters of the model. The fixed parameters describing the region are specified directly in the procedure InitialInfo; the dimensions of the parameters are noted in braces.

```
area                :=  100;   {km2}
fish_carry_capacity :=  100;   {t fish/km2}
max_fish_capacity   :=  area * fish_carry_capacity;  {t fish}.
```

Parameters that are not likely to be changed very often are entered as system parameters in the interactive query:

```
ParamQuestion[1]  :=  'max fish growth rate: (1) [1/year]';
ParamQuestion[2]  :=  'operating costs: (50000) [$/(boat.year)]';
ParamQuestion[3]  :=  'boat price: (100000) [$/boat]';
ParamQuestion[4]  :=  'boat lifetime: (15) [year]'; .
```

Parameters that are uncertain and/or variable (for example, fish price) or that have special significance as decision parameters (e.g. investment fraction, boats) are entered as scenario parameters in the interactive query:

```
ScenaQuestion[1]  :=  'max fish catch: (100) [t fish/(boat.year)]';
ScenaQuestion[2]  :=  'fish price: (1000) [$/t fish]';
ScenaQuestion[3]  :=  'investment fraction, boats: (0.5) [-]'; .
```

The results of the parameter query must be introduced into the program using the correct identifications. Since the parameters are to remain constant for the duration of an individual simulation, they are introduced into the first simulation step in the procedure ModelEqs:

```
if Time=Start then
begin
     max_fish_growth_rate      := ParamAnswer[1];
     operating_costs           := ParamAnswer[2];
     boat_price                := ParamAnswer[3];
     boat_lifetime             := ParamAnswer[4];
     decommission_rate         := 1/boat_lifetime;
     max_fish_catch            := ScenaAnswer[1];
     fish_price                := ScenaAnswer[2];
     investment_fraction_boats := ScenaAnswer[3];
end;
```

2. State variables and initial values

The state variables and their identifications, default values, and dimensions have to be introduced in the procedure InitialInfo:

```
StateVariable[1] := 'fish: (5000) [t fish]';
StateVariable[2] := 'boats: (100) [boats]'; .
```

The corresponding variable names to be used in the model equations are introduced in the procedure ModelEqs:

```
fish  := State[1];
boats := State[2]; .
```

3. Intermediate variables

The algebraic computation of intermediate variables from parameters, state variables and other intermediate variables is inserted in procedure ModelEqs:

```
fish_density              := fish / max_fish_capacity;
fish_growth_rate          := max_fish_growth_rate *fish*(1 – fish_density);
catch_potential           := max_fish_catch * boats;
fish_catch                := catch_potential * fish_density;
catch_proceeds            := fish_price * fish_catch;
boat_maintenance          := operating_costs * boats;
net_income                := catch_proceeds – boat_maintenance;
investment_capital_boats  := investment_fraction_boats * net_income;
purchase_boats            := investment_capital_boats / boat_price;
decommission_boats        := decommission_rate * boats; .
```

Writing the equations in this order ensures that each variable can be computed from initially defined quantities (parameters, state variables) and previously computed intermediate variables.

4. Rates of change

The rates of change of the state variables have to follow the computation of the intermediate variables in the procedure ModelEqs:

```
Rate[1] := fish_growth_rate – fish_catch;
Rate[2] := purchase_boats – decommission_boats; .
```

5. Runtime information and further model information

Initial time, final time, and time step of the simulation are defined in the procedure InitialInfo:

```
Start    := 0;
Final    := 20;
TimeStep := 0.02; .
```

To assist in the assessment of the simulation results, the annual fish catch is of interest in addition to the state variables which are automatically provided as results. This additional output has to be declared as an OutVariable in the procedure ModelEqs following its computation there.

```
OutVariable[1] := fish_catch; .
```

The label and dimension of this variable must be specified in the procedure InitialInfo:

```
OutVarText[1] := 'fish_catch: [t fish/year]'; .
```

In addition, the following four statements have to be provided in the procedure InitialInfo:

```
Title        := 'FISHERY DYNAMICS';
Description  := 'Fish population and fishing fleet dynamics as function of '
                + 'management, ecological and economic parameters.';
TimeUnit     := 'years';
Author       := 'H.Bossel: Modeling and Simulation 910612'; .
```

Finally, all quantities used in the model program have to be listed (behind IMPLEMENTATION under "var") and provided with the corresponding type declarations.

This finishes the specification of the SIMPAS simulation model for fishery dynamics. The corresponding TurboPascal unit FISHERY.MOD for use with SIMPAS is fully listed in Program 4.5 (in the Program Listings Appendix). This unit also contains in procedure Summary a computation of the parameter-dependent equilibrium points (cf. Sec. 4.3.7).

4.3.2 Setting up the executable simulation program

The procedure has been explained for the pendulum model. The model should be saved as FISHERY.MOD and should be available as MODEL.PAS for the compilation with SIMPAS. After starting TURBO SIMPAS, the "destination" is changed to "disk" under the menu heading "compile," and "run" is then selected. Possible programming errors are eliminated by using the TurboPascal debugging facilities. The compiled simulation program SIMPAS.EXE is renamed as FISHERY.EXE and is then available for further investigations.

4.3.3 Standard run of the fishery model

We first investigate the model behavior using the default parameter set. From the DOS level we start the model by entering FISHERY. By repeatedly pressing RETURN without any further inputs we move through the interactive query and directly to the simulation. As the simulation results become available (under OUTPUT PRESENTATION), we first scale the results to round numbers for the plots: fish (0 to 10,000), boats (0 to 100), fish catch (0 to 5000). In this case, and in all other simulations, it is always recommended to select the zero point as the lower limit of the plot range otherwise a wrong impression of the dynamic behavior of the system may easily develop.

We first consider the time-dependent, two-dimensional plot of the results (time as the horizontal axis; fish, boats, and fish catches as the vertical) (Fig. 4.16). After an initial strong decrease from an initial value of 5000 [t fish] to a minimum of about 3000, the fish population recovers again and stabilizes in the long term at the value of about 6300. The number of boats drops rapidly from an initial value of 100 to about 37, where it stabilizes. The initial fish catch decreases from an initial value of 5000 [t fish/year] very rapidly to a minimum of about 1800 and then stabilizes at about 2300. The exact values can be read off at two year intervals from the tabular presentation of the results.

With the default parameter set, the system approaches a flow equilibrium after about a decade. Thereafter, the state variables "fish" and "boats" no longer change and the losses (by the catching of fish and the decommissioning of boats) are exactly compensated by the gains (by the net growth of the fish population and the purchase of new boats). A phase plot (with fish on the horizontal axis, boats on the vertical axis) is obtained in OUTPUT SELECTION AND SCALING by moving the diamond from "time" to "fish" and using the asterisk on "boats." It also shows clearly that the system moves rapidly to a stable equilibrium point at about 6300 t fish and 37 boats.

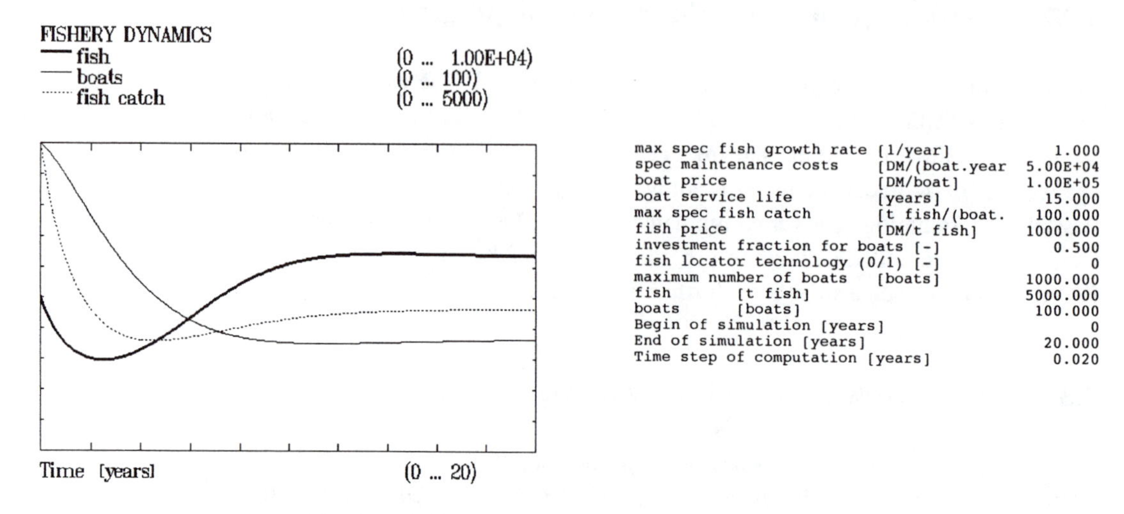

Fig. 4.16: FISHERY: Time plot of simulation results for the standard run.

4.3.4 Response to parameter changes

In order to become more familiar with the behavior of the model, the influences of the different parameters must be studied in greater detail. Individual parameters or combinations of parameters could be changed using the parameter query but this is a time-consuming process and not well suited for extensive systematic investigations. It is faster to use the possibilities for investigating the global behavior ("Global Response") and parameter sensitivity ("Parameter Sensitivity").

We therefore leave the presentation of simulation results by choosing "No, new simulation or quit" in the OUTPUT PRESENTATION selection. The next screen presents the results of the Summary procedure in the FISHERY model unit (Prog. 4.5). In it, the equilibrium points of the system are computed analytically (see Sec. 4.3.7). We note that the equilibrium values agree with our previous estimates from the simulation results.

Pressing RETURN again brings us to SIMULATION PARAMETERS. Without changing any parameters, we use RETURN again to reach MODE SELECTION. (To speed up the following simulations, run one "Simulation using Euler-Cauchy" first.) We now choose "Global Response" to study the global behavior as a function of the initial conditions. We choose "fish" for the horizontal plot axis, and "boats" for the vertical. The plot ranges chosen are the same as before: fish from 0 to 10,000, and boats from 0 to 100. Using a grid of 5*5 and normal simulation time we obtain the plot in Figure 4.17. It shows clearly that whatever the initial conditions in the given range, the system state always moves towards a stable equilibrium point at (about) 37 boats and 6300 t fish. If the initial boat number is very high, it will be reduced quickly until there is again an increase and a stabilization of the fish population. If the initial fish population is large, it will be reduced quickly to the equilibrium value by the fish catches.

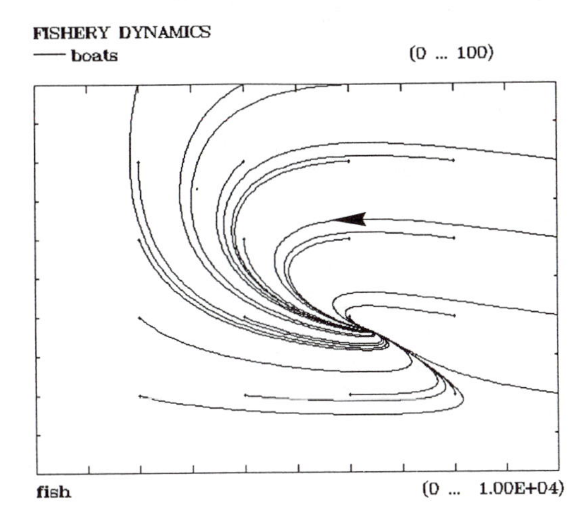

Fig. 4.17: FISHERY: Investigation of the global dynamics in state space for the default parameter set. All state trajectories move towards the equilibrium point.

We now have an overview of the global behavior for the (default) parameter values of the standard run, showing that it is determined by a stable node (degenerate focus). We are now interested in finding out how this behavior might change as individual parameters are changed. Leaving "Global Response," we return to SIMULATION PARAMETERS. The scenario parameters are immediate candidates for parameter changes, i.e. the maximum fish catch of the boats, the fish price, and the investment fraction for the purchase of new boats. In order to be better able to study the approach to the equilibrium point, we first change the initial value of boats to 25 and the "end of simulation" to 50 [years]. This allows us to study the system for a longer period, in particular when it oscillates with small damping. The initial value of 25 boats is closer to the equilibrium point we found for the default parameter set.

For studying the system response to changes in the parameters, we select under MODE SELECTION the option "Parameter Sensitivity" which takes us to the PARAMETER screen. We first study the influence of the parameter "max fish catch" (in the range from 50 to 250) [t fish/(boat·year)] by marking this parameter by an asterisk and accepting the default minimum/maximum range. For presenting the results, we select in the next screen OUTPUT VARIABLES "fish" (0 to 10,000) as horizontal variable (diamond) and "boats" (with a range from 0 to 50) as vertical variable (asterisk). In Figure 4.18 we observe the following: for a small specific fish catch (50), the boat number is reduced to zero which causes growth of the fish population to its capacity limit. In this case fishing is not economically viable; the fishermen do not have enough capital for new investments.

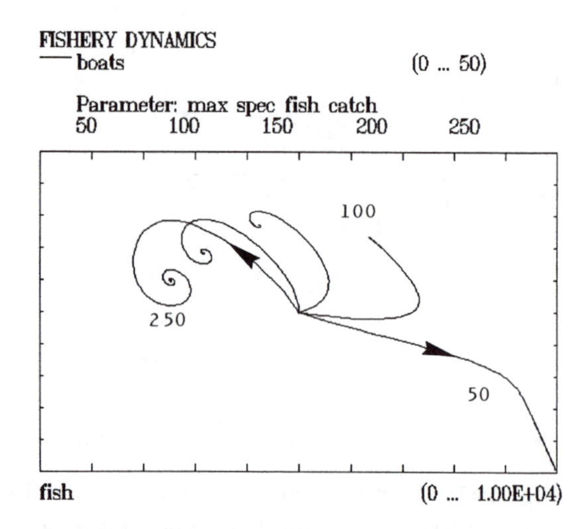

Fig. 4.18: FISHERY: Change of the global dynamics resulting from a parameter change. If the max fish catch of the boats is too small, fishing operations cease for economic reasons.

If the specific fish catch is raised (to 100), we find the previously determined equilibrium point at 6300 t fish and 37 boats which the system approaches aperiodically. As the specific fish catch is raised further (from 150 to 250), the equilibrium point changes to a smaller fish population and a smaller number of boats. In addition, there is now a strongly damped oscillation around the point of equilibrium, as evidenced by the spiralling motion of the state trajectory. The system stabilizes at different equilibrium points depending on the value for "max fish catch."

In similar fashion we can also study the sensitivity of the model behavior with respect to the fish price (in the PARAMETER list of the first "Parameter Sensitivity" screen). We select the parameter range (500 to 2500), choose the phase plot for fish and boats (now 0 to 100), and obtain the results in Figure 4.19. This now shows that a larger fishing fleet can be maintained at a higher fish price; this, however, also means a considerable reduction in the fish population.

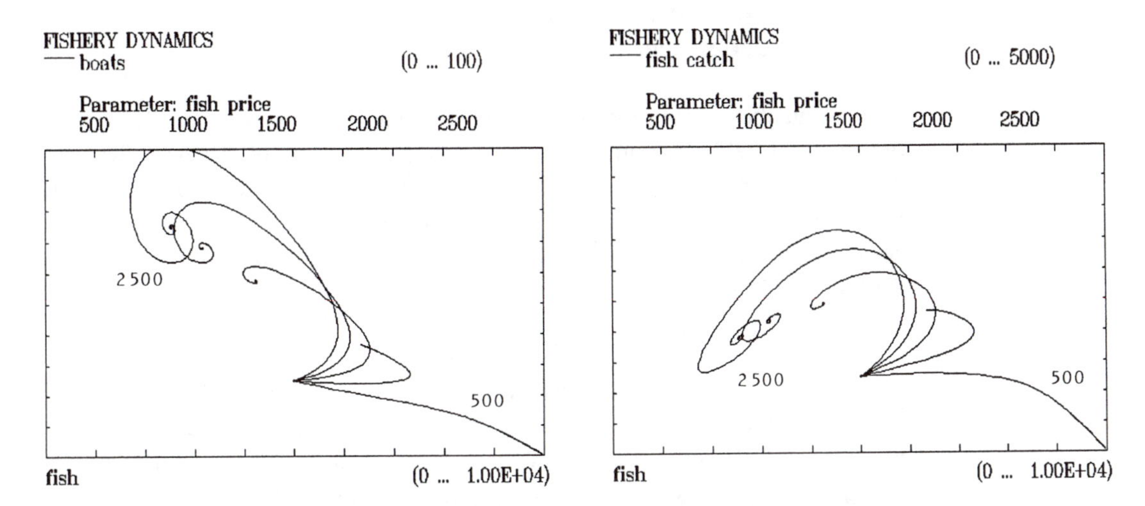

Fig. 4.19 (left): FISHERY: Response to fish price: boats as a function of fish population. If the price is too low, fishing operations cease for economic reasons.
Fig. 4.20 (right): FISHERY: Effect of the fish price: fish catch as a function of fish population. The equilibrium catch is highest for an intermediate fish price.

This brings up the question of what annual fish catches can be expected under these conditions. We therefore go back to "Parameter Sensitivity," select again the fish price (in the range from 500 to 2500) as a parameter, and select under OUTPUT VARIABLES the horizontal axis for the fish population (0 to 10,000) and the vertical axis for the fish catch (0 to 5000). The parameter sensitivity study (Fig. 4.20) now shows clearly that the (sustainable) fish catch has an optimum (of about 2500 [t fish/year]) at a fish price of about 1250 [$/t fish]: the catch under equilibrium conditions (sustainable yield) is smaller for higher as well as lower fish price. If the fish price drops below a minimum (about $900/t fish), fishing is no longer economically viable, the boat fleet disappears, and the fish population grows back to its capacity limit. (Warning: our simple model uses fictitious parameters; the results therefore cannot be compared directly to reality!)

We have only studied the sensitivity of the fishery model with respect to two parameters. Further interesting results can be obtained by working with other parameters; for example, by changing the maximum fish growth rate or the economic parameters. If the simulation model is to be used to find a favorable solution (with respect to given evaluation criteria), permissible parameter ranges can be more quickly and easily determined with the help of sensitivity studies. The analysis of the global behavior with its overview of possible state trajectories then provides a comprehensive picture of the behavioral options of the system.

4.3.5 Modification of the fishery model for density-independent harvest rate

In all simulations using the fishery model (Prog. 4.5), we occasionally observe (under unfavorable economic conditions) an "extinction" of the boat fleet but never a complete collapse of the fish population. The explanation for this can be found in the simulation diagram (Fig. 3.24) and in the corresponding model equations (in Sec. 3.6.3), or the model formulation in Program 4.5: the fish catch depends on fish density,

```
fish_catch := catch_potential * fish_density.
```

For diminishing fish density (corresponding to a decreasing fish population), the fish catch eventually goes to zero. This has corresponding economic consequences for the fishery operation: the fishing fleet is likewise strongly reduced, reducing the fish catch further. The fish population is therefore protected from complete collapse. Finally, there will be an equilibrium for fish population and boat number. This behavior can easily be demonstrated by simulations and by parameter sensitivity studies; for example, for high values of the maximum fish catch.

The implicit assumption used here, that fish are uniformly distributed over the area and are harder to catch as they get scarcer, is not realistic with respect to many fish species of economic interest. These species often occur in schools which can be located easily by modern technology. This means that the fish catch no longer depends on the (average) fish density but rather on the quality of the locator technology used and the fish catch capacity of the fleet:

```
fish_catch := catch_potential * catch_chance; .
```

This causes a fundamental change in system behavior as we shall see.

In order to modify the fishery model accordingly, we introduce two new query parameters and the "catch chance" into the procedure InitialInfo. The catch chance accounts for the fact that even for a sophisticated locator technology the fish catch will be smaller than the available catch potential:

```
catch_chance       := 0.8;
ScenaQuestion[4]   := 'fish locator (0/1): (0) [-]';
ScenaQuestion[5]   := 'max number of boats: (1000) [boats]'; .
```

The values provided by the user in response to these questions have to be introduced into the procedure ModelEqs at the first time step of simulation:

```
if ScenaAnswer[4] = 1 then
     fish_locator          := true
     else fish_locator     := false;
```

Depending on the availability or non-availability of an effective locator technology, the two different formulations for fish catch have to be introduced into the model structure in the procedure ModelEqs:

```
if not fish_locator then
      fish_catch := catch_potential * fish_density;
if fish_locator then
      if fish>0 then
            fish_catch := catch_potential * catch_chance
            else fish_catch := 0;
```

We also have to make sure (in procedure ModelEqs) that the fish population never becomes negative. (This cannot happen in the equation for the density-dependent fish catch.) In the new formulation, there is no feedback from "fish" or "fish density" to "fish_catch," and we have to state specifically that if no more fish (state[1]) are available, fish population and fish catch become zero. We also have to introduce a restriction on the maximum number of boats in order to prevent overfishing:

```
if State[1]<0 then State[1] := 0;
if boats>max_boat_number then
      purchase_boats := 0;
```

Program 4.5 already contains these changes. (It also contains a procedure Summary to compute the equilibrium points which we will discuss later.)

4.3.6 Simulation results for density-independent fish harvest

To obtain an initial overview of the behavior of the altered model, we start the FISHERY model anew and produce one (Euler-Cauchy) simulation run using the default parameter set and the parameter settings "fish locator technology" = 1 and "end of simulation" = 50. We then investigate the global behavior (fish: 0 to 10,000; boats: 0 to 100). It becomes obvious that the fish population always collapses by overfishing, irrespective of the initial conditions. Two possibilities come to mind to avoid this collapse: 1. limiting the fish catch per boat, and 2. limiting the number of boats.

Using the "Parameter Sensitivity" option, we first investigate the influence of the parameter "max fish catch [t fish/(boat.year]" (in the range from 0 to 60). The corresponding state trajectories are plotted in the phase plot for fish (0 to 10,000) and boats (0 to 125) (Fig. 4.21). The diagram shows that under these parameter conditions there is no compromise: either the fishing fleet disappears for economic reasons (for max fish catch from 0 to about 45), or the fish population disappears due to overfishing if the specific fish catch is greater than 45. (The critical value for the fish catch can be determined more exactly—about 46.1—by refining the parameter interval.)

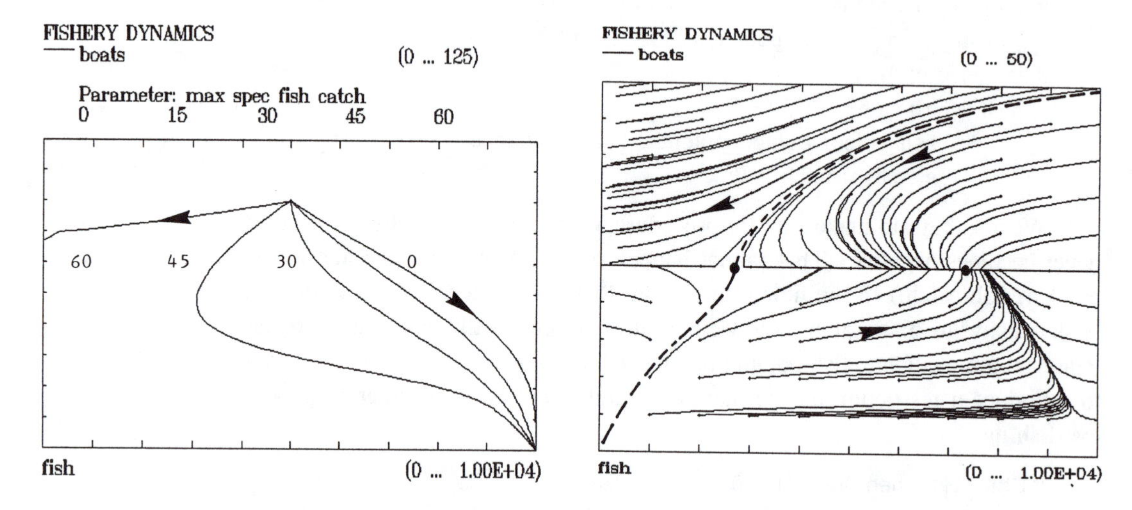

Fig. 4.21 (left): FISHERY with fish locator technology: Without a limit on fish catches, there is no compromise; either the fish population vanishes by overfishing, or the fishing operations must cease for economic reasons.

Fig. 4.22 (right): FISHERY with fish locator technology: When boat number is limited to 25, there is either an approach to stable equilibrium or collapse of the fish population; this depends on the initial state for fish and boats.

By investigating in similar fashion the effect of limiting the boat number, we find that here also the fish population always collapses (with the default parameter set) if the initial boat number is relatively large. If the initial boat number is small, the model approaches a stable state as long as the boat number is restricted to a relatively low value.

For further analysis of the obviously complex global behavior of the model system, we specify (under SIMULATION PARAMETERS) a maximum boat number of 25. We then select "Global Response." Under CHOICE OF STATE VARIABLES we select fish (0 to 10,000) and boats (0 to 50). We obtain the complex map of state trajectories in Figure 4.22. It can now be seen that there is indeed a stable region (in the lower right hand part of the diagram, below the dashed line): all state trajectories in this region approach the boat number limit (25 boats). The fish population adjusts to an equilibrium value of 7236. If the initial values are in the upper left hand part of the phase plot, above the dashed line, the fish population always collapses. Under these conditions, the boat fleet is too large with respect to the fish population. Fish regeneration cannot offset the consequences of overfishing. In this case, the collapse of the fish population is too fast to have an immediate and significant effect on fleet size; stabilization is therefore not possible.

In contrast to linear systems, we now find different stability conditions in the different regions of state space. This is a general property of nonlinear systems

As before, we are interested in the fish catch that could be maintained indefinitely and sustainably. We change (in SIMULATION PARAMETERS) the initial values for "boats" to 10 and investigate the parameter sensitivity of the "maximum number of boats" in the range from 0 to 40 boats, plotting the results for the annual fish catch as a function of time (Fig. 4.23). We now find (by narrowing the range of investigation) that for boat limits slightly below 32, a stable state with a constant and sustainable fish catch results. However, if the boat limit is raised to 32 and above, the model system collapses after about four decades as a result of overfishing. The most disturbing observation is that the maximum yield (of 2500 [t fish/year]) exactly coincides with the critical boat limit causing the collapse. For economic reasons one would attempt to operate in the vicinity of this maximum yield; however, if the permissible fish catch quota is even slightly exceeded, the system would collapse. The maximum sustainable yield, by the way, is exactly equal to that for density-dependent fishery operation, except that in that case the system stabilizes itself and overfishing cannot lead to a collapse.

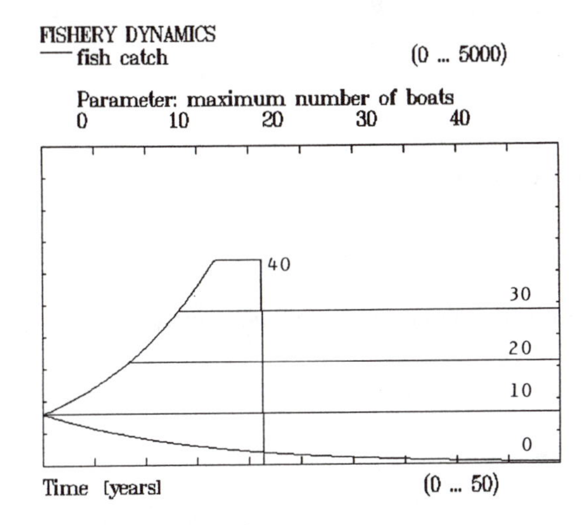

max spec fish growth rate	[1/year]	1.000
spec maintenance costs	[DM/(boat.year	5.00E+04
boat price	[DM/boat]	1.00E+05
boat service life	[years]	15.000
max spec fish catch	[t fish/(boat.	100.000
fish price	[DM/t fish]	1000.000
investment fraction for boats	[-]	0.500
fish locator technology (0/1)	[-]	1.000
maximum number of boats	[boats]	25.000
fish	[t fish]	5000.000
boats	[boats]	10.000
Begin of simulation	[years]	0
End of simulation	[years]	50.000
Time step of computation	[years]	0.020

Fig. 4.23: FISHERY with fish locator technology: Investigation of sustainable maximum yield as a function of the maximum boat number limit. The conditions for maximum yield coincide with those for collapse of the system!

4.3.7 Equilibrium points of the fishery model

In these simulation examples, the significance of the equilibrium points of the system (or model) is obvious. In order to assess whether a system is able to attain a stable equilibrium state at all, the equilibrium points and the behavior in their neighborhood have to be known: the equilibrium point is stable if all state trajectories are pointing towards it. If they are moving away from it, a stable equilibrium state is not possible. In these investigations, simulation (in particular global analysis and sensitivity analysis) can be very helpful. However, it would be easier if the equilibrium points of a system, and the stability conditions at these points, could be directly determined without time-consuming experimentation involving a large number of simulation runs. This is indeed possible with the help of the state equations.

At an equilibrium point the system is at rest; its state does not change. All rates of change of the state variables therefore have to vanish. The algebraic equations resulting from this condition can be used to determine the state values at the equilibrium point.

In Section 3.7.2 we derived the two state equations of the **density-dependent fishery model** (Eq. 3.10, see Section 3.6.2 for an explanation of the abbreviations used):

$$z_1' = a \cdot z_1 \cdot (1 - z_1/k) - f \cdot z_2 \cdot z_1/k$$
$$z_1' = c \cdot f \cdot z_2 \cdot z_1/k - e \cdot z_2$$

where

$$c = (p \cdot i)/q$$
$$e = (o \cdot i)/q + d\,.$$

Using the conditions at the point of equilibrium:

$$z_1' = 0$$
$$z_2' = 0$$

the coordinates of the three **equilibrium points** (EP) for the density-dependent fishery model can now be found.

A first equilibrium point is found at the origin:

$$z_{1EP1} = 0$$
$$z_{2EP1} = 0$$

A second equilibrium point (in the absence of boats) corresponds to the fish carrying capacity. Since "negative" boats cannot occur we have the condition,

$$z_{1EP2} = k \qquad \text{for} \quad z_{2EP2} \leq 0\,.$$

The third equilibrium point is defined by

$$z_{1EP3} = (k \cdot e)/(f \cdot c)$$
$$= (k/f)\,[(o \cdot i)/q + d]/(p \cdot i/q)$$
$$z_{2EP3} = (k \cdot a/f)\,[1 - e/(f \cdot c)]$$
$$= (k \cdot a/f)\,[1 - \{(o \cdot i)/q + d\}/\{f \cdot p \cdot i/q\}]\,.$$

If we now introduce into these relations the default parameter set, we obtain for the state coordinates at this point of equilibrium

$$z_{1EP3} = 6333 \quad [\text{t fish}]$$
$$z_{2EP3} = 37 \quad [\text{boats}]\,.$$

If the model is modified as described above in order to make the fish catch independent of the fish density, the relation $m = h \cdot b$ (catch is proportional to fish density) in Figure 3.24 or in the model equations FISHERY in Sec. 3.6.3, respectively, has to be replaced by the formulation

$$m = x \cdot b$$

where x designates the catch chance. The state equations for the **density-independent fishery model** now become

$$z_1' = a \cdot z_1 \cdot (1 - z_1/k) - x \cdot f \cdot z_2$$
$$z_2' = c \cdot f \cdot x\, z_2 - z_2 \cdot e\,.$$

Applying the equilibrium conditions $z_1' = z_2' = 0$ leads to the state coordinates of the equilibrium point for the density-independent fishery model

$$z_{1EP1} = 0$$
$$z_{2EP1} = 0\,,$$

i.e. this system does not have any "free" equilibrium point with non-vanishing state variables. However, an **equilibrium point** can be forced upon the system by fixing the boat number limit $z_{2EP} = y$. This means that the state equation for z_2 becomes superfluous (at this point). If we introduce the corresponding y into the equilibrium condition for z_{1EP}, we obtain the condition

$$a \cdot z_{1EP} \cdot (1 - z_{1EP}/k) = x \cdot y \cdot f\,.$$

This quadratic equation for z_{1EP} has the solution

$$z_{1EP3,4} = [k/(2\,a)]\,[a \pm (a^2 - 4\,a \cdot x \cdot y \cdot f/k)^{1/2}]\,.$$

Real solutions are only possible for $a \geq 4\,x \cdot y \cdot f/k$. The parameter a is the maximum specific fish growth rate. Introducing the parameter values used for the default settings, we can compute this critical growth rate

$$\begin{aligned} a_{cr} &= 4\,x \cdot y \cdot f / k \\ &= 4 \cdot \text{catch\_chance} \cdot \text{boat\_limit} \cdot \text{max\_fish\_catch} / \text{max\_fish\_capacity} \\ &= 4\,(0.8)\,(25)\,(100) / 10000 = 0.8\,. \end{aligned}$$

The fish growth rate therefore has to be greater than 0.8 for this parameter constellation in order to obtain equilibrium. For the value $a = 1$ used in the simulation, we obtain two equilibrium values for the fish population:

$$\begin{aligned} z_{1EP3,4} &= (10000/2) \cdot [1 \pm (1 - 0.8)^{1/2}] \\ &= 5000 \cdot [1 \pm (0.2)^{1/2}] \\ &= 5000 \cdot [1 \pm 0.4472]\,, \end{aligned}$$

i.e.

$$\begin{aligned} z_{1EP3} &= 7236 \qquad & z_{1EP4} &= 2764 \\ z_{2EP3} &= 25 \qquad & z_{2EP4} &= 25\,. \end{aligned}$$

The equilibrium point z_{1EP3} can be confirmed immediately by simulation of the density-indepent model (fish locator = 1) and looking at the table of simulation results. It obviously represents a stable node which is also clearly shown in Figure 4.22. In order to recognize the other equilibrium point z_{1EP4}, we have to take a closer look at Figure 4.22. In this diagram the "separatrix" separating the two regions of attraction (collapse of the fish population on the upper left hand side, survival of the fish population on the lower right hand side) crosses the horizontal line of constant boat number limit (here: 25). If the initial state is to the left of the cross-over point of boat number limit and separatrix, the state trajectory would move to the left (collapse of the fish population). If the initial point were to the right of the cross-over point, it would approach the stable equilibrium point z_{1EP3}. This cross-over point therefore must represent the second equilibrium point z_{1EP4}. It is a saddle and therefore unstable.

This conjecture can be confirmed by investigating the state space around z_{1GP4} in more detail using global analysis. It is advisable to define the interval in such a way that a simulation point coincides with the suspected equilibrium point. We therefore select for fish the range (2764 ± 100) and for boats the range (25 ± 0.5). The diagram of state trajectories in this range confirms the suspicion: we are dealing with the second equilibrium point, a saddle.

Since for this model we were able to determine the location of the equilibrium points analytically, we can also compute them for any arbitrary combination of parameters in the simulation program itself. We therefore replace the (previously empty) procedure Summary in the SIMPAS simulations model by the procedure Summary listed at the end of Program 4.5. It computes the location of the equilibrium points and prints their values to the screen after each simulation run.

4.3.8 Summary of observations using the fishery model

The fishery model has led to some insights concerning the behavior of dynamic systems in general, and economic exploitation of ecological systems in particular:

1. The investigation of the behavior of non-linear systems has to extend over the entire possible state space (global analysis of possible state trajectories) as the behavior may be dominated by several (stable and unstable) equilibrium points having separate regions of attraction.
2. The behavior, and in particular its qualitative characteristics, can be strongly parameter-dependent. If an analytical analysis is difficult or impossible (as for most non-linear systems), extensive simulation and sensitivity investigations are required.
3. Equilibrium points can be determined from the state equations (for the rates of change of the state variables): the derivatives of all state variables with respect to time must vanish at the equilibrium point.
4. The stability of an equilibrium point can be deduced from the course of the state trajectories in its neighborhood: at a stable equilibrium point, no state trajectory can leave a surface surrounding the equilibrium point in its immediate neighborhood.
5. A "slight" change of a system may cause a significant qualitative change of behavior and stability conditions. In the example of the fishery model, the originally stable system (with a prey-dependent predation rate) is changed by the introduction of fish locator technology into a system with a stable and an unstable region (prey-independent catch rate with boat number limitation).
6. The introduction of "better" technology (here: the fish locator) may have a destabilizing effect by introducing an (unnoticed) structural change into a system. While the fishery system with a prey-dependent catch rate was self-stabilizing, collapse of the system with the locator technology can only be prevented by a strict limitation of the fishing fleet—introducing catch quotas per boat is not sufficient for stabilization.
7. If the exploitation rate of an ecological (regenerative) resource ("prey") is directly dependent on the current resource supply, and if in addition the existence of the exploiter ("predator") depends *exclusively* on this resource, then a stable equilibrium (without collapse) is possible.
8. If an exploiter is able to shift to other resources as an alternative, then complete exploitation and collapse of all resources except the last one is possible. (Sustainability would therefore require restricting the exploitation of natural resources to enterprises whose economic activities are exclusively limited to this particular resource and which would themselves collapse if the resource collapsed.)

9. If the catch rate becomes independent of the prey population (here: by improved locator technology), the system can only be stabilized by strict observance of a catch limit (here: boat number and maximum fish catch) which must be kept below the regeneration rate of the prey population.
10. In this case, the maximum sustainable yield corresponds to the critical catch rate: the system collapses if the limit is slightly exceeded, or if the regeneration rate is slightly reduced. Stabilization of the system requires exploitation at a safe distance from the maximum yield.
11. In the utilization of regenerative systems, the introduction of a more efficient technology may cause destabilization and may necessitate additional efforts for stabilization without producing any advantages in the sustainable yield.

4.4 Graphic-Interactive Simulation Environment: STELLA

4.4.1 Overview of the STELLA approach

The characteristic properties and behavior of a system are determined by its structure: the previous simulation models have shown this clearly. We therefore focused on the recognition of system structure: the structural information implicit in the verbal model was laid down in the influence diagram which was then developed into the simulation diagram by specifying elements and quantifying relationships. Although a set of differential equations (with the supplementary algebraic equations) contains the same information as its equivalent simulation diagram, the graphical presentation of the model structure is intuitively much more easily understood by most people; it can be more easily changed and complemented; it is better suited as a common basis for discussion; and it is more easily and clearly communicated to others.

For this reason we concentrated on the influence diagram and the simulation diagram. From them mathematical relationships are easily and quickly deduced. Model development requires only a pencil and a piece of paper to draw diagrams; there is no need to write mathematical formulae. This is an iterative process: influence relationships are hypothesized, drawn in the diagram, impacts are analyzed, and relationships are changed or deleted. If a complex model must be developed, the constant changes and corrections of the diagrams can spoil the creative fun of model development: many simulation models owe their final shape to the fact that the developers were simply unwilling to change the complex diagram again.

This suggests the use of computer-aided design for the development of simulation models. This approach was taken in the STELLA system (**S**ystems **T**hinking, **E**xperiential **L**earning **La**boratory developed by High Performance Systems, Hanover, NH 03755, USA). STELLA was developed for the Apple Macintosh. The following system studies were performed with STELLA II.

STELLA differs from SIMPAS (and other simulation methods) by the interactive input of the structural diagram and by the interactive graphical procedures available for changing and improving the model. The structural diagram is the basis for a guided interactive query where the computer requests all as yet unspecified block names, relationships, and parameter values. In this way the user is led to specify the model until it is finally executable. As before, the mathematical relationships have to be programmed separately for each model element.

STELLA uses a model notation that is slightly different from the simulation diagrams used here. If one understands the distinctions and characteristics of both approaches, shifting from one diagram to the other should cause no problems.

The notation used in the influence and simulation diagrams of this book focuses exclusively on the influences between system elements. Each arrow indicates an influence from a sender element to a receiver element—and nothing else. This structure is first developed in the influence graph, and it is conserved in the simulation diagram. The shape of the structure therefore does not change in the course of the model development.

The notation used for the structural diagrams in STELLA goes back to the system dynamics method of J. Forrester (Industrial Dynamics 1961; Principles of Systems 1968) for which the well-known simulation language DYNAMO was developed. An important characteristic of this method is the explicit display of "flows" (rates of change) of the "stocks" or "levels" of the system (state variables). These inflows and outflows are regulated by "valves" representing the rates, where the valve setting is changed by inputs from other system elements. The diagrams therefore show two separate processes in one picture: 1. the inflows and outflows of state variables and 2. the influence relationships in the system that change these inflow and outflow rates. Although a graphical distinction is made for both processes in STELLA (double line for flows; single line for influences), the connection between influence diagram (here often called "causal loop-diagram") and STELLA diagram is not always obvious. For example, the influence of a drain on the stock from which it drains (for example, of the outflow from a bathtub on the water level in the tub) can only be inferred from the effect which this outflow has on the stock; there is no arrow in the diagram showing this influence directly (cf. Fig. 4.24). The transition from influence graph to system dynamics diagram is often confusing. For this reason we strictly adhere to the influence relationships when developing the simulation diagram.

The most significant differences between the two system representations are illustrated in Figure 4.24. The STELLA/DYNAMO diagram for the process: "A state changes as a function of its inflows and outflows" (for example, the bathtub) shows the inflow regulated by a valve, as well as the outflow also regulated by a corresponding valve. These flows flow in "pipes" from an undefined source ("cloud") into another undefined sink ("cloud") outside of the system boundary.

The system description using a simulation diagram on the other hand only shows influences: both inflow and outflow act on the state level. If both are defined as positive quantities, the outflow rate must be multiplied by (–1) before being used in the differential equation for the state. (Note that the arrows pointing into a state "box" always stand for rates of change of the state.)

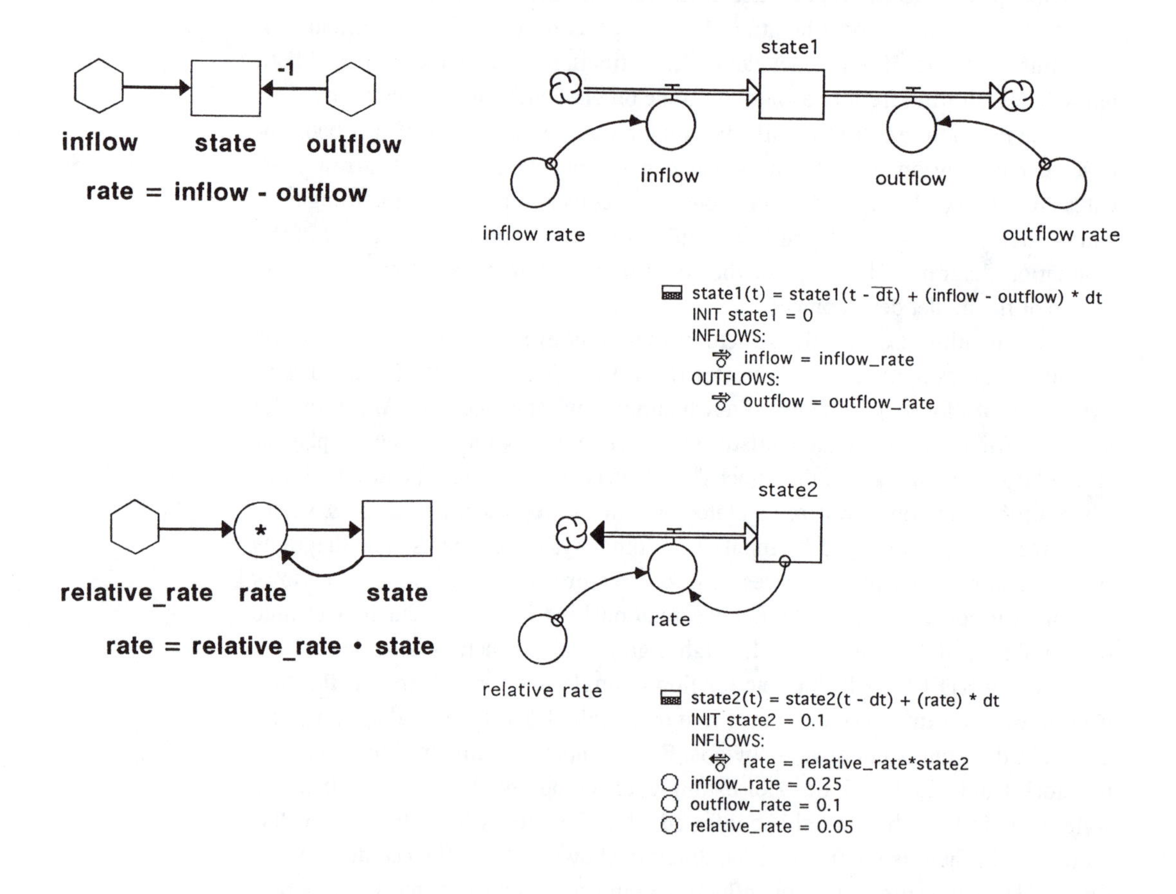

Fig. 4.24: Presentation of simple systems in the simulation diagram (left) and in STELLA (right).

In the STELLA diagram the direction of the flow arrows therefore has nothing to do with the direction of influences. It must be remembered that inflows and outflows always act on the state, independently of the flow direction shown in the STELLA diagram.

Since rates of change may sometimes be positive or negative, meaning that flows of a state variable may flow into or out of a level (box), STELLA diagrams sometimes use flow arrows in both directions ("biflows") as shown in the second example of Figure 4.24. This example shows the process "the rate of change of a state is proportional to the state and to a normal rate of change" which is the well-known process of exponential growth or decay. In the corresponding simulation diagram the influence structure is clearly discernible.

Because of the need to specify flows either as "uniflow" or as "biflow," and additionally by a dominant direction of flow, there are possibilities for errors in formulating STELLA models. These errors are often not immediately obvious in the results and are usually not easily located.

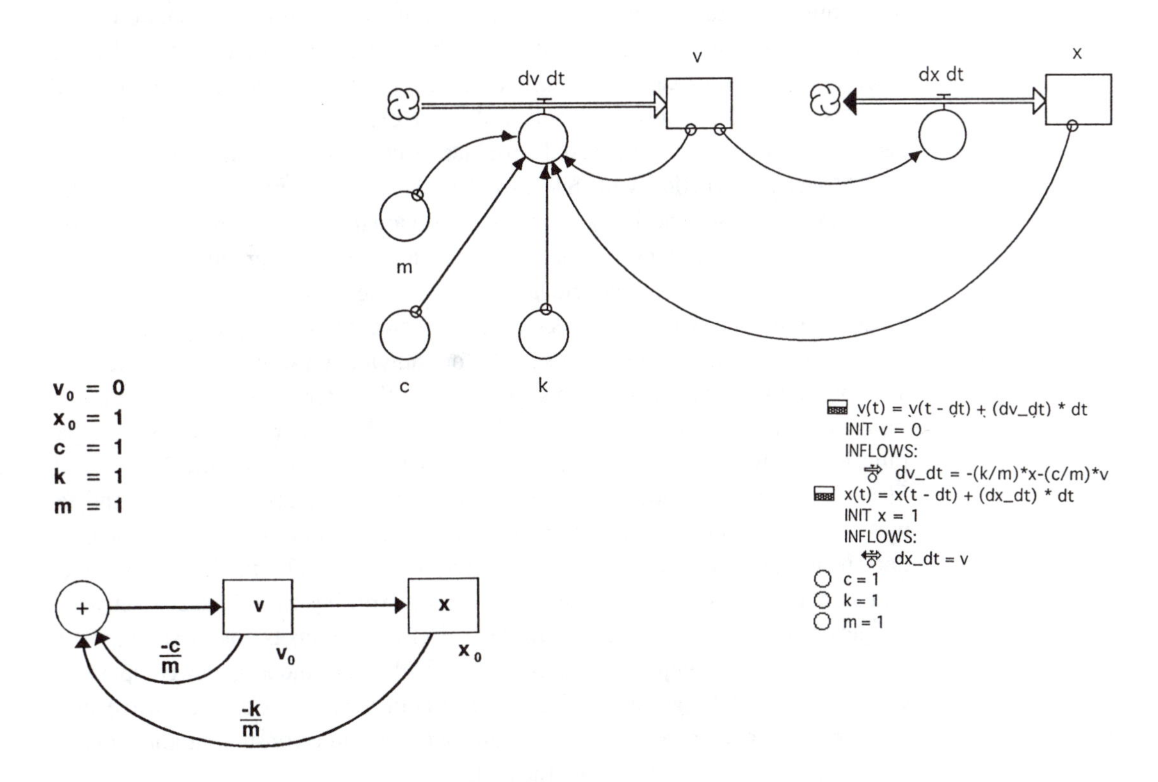

Fig. 4.25: Simulation diagram and STELLA diagram for the spring-mass-damper system.

Figure 4.25 shows the simulation diagram and the corresponding STELLA structural diagram for the spring-mass-damper system (Fig. 3.4, left). Here again, the characteristic differences between the two types of diagrams are obvious. The simulation diagram already contains all information about the mathematical relationships; only the initial values and the parameter values must be specified. In the STELLA diagram the mathematical relationships are not specified; they must be listed separately to complete the specification of the model.

In STELLA there is no distinction between the exogenous parameters or functions on the one hand and intermediate variables on the other: both are characterized by circles. Exogenous quantities have no input, however.

The model equations for the intermediate variables (circles) can be written as ordinary mathematical expressions in STELLA. For this purpose some 60 special functions and some additional logic expressions are available. Relationships that cannot be expressed analytically can be supplied by graphic (table) functions. There are also special symbols and functions for discrete models. With these programming features, the requirements for a broad spectrum of dynamic models can be met. However, requirements going beyond these features cannot be met since a programming capability using a general purpose language is not available (as it is in SIMPAS using TurboPascal). For example, STELLA models often become cumbersome since indexed variables (vectors, arrays) cannot be used.

Model construction with STELLA begins with the interactive input of the structural diagram. The blocks are then individually defined by specifying the mathematical expressions characterizing each block and providing additional explanations for the model documentation. After the complete specification of the model, input of run time parameters, and choice of the output quantities, the simulation is started. Simulation results for individual variables can be selected and can be presented as time plots, phase plots ("scatter plots"), and tables. The results of several simulation runs can be combined in one plot to investigate parameter sensitivity. Parameters and structure can be easily and quickly changed. Structural diagrams, model equations, documentation, graphical output, and tables of results can be printed out as hardcopy, thus facilitating easy and complete documentation of the model and its simulation results. The model can be saved and used again at some later time. This requires working with STELLA software: the generation of stand-alone executable (*.EXE) programs is not possible (as it is in SIMPAS). For comparable computers and identical models, the computation time is also much longer in STELLA than it is in SIMPAS. Comprehensive studies of system behavior with STELLA turn out to be time-consuming and cumbersome even for small models (e.g., the fishery model).

In the following, we implement both the rotating pendulum and fishing dynamics model in STELLA to demonstrate the features of this software.

4.4.2 Simulation of pendulum dynamics using STELLA

Working with STELLA requires the structural diagram corresponding to the STELLA conventions. We can simply translate the simulation diagram (Fig. 3.21) to STELLA notation introducing the names for the quantities that were already used in the SIMPAS-model (Prog. 4.4). The resulting STELLA structural diagram is shown in Figure 4.26.

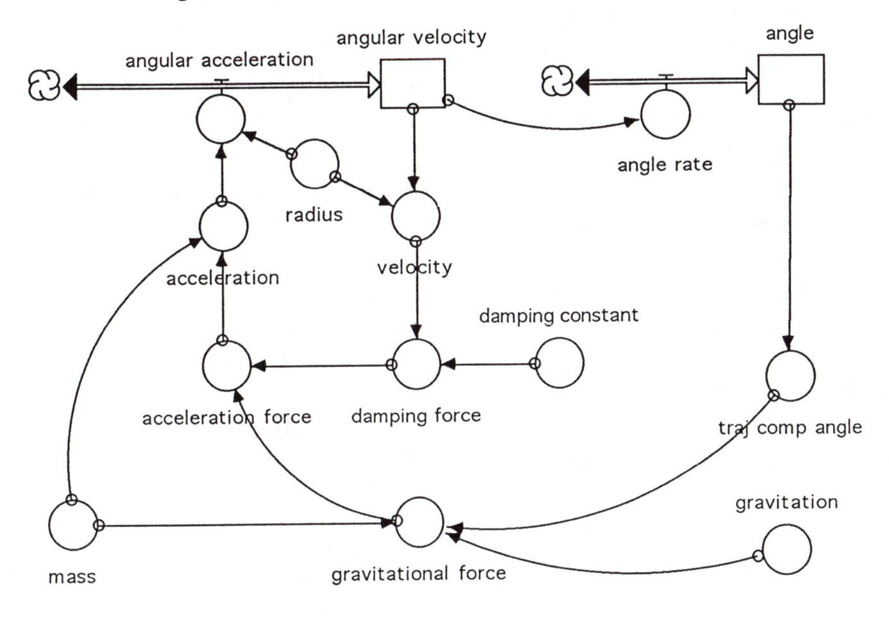


```
angle(t) = angle(t - dt) + (angle_rate) * dt
    INIT angle = 0
    INFLOWS:
        angle_rate = angular_velocity
angular_velocity(t) = angular_velocity(t - dt) + (angular_acceleration) * dt
    INIT angular_velocity = 10
    INFLOWS:
        angular_acceleration = acceleration/radius
O acceleration = acceleration_force/mass
O acceleration_force = -gravitational_force-damping_force
O damping_constant = 1
O damping_force = damping_constant*velocity
O gravitation = 9.81
O gravitational_force = mass*gravitation*traj_comp_angle
O mass = 1
O radius = 1
O traj_comp_angle = SIN(angle)
O velocity = radius*angular_velocity
```

Fig. 4.26: STELLA structural diagram for the rotating pendulum and STELLA printout of the system equations. The variable names correspond to those in the SIMPAS model PENDULUM.

In order to draw a STELLA structural diagram on the screen, the corresponding symbols are selected on the STELLA screen and are then dragged to the desired location on the screen using the mouse. The block names are also specified at this point. Following this, the block symbols are connected by arrows corresponding to the influence structure. As long as the mathematical relations applying to a block have not been specified, a question mark will appear in the block. The blocks are now individually selected (double-click). The screen then shows a dialogue box specifying the names of the inputs to the block and showing the "keyboard" of a calculator and a list of STELLA functions. Even complicated mathematical expressions can be specified by either using the calculator keyboard and clicking on the corresponding quantities, numbers, algebraic operators, or STELLA functions. For the state variables the initial values must be specified. Relationships between two quantities which cannot be formulated mathematically or logically, can be provided as "graphical" (table) functions by either specifying data pairs of the table or by drawing the function on the screen using the mouse.

Comparison of the STELLA structural diagram (Fig. 4.26) with the simulation diagram (Fig. 3.21) shows that they are quite similar. The significant difference is in the (additional) representation of flows in STELLA and the use of the circle also for exogenous quantities. Figure 4.26 also documents the model equations. The model parameters and variables are alphabetically ordered by STELLA; the state variables are listed first.

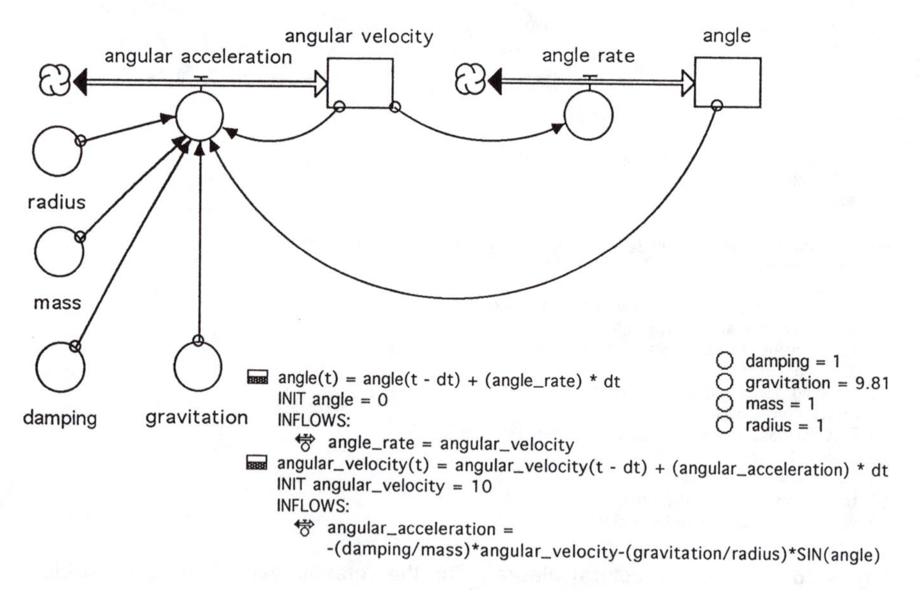

Fig. 4.27: Compact presentation of the pendulum system; STELLA diagram and system equations.

In Figure 4.26 an individual block is used for each system element. This is necessary in STELLA, if each of these quantities is to be available as a simulation result. In addition, it often makes models more transparent and easier to understand. However, the basic model structure becomes more obvious if variables are collected algebraically and are represented in only a few blocks. An example is given in Figure 4.27 which is a compact representation of the pendulum model. Here only the rates of change (and the parameters) are shown as individual blocks; the corresponding specifications for computing the rates are given in the model equations which are separately documented. Except for the particular form of presentation, the model in Figure 4.27 is identical to that of Figure 4.26. However, the feedback structure is more obvious in the more compact form. In both cases one has to make sure that the flows are correctly specified as "biflow" as shown.

Following this process of inputting and checking the structural diagram and all model equations, the simulation can be started. Run time data (begin, end, computation step, table step) must be selected first (by clicking on "run" and "time specs"). In addition, the type of output presentation and the names of the output variables must be selected (by clicking on "windows"). Minimum and maximum values for the plots can also be specified at this point. By clicking on "Graph Pad," time plots and phase plots ("Scatter Plot") can be generated. By clicking on "Table Pad," the results can be presented in tabular form and by clicking "Sensitivity," multiple runs can be generated where a selected parameter (or initial condition) is changed in a predefined range. (This corresponds to "Parameter Sensitivity" in SIMPAS.) Following this selection, the simulation can be started by clicking on "Run" (for an individual simulation) or on "S-Run" (for a sensitivity analysis) on the "Run" menu. The simulation can also be observed in an animated display of the structural diagram where the levels of the state variables change and the flows, intermediate quantities, and parameters are visualized by corresponding "dials." The complete model documentation and all results can be saved and can be called up again at some later time using STELLA.

Figure 4.28 shows the tabular presentation of the simulation results of the standard run for the rotating pendulum. Comparison with the SIMPAS results (Fig. 4.4) shows only minor differences (due to round-off). STELLA time plots for a second simulation under identical conditions are shown in Figure 4.29; this corresponds to the SIMPAS plot in Figure 4.3. The phase plot of still another run under identical conditions is shown in Figure 4.30; this corresponds to the SIMPAS plot Figure 4.5. The results of a sensitivity run using damping as a parameter are shown in Figure 4.31. This corresponds to the SIMPAS plot in Figure 4.11.

seconds	angle	angular velocity
,00	0,00	10,00
1,00	5,41	5,31
2,00	7,23	-2,51
3,00	5,65	1,21
4,00	6,68	-0,62
5,00	6,04	0,33
6,00	6,44	-0,17
7,00	6,19	0,09
8,00	6,34	-0,05
9,00	6,25	0,02
Final	6,30	-0,01
Pendulum: (1)		

Fig. 4.28: Tabular presentation of simulation results of the pendulum system (standard run) using STELLA.

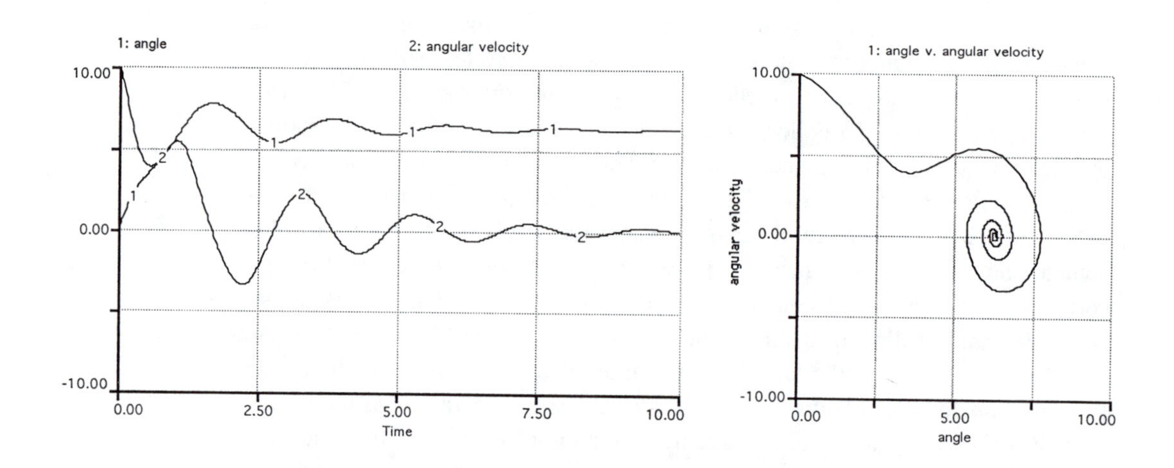

Fig. 4.29 (left): Time plots of the simulation results of the pendulum system (standard run) using STELLA.
Fig. 4.30 (right): Phase plot of the simulation results of the pendulum system (standard run) using STELLA.

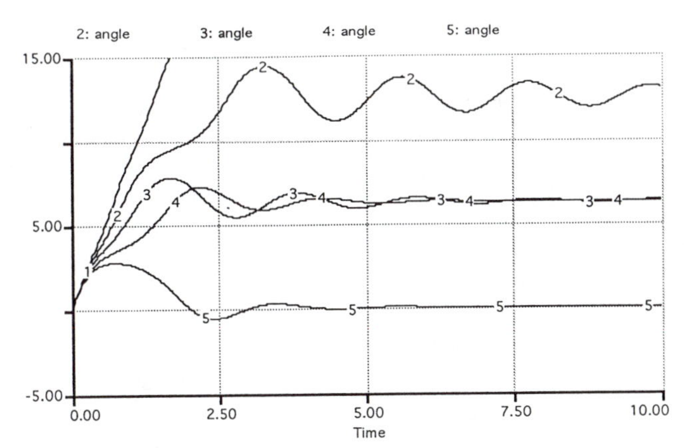

Fig. 4.31: Sensitivity run for the pendulum system with respect to the damping parameter using STELLA.

These presentations exhaust the possibilities of STELLA. Having to rerun an identical simulation again and again in order to produce the different presentations of output turns out to be quite inconvenient as it requires a new output definition as well as a new simulation run. If parameters are to be changed, the corresponding blocks in the structural diagram have to be selected by clicking and by changing them in the dialogue box. Since relatively long computation times are necessary even on fairly powerful computers, extensive sensitivity studies can usually not be made. A combined approach may be advisable: the interactive model development is done using STELLA. As soon as the model structure is established, the equations are transferred to SIMPAS. The model behavior can then be investigated fully and quickly using the features of interactive parameter change, simulation, sensitivity analysis, and global analysis available in SIMPAS.

4.4.3 Simulation of fishery dynamics using STELLA

For the STELLA version of the fishery dynamics model we can use the same variable names as used in the SIMPAS version. On the basis of the simulation diagram (Fig. 3.24), or the SIMPAS simulation program (Prog. 4.5), the STELLA structural diagram can be quickly generated (Fig. 4.32). Except for the representation of flows in the STELLA diagram, this diagram is quite similar to the original simulation diagram. However, the model formulation which is already contained completely in the simulation diagram, must be provided separately in STELLA. The corresponding STELLA equations are also listed in Figure 4.32.

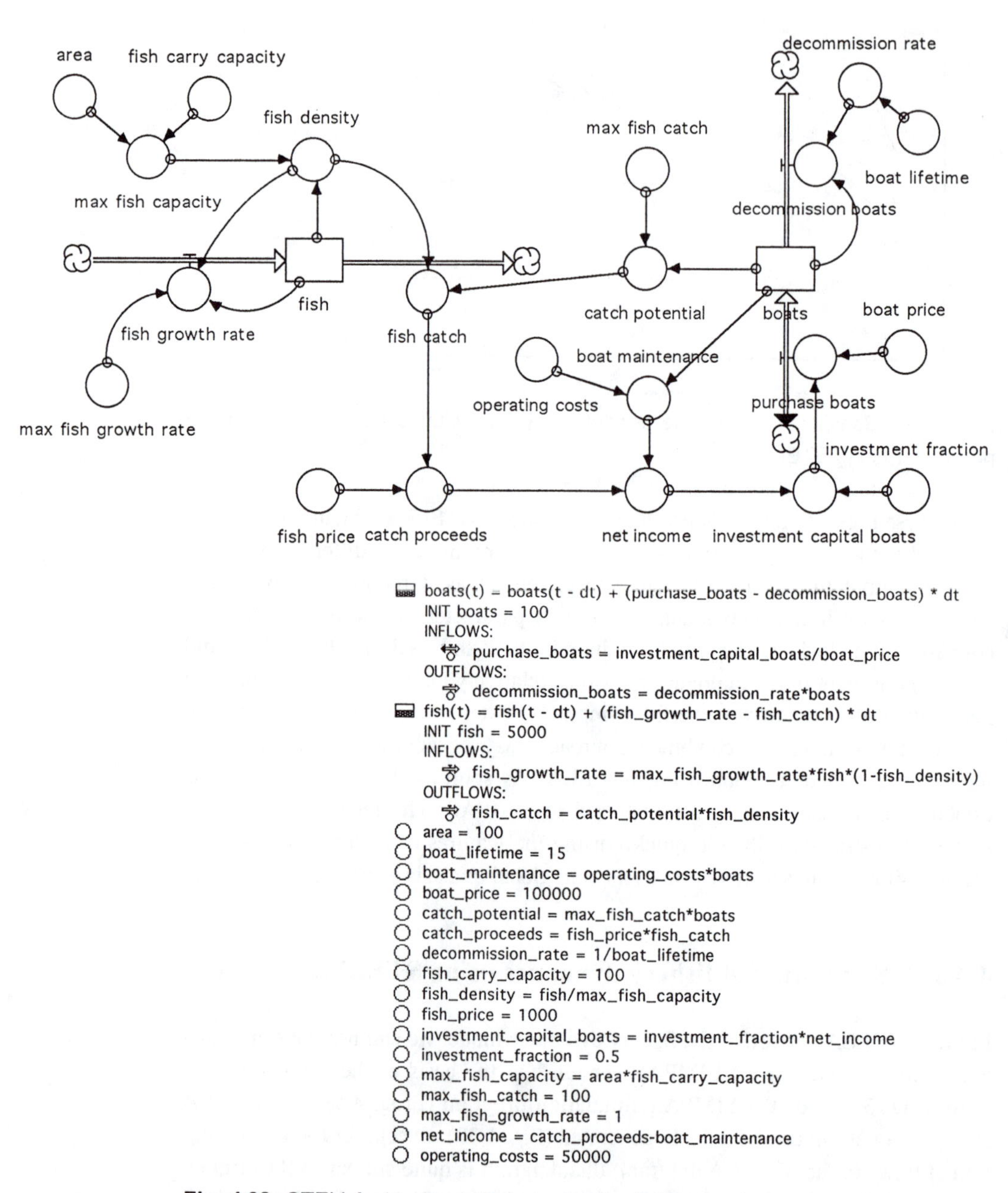


```
boats(t) = boats(t - dt) + (purchase_boats - decommission_boats) * dt
INIT boats = 100
INFLOWS:
   purchase_boats = investment_capital_boats/boat_price
OUTFLOWS:
   decommission_boats = decommission_rate*boats
fish(t) = fish(t - dt) + (fish_growth_rate - fish_catch) * dt
INIT fish = 5000
INFLOWS:
   fish_growth_rate = max_fish_growth_rate*fish*(1-fish_density)
OUTFLOWS:
   fish_catch = catch_potential*fish_density
area = 100
boat_lifetime = 15
boat_maintenance = operating_costs*boats
boat_price = 100000
catch_potential = max_fish_catch*boats
catch_proceeds = fish_price*fish_catch
decommission_rate = 1/boat_lifetime
fish_carry_capacity = 100
fish_density = fish/max_fish_capacity
fish_price = 1000
investment_capital_boats = investment_fraction*net_income
investment_fraction = 0.5
max_fish_capacity = area*fish_carry_capacity
max_fish_catch = 100
max_fish_growth_rate = 1
net_income = catch_proceeds-boat_maintenance
operating_costs = 50000
```

Fig. 4.32: STELLA structural diagram for the fishery system and STELLA printout of the system equations. The names of quantities correspond to those in the SIMPAS model FISHERY.

Using the much more compact differential equations (Eq. 3.10) in Section 3.7.2, a more compact system representation (analogous to Fig. 3.26) would be possible showing a "clearer" and "cleaner" system structure. Note that the rate "purchase_boats" must be specified as "biflow" in the manner shown to permit accelerated decommissioning of boats by "negative" purchases in case of negative net income.

The results of the standard run are shown in Figure 4.33. The curves and the scaling correspond to the SIMPAS plot in Figure 4.16. In the sensitivity run in Figure 4.34 the dependence of the simulation results on the specific catch rate of the boats was studied.

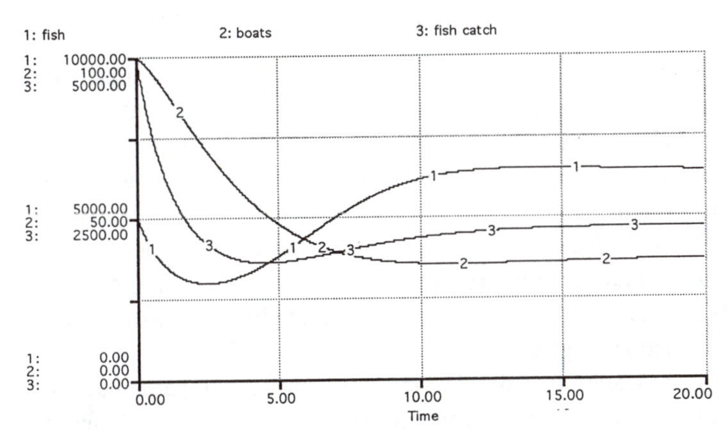

Fig. 4.33: Time plots of the simulation results for the fishery system (standard run) with STELLA.

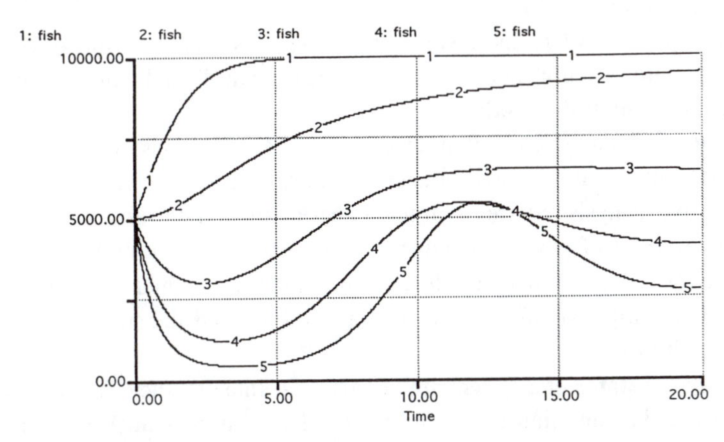

Fig. 4.34: Sensitivity runs of the fishery model with respect to the fish catch potential of the boats.

4.5 Summary of Important Results

The development of simulation models for the rotating pendulum and fishery dynamics begun in Chapter 3 was completed in this chapter by the development of executable simulation models. Using two different programming systems, these models were studied over a wide parameter and behavioral range. If we now compare the results of each model with the model purposes specified in Chapter 3, we can conclude that the original task has been achieved.

We have now gone through the process of modeling and simulation several times. Individual phases and steps of model development are summarized in Figure 4.35. The most important results of this chapter are:

1. **Simulation models** of dynamic systems consist of elements which belong to one of three categories (input quantities, state quantities, and intermediate quantities including rates of change) and can only connect in certain ways. The requirements for developing and **processing** different simulation models are therefore **similar**.
2. Independently of the specific model, it is therefore possible to develop **general purpose simulation software**. A number of such programming systems are available for different computer systems and applications; many employ their own simulation language.
3. For practical application **two simulation approaches** are of particular interest: a. simulation systems using common general programming languages with all their possibilities, and b. simulation systems using the possibilities of interactive graphics and computer-aided design for model development. Both approaches were used here.
4. A characteristic of most realistic simulation models is their **non-linearity**. These models are (usually) not open to numerical analysis and can only be investigated by **computer simulation**.
5. The **behavior** of such systems often depends in surprising ways on parameter settings, initial values, environmental inputs, and structural changes. Comprehensive analysis requires a large number of simulation runs covering the whole spectrum of possible parameter constellations and initial values. Simulation software should therefore not only allow individual simulation runs but also multiple simulations for **global analysis** and the study of **parameter sensitivities**.
6. The **equilibrium states** of a system are of particular interest; they can be determined from the condition that the rates of change must vanish at such points:

   $$dz/dt = \mathbf{0} .$$

7. The stability or instability of an equilibrium point follows from the dynamic behavior in the neighborhood of the equilibrium states, i.e. the state trajectories. Often a linearization of the original system is possible near the equilibrium point which then allows mathematical analysis of the stability of the locally valid linear substitute system.

Often, a system study will end with the development of a model and the completion of a number of simulation runs. However, in our work so far we have found points of departure for further investigations. These questions will be dealt with in the following chapters:

1. How could system behavior—in particular in the neighborhood of equilibrium points—be described analytically in order to obtain generally valid conclusions without having to conduct extensive simulations?
2. Is it possible to make statements about the stability behavior on the basis of the state equations? Since we are particularly interested in finding the circumstances under which a system may become unstable, the behavior must be investigated in particular in the neighborhood of equilibrium points.
3. Is it possible to find ways to change the structure of an unstable or marginally stable system in such a way that it will show stable behavior even under perturbation? For example, is it possible to change the pendulum system in such a way that it will also be stable in its upper dead center?
4. How do systems behave if they are subject to constant, and in particular periodic, influences from the system environment? How do these inputs modify the characteristic dynamics of a system (which are determined by system structure)? Is it possible to destabilize an otherwise stable system by external influences?
5. How can a complex system be controlled in such a way that it will produce an "optimal" behavior, where optimum must be measured by some given criteria? For example, how must a complex system of regenerative resources (fish population) be managed in order to avoid collapse, maintain species variety and still obtain sustainable high and constant yields?
6. Can dynamic systems show behavior that is qualitatively different from that which we have so far encountered in the simulations? What are chaotic systems? How is it possible that occasionally small structural changes cause radical qualitative changes in behavior?

Phases of model development

1. **Development of the model concept**
   - State problem
   - Define model purpose
   - Define a reference behavior pattern
2. **Model construction**
   - Write down detailed verbal description of system (verbal model)
   - Define system boundary
   - Identify important subsystems and their influence relationships
   - Develop influence structure
   - Determine state variables
   - Specify system elements and their functions
   - Identify and quantify influence relationships
   - Recognize feedbacks
   - Define and quantify exogenous parameters and influences
   - Choose initial conditions and free parameters for the reference behavior
   - Program, using a suitable simulation method
3. **Model testing**
   - Validate the model structure by reference to the system structure and the model purpose (structural validity)
   - Eliminate programming errors
   - Generate the reference run
   - Test over the possible parameter and behavioral range (plausibility and robustness)
   - Validate behavioral modes (behavioral validity)
   - Compare to empirical observations (time series) (empirical validity)
   - Investigate parameter sensitivity
   - Run scenario simulations (using realistic parameter combinations)
   - Test influences of inputs and control measures
   - Investigate consequences of necessary or possible structural changes
   - Check agreement with model purpose (application validity)
4. **Communication of results**
   - Condense models to "essential" structure
   - Identify behavior-determining feedback loops
   - Understand and substantiate the behavior modes

Fig. 4.35: Phases of model development and model testing.

CHAPTER 5

CHOICE AND DESIGN

5.0 Introduction

In the previous chapters we went in steps through the process of model construction and simulation, from the initial definition of model purpose to the simulation runs. At the end of this process, a simulation program is available, a tool for studying the entire behavioral spectrum of the model as a function of parameter choice. Assuming validity of the model, the development and the use of this tool can increase our understanding of the system enormously. Simulations with the model lead to new information and new insights about the system.

However, these results may also lead to the realization that the system behavior is unacceptable, unsatisfactory, or merely difficult to comprehend. For example, unstable behavior may appear even under "normal" conditions, state values may stay far below requirements, or the multitude of free parameters may frustrate the systematic search for good solutions. These problems will appear particularly if the goal was not just the simulation of the behavior of an existing system but the improvement of system performance by input or system changes.

In working with systems or models, we are dealing with three basic components: system S, system input I, and system output O. These three components define the three **fundamental tasks of systems analysis** (Fig. 5.1):

1. **Path analysis:** to determine system *output* for a given system and given system input.
2. **Policy analysis**: to determine system *input* for a given system and a desired output.
3. **System design**: to determine system *structure* for a given input and a desired output.

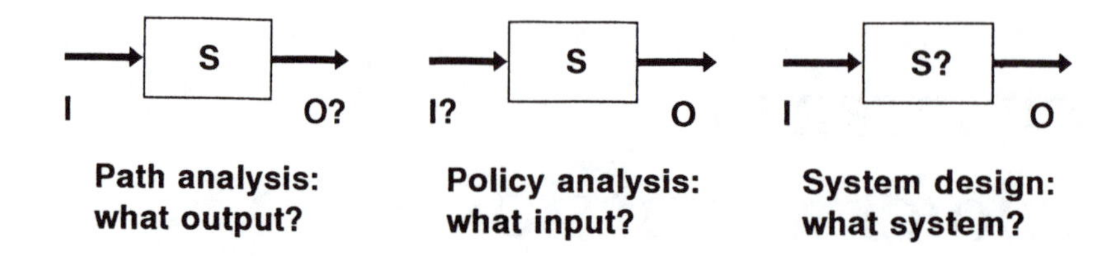

Fig. 5.1: The three fundamental tasks of systems analysis: path analysis, policy analysis, and system design.

1. **Path analysis**: In this case the model of a system is available, the constant or time-dependent inputs are known (or are assumed as "scenarios"), and the behavior of the system under these conditions is to be determined. Usually the input parameters and scenarios will be changed over a broad range of conditions to compute the resulting development paths and to compare and evaluate the outcomes. (Example: analysis and comparison of the global population and pollution development for different scenarios concerning the type, time, and vigor of birth control measures, resource conservation, and limitation of consumption.)
2. **Policy analysis**: In this case the model of a system is available, its desired performance is prescribed (perhaps as a function of time), and the constant or time-dependent inputs that would produce the desired result are sought. In technical or economic systems, optimization is often a goal: finding parameter settings (possibly time-variant) that would lead to "optimal" system behavior (in terms of given evaluation criteria). (Example: finding a fish harvest strategy leading to maximum sustainable and economic yields—without collapse of the fish population.)
3. **System design**: In this case the constant or time-dependent inputs are known, the desired performance of the system is prescribed, and a system has to be designed or changed to deal with these inputs and meet the performance criteria. In particular, the task may be to modify an existing system to enable it to produce "better" behavior than before. (Example: stabilizing the rotating pendulum at its upper (unstable) dead center by an additional negative feedback.)

In this chapter we will deal with these three aspects using the examples mentioned. For these investigations we now have available the simulation models WORLDSIM, FISHERY, and PENDULUM that were developed earlier and can now be easily investigated, changed, or expanded using SIMPAS or STELLA.

In these three tasks, evaluation of the system behavior and of the results is of central importance. What is meant, for example, by "unacceptable development," "optimal yield," or "good stability behavior?" Such evaluation results can only be taken seriously if they do not simply represent the intuitive assessment of the observer but if they are instead founded on a formal and reproducible systematic evaluation. This requires that **evaluation criteria** and evaluation procedures be clearly and completely defined. We will first deal with this aspect in Section 5.1.

In Section 5.2 the concepts developed for evaluation of system performance will be applied to **path analysis** and the comparative assessment of development paths of the global model. A method will be presented for the systematic comparison of simulation runs by assessing impacts on the functional interests of the system.

In Section 5.3 we will use the fishery model to demonstrate the basic approach of **policy analysis** to determine policy inputs which would produce desired, and possibly "optimal", outcomes.

In Section 5.4 we will demonstrate the concept of **system design** by devising a control system for the unstable rotating pendulum that is able to stabilize the system at its upper dead center. Different control functions (i.e. structural modifications) can be used for this purpose, and the question of "optimal control" poses itself here also.

In Section 5.5, important results of this chapter will be summarized.

5.1 Criteria and Evaluation of System Behavior

5.1.1 Orientors, indicators, and criteria

The concepts developed in this chapter apply not only to models but to systems in general. In the following, we therefore speak of systems and system behavior.

The evaluation of system behavior and simulation results requires the use of some sort of value criteria. In a first *ad hoc* assessment we will often classify one result as "good," another one as "bad," without being aware of specific evaluation criteria. Intuitive evaluations of this kind cannot be substantiated, are not reproducible, and are often not understood by others. If based on experience or tradition, they may serve for a first orientation and classification or as an initial search heuristic for sorting possible alternatives. However, for systematic comparative evaluation or in the search for "optimal" solutions, clearly defined evaluation criteria and reproducible evaluation procedures are mandatory.

The assessment of system development always requires two things:

1. **Evaluation criteria** must be available for all system aspects of interest.
2. The **system state** reflecting all system aspects of interest must be known at the point in time for which the evaluation is to be made.

Evaluation criteria and state description therefore must directly complement each other: an incomplete state description, for example, where the state cannot be evaluated with respect to a criterion recognized as crucial is just as inadmissible as an incomplete set of criteria which, for example, does not permit recognition of an existential threat to the system.

The set of criteria that are relevant for the evaluation of system development will be referred to as **orientors**. This term is used to make clear that we are dealing with criteria that systems (or their managers) use to orient their decisions and actions regarding the system. Orientors are aspects, terms, or dimensions (like "freedom") that designate important criteria or qualities of system survival and development but not the degree to which they must be satisfied.

To evaluate system development with respect to the relevant orientors, **indicators** describing the system state must be mapped on the orientors in order to compare the actual system state with the corresponding goal values in "orientation space."

In the evaluation of system development, three different types of normative criteria play a role:

1. **Constraints** that limit state variables, system variables, final system states, the duration of processes, or control inputs to permissible ranges.
2. **Quality measures** that allow differentiating "better" from "worse" state developments and may also allow the search for "optimal" solutions.
3. **Weights** that allow an aggregated evaluation if several quality measures of different degrees of importance must be applied simultaneously.

The assessment of the behavior of real systems is almost always complicated by the fact that several constraints and quality measures must be considered simultaneously. While there can be no compromise for constraints which must be strictly observed, trade-offs are possible if quality measures are involved. In that case, the evaluation result depends on the relative weights assigned to the different quality criteria.

System variables or functions of system variables for which goals can be defined and compared to current values are referred to as **indicator variables (indicators)**. It is certainly possible that the criteria considered to be relevant for the evaluation will not have counterparts in the routinely observed system variables or in the variables available in the simulation model. In that case, additional indicators have to be observed to assess the performance of the real system, or the necessary indicator variables must be constructed from the available variables. In simulations it may be necessary to introduce additional model formulations from which the necessary indicators can be computed.

An example for clarifying these concepts: in the evaluation of global development (of the real world or of a "global model"), orientors and the corresponding indicators with respect to the following aspects will play a role: life expectancy, nutrition, health risks from pollution, state of the environment, material welfare, resource consumption, and social progress. In order to reach a comprehensive assessment, indicators have to be defined with corresponding goal values. The orientor satisfaction can then be checked by comparing current indicator values to goal values. (The following formulations are to be understood as examples only.)

Orientor: Life expectancy.
Goal: "Average life expectancy should not be less than 60 years."
Indicator: Current life expectancy.

Orientor: Nutrition.
Goal: "Daily uptake of nutrient energy should not be less than 8000 kJ/day per person."
Indicator: Current average daily food energy consumption per person.

Orientor: Health threat by pollution.
Goal: "Nitrate load of drinking water may not exceed 50 mg/l".
Indicator: Maximum of the (seasonally fluctuating) nitrate concentration.

Orientor: State of the environment.
Goal: "The regenerative capability of the environment must be preserved."
Indicator: Sustainability of essential environmental functions, expressed by a weighted measure describing development of key populations (plants, animals).

Orientor: Material welfare.
Goal: "Attaining a (defined) level of real income in the shortest time."
Indicator: Time integral of the difference between income goal and actual income. This time integral should be minimized.

Orientor: Resource consumption.
Goal: "Minimize consumption of non-renewable resources in the long term."
Indicator: Time integral of the annual fossil fuel consumption. This time integral should be minimized.

Orientor: Social progress.
Goal: "Rapid attainment of a (certain) life expectancy and a (certain) real income."
Indicator: Time integral of the weighted sum of the (normalized) discrepancies between the current and the desired state. This time integral should be minimized.

In these examples we again encounter the three different kinds of normative criteria:

Constraints (thresholds which should not be crossed; as here, for example, for life expectancy, nutrition state, state of the environment, and health threat from pollution); **quality measures** (minimization or maximization of time integrals; as here, for example, for material wealth, resource consumption, and social progress); and **weights** (different relative emphasis on different criteria; as here, for example, for state of the environment and social progress).

Where do the criteria for the evaluation of system development originate? They obviously have an influence on the kind and extent of system observation or model construction: if criteria sets differ, the perspectives of observation, description, and evaluation will also differ. The selection of criteria is determined on the one hand by the immediate interests of the investigator, and on the other hand by his/her understanding of the system and the problem. It is rarely unique. Especially for crucial system investigations it is absolutely necessary that all of those interested in a solution—and in particular if they belong to different (political) persuasions—should agree on the same evaluation procedure and on the same set of evaluation criteria. This should be true even if each evaluator assigns to them different (but openly stated) values and weights. Only in this way can misunderstandings and useless battles over evaluation results be avoided.

Sometimes, criteria choice is not an issue. This is the case in particular for systems where states and rates of change are measured in the same "currency." In economic studies, for example, states are measured in terms of monetary values (investment capital, reserves, orders) and rates of change in money flows (proceeds, payments, depreciation, investment rate). In production processes and other technical processes, stocks of energy and/or material and their rates of change (power, material flows) play a role. In these cases it is possible to formulate indisputable evaluation criteria (with respect to the system performance) such as maximizing profit or minimizing energy and/or resource consumption.

Often, however, such "one-criteria assessments" are mistakenly applied to systems whose criteria space is multi-dimensional. In such cases, system performance cannot be evaluated in terms of a single "system currency." Examples are the attempts to evaluate the ecological impacts of road construction (for example, loss of a forest) by economic means, i.e. by monetarizing the "recreation value" (using the potential increase of life income resulting from better health of hikers) or the value of a plant species threatened by extinction (in terms of its potential economic contribution as a pharmaceutical product).

Many reject such cost-benefit analyses on intuitive grounds without being able to justify their rejection; they are often enough reproached as "irrational." In the following, we therefore want to focus on the system-theoretic side of the evaluation of system behavior. This will demonstrate why one-dimensional evaluation will in general not be permissible and can only legitimately be applied in special cases (as, for example, in the assessment of economic or resource use efficiency mentioned above).

5.1.2 System behavior and orientation theory

Systems always exist in an environment. They receive inputs from the environment, and they affect the environment by their outputs (Fig. 1.2). When discussing the general properties of systems in Section 1.3 and in particular the definition of the system boundary in Section 1.3.3, we noted that the system boundary should be drawn where the environment is relatively unaffected by the system. This excludes (significant) feedbacks from the system to itself via the environment. (Where they exist, they must be counted as part of the system and must therefore be inside of the system boundary.)

This view of system environment and system boundary means that systems are subjected to the properties and characteristics of their system environment without being able to change them significantly and rapidly. The system therefore has to cope with whatever environment in which it finds itself.

If systems have existed in their environment for a long time (such as plant or animal species that have persisted for hundreds of millions of years; or the family that has organized human life for hundreds of thousands of years), they have obviously been successful in coping with their specific environments. But this means that the system must possess properties that in some way reflect properties of the environment. The environment therefore shapes the system, and only systems that are adapted to their environment will survive in the long run.

The same basic idea applies to the design of systems. If a new system (a product, an institution, a company) is to survive in its environment, it must have been designed to cope successfully with the properties of that environment.

While these observations may seem trivial, they take on profound significance if we now take a closer look at the fundamental properties of system environments. It turns out that system environments have some fundamental properties that force upon systems certain "design criteria" to insure they will persist. These "design criteria" are termed "basic orientors:" criteria which must be fulfilled if a system is to survive and develop in its environment in the long term and which must be considered in the orientation of system behavior and development.

In order to derive these basic orientors, we therefore have to look first at the fundamental properties of system environments. On first look, system environments seem to have nothing in common: the environment of (A) a forest ecosystem is completely different from that of (B) an industrial company, or of (C) a moon lander. And yet, when we look at the properties of these different environments, we find common characteristics:

1. An environment is characterized by a **normal environmental state**. The actual environmental state can vary around this state in a certain range.

Example A: A Central European forest exists in an environment characterized by a mean annual temperature of about 10 degrees Celsius (range: -20 to +30 C), 800 mm annual rainfall (range: 500 mm to 1100 mm), nutrient supply depending on site properties, etc.

Example B: An industrial company in a small town in Germany has to deal with specific economic, social, cultural (language, schools, job training, attitudes, etc.), legal, and political environments.

Example C: A moon lander has to operate in an environment of low gravity, no atmosphere, and severe temperature differences between lunar day and night.

2. An environment is characterized by **scarce resources**. Resources required for a systems's survival are not immediately available when and where needed.

Example A: The forest ecosystem has to grow an extensive fine root system to collect scarce water and nutrients in the ground and a leaf canopy to harvest sunlight and CO_2. Sufficient sunlight is not available in the winter.

Example B: The company needs water, electricity, raw materials, loans, workers, all of which can only be secured with considerable effort.

Example C: The moon lander will not find any fuel on the lunar surface but there is sunlight during the lunar day to charge its batteries.

3. An environment is characterized by **variety**. A large number of very different processes occur constantly or intermittently.

Example A: The forest environment is affected by day and night, summer and winter, rain and snow, frost and drought, insect pests, animals, wood cutters, and many other very different processes.

Example B: The company has to exist in an environment of various sources of materials and energy, competitors, customers, different rules and regulations, alternative means of transportation and production, employees with very different training and personalities, etc.

Example C: When roving the moon surface, the moon lander encounters different surface materials, small and large rocks, hills and craters, uphill and downhill climbs.

4. An environment is characterized by **variability**. Although there is a normal environmental state, it fluctuates in random ways, and the fluctuations may occasionally take it far from the normal state.

Example A: The forest ecosystem may be stressed by unusual frost in early June, a sudden insect pest outbreak, a long drought, or pollution due to an industrial accident.

Example B: The company may be hit by a recession, a stock market crash, a sudden jump in oil prices, an unexpected competitor, or a change in government that imposes stricter environmental regulations.

Example C: While the environment on the moon is hardly variable, the conditions the moon lander finds itself in may be quite variable; for example, if it lands on its side after a maneuver or if an electrical connection shakes loose.

5. An environment undergoes **change**. The normal environmental state may gradually or abruptly change with time.

Example A: In Europe and elsewhere, the climate has changed considerably since the last ice age, and now there is an accelerated change due to greenhouse gases entering the atmosphere from anthropogenic sources. This changes the environment of forest ecosystems.

Example B: The economic, social, and technological environments of a company show gradual long-term trends (shift to service economy; changing family size, incomes, and living conditions; introduction of computers and industrial robots; integration in the European common market).

Example C: Engineers can at least forget about this one: the moon lander will be designed under the assumption that during its lifetime, conditions on the moon will not change.

6. An environment is populated by **other systems**. The behaviors of these other systems introduce changes in the environment that affect the system.

Example A: The forest ecosystem has to interact with browsing, pollinating, seed-dispersing animals; with agricultural, river, and mountain ecosystems at its edges; with urban settlements and transportation systems; with industries and their pollution.

Example B: The company vitally depends on interactions with, and actions of, suppliers and customers, city officials and politicians, competitors, and bankers.

Example C: The moon lander will have to coordinate its activities with ground control and perhaps other moon-roving research vehicles.

What does this tell us about the orientors of system evolution and system design? If a system is to survive and thrive in an environment characterized by these fundamental properties:

1. normal environmental state
2. scarce resources
3. variety
4. variability
5. change
6. other systems,

then it must either have been forced to pay attention to these properties of the environment during its evolution, or it must be designed to adequately cope with them.

Before we turn to derivation of the proper design criteria, or orientors, let us look again at the fundamental structure of systems. This will help us to determine where in a system the proper responses to environmental challenges will have to be located, and what the options are.

The general form of the state equations describing a continuous dynamic system (including ecosystems) is (Sec. 3.2.2, Eq. 3.1, 3.2)

$$dz/dt = \mathbf{f}(\mathbf{z}, \mathbf{u}, t)$$
$$\mathbf{v} = \mathbf{g}(\mathbf{z}, \mathbf{u}, t) ,$$

where $\mathbf{z}$ is the state vector, $\mathbf{u}$ the vector of environmental inputs, $\mathbf{v}$ the vector of system outputs, t the time, $\mathbf{f}$ the (vector) state function, and $\mathbf{g}$ the (vector) output function. The generic system diagram for any dynamic system corresponding to this formulation is shown in Figure 3.7.

The externally observable behavior of the system consists of output $\mathbf{v}$ which is determined partially by (algebraic or logic) transformation of external inputs $\mathbf{u}$ and partially by the internal state vector $\mathbf{z}$ of the system. The state vector $\mathbf{z}$ itself is partially a result of external inputs $\mathbf{u}$ and partially a result of the system state $\mathbf{z}$ feeding back on itself. State function $\mathbf{f}$ and output function $\mathbf{g}$ may in addition be functions of time (e.g. aging).

These basic relationships apply to any dynamic system from simple mechanical control systems to highly complex and intelligent systems such as ecosystems or human individuals and organizations.

The behavior of state-determined systems (to which this formulation applies) is only partially the result of external influences **u**. The eigendynamics of the state variables **z** provide these systems with a certain degree of autonomous behavior. The system development is therefore dependent on the system itself. This means that the state function **f** as well as the output function **g** must implicitly contain the coping capability of the system. Put differently, in a "successful" system, the system structure (**f** and **g**) reflects the opportunities and dangers of long-term system development. The following discussion is therefore based on the concept that a "successful" state-determined system must in the course of its evolutionary development have been provided with a combination of state function **f** and output function **g** that allows it to fulfill its characteristic system purpose in the long term.

In order to fulfill this system purpose, any particular system must be able to persist in its environment. This means that it must not fail to fulfill its normal functions under "normal" conditions, possibly breaking down and being destroyed in the process. On the contrary, one should expect more "intelligent" systems would be able to cope with new, significantly different environmental conditions without breaking down.

This self-evident requirement of persistence leads to concrete and distinct requirements of behavior and performance of dynamic systems. These can be formulated as general "design criteria" of dynamic systems. Because of their fundamental significance for systems and system behavior, these criteria are referred to as "**basic orientors**." As will be shown, it is of crucial importance that proper attention be paid by the system (consciously or unconsciously) to each of these basic orientors: continuous neglect of any of the orientors will be punished by a loss of viability and by threat of extinction.

Let us now derive the "design criteria" or orientors corresponding to these six fundamental properties. In the following, we will refer to "system interests" even in the case of systems that are unable to reflect on their system state. These are interests that an observer can ascribe to the system as long as it fulfills its system purpose.

5.1.3 Existence in the normal environmental state

As we have seen above, any particular system environment is characterized by a "normal" environmental state. If a system is to exist in this environment and to fulfill its system purpose, it must be assured that it can function under these conditions. This requires maintaining its state variables within certain limits:

$$\mathbf{z}_{min} \leq \mathbf{z} \leq \mathbf{z}_{max} \,.$$

This leads us to a first basic orientor **existence**. Attention to this orientor is necessary to insure the immediate survival of the system.

The existence orientor leads to three distinct requirements concerning the system structure and its properties:

1. The system must have a protective enclosure to filter or exclude any threats **u** from the environment which would be able to move the system state **z** (spontaneously) out of its safe range (a requirement on **f**).
   Example: The pressurized cabin of an aircraft protects passengers from the low pressure, lack of oxygen, and low temperature at high altitudes.
2. The system structure **f** itself must not allow any system state threatening the system's existence.
   Example: The feedback process regulating the heart beat of infants must not fail (as in sudden crib death).
3. The system must not exhibit any self-destructive behavior (a restriction on the behavior function **g**).
   Example: Suicide.

5.1.4 Effectiveness in securing scarce resources

System processes require energy, material flows, and information from the environment. These resources are usually not freely and abundantly available—systems have to expend considerable effort to secure them. Systems also have to exert influence on the environment (output **v**) to fulfill their system purpose in other ways (for example, to defend themselves). These efforts must be balanced by appropriate gains.

This leads to a second orientor **effectiveness**. It requires that in its efforts to secure resources from and to exert influence on its environment, the system should on balance be effective. The results achieved must be worth the cost. In the long run, no system can spend more than it absorbs. The better the ratio of result achieved versus effort expended, the higher the "efficiency." This balance need not be positive at each instant. It is merely required that it be positive over the long term. Deficits can be absorbed better and longer if there are large reservoirs in the system. It should be noted that the normal concern of a system would not be maximum efficiency of all processes but rather the overall effectiveness of interactions with the environment and the long-term cost/benefit balance.

Consideration of the elementary system structure in this context shows that the effectiveness orientor again has an internal component (with respect to the internal dynamics of the system described by the state function **f**) and an external component (with respect to the external behavior of the system described by behavior function **g**):

1. The internal system structure (i.e. the state function **f**) must allow efficient use of the resources required for system existence and system development.
   Example: A company must not spend more on salaries and dividends than it earns from selling its products, leaving nothing for research and new investments.
2. The behavior **v** (i.e. the behavior function **g**) of the system must be efficient with respect to its influence on the system environment (efficient resource acquisition; efficient, timely interaction with the environment).
   Example: The amount of energy collected by a hummingbird as nectar from flowers must significantly exceed the energy expended for flying and hovering, and other life-supporting activities.

5.1.5 Freedom of action to cope with environmental variety

Environmental variety poses problems for a system. In general, the environmental variety confronting the system will be vastly greater than its behavioral variety. The system should be able to cope adequately with all challenges posed by the environment. But the more variety there is in the environmental challenges, the more variable must its own responses be unless it can simply evade the challenges with which it cannot cope. As Ashby put it (1956): "Only variety can cope with variety."

This leads to the third basic orientor **freedom of action**. It requires that the system must have the ability to cope in various ways with the various challenges that confront it. In principle, there are different possibilities for achieving this, again with an internal and an external component:

1. The system reacts with an appropriate state response from its state function repertoire **f**.
   Example: A gosling will walk on land but will paddle and swim when thrown into the water.
2. By its behavior **g**, the system attempts to change the environmental input into something for which it has an appropriate response.
   Example: A person surprised outdoors by cold weather will try to reach a house, don a warm coat, or build a fire.
3. Choice of another environment, or finding a more appropriate niche in it.
   Example: Emigrating, finding and occupying an ecological niche, or changing jobs.

5.1.6 Security to protect from environmental variability

In general, environmental effects on a system not only have a certain variety, they are also subject to random changes with time, may occasionally and suddenly be outside of the normal range, and are generally uncertain and unstable.

The system state cannot be allowed to depend in critical ways on unpredictable fluctuations in the environment. In particular, the system should be safe from continuous existential threats posed by fluctuating environmental conditions.

This leads to the fourth basic orientor **security.** It requires that the system must have the ability to protect itself from the detrimental effects of the variable and unpredictable environmental conditions. The security requirement has two aspects:

1. **Relative autonomy** from unstable environmental factors.
2. **Stability of** those **environmental factors** on which the system continues to depend.

The system has to cope with these different threats to its "security" requirement with appropriate but basically different measures. These measures either aim at internal changes in the system itself or at particular changes in the external environment:

1. Broad decoupling from unstable environmental factors by (partial) isolation, filtering, or saturation effects. This amounts to corresponding adjustments of the state function **f**.

   Example: Building a house to protect oneself from unstable and adverse weather conditions.
2. The creation of buffers and reservoirs to cushion overloads and supply bottlenecks, i.e. an extension of the system structure by additional state variables with large time constants. This is equivalent to a change of system structure **f** and/or system parameters.

   Example: Food, water, and fuel supplies to tide over the winter or in times of war.
3. The creation of a self-stabilizing structure (feedback control, control parameters) and protection against a sudden shift of the system trajectory into an unstable region of attraction. This again amounts to a change in **f**.

   Example: Speed control by flyweight governor on steam turbines or diesel engines.
4. Defusing of potentially dangerous environmental threats or establishment of appropriate defense mechanisms. This is equivalent to a change of the environmental inputs **u** by purposeful behavior **v**.

   Example: Construction of dikes or extermination of wolves and other large predators.

5.1.7 Adaptivity to deal with environmental change

If a system cannot escape the threatening influences of its environment, the one remaining possibility consists in changing the system itself in order to cope better with the environmental impacts. These changes may be necessary to ensure survival and development but they may also be required to maintain the integrity of the system, even at the cost of changing identity.

This leads to the fifth orientor **adaptivity**. It requires that the system should be able to change its parameters and/or structure in order to generate more appropriate responses to challenges posed by the environment. Basically there are two possibilities to satisfy the adaptivity requirement:

1. Modification of behavior $\mathbf{g}$ such that for a given state $\mathbf{z}$ a different behavior $\mathbf{v}$ results which is better adapted to the altered environmental inputs $\mathbf{u}$.
   Example: Befriending an old foe after he/she has become influential.
2. Modification of the state function $\mathbf{f}$ such that for the given (changed) environmental input $\mathbf{u}$, a more appropriate state vector $\mathbf{z}$ results.
   Example: Conversion of the energy supply system to renewable energy sources and better energy use efficiency.

While a change of behavior $\mathbf{g}$ does not change the system itself, a change of the structure of the state function $\mathbf{f}$ is equivalent to a fundamental change of the system. The state function $\mathbf{f}$ is the core of the system and determines its specific identity. Following such a qualitative change, the system is no longer identical to the old one; its identity has been changed.

Adaptive change often offers the only possibility to cope with changing environmental conditions. One has to clearly distinguish between parameter change and structural change. Parameter change usually secures the system only for a limited time period or when environmental changes are small. By contrast, structural change permits evolutionary adaptation to larger environmental shifts. In most cases, this will be a coevolutionary process, since system change also results in a changing influence on the environment (cf. Sec. 1.3.12 and Fig. 1.4).

The adaptivity orientor is a reminder that in a changing environment the system must have the ability to fundamentally change its structure and behavior. The details of this process remain unspecified. It requires the capacity for self-organization. Certain conditions facilitate adaptivity:

1. internal diversity and variety
2. multiple-use structural elements
3. redundant, physically different processes
4. decentralization and partial autonomy
5. memory as information storage from which there can be learning
6. timely assessment of available alternatives and necessary change processes.

5.1.8 Regard for other systems in the environment

The analysis so far has been limited to an isolated system coping with its environment. This is a rare situation in reality. Usually there are other systems present whose behaviors affect the system. Sometimes it may be possible to treat other systems simply as part of the overall environment (as someone on his way to work might treat the thousands of individuals around him in the subway system) but often the behavior of one system is very much influenced by the behavior (even if only anticipated) of another system (as in a personal relationship).

This leads to the sixth orientor **regard for other systems**. It acts as a reminder of the need to pay at least some attention to behavior and orientors of certain other systems.

Almost always the external inputs $\mathbf{u}$ of a system are composed of the influences $\mathbf{u}_u$ of the "unsystemic" environment and influences $\mathbf{u}_p$ resulting from the behavior $\mathbf{v}_p$ of other systems in the common environment:

$$\mathbf{u} = \mathbf{u}_u + \mathbf{u}_p(\mathbf{v}_p) .$$

In general, a system will therefore have to respond to the behavior of other systems. In principle, this does not change our concept of system behavior: the impacts of another system are merely part of the total vector of environmental influences $\mathbf{u}$ and must be processed in the same way.

However, it may be that the behavior of certain systems in the environment is of special significance to a system. That behavior may therefore be accorded specific attention and may evoke specific behavioral response. This means, in effect, that the system takes into account the orientors of the other system, at least partially: a mother watching her small child is concerned about his security, for example. The fact that other systems may be of importance to a particular system and that their behavior may have to be taken into consideration (by a corresponding modification of behavior) is particularly obvious for systems that are equipped with certain anticipatory capabilities (like higher animals). Regard for other systems is not always driven by self-interest. In animals, one may sometimes have to revert to a "genetic egotism" hypothesis in order to explain occasionally self-destructive behavior which helps to preserve the species.

Regard for others requires 1. (conscious or unconscious) recognition of another system in its specific situation, and 2. (conscious or unconscious) anticipation of at least the further short-term development of that system.

However, it only makes sense to talk about "regard for others" if the attention extends beyond the normal attention accorded one of many environmental factors, i.e. if the other system's interests are considered with a certain weight.

The orientor **regard for others** may influence system behavior in two ways:

1. The system must be able to recognize a situation involving another system, and an appropriate response must be available. Such selective perception and response requires a corresponding state function **f** with the ability to differentiate, discriminate, and weigh observations and actions.
   Example: Different responses of an animal depending on whether it is approached by a sibling or a predator.
2. To turn different (internal) responses into appropriate action, different behavioral patterns and possibilities must be in the repertoire **g** .
   Example: Ants are capable of individual action as well as coordinated social action involving hundreds of individuals.

Without getting into a discussion of origins and modes of such "considerate" behavior, we merely note here that "regard for other systems" is an additional basic orientor playing a significant role in system orientation. In conscious actors (like humans and their organizations), who have (or could have) knowledge and perceptions about the impacts of their actions on current and future systems, "regard for others" is always a basic orientor dimension even if in the end no consideration is taken of others. The relative weight of the interests of others in behavioral decisions is a matter of normative valuation.

This brings us to ethics. Humans are free to choose how much weight they accord the interests of other systems that may be affected by their decisions. In principle, this can reach from naked egotism to self-sacrifice in altruism: as conscious beings we have the privilege, and the obligation, to choose.

There is evidence that this choice is not quite as open as is often claimed: if we value humanity, natural environment, and human culture as "permanent systems" that are worth preserving, we have no logical choice but to stand up for the interests of present and future subsystems: other people, other countries, other species, ecosystems, and future generations.

5.1.9 Basic orientors, orientation, and evaluation of system behavior

The previous discussion leads to the conclusion that the fundamental properties of a system's environment force it to pay attention to certain basic principles. These "basic orientors" play a role in the evolution and development of a system and in its behavior, and they must be observed in designing a system if it is to be successful in the long run. There is a one-to-one relationship between the properties of the environment and the basic orientors of systems (Fig. 5.2):

property of environment	basic orientor
normal environmental state	existence
scarce resources	effectiveness
variety	freedom of action
variability	security
change	adaptivity
other systems	regard for other systems

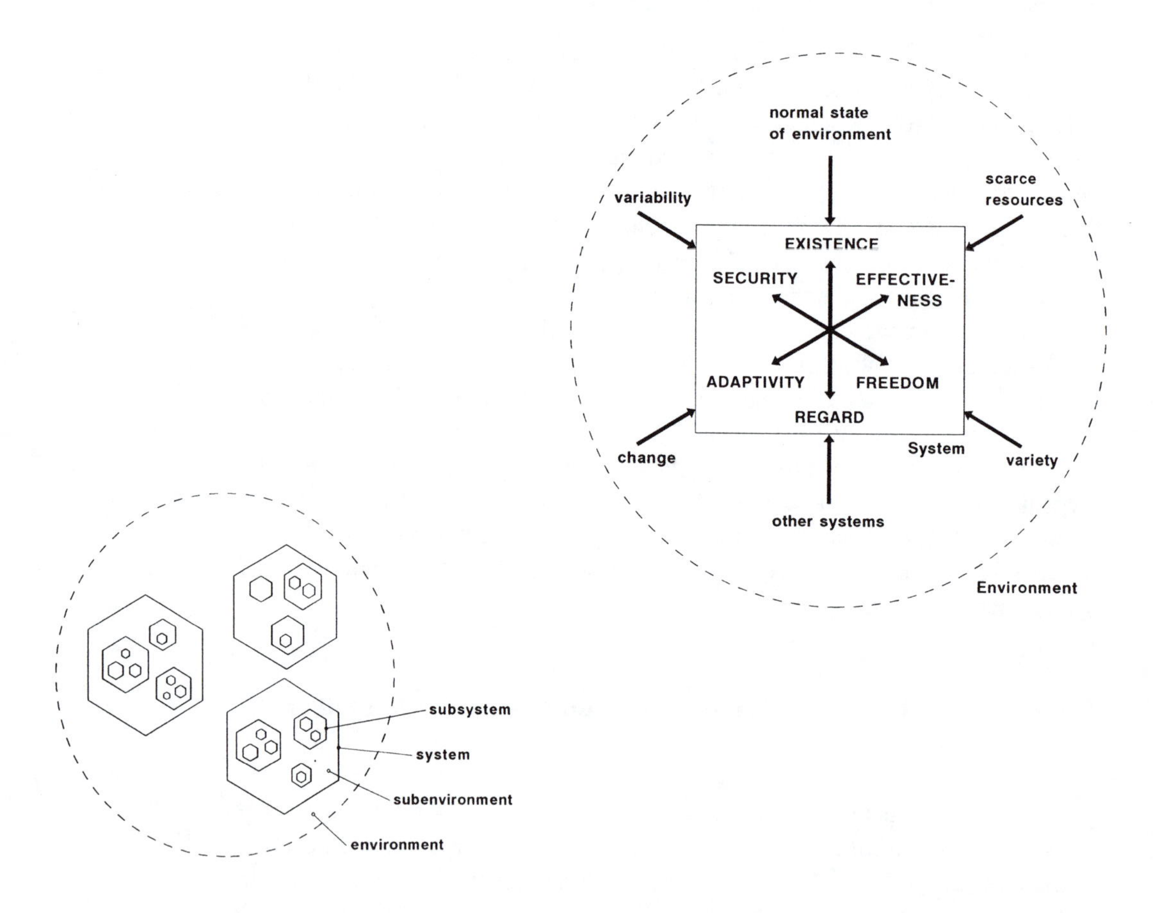

Fig. 5.2: Properties of the environment and corresponding orientors of the system (top). Systems (and subsystems) may share a common environment; each system must pay attention to its own set of orientors (bottom)

With regard to system studies this leads to the following conclusions:

1. In long-existing "proven" systems (organisms, ecosystems, social organizations), one can assume that in the course of their evolution, and in interaction with their environment, the systems have evolved in such a way that adequate satisfaction of basic orientors is assured.
2. When designing a new system, the system structure and elements, i.e. the state function **f** and the behavioral function **g**, must be "constructed" in such a way that in interacting with the system environment, adequate satisfaction of basic orientors is achieved.
3. Each of the orientors stands for a unique requirement. Attention must therefore be paid to all of them, and the compensation of deficits of one orientor by over-fulfillment of other orientors is not possible. Comprehensive assessments of system behavior and development therefore must be multi-criteria assessments.
4. In "unconscious" systems, orientation with respect to the basic orientors is forced upon the systems by the environment—non-attention to basic orientor aspects means development handicaps and evolutionary disadvantages and extinction of the system in the long run.
5. "Conscious" systems can apply different (ethical) weights to the basic orientor categories, to derived orientors, and to the interests of other systems. This causes differences in behavior. However, system existence, viability, and development require a minimum of coordinated basic orientor satisfaction even in this case.
6. Path analysis, policy analysis, and system design have to take the orientor satisfaction of the affected systems into account.

How do we translate this into a practical approach for use in path analysis, policy analysis, and system design? For the analysis of a specific system in its environment, we have to find indicator variables which can be used as measures of actual orientor fulfillment, and we have to define mapping functions for each that tell us how much the current system state contributes to the satisfaction of a particular orientor. We will usually find that one indicator can simultaneously contribute to more than one orientor. The practical use of this tool requires real-structure modeling. Note in the following analysis that the mapping of functional requirements on certain basic orientors is not always unique. However, the primary objective must be to consider all relevant aspects of orientation, and this is assured by the approach used.

Example: Assume we wish to devise an orientor assessment procedure for national development. For individuals, we recognize the following mappings from indicators on orientors:

1. "Life expectancy" maps on "freedom of action."
2. "Income" maps on "effectiveness" (weakly), "freedom of action" (strongly), "security" (weakly), and "adaptivity" (weakly).
3. "Nutrition" maps on "existence" (strongly), "freedom of action (weakly), and "security" (weakly).
4. "Education" maps on "effectiveness," "freedom of action," and "adaptivity"; etc.

In these mappings, it must be clearly stated for which value of the indicator the respective orientor satisfaction falls below the required minimum. In this case, a "red light" has to go on at this orientor, irrespective of positive contributions from other indicators. For example, if "nutrition" falls below the minimum requirement, the red light of the "existence" orientor should focus all attention on this deficit, no matter how excellent the conditions may be with respect to the other orientors.

The orientor "regard for others" must be dealt with by introducing the orientor systems of other systems explicitly; for example, in addition to those of the individual, those for:

1. "the nation"
2. "industry"
3. "the environment"
4. "other countries"
5. "future generations".

These systems will be assigned different weights in the assessment, where the weights reflect ethical positions.

In orientor assessments of complex systems, it is often advantageous to introduce several levels of orientors by splitting the more general orientor into more specific aspects. For example, in an assessment of national policy, the "effectiveness" orientor would have different aspects requiring different indicators:

1. "economic effectiveness"
2. "government effectiveness"
3. "technological effectiveness"
4. "legal effectiveness," etc.

The concepts of orientors and orientor assessment can be summarized by the "orientor star" of Figure 5.3: each of the six basic orientors represents a fundamentally different aspect of system orientation. A certain minimum of each of these dimensions must be met in order to guarantee survival, viability, and adaptive evolution of the system. A deficit with respect to one basic orientor cannot be compensated by a surplus of others. A system is not viable in the long run if even one of the orientor dimensions remains deficient.

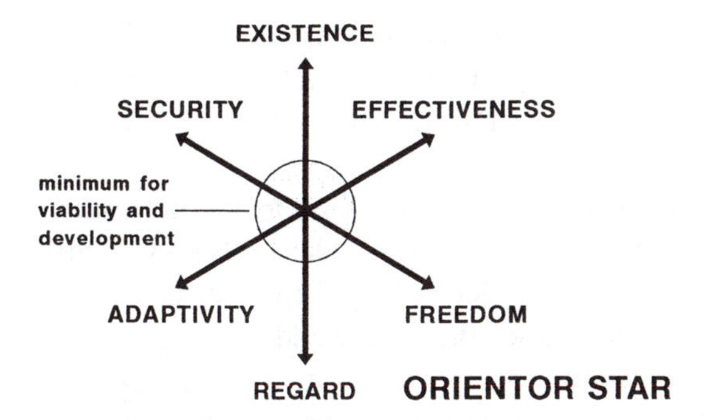

Fig. 5.3: Orientor star: If a system is to be viable in the long run, a minimum satisfaction of each of the basic orientors must be assured.

A comparative assessment of the (subjective) benefit of better orientor satisfaction with respect to the different orientor categories is possible only if the necessary minimum of *each* of the basic orientors is guaranteed. This may then result in a decision to forego improvements with respect to one orientor in favor of improving performance with respect to another. This is a subjective assessment in which the normative weights of the orientor criteria play a role.

In the assessment and orientation of system behavior we therefore deal with a two-phase assessment process where each phase is different from the other:

1. First, a certain minimum satisfaction must be guaranteed separately for each of the basic orientors. As long as this is not the case, the system's attention will have to be focused on the remaining orientor deficits.
2. Only if the required minimum satisfaction of all basic orientors is guaranteed is it permissible to use a weighted sum of surplus orientor satisfactions to obtain an overall quality or satisfaction index which can then be maximized.

Here again we come across **constraints** (which must be observed strictly and individually and cannot be balanced one against the other); **quality measures** (where surplus satisfactions can be computed in terms of a common "currency"—for example, "utility"—and be balanced one against the other); and **weights** (determining the relative importance of different contributions in aggregated quality measures).

In applications, we deal with systems at different levels of self-organization ability and consciousness. This has to be taken into account in orientor assessments.

Systems with invariant structure and constant parameters (most technical systems, organisms without learning capability) have no freedom of choice, and hence orientor assessment only plays a role in the design or evolution of such systems.

Self-organizing systems can change parameters and structure, and this change will usually be made by (conscious or unconscious) reference to the basic orientors. This will either be enforced by reality (trial and error), or it may be shaped by anticipation of the environmental response.

Conscious actors (humans, organizations, institutions) may decide differently than orientor assessment of their interests would indicate. However, such assessments can still be a valuable decision aid for finding likely responses.

In the following simulation examples for the "miniworld," the fishery system, and the stabilization of the rotating pendulum at its upper dead center, we use these fundamental considerations and the corresponding evaluation method. References on orientation theory are found in Bossel et al., 1989: p. 176-183. An application of the orientation approach to path assessment and control of a global model can be found in Bossel and Strobel, 1978: p. 191-212; an application to a development study for Mexico is described in Bossel et al., 1989: p.105-173. The relationships to human values and behavior are discussed in Bossel, 1978.

5.2 Path Analysis

5.2.1 Introduction

Path analysis of a simulation model, i.e. the study and comparative evaluation of alternative development paths, represents one important task of systems analysis. It is possible to develop an overview of behavior by many probing simulation runs but a systematic approach is always to be preferred. Orientation theory offers a framework for systematic and efficient path studies.

The first task of path analysis consists of quickly finding the most relevant development paths despite a multitude of uncertain, time-dependent, or adjustable parameters. The efficiency of this search depends on how cleverly possible parameter constellations are combined in consistent and plausible "scenarios."

The second task of path analysis is the comparative evaluation and assessment of different development paths to clarify which path (or which group of paths) should be preferred. In this phase of the work, evaluation criteria have to be introduced that reflect the existence and development interests of the system (and perhaps also the interests of its managers). For small models like the Miniworld, it may still be possible to distinguish a "good" from a "poor" solution by simple inspection of the results of simulations. For more complex models, this becomes almost impossible, and a reliable assessment requires a formal and comprehensive approach. For achieving a comprehensive evaluation procedure, the concepts of orientation theory are useful. Note that the formulation of an orientor assessment procedure is not unique: different possibilities exist. But the important point is that this approach forces us to include a set of criteria that represents the full spectrum of environmental properties and system interests. Irrespective of the actual choice of criteria and indicators, we can then be fairly certain that all aspects important to system development are included.

In this section we shall again deal with the Miniworld model of Chapter 2 (Fig. 2.14 and Eq. 2.19 in Sec. 2.2.5). We shall first program it in SIMPAS. We shall then determine criteria and indicators for assessing the system development to determine orientor satisfaction.

The computation of orientor satisfaction requires some additional programming. With this augmented program, we will then investigate two alternative scenarios of system development, each described by its own set of scenario parameters. The simulation runs generate not only the temporal development of the system variables but also that of the resulting criteria satisfactions. With this information, a comparative evaluation of the two paths becomes possible.

5.2.2 System elements and simulation model for a Miniworld

In Chapter 2 the model equations for a Miniworld system were derived. They can be used to study the interactions between population, consumption, and pollution development (Fig. 2.14 and Eqs. 2.19).

In order to make the simulation model more readable, we replace the abbreviations used in Chapter 2 by corresponding descriptive labels. Since we are more interested in the qualitative analysis of interactions and developments, we will continue using non-dimensional relative state variables while using the "year" as the relevant time unit. The following new labels will be used:

C	= capital	[-]	capital stock
N	= population	[-]	population number
P	= pollution	[-]	pollution load
P^*	= threshold	[-]	poll. absorption threshold
a	= absorption_rate	[$year^{-1}$]	normal absorption rate
b	= birth_rate	[$year^{-1}$]	normal birthrate
d	= death_rate	[$year^{-1}$]	normal deathrate
e	= pollution_rate	[$year^{-1}$]	normal pollution rate
k	= capital_growth_rate	[$year^{-1}$]	normal capital growth rate
f	= consumption_control	[-]	growth limitation
g	= birth_control	[-]	family planning effect

In order to make the model more transparent, we introduce additional labels for the intermediate variables:

quality	[-]:	environmental quality
births	[year^{-1}]:	births per year
deaths	[year^{-1}]:	deaths per year
degradation	[year^{-1}]:	environmental degradation per year
regeneration	[year^{-1}]:	environmental regeneration per year
conservation	[-]:	influence on consumption growth rate
capital_growth	[year^{-1}]:	capital growth per year

We now obtain (from Fig. 2.14 and Eq. 2.19) the following relationships. The numerical values represent the default values for parameters and initial values. They can be changed interactively in simulation runs.

1. Parameters

birth_rate = 0.03
death_rate = 0.01
threshold = 1
absorption_ratc = 0.1
capital_growth_rate = 0.05
pollution_rate = 0.02
birth_control = 1.0
consumption_control= 0.1

2. Initial values of state variables

population_0 = 1
capital_0 = 1
pollution_0 = 1

3. Algebraic intermediatc variables

quality = threshold/pollution
births = birth_rate · population · quality · capital·birth_control
deaths = death_rate · population · pollution
degradation = pollution_rate · population · capital
conservation = 1 – consumption_control · pollution · capital
capital_growth = capital_growth_rate · capital · pollution · conservation
if quality < 1 then renewal = quality
if quality ≥ 1 then renewal = 1
regeneration = absorption_rate · pollution · renewal
consumption = capital

4. State equations

d(population)/dt = births – deaths
d(capital)/dt = capital_growth
d(pollution load)/dt = degradation – regeneration

5. Run time parameters

Start = 0
Final = 500
TimeStep = 0.2

The STELLA structural diagram for this model is shown in Figure 5.4. Here again we have to take care to introduce the correct direction of flows ("biflow" for "capital_growth!") The SIMPAS model unit WORLDSIM is shown in Program 5.1. Figure 5.5 presents the simulation result for the standard run (cf. Fig. 2.16a).

5.2.3 Criteria and indicators of system development

To find out what the results of different simulation runs "mean" for the system, they have to be mapped on the orientors. In accordance with the previous discussion, two aspects have to be considered:

1. We must make sure that the necessary minimum level of orientor fulfillment is achieved for each individual orientor.
2. We must determine the total quality of orientor satisfaction (for individual orientors and some aggregated quality measure).

In comparing different paths of system development, the most favorable path will be the one for which a. the minimum conditions are always satisfied for all orientors, and b. the overall orientor satisfaction is better.

To perform these additional computations with the WORLDSIM model, we have to define an evaluation procedure and introduce it into the model. The procedure described in the following makes no claim to completeness, and the WORLDSIM model itself should only be considered as a crude image of reality. We are here merely interested in demonstrating the procedure.

We first have to decide whose interests should be represented by the assessment, i.e. whose orientor satisfactions are we assessing? There are two "actors" in the model whose development is of interest to us and on whose fate the development of the whole system depends: population and environment. In formulating the evaluation procedures, we have to keep their respective survival and development interests in mind. To include all aspects important for the total system, we apply the basic orientors. Note that the relationships formulated in the following are only meant as examples; different formulations are possible.

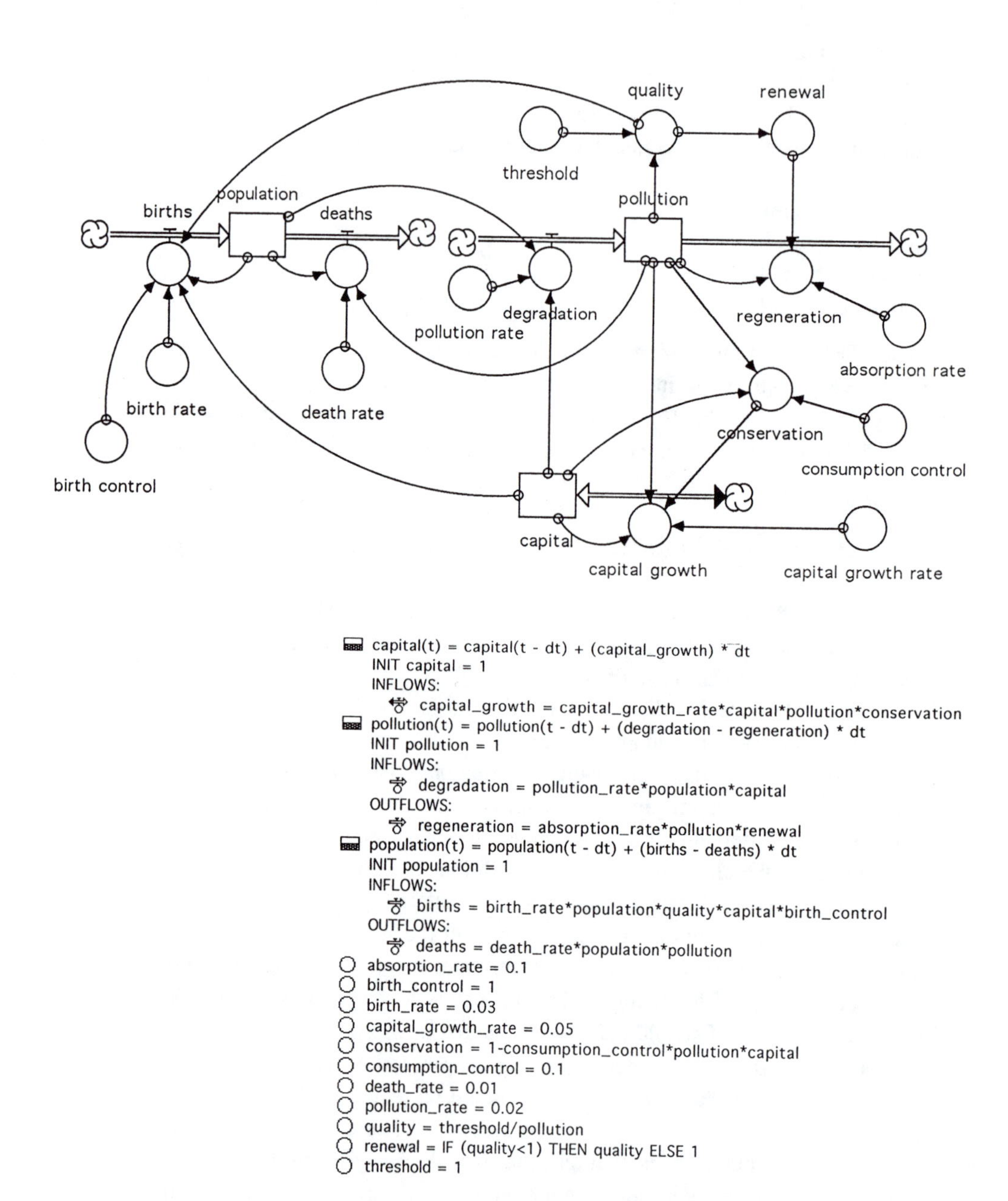


```
capital(t) = capital(t - dt) + (capital_growth) * dt
    INIT capital = 1
    INFLOWS:
        capital_growth = capital_growth_rate*capital*pollution*conservation
pollution(t) = pollution(t - dt) + (degradation - regeneration) * dt
    INIT pollution = 1
    INFLOWS:
        degradation = pollution_rate*population*capital
    OUTFLOWS:
        regeneration = absorption_rate*pollution*renewal
population(t) = population(t - dt) + (births - deaths) * dt
    INIT population = 1
    INFLOWS:
        births = birth_rate*population*quality*capital*birth_control
    OUTFLOWS:
        deaths = death_rate*population*pollution
absorption_rate = 0.1
birth_control = 1
birth_rate = 0.03
capital_growth_rate = 0.05
conservation = 1-consumption_control*pollution*capital
consumption_control = 0.1
death_rate = 0.01
pollution_rate = 0.02
quality = threshold/pollution
renewal = IF (quality<1) THEN quality ELSE 1
threshold = 1
```

Fig. 5.4: STELLA structural diagram for the Miniworld system and STELLA listing of the system equations. The labels of variables correspond to those in the SIM-PAS model WORLDSIM.

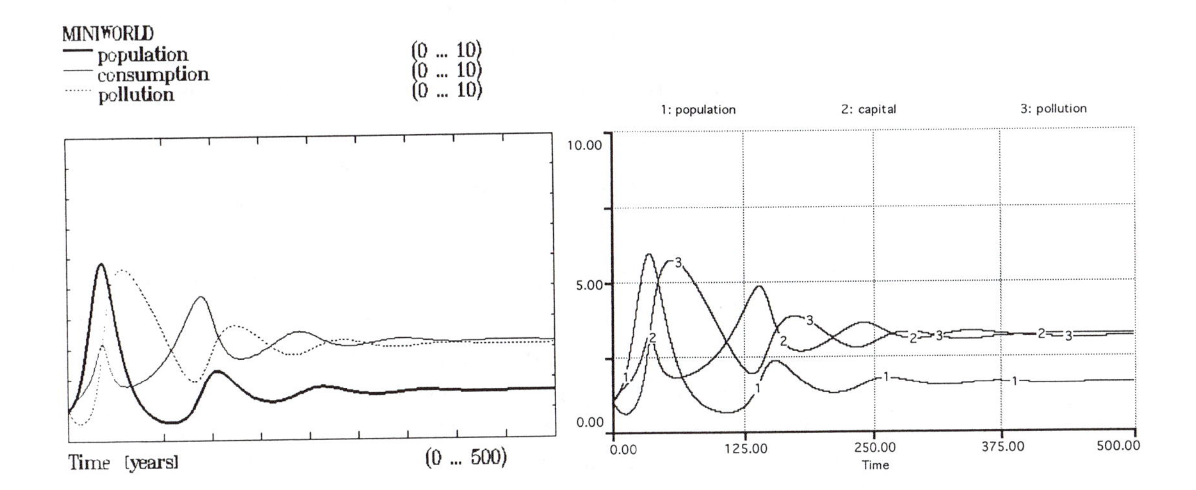

Fig. 5.5: WORLDSIM: Time plots of the simulation results for the standard scenario of the Miniworld Model in SIMPAS (left) and STELLA (right). The results correspond to those of the simple simulation model GLOBSIM in TurboPascal (Fig. 2.16a).

Existence: The existence of a population is threatened if its number falls below a critical value. Similarly, the environment will lose its identity (as the environment with which we are familiar) if its quality drops below a certain minimum. Since we are using relative quantities in the model, where "1" corresponds approximately to normal conditions, we might formulate:

"Existence is threatened if population falls below 0.1."
"Existence is threatened if environmental quality falls below 0.1" .

We therefore define the constraints:

population > 0.1 !
quality > 0.1 !

The exclamation mark is to be read as "should be." If these indicator values drop below the values given, an intolerable threat to the orientor "existence" should be reported.

Effectiveness: As measures of effectiveness, we may use here for example the ratio of environmental regeneration to degeneration (which indicates the "vigor" of the environment), as well as the ratio of environmental quality to industrial capital (which is a measure of how strong and effective the environment remains despite industrial activity) . We could formulate:

"The regeneration rate must never fall much below the degeneration rate."
"Environmental quality must remain strong with respect to consumption."

We therefore use the constraints:

regeneration/degeneration	> 0.95 !
quality/capital	> 0.4 !

Violation of the orientor "effectiveness" should be reported if these indicator variables drop below the values given.

Freedom of action: Assuming that the consumption level provides certain material opportunities, it can be used as an indicator for corresponding orientor satisfaction. The freedom of action is reduced if the environmental quality is poor (and its improvement requires resources). Also, a low life expectancy must be interpreted as restricting freedom of action. We could formulate:

"Consumption must exceed a certain level to assure freedom of action."
"Environmental quality must be high to avoid negative effects on freedom."
"Mortality must be small to allow individuals reasonable freedom of action."

We therefore use the constraints:

consumption	> 0.8 !
quality	> 0.5 !
deaths/population	< 0.02 !

The last condition corresponds to an average life expectancy of 50 years. Violation of the constraints corresponds to violation of "freedom of action."

Security: If the number of deaths is greater than the number of births, the population tends to die out. This is a threat to security. On the other hand, it is also a threat to security if the population grows too fast. Another threat to security follows from growing degradation of the environment. We can formulate:

"Deaths and births should be approximately equal."
"Births and deaths can only deviate a little from each other."
"Security must not be threatened by a lack of environmental regeneration."

We therefore define the constraints:

deaths/births	> 0.9 !
deaths/births	< 1.1 !
regeneration/degeneration	> 0.95 !

Violation of these conditions represents a threat to "security" and must be reported. Note that the last condition looks identical to that under "effectiveness," but it represents a different aspect ("security").

Adaptivity: The system will be better able to adapt if, for example, the environmental quality is relatively good, the population not too large, the growth of the capital stock small or negative, and the industrial capital stock sufficiently large. We can formulate this as:

"Adaptivity requires a relatively high environmental quality."
"Adaptivity will be threatened if the population is too large."
"Adaptivity is threatened if consumption growth is too high."
"Adaptivity requires a certain minimum level of industrial capital stock."

We define the corresponding constraints:

quality > 0.5 !
population < 4 !
capital_growth < 0.02 !
capital > 0.5 !

Threats to the orientor "adaptivity" should be reported if indicators violate these constraints. Note that the condition on "quality" has another meaning as that under "freedom of action."

Consideration of others: In the model, the environment is protected against complete destruction since this would also mean a collapse of the population. It may also be that in addition, the conservation of the environment as a value in itself is desired for ethical reasons going beyond utilitarian considerations. We may formulate this as

"The environment must be preserved in reasonably good shape,"

and write this as a constraint:

quality > 0.5 !

For smaller values of the indicator, a violation of the orientor "attention to others" will be reported. An identical formulation has been used with respect to "effectiveness" and "adaptivity," but remember that in each case it means something different to the system.

These constraints are inserted into the WORLDSIM model by using table functions (see model WORLDVAL, Prog. 5.2). In these table functions, the simplifying assumption is made that if the constraints are met, the contribution to orientor satisfaction increases linearly from 0 to a maximum value of "1" as conditions improve. Note that occasionally several constraints have to be met simultaneously in order to provide a minimum of orientor satisfaction. If even one of these constraints is not met, the orientor remains unsatisfied. As an example for the use of table functions for the computation of contributions to orientor satisfaction, Table Function No. 8, the mapping of the deaths/births ratio on "security" is shown in Figure 5.6.

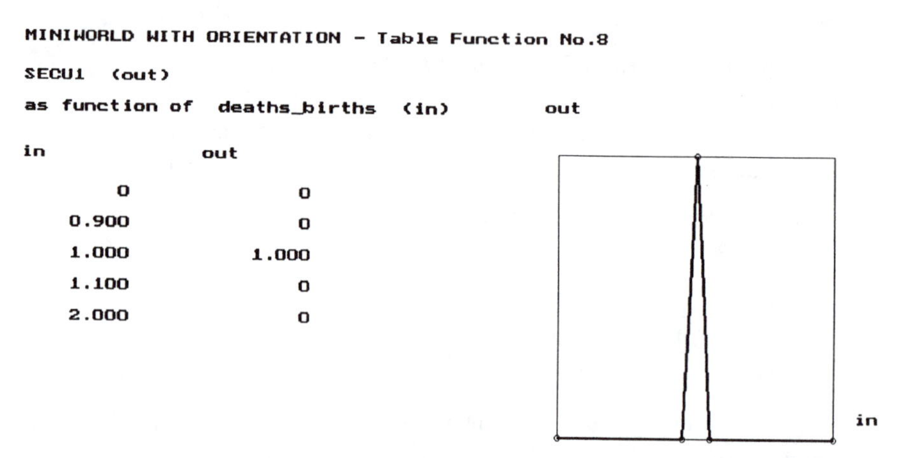

Fig. 5.6: WORLDVAL: Table function for the relationship between the deaths/ births ratio and the orientor "security."

These mappings of indicators on orientors are used in the program to report violations of individual orientors. If the required minimum of all orientors is assured, then we no longer have to worry about individual orientor satisfactions and can try to improve overall orientor satisfaction by raising orientor fulfillment beyond the essential minimum.

The shape of these table functions seems crude, and one may be tempted to formulate "smoother," more "accurate" functions. Actually, the use of precise numerical values in these functions is inappropriate since we are dealing with "fuzzy" mappings. For example, the last constraint (quality > 0.5 !) is really only a poor numerical translation of what was actually meant. ("The environment must be preserved in reasonably good shape.") Fuzzy mathematics provides a better way of expressing such relationships. "Reasonably good shape" could be expressed (for example) by a 20 percent "membership" in "poor shape," 80 percent "membership" in "good shape," and "poor shape" and "good shape" could again be defined by some fuzzy assignment of corresponding indicator values. Fuzzy descriptions are therefore more appropriate for orientation and evaluation processes, and they have proved to be quite successful in the control of complex processes (Negoita, 1985). However, while more appropriate, the use of fuzzy descriptions is not necessary, and the approach we use here is quite adequate in many applications.

To obtain a measure of overall orientor satisfaction, we simply sum up the weighted orientor contributions and divide the sum by the number of orientors. The weights represent assessments of the relative importance of each orientor dimension for the system development.

$$\begin{aligned} \text{SATIS} &= (1/6)\cdot(W_{EXIS}\cdot \text{EXIS} + W_{EFFI}\cdot \text{EFFI} + W_{FREE}\cdot \text{FREE} \\ &\quad + W_{SECU}\cdot \text{SECU} + W_{ADAP}\cdot \text{ADAP} + W_{SOLI}\cdot \text{SOLI})\,. \end{aligned}$$

Overall satisfaction is here denoted by SATIS (satisfaction). The W_n are weights (in WORLDVAL, all $W_n = 1$). The orientors are abbreviated by the following labels: **existence** (EXIS), **effectiveness** (EFFI), **freedom of action** (FREE), **security** (SECU), **adaptiveness** (ADAP), **and regard for others** (SOLI for "solidarity"). This computation of overall satisfaction is inserted into the WORLDVAL Model (Prog. 5.2).

The individual orientor contributions as well as the value for overall satisfaction are normalized to "1" by dividing by the number of individual contributions. In the present case we have, for example, three individual contributions (FREE1, FREE2, FREE3) to the satisfaction of the "freedom of action" orientor. The orientor satisfaction FREE is therefore calculated from:

$$\text{FREE} = (1/3)\cdot(\text{FREE1} + \text{FREE2} + \text{FREE3})\,.$$

If even one of these contributions is not satisfied (equal to zero), the total orientor satisfaction will be set to zero. The formulation above is therefore valid only if all of the contributions are greater than zero which can be expressed by

$$\text{FREE1} \cdot \text{FREE2} \cdot \text{FREE3} > 0\,.$$

All other orientor contributions are treated similarly (see corresponding program statements in model WORLDVAL, Prog. 5.2).

For this particular application, the mapping of the indicator states on the orientors and finally on the overall orientor satisfaction of the system is summarized in Figure 5.7. The exact relationships are found in Program 5.2. The scheme is typical for assessments of this nature. Note that individual indicators may load simultaneously on different orientors—the same indicator may at any time even provide positive contributions to some, and negative contributions to other orientors. Conversely, the satisfaction of individual orientors is usually dependent on acceptable states of different indicators. In more complex assessments, a whole hierarchy with several orientor levels may have to be developed (Bossel et al. 1989).

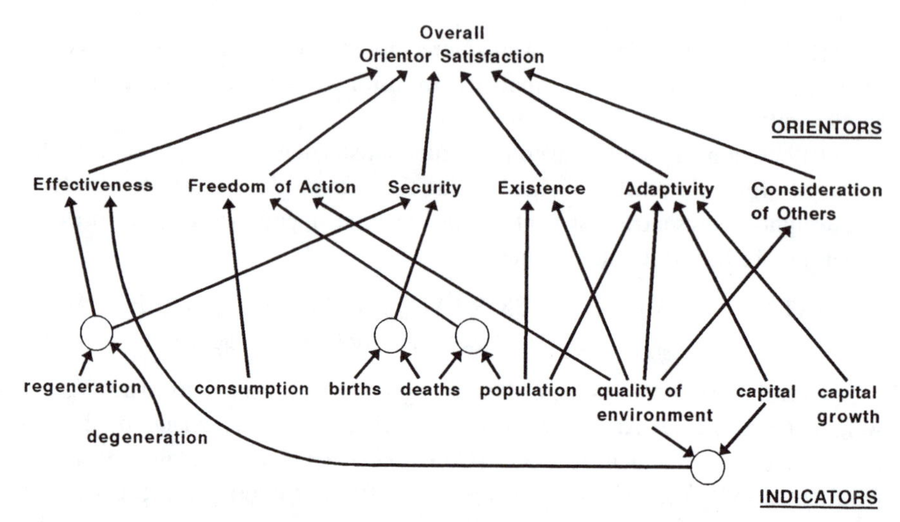

Fig. 5.7: Mapping relationships as used in the WORLDVAL model for relating system indicators to system orientors.

5.2.4 Scenarios and simulation runs

The behavior of systems is partially determined by parameters and inputs from the system environment, both of which may be functions of time. While system parameters are generally well-known (or are assumed to be known system properties), this is not true for inputs from the system environment. These are by definition not influenced by the system itself; statements about their magnitude and time development must therefore come from sources other than the system description. In general, the only option is to make well-founded assumptions concerning their future development. However, the set of assumptions about inputs from the system environment must be plausible and internally consistent. Arbitrary combinations of parameters make no sense. The sheer number of possible combinations demands a bundling of assumptions.

We denote as "scenario" a set of internally consistent and plausible assumptions concerning the future development of exogenous quantities. The formulation and application of scenarios are essential for using complex simulation models efficiently. They should be developed with care.

It is customary to denote scenarios by aggregate, highly descriptive labels. A first scenario candidate is always the (surprise-free) continuation of current conditions (standard scenario, reference scenario, status-quo scenario, business-as-usual scenario). Other scenarios usually explore specific policy combinations.

In our simple Miniworld model we introduced four scenario parameters:

1. birthrate
2. consumption rate
3. birth control
4. consumption control.

Just these four parameters would allow a huge number of possible combinations, most of which would not be plausible (for example, low birthrates in the absence of birth control). On inspection of the computation of "births" in Program 5.2, we note that birthrate and birth control both appear as factors. It is therefore sufficient to change only one of these parameters. We therefore set birth control = 1 (no influence) and restrict ourselves to assumptions concerning the remaining three parameters.

To show some fundamental behavioral trends of the model, we formulate in the following two distinct scenarios which promise to provide some insights into the breadth of the behavioral spectrum of the model under the conditions of interest to us and which allow us to demonstrate the usefulness of the orientor assessment approach.

The **reference scenario** is defined by the default parameter set of the model which should correspond to a continuation of current conditions: high birthrate, high capital growth, and little consumption control:

birthrate	= 0.03	[1/year]
capital growth rate	= 0.05	[1/year]
consumption control	= 0.1	[-]

To contrast this scenario, it would be of interest to define a scenario where the birthrate corresponds to a death rate for high life expectancy, and where further capital (and consumption) growth is constrained. For a constant population, a life expectancy of 80 years corresponds to a death rate of 1/80 = 0.0125. For equilibrium conditions, the birthrate would have to assume this value. Assume that consumption control, the inverse of which corresponds to the saturation of material consumption, be increased fivefold.

We therefore define a contrasting **constraint scenario** with reduced birthrate and strong consumption control :

birthrate	= 0.0125	[1/year]
capital growth rate	= 0.05	[1/year]
consumption control	= 0.5	[-] .

The different behavioral characteristics of both simulation runs become evident from the phase plots in Figures 5.8a and b (500 year simulations).

In the **reference scenario**, population and consumption initially increase strongly (up to a maximum population of 5.932). The high population then collapses, and the solution approaches an equilibrium value in a damped oscillation. Equilibrium is found at values for population of 1.55, capital stock of 3.22, and pollution of 3.01. If the three state variables are plotted as a function of time (Fig. 5.5), it becomes obvious that environmental pollution which follows the increase of the capital stock with a time lag, causes the collapse of the population.

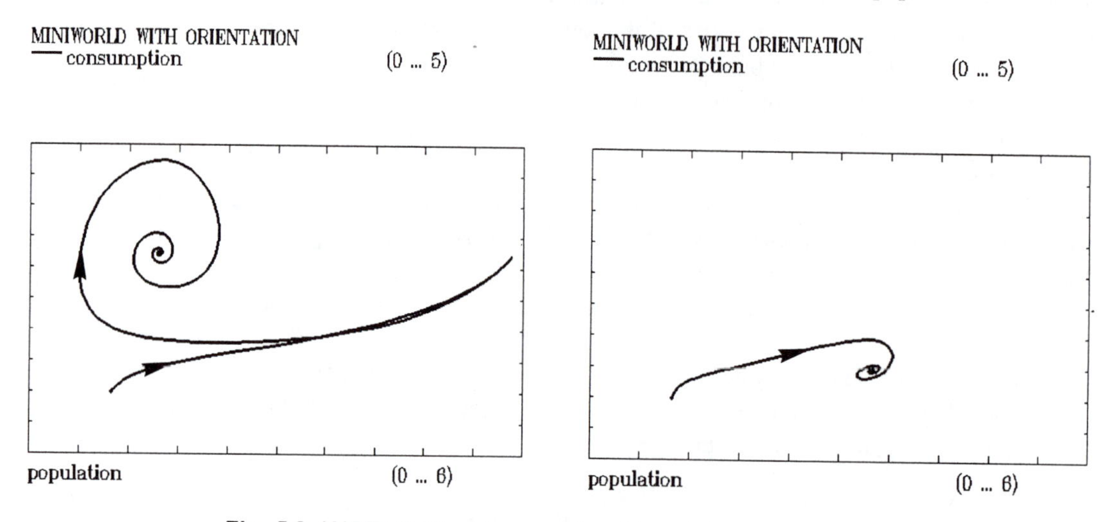

Fig. 5.8: WORLDVAL: Phase plots of the state variables population and specific consumption for the standard scenario (left) and the constraint scenario (right).

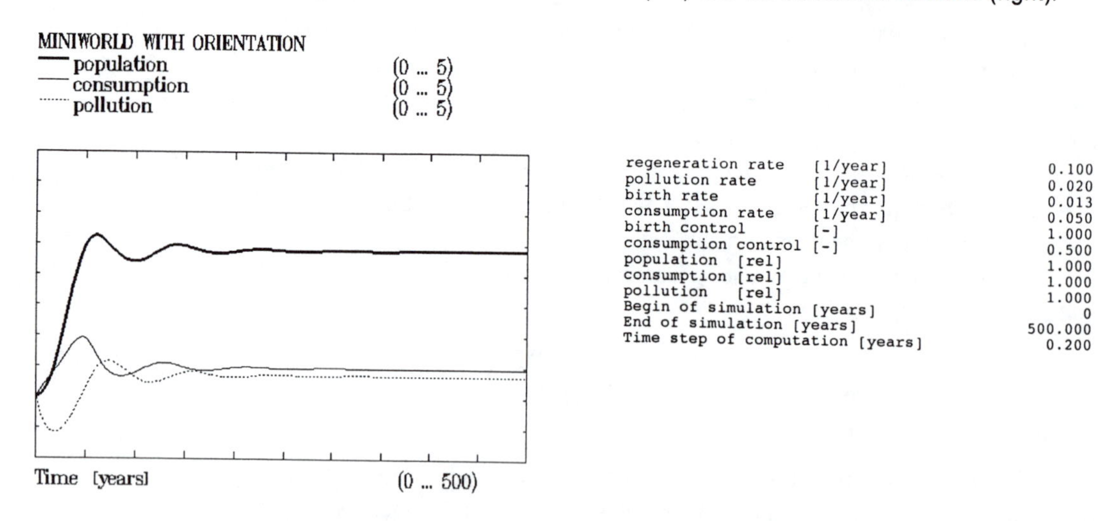

regeneration rate	[1/year]	0.100
pollution rate	[1/year]	0.020
birth rate	[1/year]	0.013
consumption rate	[1/year]	0.050
birth control	[-]	1.000
consumption control	[-]	0.500
population	[rel]	1.000
consumption	[rel]	1.000
pollution	[rel]	1.000
Begin of simulation	[years]	0
End of simulation	[years]	500.000
Time step of computation	[years]	0.200

Fig. 5.9: WORLDVAL: Time plot of simulation results for the constraint scenario.

In the **constraint scenario** (Fig. 5.9) a collapse does not occur. Population and capital stock grow continuously and then move in a highly damped oscillation to an equilibrium value of population = 3.933 and capital = 1.473. Environmental pollution exhibits only a gradual increase to an equilibrium value of 1.357.

5.2.5 Comparative evaluation of the simulation runs

For the simple global model with its few variables, comparison of the two simulation runs is still easy, and a mapping on the orientors is really not required. However, for models of higher complexity with many state and other output variables, comparative evaluation becomes difficult as long as it must be based on comparison of these variables alone. Which combination of state values must be considered "better" or "worse" than another? Which values of state variables are admissible and which are not? Does the system run into a state which endangers development and survival without this becoming obvious in the major variables?

The evaluation task is simplified if the system state is mapped on the orientors to determine current orientor satisfaction (Fig. 5.7). This allows an assessment of the state with respect to the overall interests of the system. The relationships between the model indicators and orientors were defined Section 5.2.3.

The results for both scenarios (reference scenario and constraint scenario) can be compared by printing out the orientor satisfactions (here in intervals of 50 simulation years) (Fig. 5.10a,b). In both cases the existence of the system is secure for the simulation period but the other orientor satisfactions differ enormously.

In the reference scenario all orientors except "existence" are unsatisfied for almost the entire simulation period (500 years). In the constraint scenario, however, only two of the orientors (security and adaptiveness) show some initial problems; thereafter all orientor satisfactions show acceptable values.

This result provides a clear indication that the constraint scenario would be of great advantage over the standard scenario for our Miniworld. However, this does not mean that this constraint scenario would be the best possible development in terms of orientor satisfaction. In the search for further improvements, it makes sense to compare overall system satisfaction for the different scenarios.

To search for a better solution, the SIMPAS option "Parameter Sensitivity" was used to investigate the effect of parameter changes on the overall satisfaction (SATIS) (Fig. 5.11). We find here that further improvements of overall satisfaction are possible by a further increase in consumption control (lowering the saturation level of consumption). Phase plots for the three state variables show clearly (Fig. 5.12a,b) that the states with high orientor satisfactions correspond to different equilibrium conditions of population, consumption, and pollution.

MINIWORLD WITH ORIENTATION

	existence [-]	effectiveness [-]	freedom [-]	security [-]	adaptability [-]
(a)	1.000	1.000	0.722	0	0
	0.911	0	0	0	0
	1.000	0	0	0	0
	1.000	0	0	0	0
	1.000	0	0	0	0
	1.000	0	0	0	0
	1.000	0	0	0	0
	1.000	0	0	0.236	0
	1.000	0	0	0.446	0
	1.000	0	0	0.505	0
	1.000	0	0	0.556	0
(b)	1.000	1.000	0.722	0	0
	1.000	0	0.881	0	0.639
	1.000	0.264	0.619	0.795	0.543
	1.000	0.088	0.593	0.178	0.519
	1.000	0.134	0.638	0.348	0.540
	1.000	0.134	0.610	0.490	0.529
	1.000	0.125	0.622	0.554	0.533
	1.000	0.131	0.618	0.599	0.532
	1.000	0.128	0.619	0.595	0.532
	1.000	0.129	0.619	0.596	0.532
	1.000	0.129	0.619	0.598	0.532

Fig. 5.10: WORLDVAL: Comparison of orientor satisfactions for the standard scenario (above) and the constraint scenario (below).

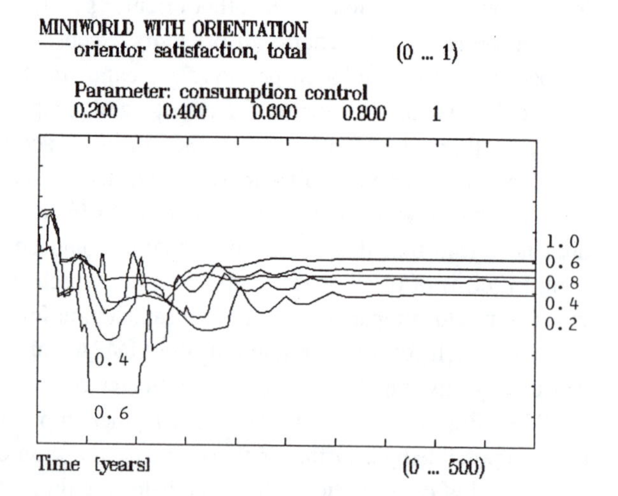

Fig. 5.11: WORLDVAL: Development of the overall orientor satisfaction in the constraint scenario as a function of the parameter "consumption control".

In the example of the Miniworld, the scenario parameters were not changed as functions of time during the simulation. For many practical investigations, this assumption of constant parameters is not realistic. It would be more realistic, for example, to assume a gradual decrease of the specific birthrates from a high level today to a lower equilibrium value later. Or to assume that the introduction of energy and resource conserving technologies changes the saturation value for consumption (*via* consumption control) over a certain period of time.

Time dependence of the parameters can be included in a simple way by time-dependent table functions. Otherwise, the procedure of producing and evaluating the simulation runs, in particular, the comparative evaluation of orientor satisfaction, is identical to the procedure just described.

In this section, we have used a simulation model to determine the possible paths of development of a system for different scenarios and to assess their respective impacts on the vital interests of the system. In the next section, we will use a simulation model to help us find the policy inputs for achieving a certain development path which is seen as desirable, usually because it is in some sense "better" or even "optimal."

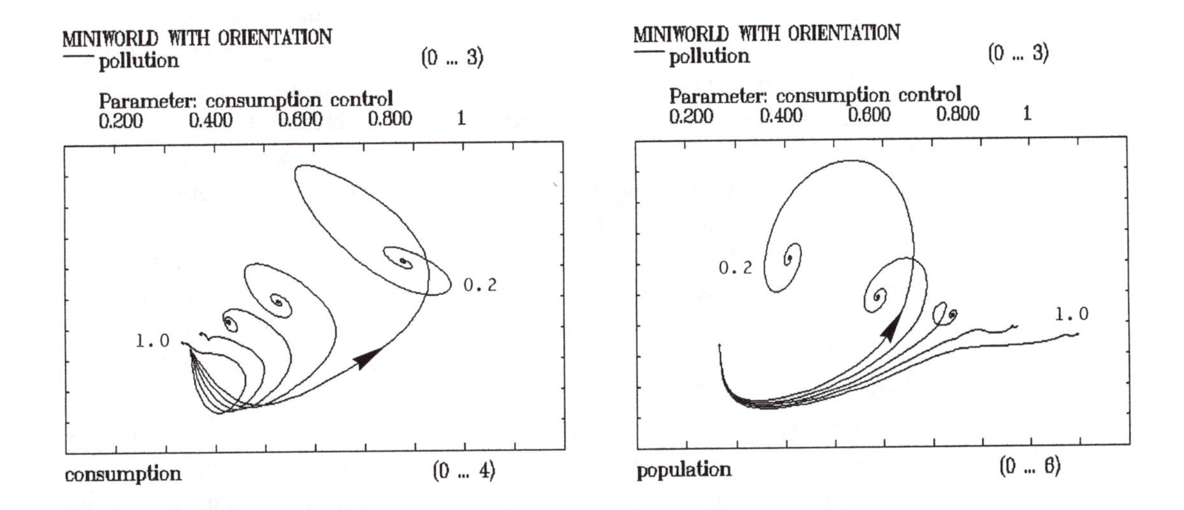

Fig. 5.12: WORLDVAL: Dependence of the points of equilibrium for pollution, consumption, and population on the parameter "consumption control" in the constraint scenario.

5.3 Policy Analysis

5.3.1 Introduction

Often, the desire to find an "optimal" solution to a control, administrative, or management problem is the reason for the development of a simulation model. In many cases it is indeed possible to define a unique quality measure whose relative level of satisfaction can be computed at any time using corresponding indicator variables from the model system. The remaining task then consists of programming a systematic search for the parameter constellation that provides an optimal solution, subject to the given constraints. A large number of numerical procedures which will not be discussed here are available for this purpose. The reader is referred to the extensive literature on optimization (Schwefel, 1981).

While the desire to find an "optimal" solution is understandable, we should also be aware of the problems and pitfalls of optimization. Briefly stated, optimization is permissible only if there is perfect knowledge about the system and its performance as a function of parameter changes and about the future development of the environment and the inputs it delivers to the system. Also, it must be guaranteed that the "optimal" set of system parameters and system inputs will not erode system viability and resilience: all system orientors must remain above their minimum. This means that optimization can be applied to problems like minimizing the costs of production or finding better control strategies for technical systems but that it cannot usually be applied to the "messier" real-world problems of social and ecological systems and their paths into an uncertain future.

Nevertheless, the basic ideas of optimization can help in finding "better" solutions in the routine business of policy analysis, provided some quality measure can be defined. The mapping of indicators on orientors and the computation of individual and aggregate satisfaction values can provide such a measure.

Different values of the quality measure will result from the different possible parameter constellations of a system. If we locate a particular parameter combination as a point in the two-dimensional parameter plane, then the quality measure can be plotted as a "height" over that point such that a "quality mountain" with "peaks" and "valleys" evolves. It is the task of the optimization program to find the highest peaks as quickly as possible (usually in n-dimensional space). Most programs first probe the entire range of parameters and then concentrate on areas where "peaks" are to be expected. The program "climbs" a peak, for example, using the method of steepest ascent following the maximum gradient until no further improvements can be found (meaning that the top has been reached). However, if one peak is found, one has to make sure that it is not just a local optimum: there may be other peaks that are even higher.

A different approach to numerical optimization (evolution strategy and genetic algorithms) is formulated following the processes of evolution (Holland, 1975; Schwefel, 1977; Kirkpatrick, 1983). In this case the parameters are changed by small random perturbations, analogous to mutations of organism. If this results in improvements of the quality measure, the successful parameters are adopted and serve as the points of departure for further search.

In many practical cases it is not only necessary to observe a number of constraints simultaneously but also several distinct quality measures that cannot be aggregated to a single criterion. (Example: maximization of profits with simultaneous minimization of ecological destruction and maximization of social benefits.) In dealing with the basic orientors, we found that they cannot be aggregated as long as the necessary minimum of each orientor satisfaction is not guaranteed. Only if this is the case can aggregated quality measures be defined; the result of the aggregation depends on the chosen weights, however. Questions of optimization under simultaneous consideration of several quality criteria are dealt with by poly-optimization (e.g. Peschel and Riedel, 1976; Steuer, 1986).

Here we shall only deal with the formulation of the optimization task using a dynamic system model, i.e. the definition of an applicable quality measure and of the constraints to be met, as well as the definition of indicator quantities to compute the value of the quality measure. With these additions to the simulation model, a systematic search for the optimum (or the optima) can then be conducted. We shall only conduct a manual heuristic search here in order to explain the procedure. To demonstrate the procedure, we shall try to find optimal management strategies for the model of fishery dynamics.

5.3.2 Constraints and quality measures for the fishery optimization

In Chapter 4 we studied two versions of the fishery model. In the first version, the fish catch was dependent on the current fish density. Overfishing eventually reduced the fish catch such that the boat fleet had to be reduced because of insufficient proceeds from the fish catch. This then led to an eventual recovery of the fish population. A complete collapse of the fish population was therefore not possible; the system is inherently stable.

In the second version, an efficient fish locator technology was introduced that allowed finding and catching of the remaining fish even if the fish density was low—until the population was reduced to a point where it could not recover. This system was inherently unstable and could only be stabilized by constraints on the number of fishing boats and on the maximum catch capacity of each boat.

Even then, a breakdown is possible if the fish population drops below a certain critical limit (for example, by natural processes). For this density-independent fishery system, the orientor "security" could therefore only be satisfied if strict adherence to the catch limits was assured, and if, in addition, the system would not be subject to the slightest natural fluctuations. Since these conditions are not realistic, there is always the risk of a sudden collapse for this type of system.

We shall now modify our original simulation program for both of these cases (FISHERY, Prog. 4.5) by adding criteria which allow us to search for a management policy which is in some sense "optimal" or at least "better." The result will crucially depend on the choice of quality criteria.

The optimization task therefore has to deal first with the question of what has to be optimized. From the standpoint of providing food for a human population, the annual fish catch should be maximized. However, if one considers a fishing company and its economic success, profit maximization would be the major concern.

The next question concerns the time horizon of optimization: a year, a decade, a century, forever? Obviously, the result depends on this decision: if the time period is limited, optimization will always lead to the result that at the end of the period all remaining resources will be exploited. However, if we demand an optimal yield "forever," i.e. "sustainability," then this means a continuous utilization of the resource at an equilibrium point where the state variables are in flow equilibrium: fish would be caught at a constant rate corresponding to their reproduction capacity. In the present case we shall aim for the sustainable solution with an indefinite time horizon.

Note that the idea of sustainability is inherent in the orientor concept we discussed in previous sections: satisfying the required orientor minima means survival and development of the system, hence sustainability.

As a condition for sustainable use of a system we should therefore introduce the sustainability condition, saying that neither the fish population nor the number of boats should change under equilibrium conditions:

$$\begin{aligned} d(\text{fish})/dt &= 0! \\ d(\text{boats})/dt &= 0! \end{aligned} \qquad (K_0)$$

We have to add the constraint that the fish population will not disappear, i.e. we must require that fish > 0 for all time.

The maximization of the sustainable catch without consideration of the economic conditions of the fishermen is certainly not a realistic solution. We could simulate this case (for example, for a state-subsidized fishery fleet) by deleting all economic computations and by coupling the boat growth rate directly to the fish catch. (The boat loss rate by aging and decommissioning remains.)

Under realistic economic conditions, by contrast, one would have to maximize the sustainable net profit. Since sustainability implies remaining at an equilibrium point (EP), the rate of profit will be constant (profit per time unit). For an optimal result at the point of equilibrium, we would then have the requirement that the profit rate (the annual net gain) should have a maximum:

$$\text{profit\_rate}_{EP} = \max! \qquad (K_1)$$

Note that the optimization problem can always be reduced to either a minimization or a maximization problem since finding the maximum of a function X amounts to finding the minimum of its negative (–X):

$$\max (X)! = \min (-X)!$$

Normally, the initial state of the investigation will be quite different from the equilibrium state reached later. There is a multitude of possible state trajectories leading to the equilibrium state. In the fishery problem, different paths would result in different developments of the profit rate which would result in different accumulated profit in thc initial phase before the equilibrium point is reached. We could therefore also formulate the optimization criterion as a time integral:

$$\int_0^T \text{profit\_rate}(t)\, dt = \max! \qquad (K_2)$$

Here T is the time period until the system state has come sufficiently close to the equilibrium point with profit rate$_{EP}$. The time period T itself could be defined as a quality criterion:

$$T = \min! \qquad (K_3)$$

In order to consider problems of food supply, a further criterion could be

$$\text{fish\_catch}_{EP} = \max! \qquad (K_4)$$

or a criterion to optimize the initial adjustment to the equilibrium point

$$\int_0^T \text{fish\_catch}(t)\, dt = \max! \qquad (K_5)$$

Depending on management interests, different formulations of quality criteria (here: K_1 to K_5) are obviously possible and meaningful. Also, several of these quality criteria may have to be optimized simultaneously. This is only possible if compromises can be made using a sum of weighted criteria.

Since the individual criteria contributions K_i may have different orders of magnitude, they first have to be normalized by reference to their possible maximum values or other reference values K_i^*, before formulating a weighted criteria sum. In order to normalize this sum, it is divided by the sum of the weights:

$$k = \Sigma_i\, w_i \cdot (K_i/K_i^*) \,/\, \Sigma_i\, w_i \qquad (5.1)$$

The (dimensionless) quality criterion k now has the order of magnitude "1." (By multiplying it with 100 this would result in an index of the order of magnitude "100"). The prescription for optimization could now be

$$k = \max!$$

If individual criteria K_j are to be minimized, the corresponding term (K_j/K_j^*) in Eq. 5.1 must be replaced by its inverse (K_j^*/K_j).

5.3.3 Extension of the simulation model for optimization studies

In order to use the fishery model to demonstrate differences between a fish catch optimization and an economic optimization and to study compromise solutions between both extremes, we first define a (dimensionless) quality index (order of magnitude "100"). (The following equations are written as Pascal statements.)

```
quality := 100 * ((catch_weight * rel_fish_catch) + (profit_weight *
      rel_profit_rate)) /(catch_weight + profit_weight);
```

For the definition of the normalized quantities "rel_fish_catch" and "rel_profit_rate" (in procedure ModelEqs), we use an annual fish harvest corresponding to the maximum fish capacity (fish population at the capacity limit):

```
rel_fish_catch  := fish_catch/max_fish_capacity;
rel_profit_rate := profit_rate/(fish_price*max_fish_capacity);
```

These relations are introduced into the model FISHERY (Prog. 4.5).

The criteria weights are introduced as scenario parameters into the interactive query:

```
ScenaQuestion[6] := 'catch weight: (0) [-]';
ScenaQuestion[7] := 'profit weight: (1) [-]';
```

Corresponding definitions of program variables have to appear in the procedure ModelEqs:

```
catch_weight := ScenaAnswer[6];
profit_weight := ScenaAnswer[7];
```

The computation of the profit rate has to be inserted into the model (ModelEqs):

```
profit_rate := net_income – investment_capital_boats;
```

The profit rate and the quality index should be available as output quantities:

```
OutVariable[2] := profit_rate;
OutVariable[3] := quality;
```

They must therefore also be defined in the procedure InitialInfo:

```
OutVarText[2] := 'profit_rate: [$/year]';
OutVarText[3] := 'quality_index: [-]';
```

With these extensions to the fishery model (model FISHOPT, Prog. 5.3), simulation runs with different weights of fish catch and profit rate can be conducted. Since we are not using a formal optimization program, we have to find optima from a multitude of runs. This can be done by employing the SIMPAS option "Parameter Sensitivity" which shows the results of five simulations in comparison.

Before we do so, we have to clarify which parameter is to be used in the optimization. In principle, all system and scenario parameters (except for the weights) are available. However, we shall assume here that except for the parameter "investment_fraction_boats" (net income fraction which is reinvested in the purchase of boats), all other parameters remain constant. We will therefore only change this one parameter and the relative weights of the fish catch and profit in search for optimal solutions.

5.3.4 Search for optimal investment when fishing without fish locators

A fishing company working under economic conditions would have to determine what percentage of its annual net profit should be invested in the annual purchase of new boats. A large number of boats means high maintenance and operating costs cutting into the profit but a larger number of boats could produce bigger catches. On the other hand, if the boat number is small, the catch rate will be correspondingly low, and the net profit will also be low.

For the following investigations we initially use the default parameter settings of model FISHOPT (Prog. 5.3). For density-dependent fish harvesting, the fish population cannot collapse even for a larger number of boats (as shown in Ch. 4). For the following, we use an initial value of 25 boats and a simulation period of 250 years.

If the equilibrium points for different investment fractions (in the plot of profit rate as a function of fish catch and boat number) are connected by a curve (Fig. 5.13a,b), a maximum of the profit rate (about 470,000 $/year) at an investment fraction of 0.25 (25%) and a boat number of 23 is obtained. In this case the fish catch (around 1800 t fish/year) is significantly below the maximum sustainable yield of 2500 which would be possible if no attention was paid to economic considerations. This larger fish catch is obtained with an investment fraction of 1 (100%) and 43 boats.

Figure 5.14a and b show how the optimum moves if the criteria "profit" and "fish catch" are weighted differently. If the fish catch is weighted five times as heavily as the profit, the optimum of the quality index shifts to a high investment fraction of > 0.9 and a correspondingly high boat number of about 45 (Fig. 5.14a). If on the other hand profit is weighted five times as high as the fish catch, the optimum of the investment fraction is at about 0.3 with a boat number of about 27 (Fig. 5.14b). It becomes obvious from these simulations that the results of the policy search are quite dependent on the definition of the quality index used. (It is always a good idea to ask for definition of the quality index if someone tries to sell you the "best" solution!).

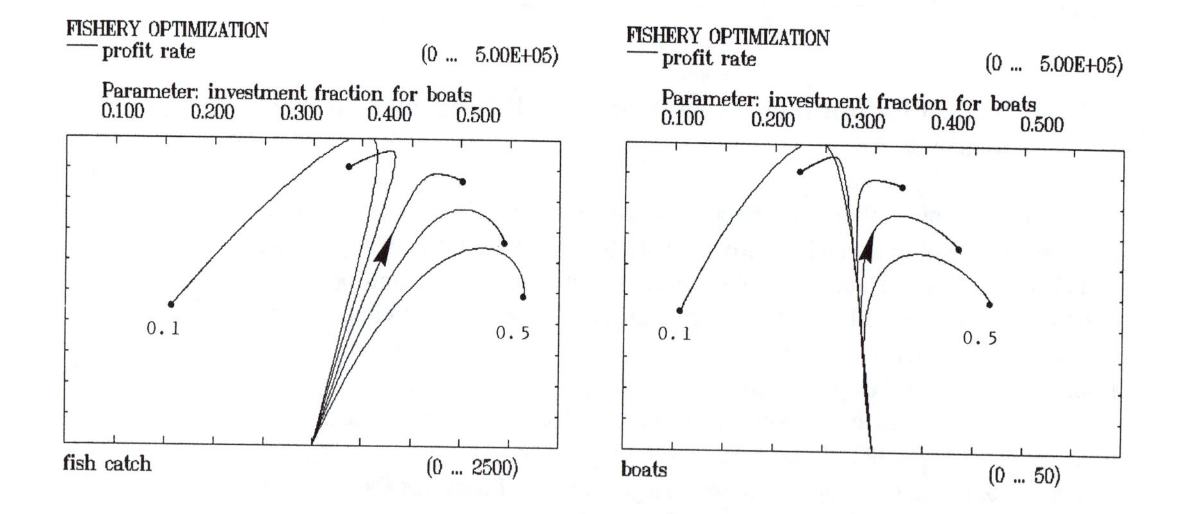

Fig. 5.13: FISHOPT without fish locator technology: Dependence of the equilibrium points of profit rate and fish catch (left) and boat number (right) on investment_fraction_boats. Maximum profit rate is found for an investment fraction of about 0.25.

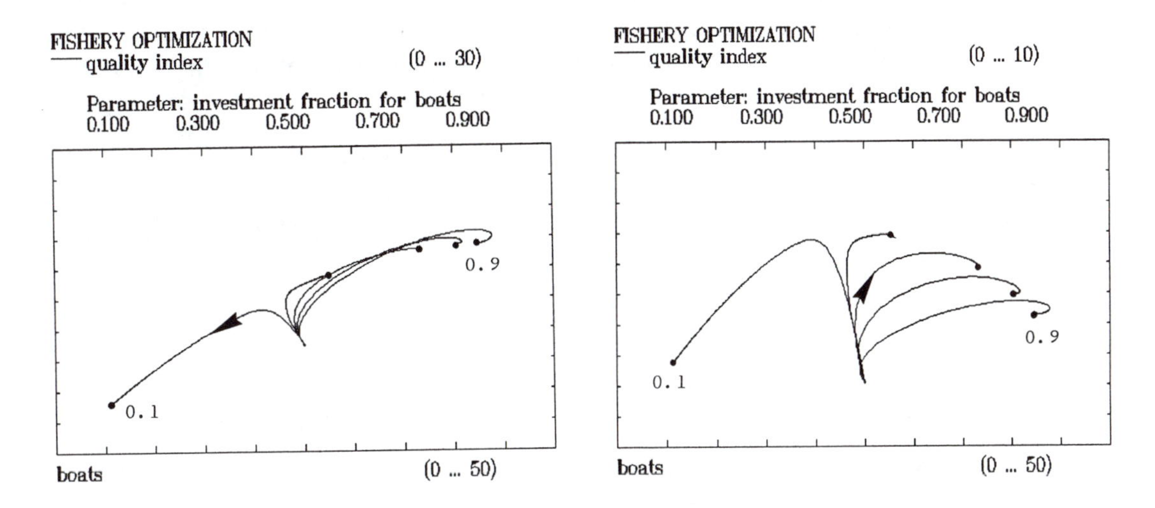

Fig. 5.14: FISHOPT without locator technology: Shifting of the optimal value for the quality index as a function of the relative weights of fish catch (left: 5, right: 1) and profit (left: 1, right: 5).

5.3.5 Search for optimal investment when fishing with fish locators

If fish locator technology is used, the behavior of the fishery system changes radically as already shown in Chapter 4. This can be seen from the phase plots in Figure 4.17 and 4.22. For (density-dependent) fishing without locator technology, the fish population cannot collapse (Fig. 4.17). For (density-independent) fishing with fish locator technology, only a strict limit on the number of boats and their annual catch can lead to an equilibrium state and sustainable yield (Fig. 4.22). If the fish population is initially small, it will collapse completely even for an initially small boat number. In the following investigations, we therefore choose a small initial value of 10 for the boat number and optimize exclusively with respect to the profit rate. The "locator technology" parameter is set to "1" and the time period of simulation to 100 years.

We first focus on finding the investment fraction for an optimal profit rate and subsequently search for the boat number limit promising maximum profit. If the boat number is initially restricted to 25, an optimal investment fraction is determined at 0.3. The subsequent investigations are conducted using this value. Plotting the profit rate as a function of fish population (Fig. 5.15) shows an increase in profit (for boat numbers from 30 to 34) which however for a boat number >32 does not lead to a stable equilibrium solution but rather a collapse of the fish population. The optimal result is achieved for a limitation of the boat number to 31, giving a profit rate of about 680,000 $/year.

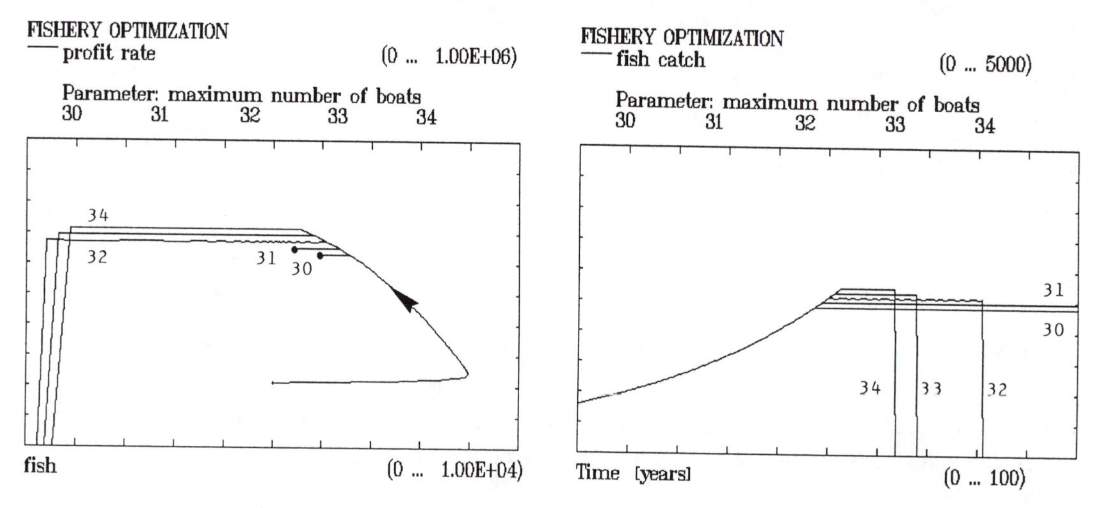

Fig. 5.15: FISHOPT with fish locator technology: Profit rate and sustainability as a function of the maximum boat number parameter. The optimal result is found exactly at the stability limit.

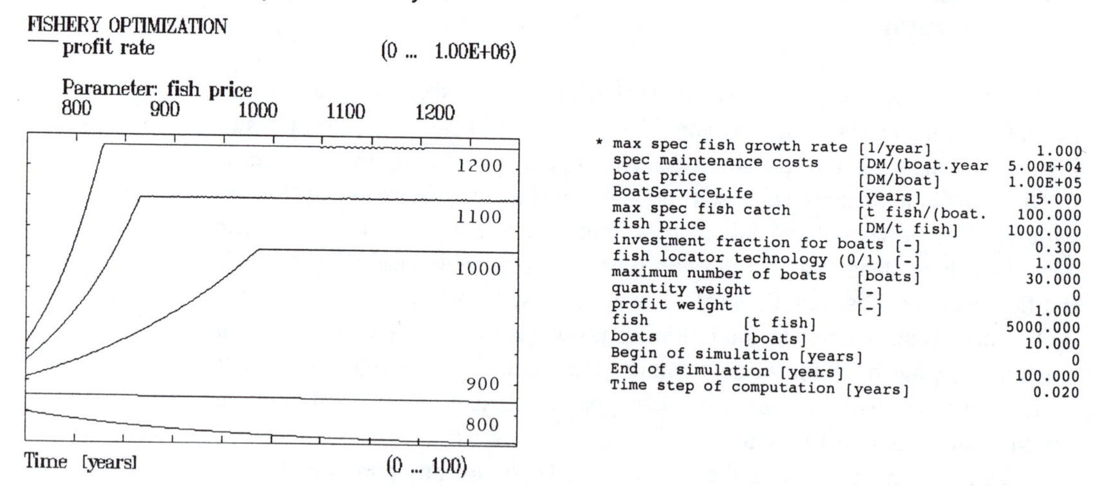

Fig. 5.16: FISHOPT with fish locator technology: Strongly non-linear dependence of profit rate on fish price.

Comparison with the optimization for density-dependent fishery (profit rate 470,000, 23 boats), shows a significantly better economic result when using locator technology but this is achieved at the cost of operating at the stability limit of the system. A small amount of overfishing, or even a fluctuation in fish reproduction, will lead to collapse of the fish population (and therefore collapse of the fishing industry). If we introduce an upper limit of 30 boats, the investigation of the profit rate as function of fish price (Fig. 5.16) shows a strongly non-linear relationship: fishing then does not pay anymore if the price is less than 800 $/t fish.

5.3.6 Optimization over a time path, and using system orientors

In the examples shown here, we have searched for optimal parameter values under equilibrium conditions without paying attention to the development path from the initial state to the equilibrium state. In reality it will usually be necessary to consider the transition from the initial state to the equilibrium state. In order to find the conditions (parameter settings) that would produce a short transition period and maximum accumulated profit, the quality index will have to be written as a weighted time integral. This integral would have to be optimized over the time period from the initial state to the desired equilibrium state (for example, a certain number of boats). For example,

$$\text{transition quality} := \int_0^T (\text{profit\_weight*rel\_profit\_rate} - \text{time\_weight})\, dt/T$$

The formulation of time-integrated quality criteria was discussed at the beginning of this section. The negative sign takes into account that the elapsed time should be minimized; the overall index must therefore be maximized.

The discussion of orientors in Section 5.1 stressed the point that a complete and "responsible" assessment of system development should take into account all of the relevant aspects of a system's interaction with the environment. Only in this way can the "interests" of the system in its long-term survival and development be correctly included in long-term planning and policy analysis. The orientor assessment can also be a basis for optimization. The task then amounts to finding maximum aggregate orientor satisfaction under the constraint that basic orientors should always at least fulfill their minimum levels.

However, in transition processes (for example, from a "growth economy" to a "sustainable economy") there may be periods where temporary violation of certain orientor satisfactions must be tolerated for a limited time in order to reach a sustainable state with high overall orientor satisfaction later. In this case the concept of optimization over a time path, discussed above, also applies.

5.4 System Design

5.4.1 Introduction

As natural systems evolve, their environment forces them to respect the basic orientors. This assures that only systems (organisms, ecosystems) with a viable, robust, and adaptive system structure (**f**, **g**) survive. However, human impacts (for example, environmental pollution) have often disturbed or even destroyed the system structure and function to the point where systems fail. For example, in agriculture and forestry where ecological systems have been massively changed by humans, constant human control, regulation, and stabilization are now necessary in order to keep these systems in the desired state (weed and pest control, irrigation, fertilization, protection from erosion).

The control and stabilization task is even more obvious and necessary in many technical systems which would be hopelessly unstable without specific structural additions such as control components or specific control systems and which could not be operated safely without them. Examples are: aircraft, chemical reactors, and nuclear power stations. It is not always possible to achieve inherent stability that would remain active even after loss of all power sources (as for example, in aircraft where it is achieved by careful balancing of the different forces and moments).

In still another area, society is constantly creating new and often large institutions, organizations, and legal systems or is forced to modify existing ones. These are systems which have memories and stocks and therefore have characteristic eigendynamics which may be at odds or even counteract the processes they are supposed to facilitate. Examples are: new international organizations, environmental laws, and political and economic reorganization in many countries.

In these cases new systems have to be designed, or existing systems have to be modified, to ensure safe and stable operation while fulfilling certain criteria.

It is certainly possible to improve the performance of systems by structural modifications, and in particular it is even possible to stabilize unstable systems by suitable additions to the system structure. It is the task of system design to devise better systems or to modify existing ones to better meet specified criteria of performance and stability. In particular, the important task of analysis and synthesis of control systems is handled by control systems theory.

The task of stabilizing a system by structural additions is directly related to the basic orientors. A structural addition is expected to guarantee the "security" orientor of the system by conserving its stability which also means securing its "existence" orientor. The control process itself is expected to be efficient, i.e. it should be quick and require only a minimum amount of energy and time ("effectiveness" orientor). If possible, the control system should also be able to adapt to changing environmental conditions ("adaptivity" orientor). The control process should create "freedom of action" (orientor) such that the operator (for example, the pilot or the manager) as part of the total system is not constantly occupied by stabilization tasks but can concentrate on other tasks.

In this section, we shall be concerned with stabilization of a hopelessly unstable system, i.e. the rotating pendulum (discussed in Ch. 3, 4) at its upper dead center. We will first design a possible structural addition that could lead to stabilization. For this new structure, we then derive the equations of motion which turn out to be highly non-linear. Assuming that we will be successful in solving the stabilization task and that the pendulum will then remain in the neighborhood of the upper dead center, we can simplify the equations of motion significantly by linearization. We then design a control function that depends on the state variables of the system; initially the corresponding control parameters will be left unspecified. After these steps, we are equipped with the necessary equations for the corresponding simulation model. In multiple simulations, we now search for combinations of control parameters producing good stability behavior. In this search the introduction of a quality criterion for the efficiency of control (minimum energy consumption) is of some value. The whole process is typical of the tasks to be solved by system design.

5.4.2 Stabilization by a modified system structure: system equations

From experience (balancing a broom) we know that an unstable system like the rotating pendulum can be stabilized at its upper dead center if the point of rotation is moved sideways quickly and at the right moment such that the motion counteracts and catches the falling motion. Movement of the point of rotation is therefore necessary for stabilization.

In order to stabilize the pendulum, we could mount its point of rotation on a small carriage whose wheels could be driven forward and backward by a small (battery-driven) electric motor. (We use the assumption that the pendulum can only move in one plane of rotation.) Sensors are mounted on the carriage and on the pendulum, providing instantaneous information about the system state to the controller directing the motion of the carriage. We have to determine the relevant state variables and the response of the controller to the system state.

We note that now our system has become more complicated and that the description by the system equations of the non-linear rotating pendulum is no longer sufficient. The complete system now consists of the pendulum *and* the carriage with its drive controlled by the control system. Since the carriage has a certain mass, it also has to be accelerated (or decelerated) for each corrective motion of the carriage. Carriage and pendulum have to be considered as a dynamic unit, and the equations of motion must be derived again from the beginning.

This system is shown in Figure 5.17. The pendulum of length L with pendulum mass m is supported at its lower end on a carriage of mass M such that it can rotate in one plane (paper plane). The angle of rotation β will change with time. In order to catch the rotating and falling motion of the pendulum, the carriage can be moved to the right or left in the x-direction (paper plane) by the control force u. The task is to find a control function that, in response to the rotational motion of the pendulum and the translational motion of the carriage, leads to a stabilizing control force $u(t)$.

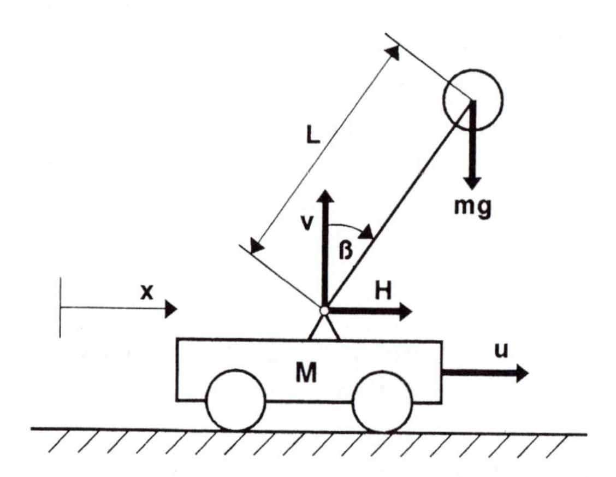

Fig. 5.17: Pendulum balancing system: Stabilization at the upper dead center by horizontal motion of the carriage.

The equations of motion are found from the (usual) condition that the sums of the forces and moments at the pendulum and at the carriage must vanish. To derive the equations, the total system is "cut" at the rotational support bearing; the reaction forces in the horizontal direction H and in the vertical direction V at this point must be considered in the corresponding equations.

The horizontal acceleration force at the carriage is equal to the control force minus the horizontal support force H caused by the pendulum. At the pendulum itself, this horizontal support force is equal to the horizontal component of the acceleration force of the pendulum's center of gravity. The vertical acceleration force at the pendulum's center of gravity is given by the vertical support force V

minus the weight force of the pendulum acting at the pendulum's center of gravity. In order to describe the dynamics completely, the sum of the moments is written around the pendulum's center of gravity: the acceleration moment is equal to the sum of the moments resulting from the vertical and horizontal support forces and their respective distance from the pendulum's center of gravity.

Forces at the carriage (horizontal only):

$$u - H = M\, d^2x/dt^2 .$$

Forces at the pendulum:

$$H = m\, d^2(x + L \sin ß)/dt^2$$
$$V - mg = m\, d^2(L \cos ß)/dt^2 .$$

Sum of moments around the pendulum center of gravity:

$$V L \sin ß - H L \cos ß = I\, d^2ß/dt^2.$$

On the right hand side, we now find the product of the moment of inertia I of the pendulum and its angular acceleration. Assuming that the pendulum mass is concentrated in a sphere of radius r and the connecting rod itself is weightless, we have for the moment of inertia:

$$I = (2/5)\, m\, r^2 .$$

(For a stick of mass m and length m L, we would have $I = (1/12)\, m\, L^2$.)

In the equations for the horizontal and vertical forces at the pendulum, we find the second derivatives of the center of gravity coordinates with respect to time. Carrying out the differentiations of these terms, we obtain four relatively complex equations of motion for the unknowns $ß$, x, V, and H. The non-linear expressions in the equations render mathematical analysis next to impossible.

At this point we have to remember the purpose of the investigation: the objective is not the precise description of the complex sequence of motions even for large angles of rotation but it is rather to find a controller that would keep these angles very small. Assuming that we will be able to solve this task, we introduce into the equations of motion the assumption that the angle will remain small (few degrees) as a consequence of the control action. The equations can then be simplified considerably.

Under these conditions, we can apply the fact that the sine of a small angle is approximately equal to the angle itself (in radians) and that the cosine will be approximately equal to 1. Introducing these assumptions into the system of equations results in significant simplifications. However, there are still non-linear terms. These terms are products of the angle of rotation and of its first and second derivatives with respect to time (angular velocity and acceleration).

Since according to our assumption the angle always remains relatively small, the angular velocity will also be relatively small. If two small quantities like these are multiplied, the product itself will be quite small and can therefore be neglected for the purposes of the investigation. With this we now obtain four relatively simple equations from which we can also eliminate the support forces H and V, finally leaving two differential equations for ß and x:

$$(I + m L^2)\, d^2ß/dt^2 + m L\, d^2x/dt^2 - m g L ß = 0$$
$$m L\, d^2ß/dt^2 + (m + M)\, d^2x/dt^2 = u\,.$$

Solving these equations with respect to the highest derivatives, we can write the system equations in the simple form

$$d^2ß/dt^2 = a ß + b u$$
$$d^2x/dt^2 = c ß + d u\,.$$

The following abbreviations were used in the set of equations:

$$A = I (m + M) + m ML^2$$
$$a = g (m + M) m L / A$$
$$b = - m L / A$$
$$c = - g m^2 L^2 / A$$
$$d = (I + m L^2) / A\,.$$

(The full derivation of the equations is given in Bossel, 1987/1989: p. 206-207.)

The two remaining differential equations of second order essentially state that both the angular acceleration and the acceleration of the carriage are a function of the instantaneous angle of rotation ß and the control force u. Obviously, the angle of rotation and the translation of position are state variables of the system. However, since both can only be obtained by integrating the angular acceleration and the translational acceleration twice, the angular velocity and the translational velocity are also hidden in this formulation as additional state variables.

In the system equations the control function u is so far undefined. In order to carry out the control process, the controller must be informed about the state of the system. In principle, one therefore has to assume that each of the four state variables is coupled to the controller. Assuming that u is a linear function of angle ß, angular velocity dß/dt, position x, and velocity dx/dt with control parameters $k_ß$, $k_{ßt}$, k_x, k_{xt},

$$u = k_ß\, ß + k_{ßt}\, dß/dt + k_x\, x + k_{xt}\, dx/dt,$$

then the system equations become linear and (also) allow an analytical solution. In the following we will only be concerned with simulation, i.e. the numerical integration of this system.

The derivation results in two differential equations of second order, i.e. a fourth order system corresponding to four differential equations of first order. With

$$z_1 = ß, \quad z_2 = dß/dt, \quad z_3 = x, \quad z_4 = dx/dt,$$

we obtain the system

$$\begin{aligned} dz_1/dt &= z_2 \\ dz_2/dt &= a\, z_1 + b\, u \\ dz_3/dt &= z_4 \\ dz_4/dt &= c\, z_1 + d\, u \end{aligned} \qquad 5.2)$$

with the control function

$$u = k_ß\, z_1 + k_{ßt}\, z_2 + k_x\, z_3 + k_{xt}\, z_4. \qquad (5.3)$$

For the simulation we use the system equations in this form.

The procedure shown here in developing the system equations is of fundamental importance. It always has the following steps:

1. Derivation of the complete (usually non-linear) equations of motion by applying the relevant physical relationships.
2. Simplification of the differential equations assuming small perturbations from a reference state. Products of perturbations or of their derivatives then become very small and can be neglected, leading to linear differential equations.

The dynamic behavior (in the neighborhood of the equilibrium point for which the linearization is valid) can be analyzed or simulated only if the control function u has been prescribed more precisely.

5.4.3 Simulation model for the stabilized pendulum system

With the system equations (Eq. 5.2) and the assumed linear control function (Eq. 5.3) the simulation diagram can be drawn (Fig. 5.18). Using the constant parameters of the system (m, M, L, and g), the moment of inertia I of the pendulum as well as the four coefficients a, b, c, and d can be computed. These quantities must only be calculated once at the beginning of the simulation. In the simulation diagram, they appear as multiplicative weights. The control parameters are determined in the interactive query at the beginning of the simulation.

In order to investigate the response of the system to random disturbances, we also add a random disturbance force s acting horizontally on the carriage (in the direction of u); it is determined by a random function of time. The maximum amplitude of the disturbance is specified by the user during the initial query.

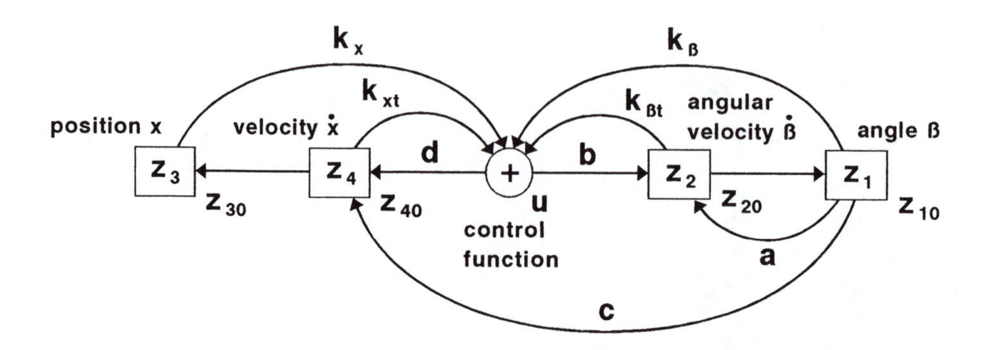

Fig. 5.18: Simulation diagram for the rotational and translational motion of the inverted pendulum and its control and stabilization.

The model unit for the simulation with SIMPAS can now be easily written down by reference to the simulation diagram (model BALANCER.MOD, Prog. 5.4). For a better assessment of the control performance, we also compute the instantaneous control power as the product of control force and translational velocity (u·xt). The integral of this power over time provides the total control work (energy) expended; it is computed as an additional state variable.

1. Parameters

g	=	9.81	gravitation constant	[m/s^2]
L	=	1	pendulum length	[m]
mw	=	1	carriage mass	[kg]
rp	=	0.05	radius of pendulum mass	[m]
kw	=	100	angle feedback factor	[N/rad]
kwt	=	30	angular velocity feedback factor	[N/(rad s)]
kx	=	3	position feedback factor	[N/m]
kxt	=	10	velocity feedback factor	[N/(m/s)]
mp	=	1	pendulum mass	[kg]
amp	=	0	disturbance amplitude	[N]

Computed constants:

ii	=	$2 \cdot mp \cdot rp^2/5$	moment of inertia for sphere
aa	=	$ii \cdot (mp+mw)+mp \cdot mw \cdot L^2$	
a	=	$g \cdot (mp+mw) \cdot mp \cdot L/aa$	
b	=	$-mp \cdot L/aa$	
c	=	$-g \cdot mp^2 \cdot L^2/aa$	
d	=	$(ii + mp \cdot L^2)/aa$	

2. Initial values of the state equations

z_1	=	w	=	0.2	angle	[rad]
z_2	=	wt	=	0	angular velocity	[rad/s]
z_3	=	x	=	0	position	[m]
z_4	=	xt	=	0	velocity	[m/s]
z_5	=	ar	=	0	control energy	[Nm]

3. Algebraic intermediate variables

s	=	amp·(2·random–1)	disturbance function	[N]
u	=	kw·w + kwt·wt + kx·x + kxt·xt	control function	[N]
wtt	=	a·w + b·(u+s)	d^2w/dt^2	[rad/s²]
xtt	=	c·w + d·(u+s)	d^2x/dt^2	[m/s²]

4. State equations

dz_1/dt = wt
dz_2/dt = wtt
dz_3/dt = xt
dz_4/dt = xtt
dz_5/dt = abs(u·xt)

5. Run time parameters

Start = 0
Final = 10
TimeStep = 0.01

In order to obtain a clearer picture of the stabilization process and the influences of the different parameters, we supplement the standard presentations of results in SIMPAS by an additional graphic animation. The procedure Summary in the unit Model which we have not used so far is intended to be used for program additions of this kind.

In order to match the graphic presentation to different screens, we first determine the graphic setting used in SIMPAS (using GetGraphMode) (see Prog. 5.4 for BALANCER.MOD). Using SetGraphMode, the screen graphics are activated. The scaling factors sx and sy for the screen graphics are found next. In programming the graphics, all lengths in the x and y direction are given in percent of the screen width and height and are then multiplied by the scale factors sx and sy.

The horizontal plane, on which the carriage moves with the pendulum, is drawn first as a horizontal line. Scaling ticks are drawn on the line. In the upper part of the screen, the currently used control function and the necessary control energy are indicated.

The values of the state variables (angle w = ß and position x) computed during the simulation are then read in a "repeat...until i=250" loop from the result file (see Sec. 4.1.4). They determine the position of the pendulum support and the center of gravity of the pendulum mass. With this information, the carriage and the pendulum can be drawn at their instantaneous position. In order to obtain an (approximate) presentation in real time, each image is delayed by 15 milliseconds before it is erased and a new picture is drawn. The animation is repeated until the user stops it by pressing a key to return to the simulation program.

5.4.4 Simulation runs and the search for "good" control parameters

The default values of the model correspond to an acceptable solution of the control problem for the inverted unstable pendulum. This is best seen by viewing the animation. After calling up the (compiled) model, we move by repeated pressing of RETURN from "Simulation" (under MODE SELECTION) to the "Continue OUTPUT PRESENTATION?" screen to which wc rcspond with "No, new simulation or quit." An animated image of the balancing motion of the inverted pendulum and the control carriage is then shown on the screen. The continuous repetition of this animation must be interrupted by pressing a key. The SIMULATION PARAMETERS screen then appears, and we can again change individual parameters to produce a new simulation and animation.

By experimenting with the four control parameters, one finds that stable solutions can only be obtained for certain ranges of the different parameters. Even within these limits, changes of the control parameters will result in significantly different behavior of the system. Changing the pendulum mass leads to a significant change of the dynamics. If random disturbances of the carriage movement are introduced as a "scenario," the control system becomes increasingly less able to cope with the disturbances as the disturbance strength is increased. Finally it can no longer stabilize the system.

The control behavior is best observed in the phase plot of angle and angular velocity (Fig. 5.19). In this particular plot the influence of the velocity feedback parameter on the behavior was studied (using the option "Parameter Sensitivity"). In all of these cases the angle is reduced to zero fairly quickly while the displacement may become quite large and will only be slowly reduced to zero. (The response is particularly slow for $k_{xt} = 5$.)

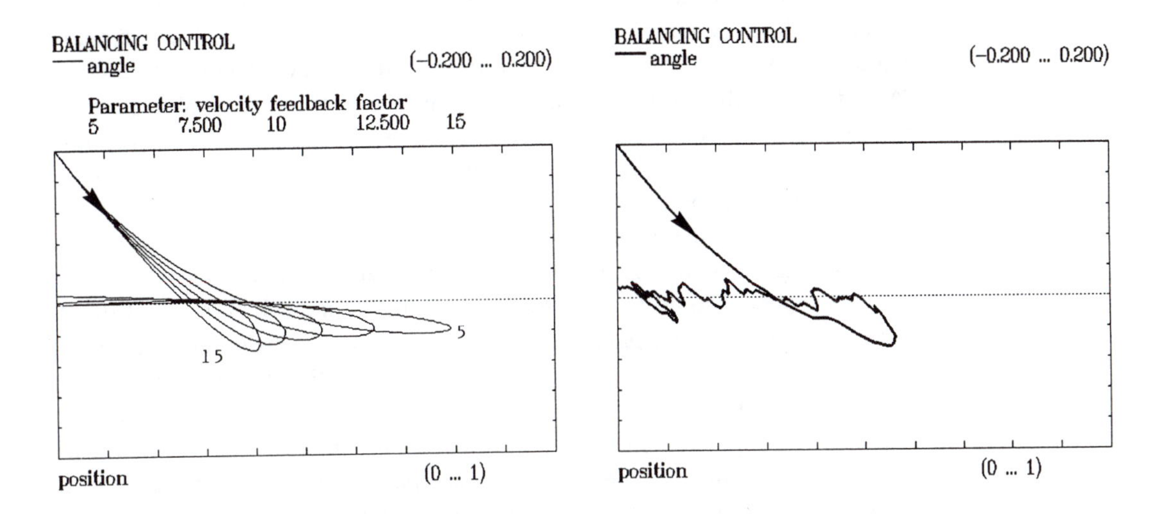

Fig. 5.19 (left): BALANCER: State diagram for position and angle as function of the velocity feedback parameter. The position control is significantly slower than the angle control.

Fig. 5.20 (right): BALANCER: Influence of random disturbances in the phase plot. Stabilization becomes more difficult as the disturbance amplitude increases.

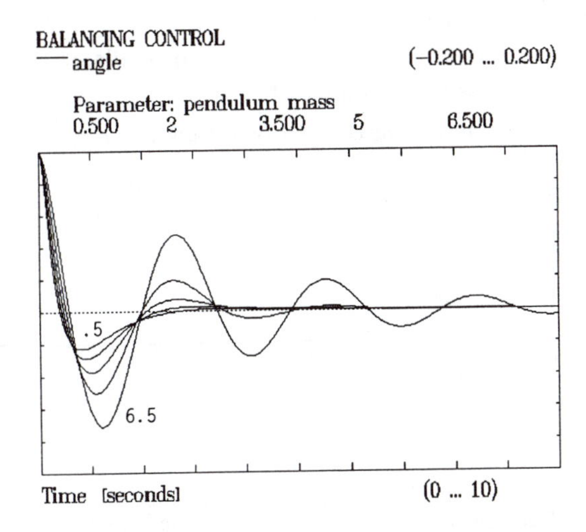

Fig. 5.21: BALANCER: Influence of pendulum mass on the stabilization of the angular motion. As mass increases, the angular deflections become larger and the motion more difficult to control.

If a random disturbance force is introduced, the basic response corresponds to that of the unperturbed system as long as the disturbance amplitude remains reasonably small (Fig. 5.20). Obviously, the controller not only fulfills the original task to balance the falling pendulum and return it to the vertical position but it can also cope with additional disturbances within certain limits. In most applications this has to be expected of a control system.

The influence of the mass is shown in the time plot of the angle (Fig. 5.21): as mass increases, the pendulum swings become larger. Beyond a critical mass, the system can no longer be stabilized (with constant control parameters).

The option "Parameter Sensitivity" also provides a quick overview of the effects of different control parameters. In Figure 5.22 and b, the influence of the feedback factors for angle and angular velocities on the time behavior of the angle is investigated. In Figure 5.23 a and b, the time curves for the carriage position are shown as function of the feedback factors for position and velocity. If we now select from these diagrams the most favorable parameter values (for quick damping) in each case, we find that the corresponding control function is destabilizing. This suggests that the parameters cannot be optimized independently of each other.

In searching for "good" or even "optimal" settings for the control parameters, we have to decide which criteria to use in the assessment. Should the pendulum be returned to a vertical position as quickly as possible? Should the carriage be returned to the zero position quickly? Should both be achieved simultaneously, or can we tolerate that the position control may take longer than the control of the angle (as in the examples)? Is the system allowed to overshoot and oscillate? Is the energy necessary for the control process to be minimized?

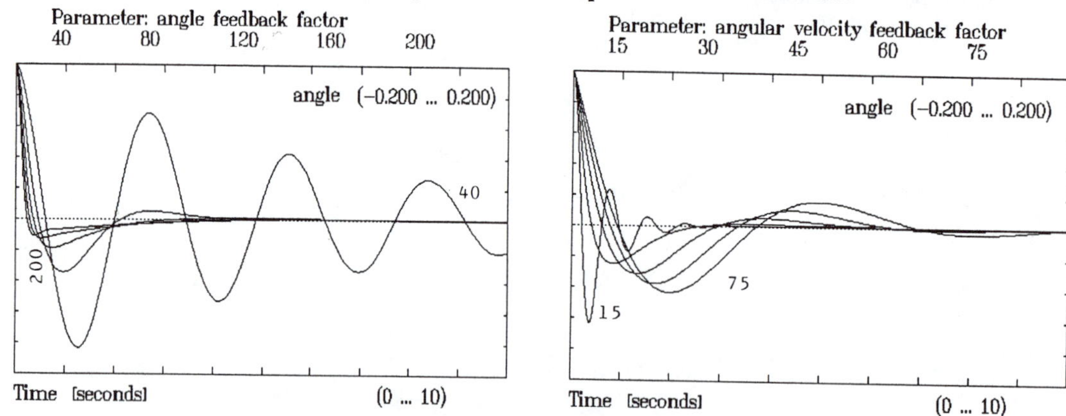

Fig. 5.22: BALANCER: Influence of the feedback parameters for angle (left) and angular velocity (right) on the rotating motion. Stable solutions only appear in a certain parameter range; the control parameters can only be optimized as a set.

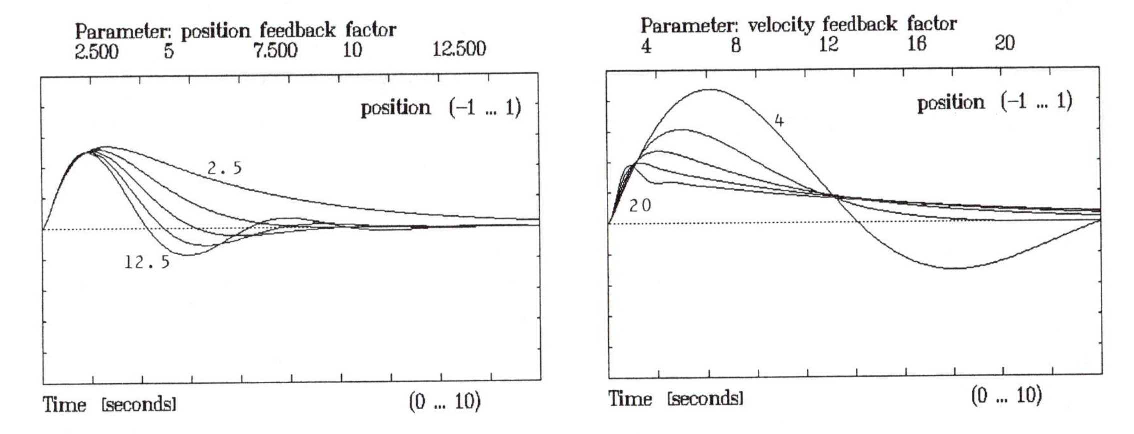

Fig. 5.23: BALANCER: Influence of the feedback parameters for position (left) and velocity (right) on the position. Stable solutions only appear in a certain range.

In order to be better able to compare different control solutions, criteria must be formulated that can be used to measure and compare the relative success of the control process. *Examples:* time constant (initial slope) of the returning motion, overshoot measure, damping, and energy expended. Since in many control processes it is essential to minimize energy use, we here use for the comparison of different control functions the energy necessary for stabilizing the initial disturbance, i.e. the control energy (control work) as time integral of the control power.

Figures 5.24 and b clearly show that as a function of the choice of the control parameters, significant differences in the energy consumption of the control system can be found. By changing the angular velocity feedback factor k_{wt} (Fig. 5.24a), we obtain the following energy consumptions (control work) for the stabilization of the initial disturbance of 0.2 rad:

k_{wt}	20	30	40	50	60	
control work	1.857	1.078	1.133	1.491	1.969	[Nm] .

Here an optimum value of 1.078 for the control work appears for $k_{wt} = 30$; for $k_{wt} = 20$ and $k_{wt} = 60$ energy consumption is almost twice as high. Large differences also appear as the velocity feedback factor k_{xt} is changed (Fig. 5.24b).

Further search for the energetically best solution shows that with the choice of control parameters $k_w = 100$, $k_{wt} = 30$, $k_x = 1$, and $k_{xt} = 7.5$, energy consumption can be lowered to 0.707 without significantly changing quality and dynamics of the control. Obviously, investigations of this kind have considerable significance in many technical applications; with the help of simulation and the use of corresponding evaluation criteria they can be easily carried out.

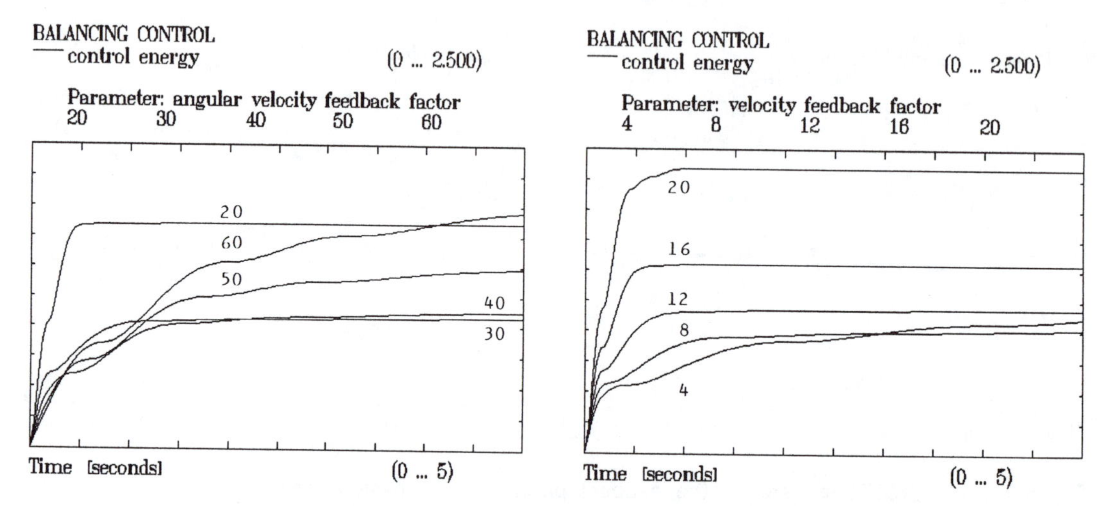

Fig. 5.24: BALANCER: Optimization of the control parameters for angular velocity (left) and translational velocity (right) by minimization of the control work necessary for stabilization.

5.5 Summary of Important Results

The original task to better understand the system and to "improve" its response is often only partially solved by the construction of a simulation model and the production of several simulation runs. The desire to find a reproducible answer to the question: What would be the "better" solution?, leads to the definition of evaluation criteria and the selection of an evaluation procedure. The evaluation problem is not only relevant to the assessment of simulation results but also to the assessment of the behavior of real systems. It is not only relevant for the observer but it also concerns the development of the system itself and its ability to cope with its environment.

We have dealt with these questions in this chapter. It became obvious that the fundamental structure of a dynamic system and its embedding in its system environment pose certain requirements that must be fulfilled if the system is to remain viable, functioning, and adaptive. The "basic orientors" that can be derived from these requirements can serve as a framework for assessments when evaluating alternative development paths, searching for better policies or even "optimal" solutions, trying to stabilize unstable systems, and designing new systems. These aspects were investigated using the three simulation models as examples. The most important results:

1. Every dynamic system interacting with its environment can in principle be described by a **state function** **f(z, u, t)** and a **behavioral function** **g(z, u, t)** that establish the internal system relationships as well as the interactions with the system environment (Fig. 3.7).
2. These system relationships imply certain **requirements** for the "design" of the system if it is to be viable in its system environment. These requirements are denoted as "basic orientors." They correspond to the **fundamental properties of the environment:** e.g. the normal state of the environment, scarce resources, variety, variability, change, and other systems.
3. The **basic orientors** of open systems (interacting with their environment) are: **existence**, **effectiveness, freedom of action, security, adaptiveness, regard for others.**
4. These basic orientor requirements are **independent** of each other and are not mutually substitutable.
5. A system can permanently exist and thrive in its environment only if each of the basic orientors is given a minimum of attention and fulfillment.
6. Reducing the evaluation problem to a single quality criterion or to a set of criteria not completely covering the basic orientor dimensions is admissible only if the necessary **minimal satisfaction of all basic orientors** is guaranteed.
7. The **comparative evaluation** of different system developments (development paths for different scenarios) should be based on basic orientor assessments.
8. **System design and optimization** also have to be guided by basic orientor considerations.
9. **System stabilization** by system change (structural augmentation) can be used to eliminate an existence-threatening deficit in orientor satisfaction (in particular, "security").
10. The assessment of system development requires the observation of **indicators**, i.e. certain system variables for which goal values must be defined.
11. The selection of the indicators must be **complete**, i.e. it must be able to provide a comprehensive picture of the instantaneous system state.
12. For the evaluation and assessment of system states, **orientors** must be defined, i.e. a set of relevant evaluation criteria.
13. In the evaluation process, the **system state** (expressed by the current values of the indicators) **is mapped on the "system interests"** (expressed by the orientors).
14. The **selection of the orientors** must be complete with respect to the interests of the system (and of the system managers).

15. For the assessment of system development, **three kinds of criteria** are required which differ fundamentally in application: **constraints, quality measures,** and **weights**.
16. As long as **constraints** are not met, **attention** has to focus on these **gaps**. Compensation of a deficit by overfulfillment of other constraints is not permissible. This is particularly true for the basic orientors.
17. **Quality measures** that allow the summation of satisfactions with respect to different criteria can only be applied if all constraints are satisfied (e.g. satisfaction of minimum basic orientor requirements).
18. The **weights** of criteria reflect subjective assessment of their relative significance.
19. **Scenarios** are plausible assumptions concerning the bundle of future exogenous influences that may determine the development of a system.
20. **Development paths** resulting from different scenarios can be evaluated and compared by systematic mapping of the indicators of system state on the relevant orientors.
21. Mapping of **system development** on the orientors provides quick and early recognition of developments threatening the system.
22. The **search for better solutions**, and in particular **optimization**, requires the definition of a quality measure which may consist of the weighted contributions of different criteria.
23. The **result of optimization** always depends on the definition of the quality measure, the selection of criteria considered, and their relative weights.
24. Systems exhibiting an initially unsatisfactory or unstable behavior may be improved in their response or stabilized in their behavior by specific **structural changes** (i.e. feedback).
25. Investigations of **control and stabilization** require knowledge of the system equations of the (changed) system.
26. In order to investigate the behavior of a controlled system in a narrow state range, the complete **state equations** can usually be significantly **simplified** (linearization, neglect of small quantities).
27. Controlled behavior depends on the choice of the **control function** and its **control parameters**.
28. For the (comparative) **evaluation of control behavior**, quality measures should be constructed as weighted sums of individual criteria. The evaluation result depends on the weights chosen.
29. The (unavoidable) subjective elements of evaluation procedures become more transparent if **formal evaluation procedures** with clearly defined criteria are used.

CHAPTER 6
SYSTEMS ZOO

6.0 Introduction

Up to now we have dealt almost exclusively with only three simulation models which we used to demonstrate the methods of model development, programming, simulation, behavioral analysis, and system management: the world model, the rotating pendulum, and fishery dynamics. The three models are described by non-linear differential equations with two or three state variables. Although relatively simple, they show rather complex behavior that strongly depends on parameters and initial values. In order to obtain an overview of the possible behavioral modes, equilibrium states, and stability regions, we investigated the dependence of behavior on parameters and initial conditions using the "parameter sensitivity" and "global analysis" options. Using these three models, we demonstrated the most important aspects of non-linear (and linear) dynamic systems.

While the approach can be generalized to the investigation of other dynamic systems, this is hardly true for the results found in different cases. In contrast to linear systems which all belong to the same type and show the same generalizable behavioral modes, even small structural differences may cause fundamentally different behavior in non-linear systems. Each non-linear system usually has its own unique characteristics.

To present interesting and typical representatives of different system types with their characteristic structures and behaviors, some fifty simulation models were collected in the "systems zoo." These simulation models on the accompanying diskette are embedded in the SIMPAS software which allows comprehensive investigations into their behavior.

An attempt was made to include rather common systems in this selection, i.e. system structures that are encountered again and again in modeling real systems and describe important processes of our everyday experience ("everyday dynamics"). Mathematical curiosities that have no counterpart in the real world were not included. The collection is limited to systems with at most four state variables. Many of the "system animals" included in the systems zoo are found as subsystems in more complex systems. This has already been pointed out during the development of the three models in the first part of the book which contain important elementary structures (exponential growth or decay, logistic growth, delay, oscillation, predator-prey relationship). The acquaintance with these elementary systems can help in understanding much more complex systems.

The systems zoo contains mostly "generic" models where the state variables were normalized by relating them to a reference value, thus producing results of order of magnitude "1." This facilitates comparison of systems having the same structure but very different parameters. In Section 3.7 it was shown how model equations can be converted from the general to the normalized form and *vice-versa*. The models of the systems zoo therefore apply also to other contexts.

On the following pages, the complete documentations for each of the 43 new models of the systems zoo are given. (In addition, seven models were documented in the previous chapters.) All 50 models can be found as run time versions in the systems zoo on the accompanying diskette. The systems zoo is run on the DOS level by calling SIMZOO. The models can be individually selected and are then executed in the familiar SIMPAS environment. However, the documentation on the following pages is independent of SIMPAS. It contains all necessary information to program the models in any programming or simulation language.

Each model documentation contains the following points:

0. Identification
1. Description
2. Occurrence
3. Structural characteristics
4. Behavioral characteristics
5. Critical parameters
6. Reference run
7. Suggestions for further study
8. References
9. System diagram and system equations
10. SIMPAS-output for:
    a. Simulation results in graphical form
    b. Parameter values (default values)
    c. Equilibrium points

The models are ordered with respect to the number of state variables (first figure). Section 6.1 presents **elementary systems with one state variable**:

101 SIMPLE INTEGRATION: CHANGE OF STOCK AND CONCENTRATION
102 EXPONENTIAL GROWTH AND DECAY
103 EXPONENTIAL DELAY OF FIRST ORDER
104 TIME-DEPENDENT EXPONENTIAL GROWTH
105 BIRTH AND DEATH: SIMPLE POPULATION DYNAMICS
106 OVERLOADING A BUFFER
107 LOGISTIC GROWTH WITH CONSTANT HARVEST RATE
108 LOGISTIC GROWTH WITH PREY-DEPENDENT HARVEST
109 DENSITY-DEPENDENT GROWTH (MICHAELIS-MENTEN)
110 DAILY PHOTOPRODUCTION OF VEGETATION

All of these processes play an important role in our environment and every day experience. Oscillations cannot occur in these one-dimensional systems.

Section 6.2 presents **elementary systems with two state variables**:

201 DOUBLE INTEGRATION AND EXPONENTIAL DELAY
202 TRANSITION FROM ONE STATE TO THE NEXT
203 LINEAR OSCILLATOR OF SECOND ORDER
204 ESCALATION
205 BURDEN SHIFTING AND DEPENDENCE
206 PREDATOR-PREY SYSTEM WITHOUT CAPACITY LIMIT
207 PREDATOR-PREY SYSTEM WITH CAPACITY LIMIT
208 COMPETITION
209 TOURISM AND ENVIRONMENTAL POLLUTION
210 OVERSHOOT AND COLLAPSE
211 FOREST GROWTH
212 RESOURCE DISCOVERY AND DEPLETION
213 TRAGEDY OF THE COMMONS
214 SUSTAINABLE USE OF A RENEWABLE RESOURCE
215 DISTURBED EQUILIBRIUM: CO_2-DYNAMICS OF THE ATMOSPHERE
216 STOCKS, SALES, AND ORDERS
217 PRODUCTION CYCLE
218 ROTATING PENDULUM
219 OSCILLATOR WITH LIMIT CYCLE (VAN DER POL)
220 BISTABLE OSCILLATOR
221 CHAOTIC BISTABLE OSCILLATOR

Most of these systems also play an important role in our environment and every day experience. In most of the systems, oscillations may occur on account of feedback loops linking two state variables.

Section 6.3 presents another set of **elementary systems with three or four state variables**:

301 TRIPLE INTEGRATION AND EXPONENTIAL DELAY
302 POPULATION DYNAMICS WITH THREE GENERATIONS
303 LINEAR OSCILLATOR OF THIRD ORDER
304 MINIWORLD: POPULATION, CONSUMPTION, AND POLLUTION
305 PREDATOR WITH TWO PREY POPULATIONS
306 TWO PREDATORS WITH ONE PREY POPULATION
307 BIRDS, INSECTS, AND FOREST
308 NUTRIENT CYCLING AND PLANT COMPETITION
309 CHAOTIC ATTRACTOR (RÖSSLER)
310 HEAT, WEATHER, AND CHAOS (LORENZ SYSTEM)
311 COUPLED DYNAMOS AND CHAOS
312 BALANCING AN INVERTED PENDULUM

Systems with three or more state variables may again show oscillatory behavior; in addition, some systems may produce chaotic behavior.

These systems can be simulated and analyzed extensively with the SIMPAS simulator to study their behavioral characteristics and their response to parameter changes. All models are supplied with default values that allow the demonstration of certain characteristic features. In addition, there are "suggestions for further study" under Point 7 of the model documentation.

In addition to these specific suggestions the following general suggestions apply to all models:

1. Start by studying the behavior of the reference run (with the given default values for the parameters) using the different modes of presentation of results in SIMPAS.
2. Investigate the dependence of system development on the "critical" parameters listed in the documentation (using the SIMPAS option "Parameter Sensitivity").
3. Analyze the system in greater detail in parameter ranges where significant changes of system behavior can be observed (for example, stability/instability, equilibrium/collapse) or where other interesting effects occur.
4. Investigate the global behavior in the complete (relevant) state space (x, y, z) for the reference case and/or interesting combinations of parameters using the option "Global Analysis," paying particular attention to the equilibrium points in state space. Observe the state trajectories and deduce from them stability/instability in a given state domain. (Note that "Global Analysis" is a presentation of two state variables in the phase plane. It is therefore not available for one-dimensional systems. For systems having three or more state variables it may produce results which are difficult to interpret.)

5. Calculate (analytically) the location of the equilibrium points as function of the parameters using the state equations and the condition that the rates of change must disappear at the equilibrium points ($dz/dt = 0$). Compare this theoretical result with the simulation results (for the same parameter choice).
6. Linearize the non-linear state equations at the equilibrium points and analyze the behavior of the corresponding linearized substitute system using the model of the LINEAR OSCILLATOR (of same order) by substituting the corresponding system parameters. Does the behavior of the original non-linear system near the equilibrium points agree with the properties of the substitute linearized system and its eigenvalues? (See Ch. 7 for the theoretical background).

(Suggestions 5. and 6. refer in particular to two-dimensional systems and are intended for readers with some mathematical interest and background).

A note on the use of terms: It is often necessary to distinguish between two kinds of rates of change—absolute and specific (or normalized, or relative) rate of change. If x is a state variable with dimension [dim x], then dx/dt (with dimension [dim x / dim t]) will sometimes be referred to as "absolute rate of change." It represents the incremental change of the state variable per time step. If this rate of change is "normalized" by dividing it by the state variable, it will be referred to as "specific rate of change," defined by (dx/dt)/x and having dimension [1 / dim t]. The term "rate of change" will be used in both cases if there is no danger of confusion.

An example should illustrate the difference: in 1993, the global population stood at about 5500 [million people] and grew at a net absolute rate of growth of 90 [million people / year]. Its specific rate of growth was therefore 90/5500 [1 / year] = 0.01636 per year.

For readers who wish to implement models of the systems zoo in STELLA (or DYNAMO), Figure 6.1 presents a summary of equivalent simulation diagrams and STELLA structural diagrams.

SIMULATION DIAGRAMS **STELLA DIAGRAMS**

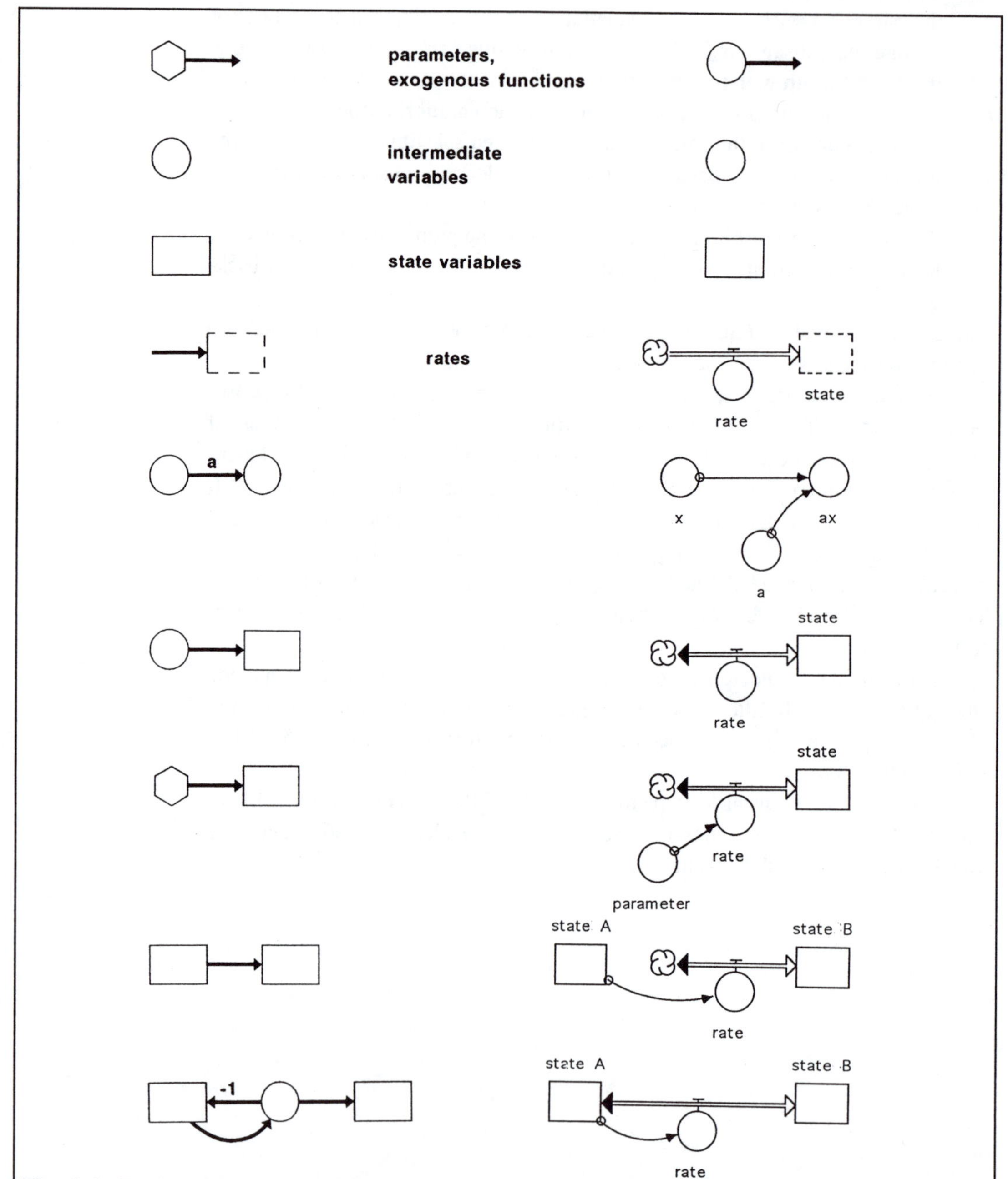

Fig. 6.1: Equivalent system structures in simulation diagram and STELLA diagram notation..

6.1 Dynamic Systems with One State Variable

101 SIMPLE INTEGRATION: CHANGE OF STOCK AND CONCENTRATION
102 EXPONENTIAL GROWTH AND DECAY
103 EXPONENTIAL DELAY OF FIRST ORDER
104 TIME-DEPENDENT EXPONENTIAL GROWTH
105 BIRTH AND DEATH: SIMPLE POPULATION DYNAMICS
106 OVERLOADING A BUFFER
107 LOGISTIC GROWTH WITH CONSTANT HARVEST RATE
108 LOGISTIC GROWTH WITH PREY-DEPENDENT HARVEST
109 DENSITY-DEPENDENT GROWTH (MICHAELIS-MENTEN)
110 DAILY PHOTOPRODUCTION OF VEGETATION

SIMPLE INTEGRATION: CHANGE OF STOCK AND CONCENTRATION **M 101**

1. **Description:** The model describes the change of a stock z as function of a time-dependent rate of change u(t) and an initial value z_0. The process occurs in any kind of stock with time-dependent continuous gains or losses. Effects of the fluctuations of the gain and loss rates are smoothed out by the stock.

2. **Occurrence:** Stocks (levels) of any kind with time-dependent inflows and outflows, e.g. populations, population classes, water reservoirs, warehouse stocks.

3. **Structural characteristics:** No feedback of the state variable z (stock, level) on the gain or loss rate dz/dt. The state change is totally dependent on the exogenously determined gain or loss rates u(t); it may be positive (gain) or negative (loss).

4. **Behavioral characteristics:** The integration accumulates the current gain or loss rates over time. The state (level) changes accordingly.

5. **Critical parameters:** The (absolute) change of state dz/dt is dependent on sign and magnitude of the rate of change u(t). (The relative state change (dz/dt)/z must be expressed in terms of the current state, and it is therefore dependent on it.)

6. **Reference run:** Input of a pulse sequence (here: pulse (1,1,1)) produces a stepwise change of state—a "staircase" function. Each pulse of strength 1 (pulse area = 1) increases the state by 1.

7. **Suggestions for further study:** Select other input functions u(t) and note the system response: single pulse, step, ramp, sine function. Confirm the analytical integration formula: constant rate of change leads to a linearly increasing state; linearly increasing rate of change leads to a quadratically growing state; the integration of a sine function leads to a cosine function.

8. **References:** -

SIMPLE INTEGRATION: CHANGE OF STOCK AND CONCENTRATION **M101**

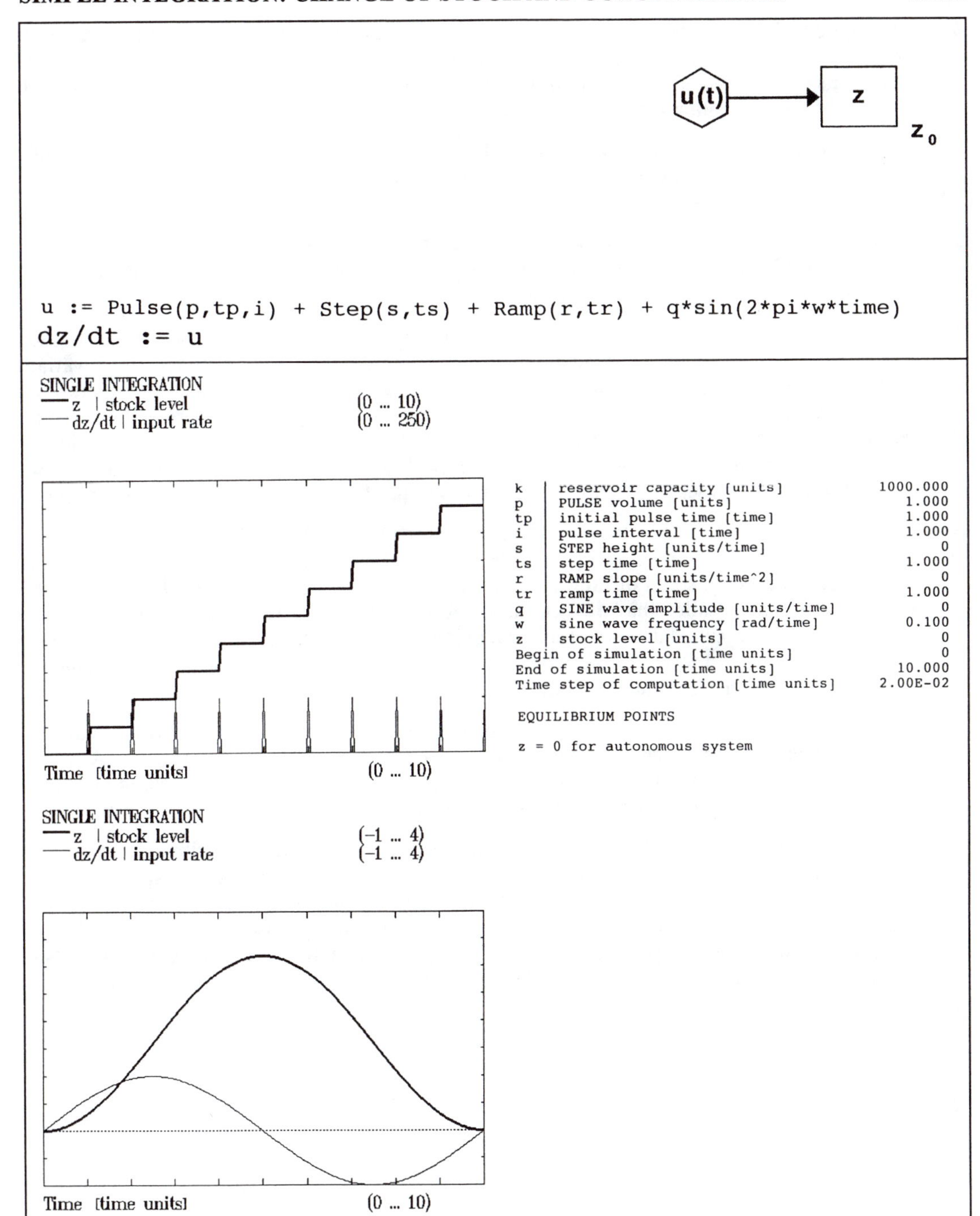

EXPONENTIAL GROWTH AND DECAY **M 102**

1. **Description:** In this system the current state z determines the rate of change of the state dz/dt; i.e. there is a feedback coupling the state back to itself (self-loop). This results in a self-generated characteristic dynamic (eigendynamic) even if no external influence is present. The growth or decay of the state follows an exponential function of time.

2. **Occurrence:** All processes where the rate of change depends on the state itself: for example, growth of a population, interest in a bank account, drain of a stock, economic growth, radio-active decay, coffee cooling in a cup.

3. **Structural characteristics:** Feedback of the state on itself (self-coupling). The feedback factor a which must be multiplied by the current state z in order to obtain the current rate of change dz/dt, represents the specific rate of change. It always has the dimension [1/time unit].

4. **Behavioral characteristics:** A system is autonomous if there is no exogenously determined gain or loss rate, i.e. $u(t) = 0$. In an autonomous system the (non-zero) state remains invariant if the feedback parameter $a = 0$. If $a > 0$, the state will grow exponentially and towards infinity with a specific rate of change a. If $a < 0$, the state follows an exponential decay with a specific decay rate $|a|$. As the time t increases towards infinity, the state then approaches the value 0. Note that for exponential growth, the increment added per unit time (the absolute rate of growth) steadily increases exponentially although the specific rate of change a remains constant.

5. **Critical parameters:** The sign of the feedback parameter determines the stability $(a < 0)$ or instability $(a > 0)$ of the state change and the system development. The magnitude of the feedback $|a|$ determines the speed of growth or decay, respectively. The inverse of the feedback parameter $T = 1/|a|$ is called the "time constant" of the system. The time it takes for the state to double or to halve is called doubling time or halving time and is given by $T_2 = \ln 2 / |a|$ or $T_2 = T \ln 2$, where $\ln 2 = 0.693\ldots$. If the specific rate a is stated in percent per year, for example, then the doubling time is therefore about 70/a.

6. **Reference run:** A value of $a = -1$ for the feedback parameter results in a quickly decaying exponential behavior. If the sign is changed $(a = 1)$, one obtains exponential growth. A typical characteristic of exponential growth is the constant acceleration of growth, reaching very high state values at the end of the simulation period (here: $z \approx 20{,}000$ for $t = 10$).

7. **Suggestions for further study:** Using the SIMPAS option "Parameter Sensitivity", investigate the time development of the state as a function of the feedback parameter a (specific rate of change) in the range from $a = -1$ to $a = +1$. Note: typical rates of change of societal processes are between 0 and 10 percent per year, i.e. $0 < a < 0.1$. In this case the time unit is "year." Find out how the state changes over a period of 100 years when different specific growth rates in this range are assumed. Verify the "formula" for the doubling time ($T_2 = 70/a$, a in percent).

8. **References:** -

EXPONENTIAL GROWTH AND DECAY

M 102

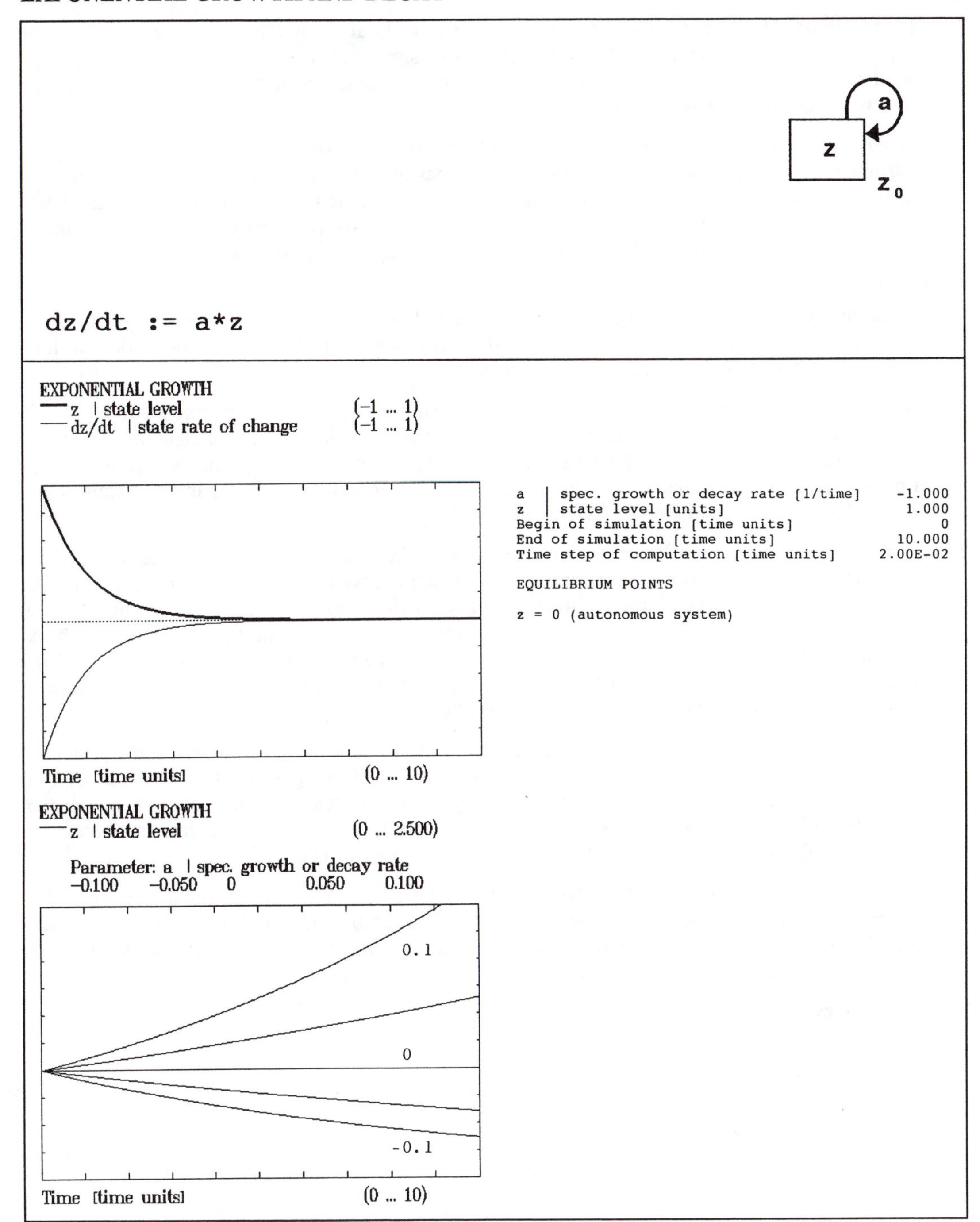

EXPONENTIAL DELAY OF FIRST ORDER — M 103

1. **Description:** The state z is subject to a constant exponential decay ("outflow" $a \cdot z$; "exponential leak") but is also constantly "replenished" by a time-dependent inflow u(t). This means a constant drain of the "history" of the system and a delayed adjustment to the input. This system structure therefore acts as a delay.

2. **Occurrence:** The process corresponds to the filling of a reservoir with simultaneous state-proportional losses due to drainage. Examples: dynamics of soil-water due to seepage and simultaneous rainfall; input of fertilizer and chemicals in soil and their subsequent chemical breakdown or leaching. In mechanical systems, the negative feedback always means a damping of motion (for example, viscous drag). Other examples are capital depreciation and the deathrate of a population.

3. **Structural characteristics:** The dynamics are caused by two different rates of change: 1. exponential decay (exponential damping) at constant decay rate (–a) and 2. a time-dependent inflow rate u(t) which is independent of the system state.

4. **Behavioral characteristics:** The state constantly suffers losses which are proportional to the current state level. Since the current state represents the time integral of past changes, the losses correspond to the "history" of the system while the gains (from the exogenously determined inflow u(t)) represent the present. The most recent changes therefore dominate in the system state; in all, the state follows the input u(t) with some delay.

5. **Critical parameters:** The magnitude of the negative feedback a is the critical parameter: the parameter a determines the equilibrium state z* which is given by the condition $dz/dt = 0$, i.e. $a \cdot z^* = u$. From this, the equilibrium state $z^* = u/a$ (with $a > 0$). The larger the magnitude of a (the larger the loss rate), the smaller is the equilibrium state z*. The inverse of the feedback parameter, $(1/a) = T$ = time constant of the system, is the most important system parameter. The larger a, the smaller T, the faster the system will adjust to a new value of the exogenous rate of change u(t), and the shorter is the delay in the system.

6. **Reference run:** The system input, a step function u(t) of constant magnitude, begins at time t = 1. Initially, the state follows this rate of change linearly (cf. integration of the step function in M 101). Since the loss rate is proportional to the current state, it will eventually reach the magnitude of the constant input. At this point a flow equilibrium z* = const results. Transition to this equilibrium value is described by the function $z(t) = (u/a) \cdot (1 - e^{-at})$.

7. **Suggestions for further study:** Determine the shape of the system response and its delay for different test functions (pulse function, step function, ramp function, sine function). Use the SIMPAS option "Parameter Sensitivity" to determine how the system response and the delay of the input signal depend on the magnitude of the feedback constant a.

8. **References:** -

EXPONENTIAL DELAY OF FIRST ORDER

M 103

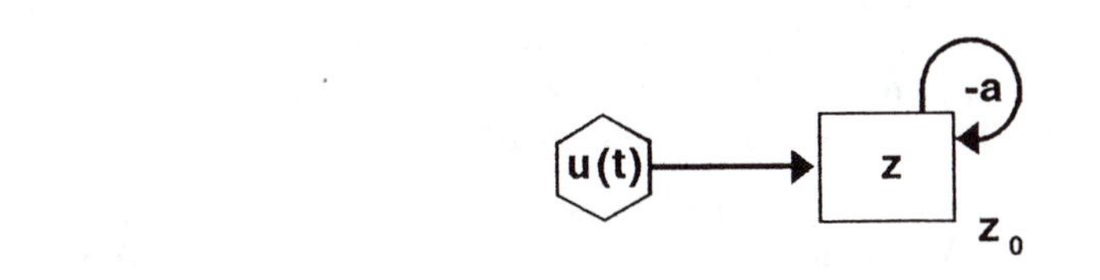


```
u := Pulse(p,tp,i) + Step(s,ts) + Ramp(r,tr) + q*sin(2*pi*w*time)
dz/dt := u - a*z
```

EXPONENTIAL DECAY
z | stock level (0 ... 1.250)
dz/dt | total change rate (0 ... 1.250)
u | exogenous change rate (0 ... 1.250)

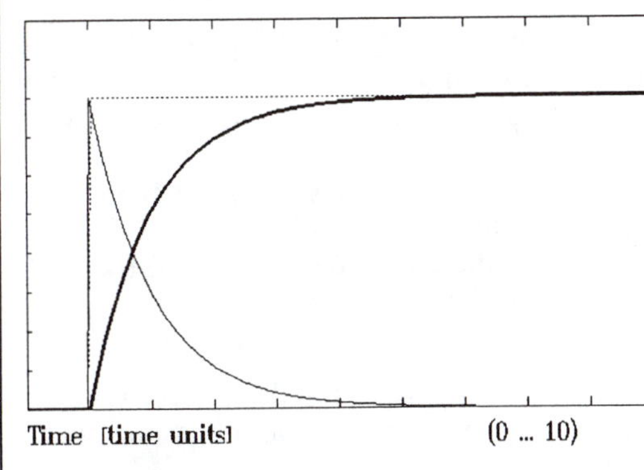

Time [time units] (0 ... 10)

a	decay rate [1/time]	1.000
p	PULSE volume [units]	0
tp	initial pulse time [time]	1.000
i	pulse interval [time]	1.000
s	STEP height [units/time]	1.000
ts	step time [time]	1.000
r	RAMP slope [units/time^2]	0
tr	ramp time [time]	1.000
q	SINE wave amplitude [units/time]	0
w	sine wave frequency [rad/time]	0.100
z	stock level [units]	0
Begin of simulation [time units]		0
End of simulation [time units]		10.000
Time step of computation [time units]		2.00E-02

EQUILIBRIUM POINTS

z = 0 for autonomous system
z = u/a for u = const

EXPONENTIAL DECAY
z | stock level (0 ... 2)

Parameter: a | spec. decay rate
0.500 1 1.500 2 2.500

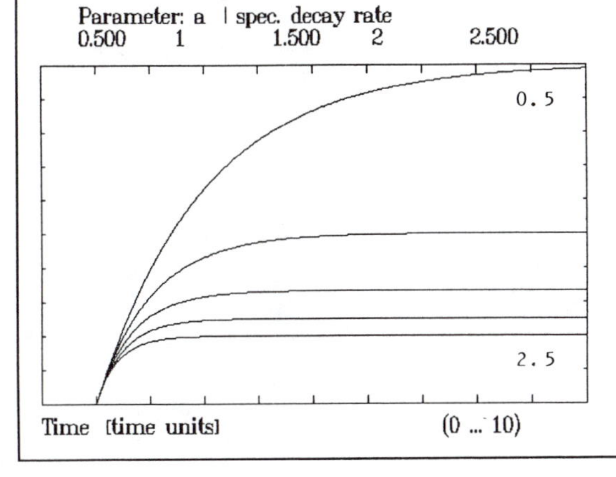

Time [time units] (0 ... 10)

TIME-DEPENDENT EXPONENTIAL GROWTH **M 104**

1. **Description:** The basic structure corresponds to that of exponential growth or decay. The growth rate dz/dt is proportional to the current state z and a specific growth rate u(t) which is a function of time. Since this specific rate of growth may change in time, the level may increase as well as decrease, depending on u(t).

2. **Occurrence:** A time-dependence of the specific growth rate is found, for example, in population development with time-dependent birthrates and deathrates as a function of the historical development of the medical system. Another example is the net primary production of plants as a function of the seasonally changing solar radiation. Economic development also is partly determined by time-dependent investment decisions.

3. **Structural characteristics:** The structure is determined by the self-coupling, where the specific rate of growth is now provided exogenously as a function of time.

4. **Behavioral characteristics:** The state z will change (increase or decrease) at a rate (speed) that depends on the exogenously determined time-dependent rate of change u(t).

5. **Critical parameters:** As in other self-loops, the sign of u(t) determines the increase or decrease of z, respectively. The absolute value of u(t) determines the speed of the state change.

6. **Reference run:** The initial growth rate (valid up to time t_1) and the final growth rate (valid after time t_2) are specified by the user. Between time t_1 and time t_2 the growth rate is linearly interpolated using the initial and final value. (This linear development of the specific growth rate u is produced by two ramp functions.) In the reference run the initial high growth rate of 0.1 [1/time unit] applies up to $t = 2$; thereafter the growth rate decreases linearly to a final value of 0.01 at time $t = 8$. Beyond $t = 8$, the growth rate remains constant at 0.01. As a result of this time development of the specific growth rate, there is initially a strong, then a decreasing, and finally a weak increase in z. The time plot shows the development for z, dz/dt, and u(t).

7. **Suggestions for further study:** Investigate different developments for the system state as a function of the scenario parameters a, b, c, and d (initial growth rate, beginning of rate change, final growth rate, end of rate change).

8. **References:** -

TIME-DEPENDENT EXPONENTIAL GROWTH M 104

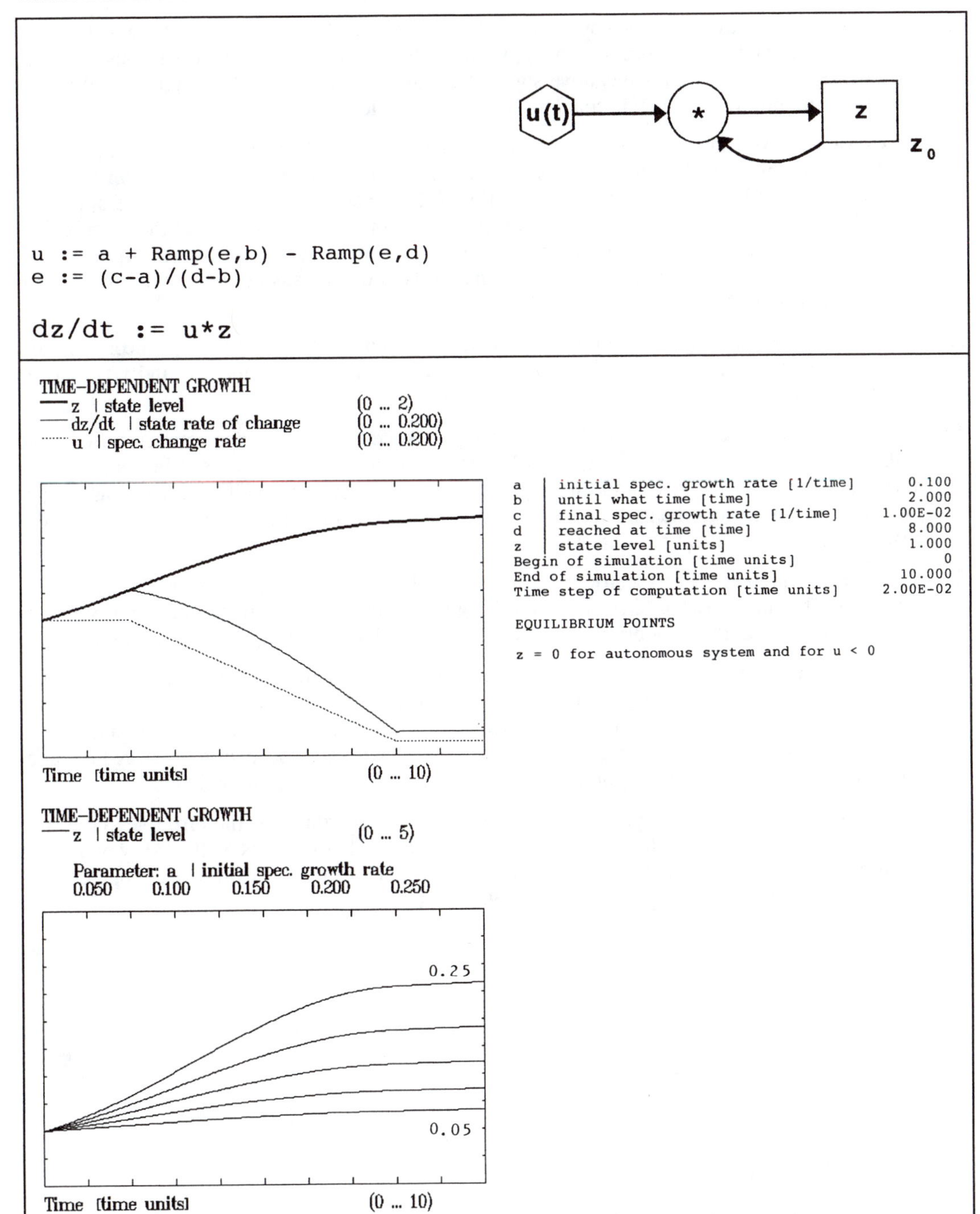

BIRTH AND DEATH: SIMPLE POPULATION DYNAMICS M 105

1. **Description:** The dynamic development of many systems is determined by simultaneous exponential growth and decay processes changing the system state z, i.e. the simultaneous existence of self-loops with positive and negative signs. A dynamic equilibrium (flow equilibrium) results if the specific inflow rate u(t) is equal to the specific outflow rate v(t).

2. **Occurrence:** An example is population development as a function of time-dependent changes of the specific birth- and deathrates. A stabilization of population occurs when the birthrate approaches the deathrate ("demographic transition"). In general, this system structure applies to state-dependent state development as a function of inflows and outflows, where the specific flow rates are time-dependent (for example, pollution concentration or CO_2 in the atmosphere). The same approach applies to the computation of the levels of orders, stocks, and capital in economics and management .

3. **Structural characteristics:** State-dependent inflows and outflows, where the specific inflow and outflow rates are determined by time-dependent exogenous functions u(t) and v(t) that are independent of each other.

4. **Behavioral characteristics:** The absolute change of state is the difference between inflows and outflows in a given time. Even at high inflow rates, the level may decrease if losses are larger than the gains. On the other hand, the level may increase even at a small inflow rate as long as the loss rate is less than the gain rate. Obviously, the net rate of change $r(t) = u(t) - v(t)$ is decisive for the development of the system.

5. **Critical parameters:** The specific net rate of growth r(t) clearly determines the time-development of the state variable z(t). By controlling one or both specific rates u(t) and v(t), one can therefore obtain different developments of the state value including flow equilibrium.

6. **Reference run:** In the reference run, birth- and deathrates remain constant from 1990 to 2000. Subsequently there is a linear reduction of the deathrate from 0.015 to 0.012 (between 2000 to 2020) as well as a linear reduction of the birthrate from 0.04 to 0.01 (between 2000 and 2050). This results in a population increase until the year 2047 with a subsequent gradual reduction of the population level. (The initial numbers are fairly typical of developing nations.)

7. **Suggestions for further study:** Start by assuming different values for the birthrate and deathrate, keeping them constant for the whole period. Explore what happens as they vary relative to each other. Then assume different time-dependent scenarios for the birthrate development and investigate the corresponding population development.

8. **References:** -

BIRTH AND DEATH: SIMPLE POPULATION DYNAMICS

M 105

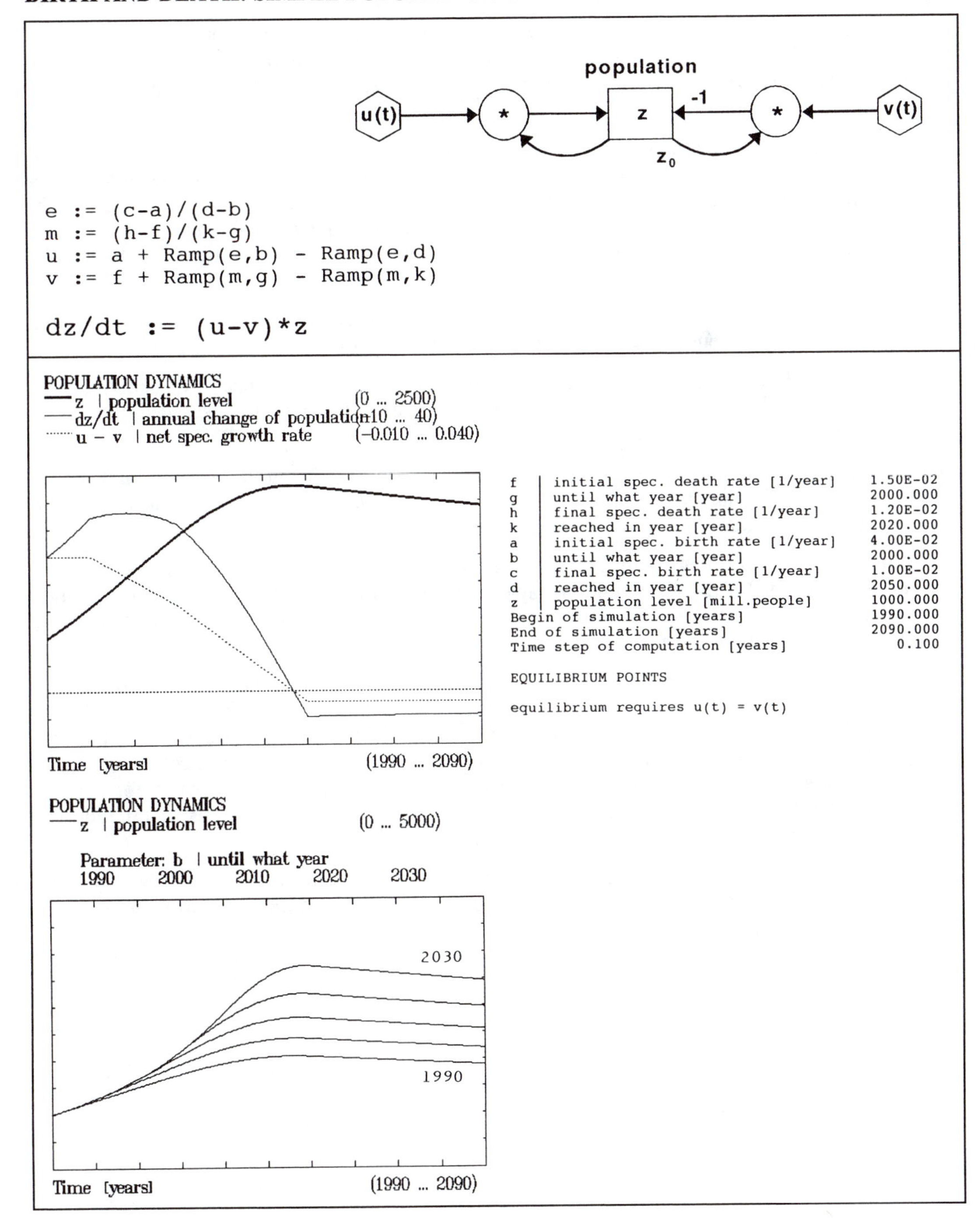

OVERLOADING A BUFFER

M 106

1. **Description:** The dynamics of this system are characterized by a slow depletion of the stock z with a specific loss rate a as long as only a part of the available storage volume is used. If the reservoir is filled beyond its capacity limit k, the surplus will flow out of the system at a much higher specific rate b.

2. **Occurrence:** Overflow of a dam after a rainstorm, fast run-off of rainwater from water-saturated soil, and slow seepage of water from soil which is not yet water-saturated. Overloading of body organs (e.g. uptake of iodine by the thyroid gland); if the capacity limit is exceeded, substances are eliminated at higher rates. Pent-up frustration giving way to an angry outbreak. This phenomenon is quite common in everyday life: "I am fed up" or "The straw that breaks the camel's back."

3. **Structural characteristics:** Normal state-dependent drainage with activation of an additional high state-dependent outflow when the capacity limit of the reservoir is exceeded.

4. **Behavioral characteristics:** Slow (exponential) loss (decomposition, decay, breakdown, leaking) as long as the state level is less than the capacity limit. If this limit is exceeded, there is an outflow of the surplus at a higher rate.

5. **Critical parameters:** The behavior of the system depends strongly on the capacity limit k and the normal specific loss rate a: if k and/or a are large, an overloading is less likely to occur. For the condition ($ka > u$), no overloading will occur.

   With respect to soil-water dynamics this means, for example, that overloading (water logging) is not to be expected if the water holding capacity of the soil is high, the rate of seepage is high, and/or the input from rainfall u(t) is small. These conditions are not satisfied for a thin clay layer after a heavy rainstorm, for example .

6. **Reference run:** If the capacity is relatively low (here $k = 1$), an input pulse u(t) at the time t produces an input which temporarily raises the level beyond the capacity k, therefore leading to a brief "flood." If the capacity is increased (to $k = 2$), a "flood" will not occur under otherwise equal conditions; there is only a small change in the normal outflow rate. The constancy of outflows caused by the storage capacity of the system can be observed in many springs, for example which often have an almost constant rate of flow despite strong fluctuations in rainfall. The same function is served by dams in river systems.

7. **Suggestions for further study:** Study the consequences of repeated "rainstorms" by using a sequence of pulses. Determine the influence of the pulse interval and the pulse area on the results and on the "flooding" of the system.

8. **References:** -

OVERLOADING A BUFFER

M 106

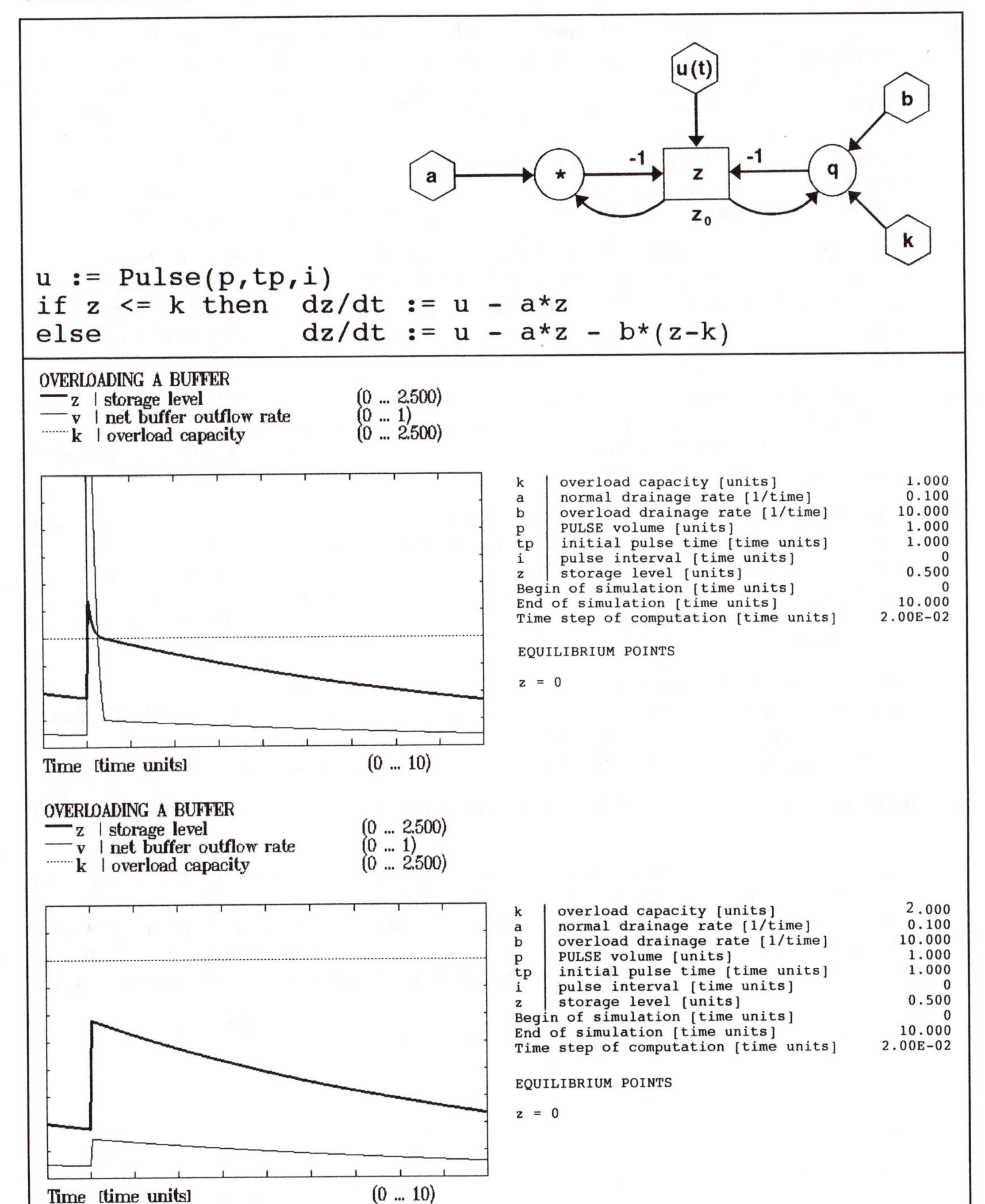

LOGISTIC GROWTH WITH CONSTANT HARVEST RATE M 107

1. **Description:** The logistic growth process is of fundamental importance; it can be found in almost all fields: ecology, economy, technology, etc. At small levels of the state variable z growth occurs almost exponentially. However, as the level approaches the capacity limit k, the negative feedback of the system increases in strength until it finally reduces net growth to zero as the level reaches the capacity limit. The time development of the state variable has a typical S-shaped form. If the stock (e.g. a population) is harvested at a constant (absolute) harvest rate, a dynamic equilibrium will establish itself if the harvest rate is small enough. However, if the harvest rate exceeds a critical value, there will be an unavoidable collapse of the population.

2. **Occurrence:** Growth and harvesting of animal and plant populations at a constant (absolute) harvest rate. Economic development with fixed taxes or duties and increase of taxes beyond a critical level. Burdening of an ecosystem with pollution beyond its capacity for regeneration, leading to ultimate collapse. Saturation processes of any type (product sales, innovation, human settlement,etc.).

3. **Structural characteristics:** An exponential growth process (positive self-loop) is multiplied by a variable factor f which approaches 0 as the state level approaches the capacity limit: $f = (1-z/k)$. For this model, the harvesting rate is assumed to remain constant and to be independent of the level of the state variable z.

4. **Behavioral characteristics:** As long as the state level is small, the saturation term f has little influence on the growth rate dz/dt: the level grows (almost) exponentially. As the level approaches the capacity limit, the growth rate is reduced to 0. Without a harvest, the equilibrium is at the capacity limit k. At constant harvest rate h, a stable equilibrium results as long as $h < a \cdot z\,(1 - z/k)$. The maximum possible harvest rate $h = a \cdot k/4$ is exactly at the stability limit. If it is exceeded only minimally, the system collapses.

5. **Critical parameters:** The critical parameter of the system is the harvest rate h. If a critical limit is exceeded, the system collapses. The system has two equilibrium points of which only one is stable. If a negative value is chosen for k (k then has no longer a meaning of a capacity limit), the state grows to an infinite value in finite time ("finite escape time;" $h = 0$).

6. **Reference run:** For $k > 0$ and subcritical harvest rate the system shows logistic growth without collapse.

7. **Suggestions for further study:** Using the option "Parameter Sensitivity," investigate the development of a population for different growth rates a, starting with different initial conditions, at harvest rates from $0.1 < h < 0.9$. In each case, determine the stable equilibrium point analytically (if it exists), and use simulation to confirm it. Use the system to study finite escape time with $h = 0$ and $k = -1$, and use the "Parameter Sensitivity" option to study the behavior for $0.1 \leq a \leq 0.5$.

8. **References:** Luenberger, 1979: p. 317-319; Richter, 1985.

LOGISTIC GROWTH WITH CONSTANT HARVEST RATE

M 107

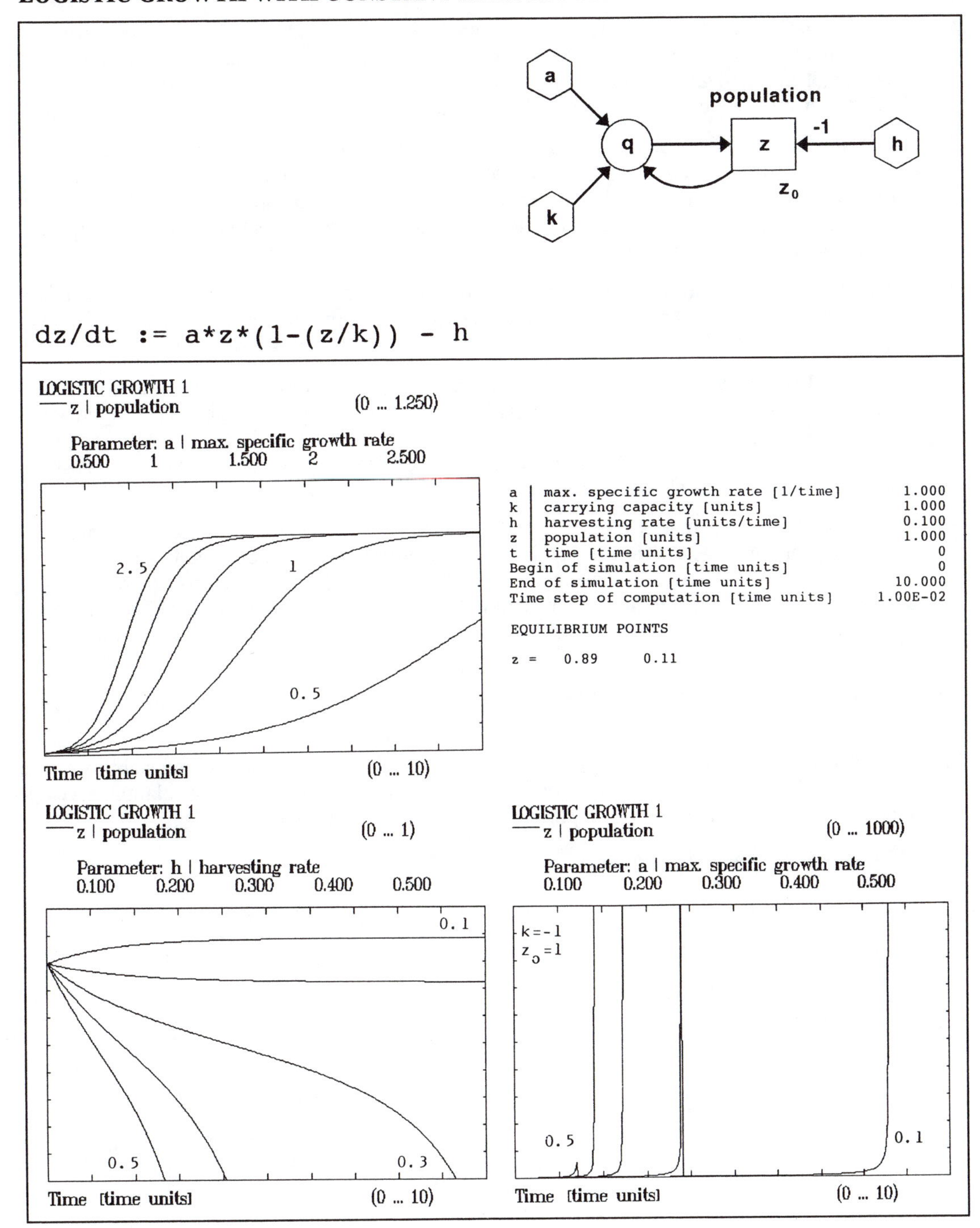

LOGISTIC GROWTH WITH PREY-DEPENDENT HARVEST — M 108

1. **Description:** A population z growing logistically is harvested at a rate which is proportional to the current level. The specific harvest rate g may be an arbitrary (positive) constant. With this type of harvesting, a collapse of the system cannot occur. The system state approaches an equilibrium value which is a function of the specific harvest rate.

2. **Occurrence:** Growth of populations if the harvest rate is state-dependent (or if it depends on the population density). An important application in fishing: without fish locator technology, the fish catch decreases as the population density decreases. The population can therefore not collapse. Economic example: taxes that are dependent on the currently active production capacity. Saturation processes of any kind.

3. **Structural characteristics:** The exponential growth process $(a \cdot z)$ with growth rate a is modified and eventually reduced to 0 by the capacity-dependent saturation term $(1 - z/k)$ as the level approaches the capacity limit k. The harvest $(g \cdot z)$ depends directly on the current level.

4. **Behavioral characteristics:** Logistic growth with S-shaped saturation. If the population level is small there is only a small harvest; overexploitation is therefore not possible. (As z approaches 0, the product $g \cdot z$ also approaches 0.) Equilibrium establishes itself at $z = k \cdot (1 - g/a)$, if $g \leq a$. The maximum specific harvest rate is given by $g/a = 0.5$; this leads to maximum possible harvest rate $g \cdot z = a \cdot k\,(1 - 0.5)/2 = a \cdot k/4$.

5. **Critical parameters:** In this case the specific growth rate a is not critical for the existence of the population but it determines the population at equilibrium and the harvest rate. For maximum harvest yield, the specific harvest rate should correspond to half the growth rate a. The maximum harvest is equal to the case of state-independent constant harvest (M 107) but here the maximum harvest corresponds to a stable condition; there is no threat of collapse. For $g = 0$ and $k < 0$ the system again produces finite escape times.

6. **Reference run:** For the assumed specific harvest rate of $g = 0.1$ a stable equilibrium results at $z = 0.9$.

7. **Suggestions for further study:** Compare the behavior, the possible harvest and stability to the system M 107 LOGISTIC GROWTH AT CONSTANT HARVEST. Investigate the population development for different initial conditions, growth rates a, and harvest rates g.

8. **References:** Richter, 1985.

LOGISTIC GROWTH WITH PREY-DEPENDENT HARVEST **M 108**

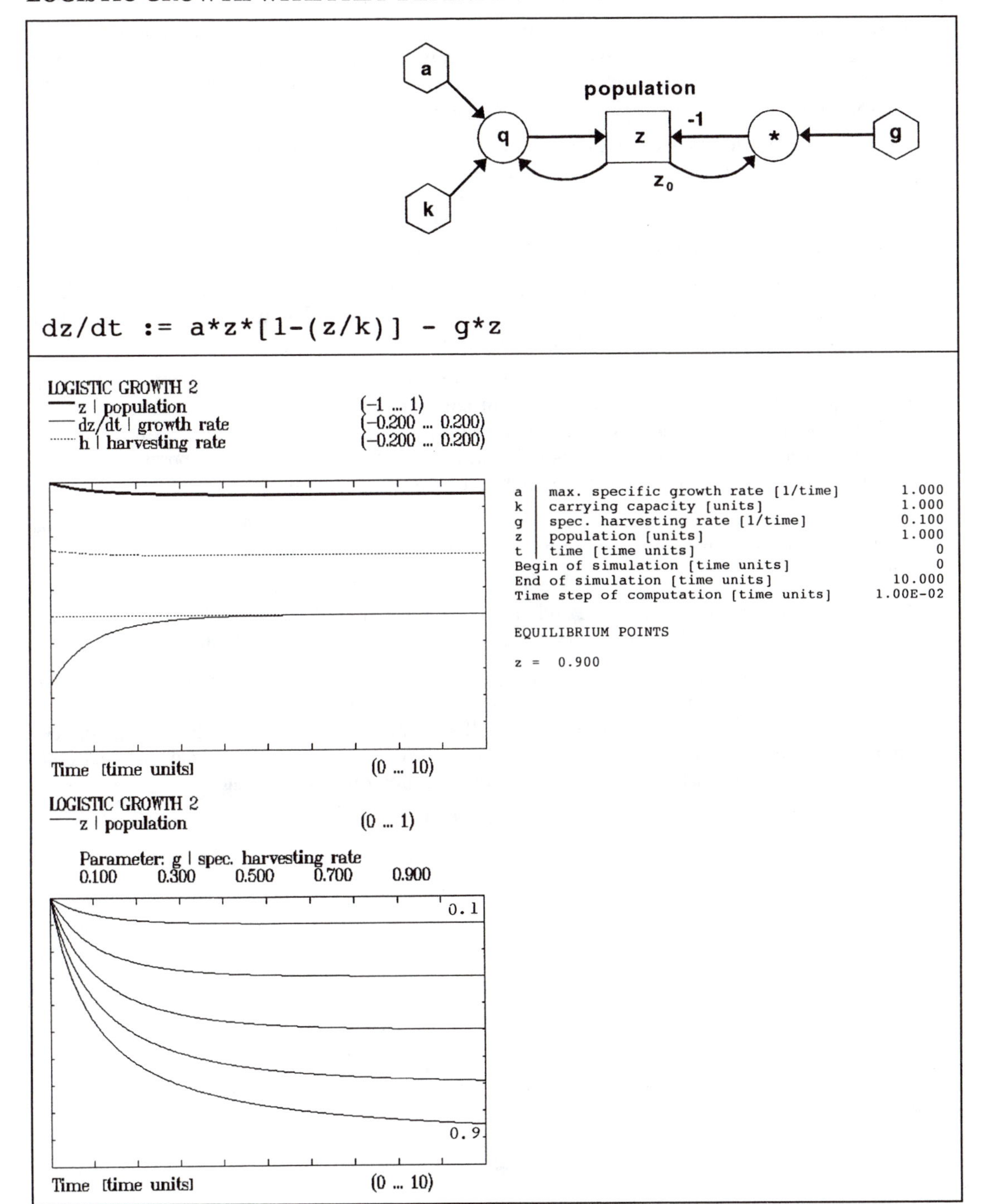

DENSITY-DEPENDENT GROWTH (MICHAELIS-MENTEN) M 109

1. **Description:** In this system formulation, a saturation term of the Michaelis-Menten form $(z/(z+c))$ is used to produce an S-shaped saturation function. The remaining structure is similar to that of the logistic growth process with an exponential growth loop $a \cdot z$ and a harvest (or the corresponding mortality) $g \cdot z$ depending on population density.

2. **Occurrence:** Michaelis-Menten kinetics are used to describe certain growth processes in biology and ecology (e.g. plant growth as a function of carbohydrate or nutrient availability, predator growth as a function of prey availability), certain chemical processes depending on the concentration of chemical compounds, and other saturation processes. In contrast to logistic growth which saturates at a capacity limit k, this description contains no explicit capacity limit.

3. **Structural characteristics:** The saturation term $(z/(c+z))$ is 0 for $z = 0$, 0.5 for $z = c$, 1 for z approaching infinity. The parameter c is therefore called the "half saturation constant," as half of the maximum saturation effect occurs when $z = c$.

4. **Behavioral characteristics:** In the system formulation shown here, linear growth results even for z approaching infinity if the system is not harvested (or no deaths occur). A saturation is only possible if $g > 0$. The condition for equilibrium is $z = (c \cdot a/g) \cdot (1 - g/a)$. Since the harvesting is proportional to the current population level, overexploitation is not possible; the system will not collapse even at high specific harvest rates g as long as $g < a$.

5. **Critical parameters:** Saturation occurs only if $g > 0$. The system cannot collapse as long as $g < a$. The equilibrium state is more critically dependent on the ratio a/g than the logistic system.

6. **Reference run:** For $g = 0.5$, $c = 1$, and $a = 1$, the saturation value is $z = 1$. Starting with an initial value $z_0 = 0.01$, the level grows in an S-shaped curve towards the final value $z = 1$.

7. **Suggestions for further study:** Analyze the behavior of the system for different specific growth rates a at constant values of $g \neq 0$. Investigate the behavior for different g at constant a. Study the behavior for different values of the half saturation constant c. Analyze the global behavior for different initial conditions by using the SIMPAS option "Global Analysis" and plotting the results with z as vertical coordinate and t as horizontal coordinate.

8. **References:** Richter, 1985: p. 91-103, 175-179, 182-187.

DENSITY-DEPENDENT GROWTH (MICHAELIS-MENTEN) M 109

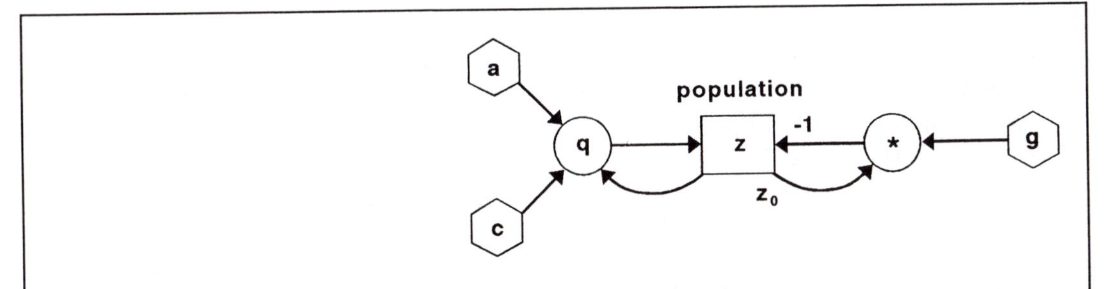


```
dz/dt := a*z*[1-(z/(c+z))] - g*z
```

DENSITY-DEPENDENT GROWTH
— z | population (0 ... 1.000)
— dz/dt | growth rate (0 ... 0.100)

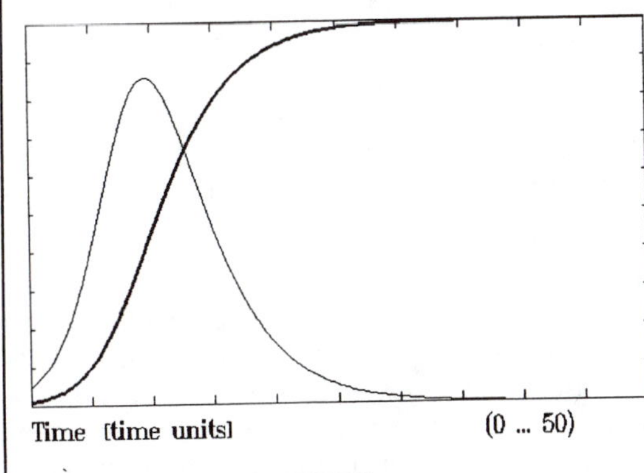

a	max. growth rate [1/time]	1.000
c	half-saturation constant [units]	1.000
g	harvesting rate [1/time]	0.500
z	population [units]	1.00E-02
t	time [time units]	0
Begin of simulation [time units]		0
End of simulation [time units]		50.000
Time step of computation [time units]		1.00E-02

EQUILIBRIUM POINTS

z = 1.000

DENSITY-DEPENDENT GROWTH
— z | population (0 ... 1.000)

Parameter: a | max. specific growth rate
0.600 0.700 0.800 0.900 1

1

0. 6

Time [time units] (0 ... 50)

DAILY PHOTOPRODUCTION OF VEGETATION M 110

1. **Description:** The model computes the daily course of photoproduction of a leaf canopy as a function of solar radiation and light attenuation in the different layers of the leaf canopy. Solar radiation is a function of the geographic latitude, the calendar day, and the time of day.

2. **Occurrence:** The model applies to the vegetation of all terrestrial ecosystems, i.e. to forests, meadows, fields, bushland, etc.

3. **Structural characteristics:** The incoming solar radiation s is computed from the solar elevation e which depends on the geographic latitude l, solar declination d, and atmospheric turbidity (c, a). The solar declination d depends on the calendar day n. The radiation energy x received during one day is the time integral of the instantaneous radiation power s. The day length y is the time integral of light hours. The net production rate q of the leaf canopy is found by analytical integration of leaf production in i leaf layers, after accounting for leaf respiration r. The production of each leaf layer depends on light attenuation k by higher leaf layers and the photoproductivity (p, m) of individual leaves. Integration of q over time gives the daily net production z of the canopy layer.

4. **Behavioral characteristics:** Corresponding to the solar position, there is a sinusoidally increasing radiation with a maximum at noon. The beginning and end of the radiation correspond to the time of sunrise and sunset. Photoproduction in the leaf canopy follows the radiation distribution. However, it has a broader shape as full leaf production already occurs at low radiation values (light saturation effect). Some of the photoproduction during daytime is used for respiration at night; the production curve therefore has a negative contribution during the night. In summer, high production is possible in the polar latitudes as a result of the long daylight hours.

5. **Critical parameters:** Leaf respiration r accounts for relatively high energy losses, especially during days with shorter daylight hours. At lower levels of the canopy, leaves receive much less light (light attenuation k). If their net production is negative, they are shed by the plant. This leads to a maximum leaf area index i of about 5 (leaf area index = square meter of leaf area per square meter of surface area). On account of the length of the daylight period, the elevation-dependent solar radiation, and seasonal time there is a considerable effect of the geographic latitude l. The shape of the light sensitivity curve is important for production: the parameter p determines the maximum photoproduction at light saturation while the parameter m characterizes the initial slope of the photosynthesis curve at low levels of radiation.

6. **Reference run:** The results given are for day 173 (i.e. June 22, summer solstice) and for 50 degree northern latitude (Frankfurt, Kiev, Vancouver, Winnipeg). The assumed leaf layer index of i = 5 corresponds to a broadleaf canopy. Canopy production is high during most of the 16 daylight hours.

7. **Suggestions for further study:** Analyze the photoproduction of the leaf canopy on other days of the year as well as for other geographic latitudes. Find out (using the SIMPAS option "Parameter Sensitivity") how the net photoproduction is affected by a change of leaf area index i (from 1 to 10). Which i yields the maximum net production? Study the role of the leaf respiration r and of the photosynthesis parameters m and p.

8. **References:** France and Thornley, 1984 : p. 114-121; Richter, 1985: p. 164-172.

DAILY PHOTOPRODUCTION OF VEGETATION M 110

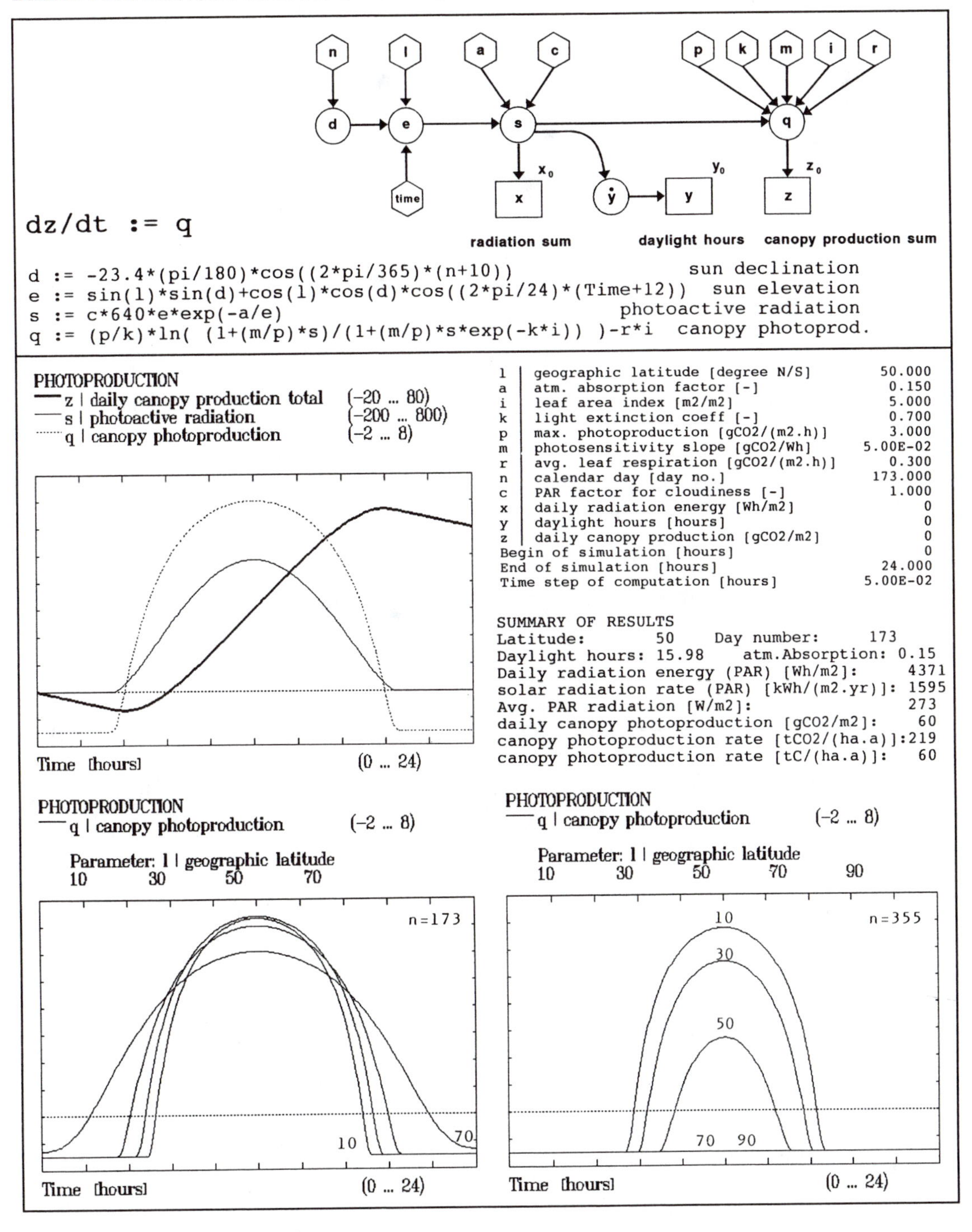

6.2 Dynamic Systems with Two State Variables

201 DOUBLE INTEGRATION AND EXPONENTIAL DELAY
202 TRANSITION FROM ONE STATE TO THE NEXT
203 LINEAR OSCILLATOR OF SECOND ORDER
204 ESCALATION
205 BURDEN SHIFTING AND DEPENDENCE
206 PREDATOR-PREY SYSTEM WITHOUT CAPACITY LIMIT
207 PREDATOR-PREY SYSTEM WITH CAPACITY LIMIT
208 COMPETITION
209 TOURISM AND ENVIRONMENTAL POLLUTION
210 OVERSHOOT AND COLLAPSE
211 FOREST GROWTH
212 RESOURCE DISCOVERY AND DEPLETION
213 TRAGEDY OF THE COMMONS
214 SUSTAINABLE USE OF A RENEWABLE RESOURCE
215 DISTURBED EQUILIBRIUM: CO_2-DYNAMICS OF THE ATMOSPHERE
216 STOCKS, SALES, ORDERS
217 PRODUCTION CYCLE
218 ROTATING PENDULUM
219 OSCILLATOR WITH LIMIT CYCLE (VAN DER POL)
220 BISTABLE OSCILLATOR
221 CHAOTIC BISTABLE OSCILLATOR

DOUBLE INTEGRATION AND EXPONENTIAL DELAY M 201

1. **Description:** The input u(t) is integrated twice with respect to time. Each integrator has a self-loop with negative sign (damping). Both feedback loops have the same magnitude $|a|$. If $a > 0$, these exponential damping processes will have a delaying effect (second order delay).

2. **Occurrence:** The double integration can be found in many important physical processes. For example, in mechanics the integration of the acceleration over time produces the velocity of a mass; a further integration of the velocity over time produces the position of the mass. An equivalent formulation is the time integration of the acceleration force ($m \cdot dv/dt$) to yield the impulse (mv); a second time integration then results in the kinetic energy ($mv^2/2$). A structurally equivalent double integration also follows from summing the voltage drops in an electrical system consisting of capacitor, resistor, and inductance.

3. **Structural characteristics:** If the self-couplings are absent ($a = 0$), the first integrator y will yield the single, the second integrator x the double integration of the input function u(t). If exponential damping is present at both integrators ($a > 0$), there will be at each instant a partial "loss" (leakage) of the state as accumulated in the course of the system development. This results in a delaying effect (cf. Model M 103 EXPONENTIAL DELAY).

4. **Behavioral characteristics:** The undamped integration ($a = 0$) corresponds to the analytical integration rules. If both integrators have damping ($a > 0$), the original signal which was modified and delayed in the first integrator, will be modified and delayed again in the second integrator.

5. **Critical parameters:** For a strong negative feedback (large absolute value of a) the delay effect is small since the time constant of the system $T = 1/a$ is small. The damping factor a determines the equilibrium values at the two integrators for a constant input u = const. One obtains $y^* = u/a$, $x^* = u/a^2$.

6. **Reference run:** The input function is a step function starting at $t = 1$. Both integrators have damping $a = 1$. The results initially show a linear increase of the level of the first integrator y and a parabolic increase of the second integrator x. Eventually, both integrators saturate at the equilibrium values y^* and x^* given above. The input signal was delayed twice. The delay time over both integrators follows from the sum of both time constants $T_D = 1/a + 1/a = 2/a$.

7. **Suggestions for further study:** Confirm the statements concerning the equilibrium values (under Point 5) for different (positive) values of a. Study the response of the system to the different test functions (pulse, step, ramp, sine). Construct your own test function u(t) having some interesting shape as a sum of test functions pulse, step, ramp, and sine. Use ($a = 0$) to produce the first and second numerical integral of this function. Use several simulations to determine how the input u is distorted and delayed for small and large damping factors a.

8. **References:** -

DOUBLE INTEGRATION AND EXPONENTIAL DELAY

M 201

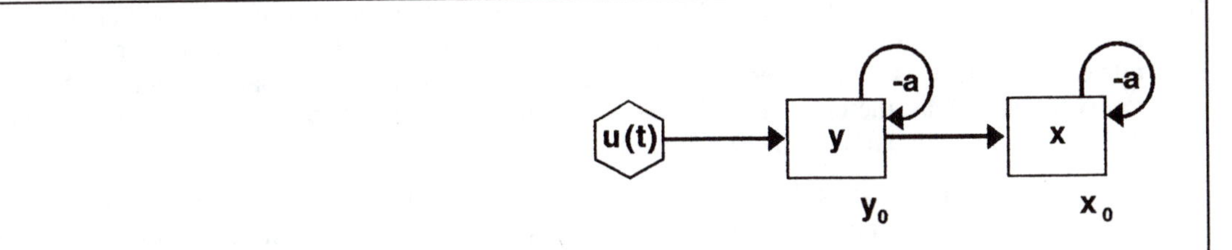


```
u := Pulse(p,tp,i) + Step(s,ts) + Ramp(r,tr) + q*sin(2*pi*w*time)
dx/dt := y - a*x
dy/dt := u - a*y
```

DOUBLE INTEGRATION
x | state1 (0 ... 1.250)
y | state2 (0 ... 1.250)
u | exogenous change rate (0 ... 1.250)

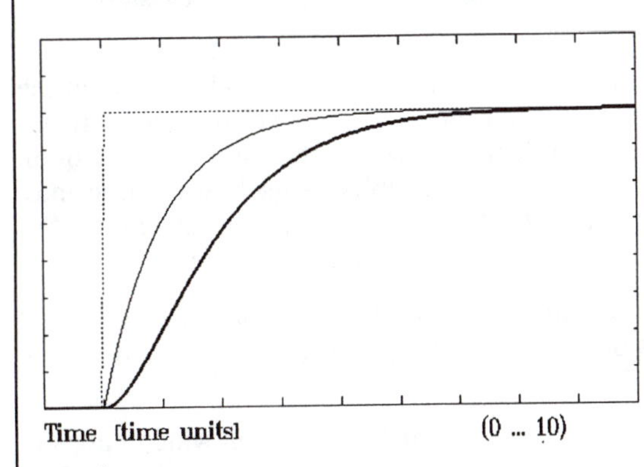

a	spec. decay rate [1/time]	1.000
p	PULSE volume [units]	0
tp	initial pulse time [time]	1.000
i	pulse interval [time]	1.000
s	STEP height [units/time]	1.000
ts	step time [time]	1.000
r	RAMP slope [units/time^2]	0
tr	ramp time [time]	1.000
q	SINE wave amplitude [units/time]	0
w	sine wave frequency [rad/time]	0.100
x	state1 [units*time]	0
y	state2 [units]	0
Begin of simulation [time units]		0
End of simulation [time units]		10.000
Time step of computation [time units]		2.00E-02

EQUILIBRIUM POINTS

depend on input; x = y = 0 for autonomous system
y = u/a, x = u/a² for u = const

DOUBLE INTEGRATION
x | state1 (0 ... 5)

Parameter: a | spec. decay rate
0.400 0.800 1.200 1.600 2

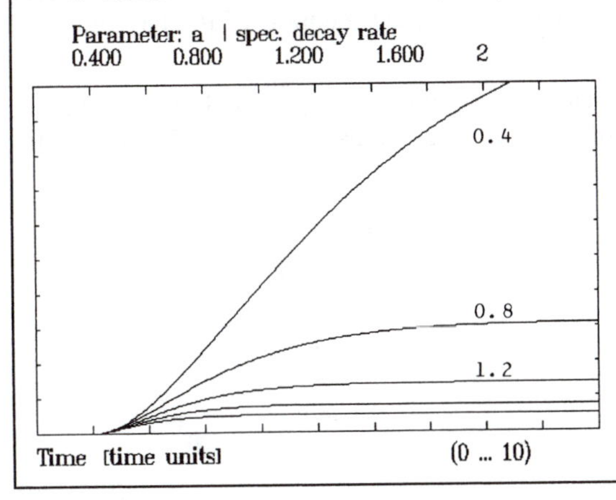

TRANSITION FROM ONE STATE TO THE NEXT **M 202**

1. **Description:** The model describes the gradual transfer of the contents of one reservoir into the next. The transition rate is proportional to the content of the first (donor) reservoir. The loss rate of reservoir y is equal to the gain rate of reservoir x. The process may continue at the next reservoir; both may therefore have losses which are proportional to their corresponding level (self-couplings with damping factors b and a).

2. **Occurrence:** This process describes the transitions between age classes in population models, seepage through different soil layers (water, nutrients, chemicals), energy transfers in food chains, etc.

3. **Structural characteristics:** Transition losses at the donor reservoir have to be subtracted there and are then added to the receiver reservoir. The transition rate is proportional to the content of the donor reservoir. However, there are other possibilities (not implemented here): transition rates may also be determined by the receiving reservoir or by a combination of the influences of the donor and receiver reservoirs as, for example, in the predator-prey system. Note: the (normalized, specific) transition rate b can be used to compute the average residence time in the donor reservoir y. If, for example, $b = 0.1$, 1/10 of the population will leave the donor reservoir per time unit. The average residence time is therefore the inverse of the transition rate, here 10 time units.

4. **Behavioral characteristics:** The response of the first integrator y corresponds exactly to the exponential delay with a feedback factor b. For a constant input u, the level will therefore adjust to an equilibrium value of $y^* = u/b$. The second integrator also has the basic structure of an exponential delay. The equilibrium value of $y^* = u/b$ which establishes itself at the first integrator y will eventually lead to an equilibrium value of $x^* = u/a$ at the second integrator. For a time-dependent input $u(t)$, the system will again produce a delay effect at each level.

5. **Critical parameters:** The transition rates between the reservoirs (here b) and the loss rates of the reservoirs (here b, a) determine the rates of change of the system, the equilibrium states, and the time constants and delays.

6. **Reference run:** The input function of the first integrator is a unit step function which jumps to a value of 1 at time $t = 1$. At transition rates of $b = 0.5$ and $a = 0.25$, the corresponding equilibrium values of the two integrators are found as $y^* = 2$ and $x^* = 4$. The transition rates chosen in this case mean that $b = 1/2$ of level y will be lost per time unit, while the state variable x loses $a = 1/4$ of its level per time unit.

7. **Suggestions for further study:** Study the behavior of the state variables y and x for different values of b and a. Investigate how different test functions (in particular sine functions) are processed by the system.

8. **References:**

TRANSITION FROM ONE STATE TO THE NEXT

M 202

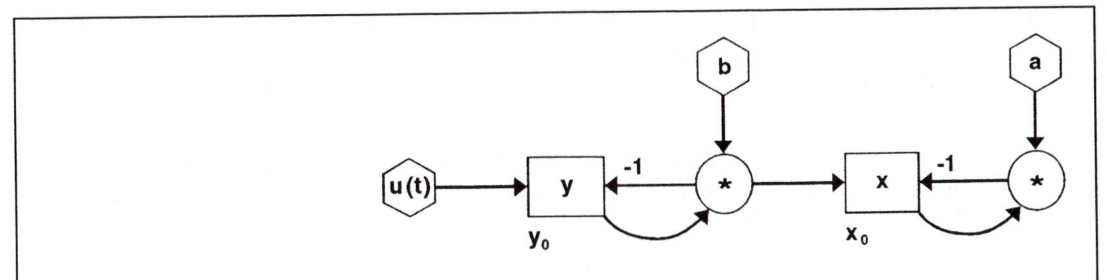


```
u := Pulse(p,tp,i) + Step(s,ts) + Ramp(r,tr) + q*sin(2*pi*w*time)
dx/dt := b*y - a*x
dy/dt := u    - b*y
```

STATE TRANSITION
x | state1 (0 ... 5)
y | state2 (0 ... 5)
u | exogenous change rate (0 ... 5)

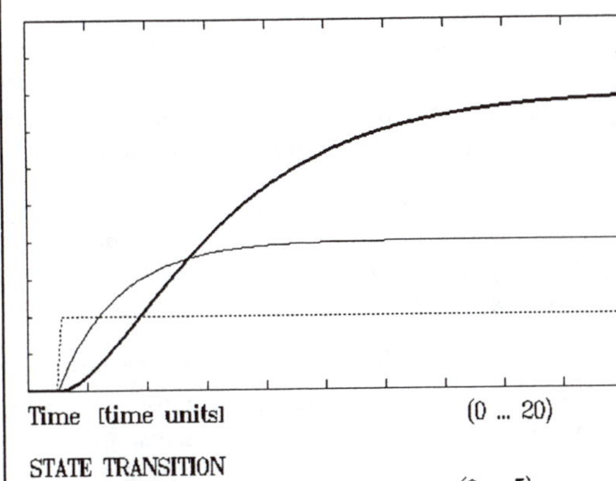

Time [time units] (0 ... 20)

STATE TRANSITION
x | state1 (0 ... 5)

Parameter: b | transition rate from y to x
0.200 0.400 0.600 0.800 1

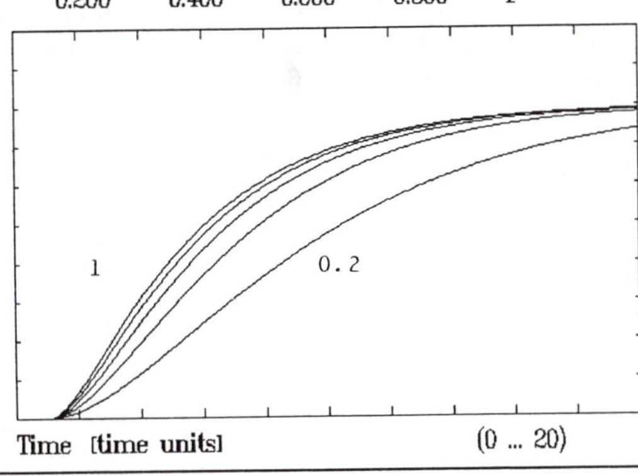

Time [time units] (0 ... 20)

b	transition rate, y to x [1/time]	0.500
a	loss rate from x [1/time]	0.250
p	PULSE volume [units]	0
tp	initial pulse time [time]	1.000
i	pulse interval [time]	1.000
s	STEP height [units/time]	1.000
ts	step time [time]	1.000
r	RAMP slope [units/time^2]	0
tr	ramp time [time]	1.000
q	SINE wave amplitude [units/time]	0
w	sine wave frequency [rad/time]	0.100
x	state1 [units*time]	0
y	state2 [units]	0
Begin of simulation [time units]		0
End of simulation [time units]		20.000
Time step of computation [time units]		2.00E-02

EQUILIBRIUM POINTS

depend on input; x = y = 0 for autonomous system
x = u/a, y = u/b for u = const

LINEAR OSCILLATOR OF SECOND ORDER **M 203**

1. **Description:** A feedback loop with coupling parameters b and c links both state variables y and x. This may cause the system to oscillate. Depending on the signs of the self-loop parameters d and a, the response may be damped or amplified. In addition, other (aperiodic) responses are possible.

2. **Occurrence:** Systems of this nature which may show oscillations, are found in many different areas: mechanical spring-mass-damper system; pendulum (for small angle); electrical circuit; interaction of market, capital, and production.

3. **Structural characteristics:** The behavioral characteristics are crucially determined by the feedback loop linking both state variables.

4. **Behavioral characteristics:** Depending on the eigenvalues λ_1, λ_2 of the system matrix, several distinct modes of behavior, which may be stable or unstable, may occur: source and sink, focus and center, saddle, node, line source, line sink (cf. Ch. 7.3.13). By a corresponding selection of the coefficients a, b, c, and d of the system matrix, each of these modes can be generated with the model. The system is linear and in each case (except for the line source and sink) has only a single equilibrium point which for autonomous motion ($u = 0$, unforced motion) is independent of the initial conditions and is always given by $x = 0$ and $y = 0$.

5. **Critical parameters:** A small change in one or several of the system parameters may result in a qualitative change of behavior since it may cause a shift of the eigenvalues to another region in the complex number plane. Eigenvalues in the right half of the complex number plane (positive real parts) always indicate instability; eigenvalues with imaginary parts indicate oscillations.

6. **Reference run:** For the system parameters $a = 0$, $b = 1$, $c = -1$, and $d = -1$, the resulting eigenvalues are $\lambda_{1,2} = -0.5 \pm 0.866i$. Both eigenvalues are complex conjugates; the imaginary part of both eigenvalues indicates oscillation, the real part with the negative sign means damping. The characteristic frequency (natural frequency) is $x_0 = (|b \cdot c|)^{1/2}/(2\pi) = 0.1592$.

7. **Suggestions for further study:** As damping is increased, the oscillation of the system will eventually disappear. At which value of d does the periodic motion change into aperiodic motion? How are response and stability of the system affected by applying different test functions (pulse, step, ramp) as inputs? What is the consequence of changing the sign of parameter c for the behavior of the system? Try to generate all the different responses listed under Point 4 by changing the parameters in the system matrix. (Consult Sec. 7.3.11 – 7.3.15.) Note: for all runs use $a = 0$ and values of either $+1$ or -1 for b, c, and d. Study the response (in particular resonance) of the system to forced oscillation by using the sine test function and the default parameter values in the program. Note: plot the amplification as a function of frequency in a log/log plot (cf. Fig. 7.8).

8. **References:** Linear systems theory, in particular control systems theory, for example: Luenberger, 1979.

LINEAR OSCILLATOR OF SECOND ORDER **M 203**

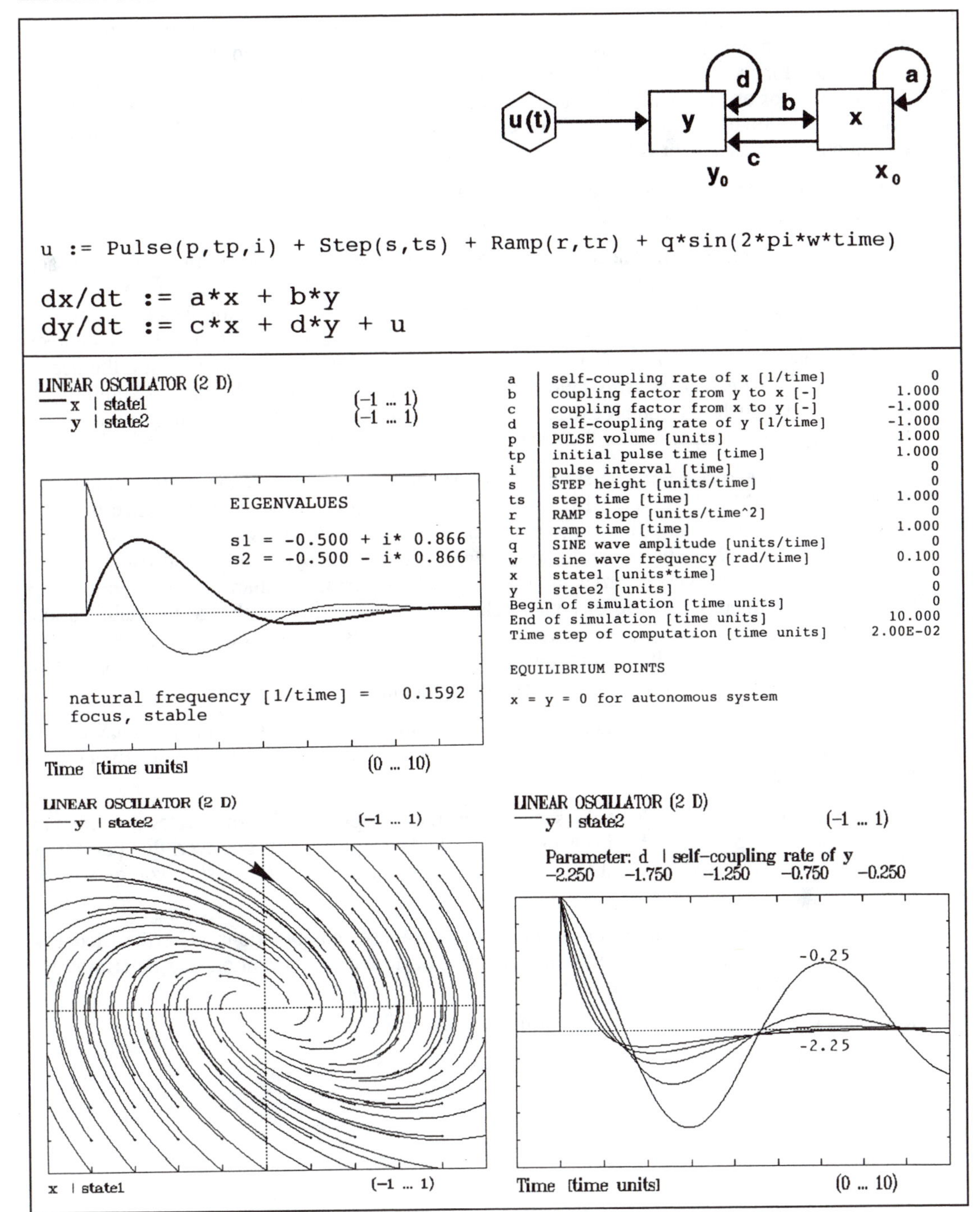

ESCALATION

M 204

1. **Description:** The model describes the mutual influence of two competing actors X and Y on each other, each of which controls a corresponding state variable x and y which is behaviorally relevant for the other actor. Examples: aggression, armament, advertising, status symbols. The behavior of one actor therefore depends on the behavior of the other and in particular on his/her perception of this behavior. Exaggeration of the potential of the other leads to spiraling escalation where each actor is forced to increase his or her own potential further.

2. **Occurrence:** Armament race, status symbol dynamics, competition of communities, sports, price and wage spiral, family conflicts, and advertising battles.

3. **Structural characteristics:** The system is explained here with reference to the armament race. Actor Y determines his investment in arms by comparing his own arms level y to that of his adversary X, exaggerating (or diminishing) the true ratio (x/y) by a perception factor q. This factor is equal to 1 if the ratio x/y is represented truthfully. The armaments of actor Y depreciate at the rate $(-b \cdot y)$. The rate of arms purchases is then determined by multiplying the depreciation rate by the perceived arms ratio: $b \cdot y \cdot q \cdot (x/y) = bqx$. Actor X behaves similarly. In addition, there may be additional exogenous rates of arms level increase h and g of actors X and Y, which are independent of the respective armament levels.

4. **Behavioral characteristics:** The system is linear; its elementary structure corresponds to the linear oscillator (Model M 203). However, oscillations cannot normally occur since both the coupling parameters $(p \cdot a, q \cdot b)$ are normally positive. Stable solutions are only possible if $p \cdot q < 1$, i.e. if one or both of the actors do not exaggerate the threat by the other and tend to take it less seriously. If, however, in order to "play it safe" the threat is somewhat exaggerated, this will lead to continuous increase of both levels of armament with no chance for stabilization. Similarly, if one actor exaggerates greatly, even if the other is trying to de-escalate the race, the aggressive actor can still pull the other one into an arms race. The race will end if one or both of the actors "play down" the threat by the other enough to make $p \cdot q < 1$.

5. **Critical parameters:** The exaggeration factors p and q are decisive for the behavior and stability of the system. If $g \cdot h > 0$, stability is possible only if $p \cdot q < 1$. If $p \cdot q = 1$, then stability is only possible if $g \cdot h = 0$.

6. **Reference run:** Both actors have the same parameters except that X exaggerates the threat by Y by 10 percent: $p = 1.1$. This leads to an unstable development. If, however, this perception of the threat by Y is reduced to a value under 1 (for example, $p = 0.5$), stabilization will result.

7. **Suggestions for further study:** Investigate other parameter combinations. Use the Model M 203 LINEAR OSCILLATOR with the parameter values selected for escalation and determine the eigenvalues and the stability of the system.

8. **References:** Luenberger, 1979: p. 206-209.

ESCALATION

M 204

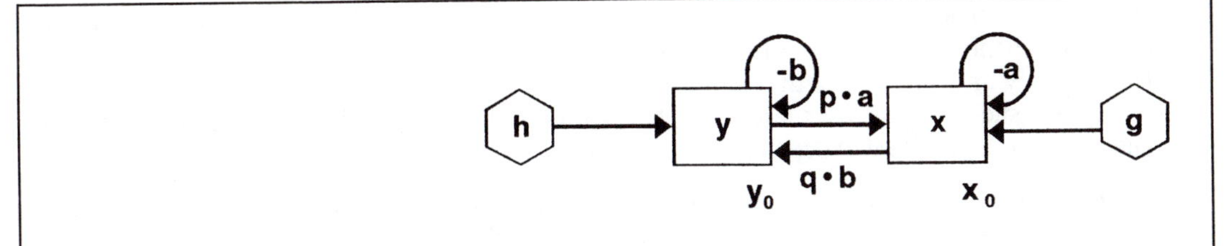


```
dx/dt = p*a*y - a*x + g
dy/dt = q*b*x - b*y + h
```

ESCALATION
— x | armament level of X (0 ... 10)
— y | armament level of Y (0 ... 10)

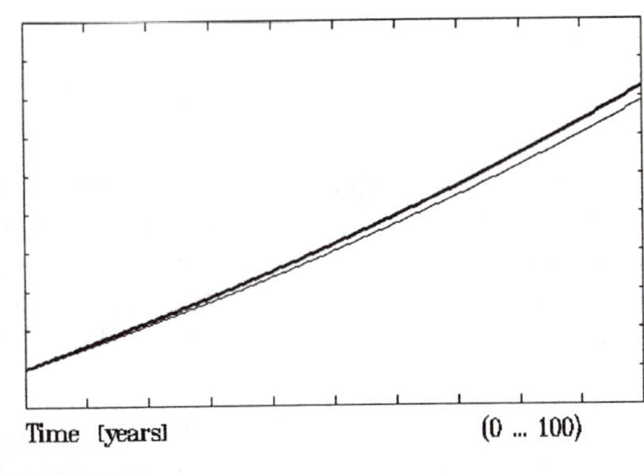

Time [years] (0 ... 100)

a	arms obsolescence rate of X [1/time]	0.100
b	arms obsolescence rate of Y [1/time]	0.100
p	X exaggerates Y by factor [-]	1.100
q	Y exaggerates X by factor [-]	1.000
g	auton. arms buildup rate, X [arms/yr	5.00E-02
h	auton. arms buildup rate, Y [arms/yr	5.00E-02
x	armament level of X [arms]	1.000
y	armament level of Y [arms]	1.000
Begin of simulation [years]		0
End of simulation [years]		100.000
Time step of computation [years]		0.100

EQUILIBRIUM POINTS

x, y = -10.500, -10.000
unstable

ESCALATION
— x | armament level of X (0 ... 2)
— y | armament level of Y (0 ... 2)

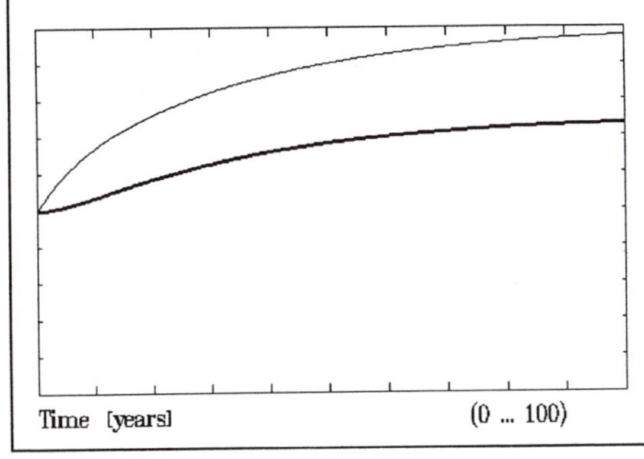

Time [years] (0 ... 100)

a	arms obsolescence rate of X [1/time]	0.100
b	arms obsolescence rate of Y [1/time]	0.100
p	X exaggerates Y by factor [-]	0.500
q	Y exaggerates X by factor [-]	1.000
g	auton. arms buildup rate, X [arms/yr	5.00E-02
h	auton. arms buildup rate, Y [arms/yr	5.00E-02
x	armament level of X [arms]	1.000
y	armament level of Y [arms]	1.000
Begin of simulation [years]		0
End of simulation [years]		100.000
Time step of computation [years]		0.100

EQUILIBRIUM POINTS

x, y = 1.500, 2.000
stable

BURDEN SHIFTING AND DEPENDENCE

M 205

1. **Description:** The model describes the accelerating loss of the ability for self-help and the increasing dependence on foreign help. This development may come about if an actor is able to partially rely on foreign help in addition to his own efforts at maintaining a system state. The more he or she does this, the more his or her ability for self-help (self-help capacity) will erode, and the more he or she will depend on foreign help.

2. **Occurrence:** Examples for this dynamic development are found in drug addiction, foreign aid, and the dependence of individuals, communities, or whole industries on state subsidies.

3. **Structural characteristics:** Despite some erosion r, the system state x is moved towards a state goal g partially by foreign aid f (with a specific help effect b), and partially by self-help (self-help capacity y and specific self-help effect a). The self-help capacity y erodes with a specific rate s and has to be renewed and improved at a specific rate c by constant use.

4. **Behavioral characteristics:** If foreign help f is available, the use of the self-help capacity will be reduced accordingly. This decreases the efforts for renewal and improvement of the self-help capacity and therefore also its possible contribution to the improvement of the system state x. If the foreign aid is too high, the self-help capacity finally erodes completely, and the system becomes completely dependent on foreign help.

5. **Critical parameters:** The foreign aid fraction f is critical. If it remains below a certain value, the self-help capacity will be maintained. If it climbs above a critical value, the self-help capacity will collapse.

6. **Reference run:** For subcritical foreign help ($f = 0.5$) the model shows maintenance and even some strengthening of the self-help capacity. The system then reaches a stable equilibrium state independent of the initial values of the state variables. (Use the option "Global Analysis" to study this.) For supercritical foreign aid ($f = 0.7$) however, there will be increasing erosion and destruction of the self-help capacity. The stable equilibrium state is then found for $y = 0$, i.e. for total reliance on foreign help.

7. **Suggestions for further study:** Using the option "Parameter Sensitivity," find the critical foreign aid fraction f where the self-help capacity will not completely disappear (for the parameter set of the reference run). Study the influence of the other parameters on the system response using the option "Parameter Sensitivity."

8. **References:** -

BURDEN SHIFTING AND DEPENDENCE

M 205

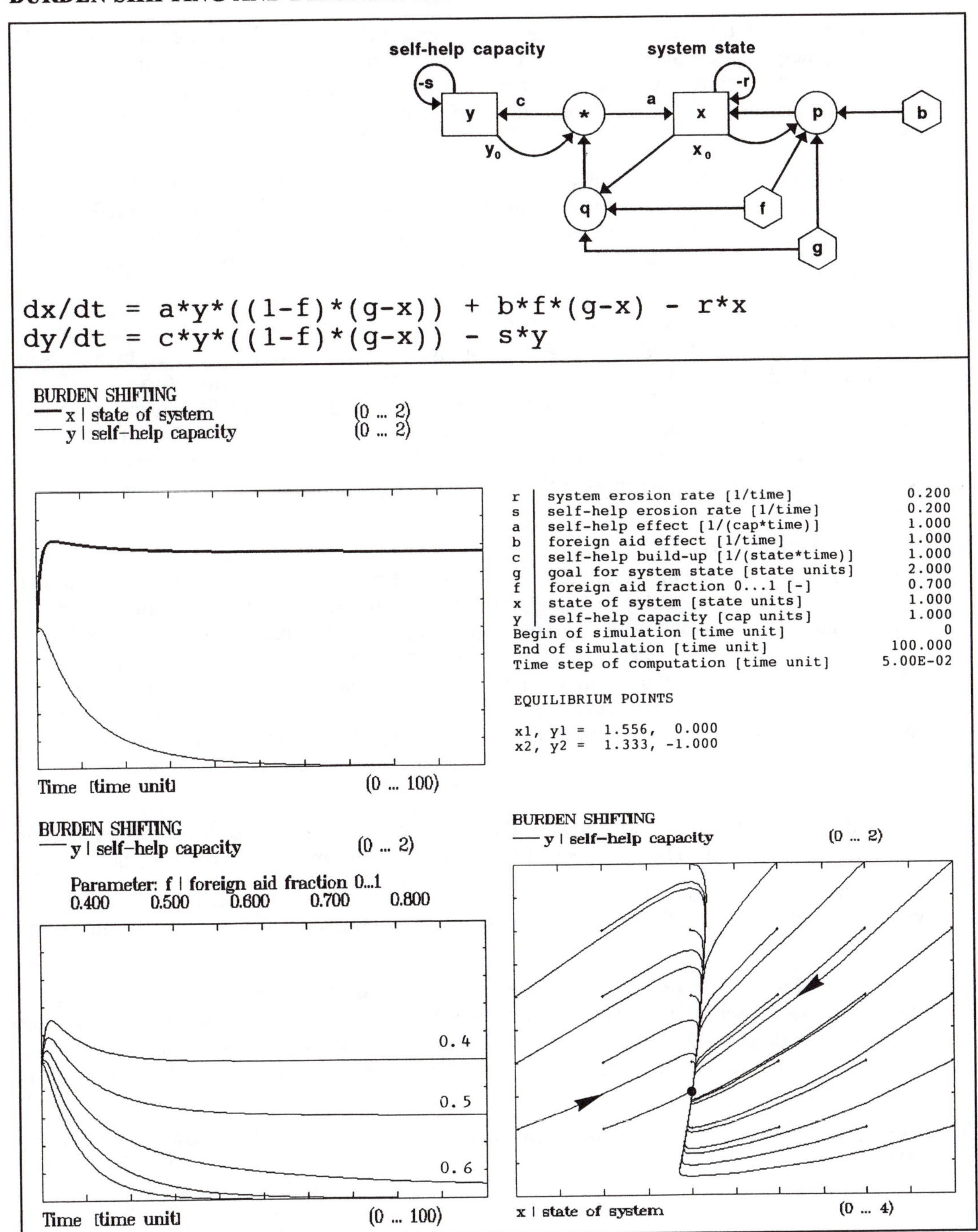

PREDATOR-PREY SYSTEM WITHOUT CAPACITY LIMIT M 206

1. **Description:** The interaction between predator y and prey x means losses for the prey population and (energy) gains for the predator. If the prey population, despite regeneration, shrinks to a very low level, then this will also reduce the energy supply to the predator population and impede its further growth. These interactions may lead to oscillations of the predator and prey populations.

2. **Occurrence:** The model is a general description of processes where a "predator" depends on "prey," and where the losses of the prey mean gains for the predator. Examples: predator-prey interactions in ecosystems (rabbits and foxes, fish and sharks, fishing boats and fish, sawmills and forests). In the socio-economic domain we also find this fundamental structure (slavery, producer and consumer, etc.).

3. **Structural characteristics:** The amount of prey seized depends on the amount of prey available x and on the number of predators y. Predation losses are therefore characterized by a non-linear term $x \cdot y$. Corresponding losses are subtracted from the prey population ($b \cdot x \cdot y$, loss factor b) and are added as gain to the predator population ($c \cdot x \cdot y$, gain factor c). The losses of the prey population are offset by the net growth rate of the prey population which is proportional to the population level and the specific net growth rate a. The predator needs the prey (energy gain) to compensate for its normal respiration losses (specific respiration rate d) and thereby maintain its life processes.

4. **Behavioral characteristics:** If the system is started with low initial values for the predator and prey populations, there will initially be a growth of the prey population in an exponential pattern. As a result of the quickly growing prey supply, there will be a corresponding increase in the predator population and of the amount of prey caught. This causes a subsequent decline in the prey population. The lack of prey eventually reduces the predator population again. This process leads to (undamped) oscillations around the equilibrium point of the system. A complete disappearance of the prey population is not possible as an alternative source of food is not available to the predator, and its population has to decrease with the diminishing prey population. In this system of total dependence, the prey population is therefore protected from complete extermination.

5. **Critical parameters:** The parameters of the system effect the speed of the processes and the frequency of oscillation. The amplitude of oscillation is determined by the initial values.

6. **Reference run:** For this normalized system $a = b = c = d = 1$. This system has an undamped oscillation with a period of about 10 time units whose amplitude depends on the initial conditions of the system.

7. **Suggestions for further study:** Investigate the influence of the different parameters using the option "Parameter Sensitivity." Which parameters determine the equilibrium state? Replace the normalized parameters of the reference run by more realistic parameters. The order of magnitude of the parameters will differ by about 10^3. Does this change the fundamental behavior of the system? (cf. Sec. 3.7).

8. **References:** Lotka, 1925; Volterra, 1926; Richter, 1985; Wissel, 1989;: Bossel, 1985: p. 91-102.

PREDATOR-PREY SYSTEM WITHOUT CAPACITY LIMIT — M 206

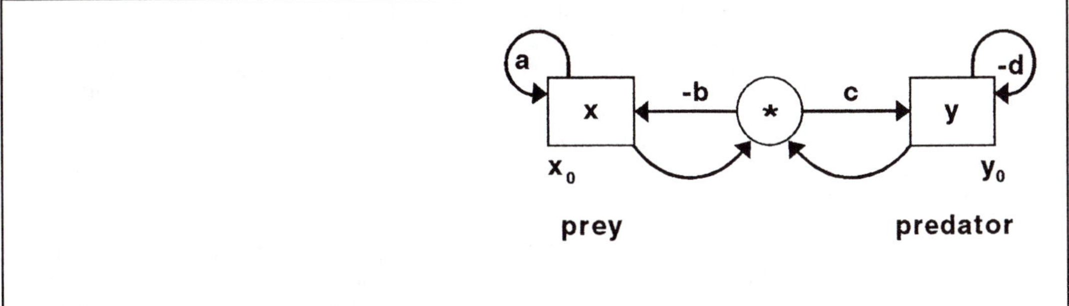


```
dx/dt = a*x   - b*x*y
dy/dt = c*x*y - d*y
```

PREDATOR PREY SYSTEM 1
— x | prey (0 ... 10)
— y | predators (0 ... 10)

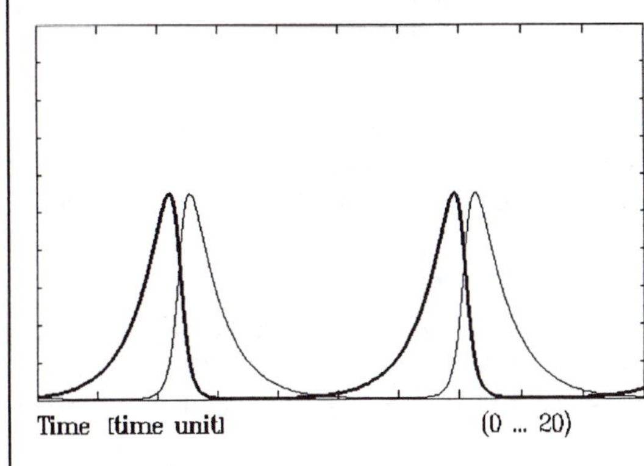

Time [time unit] (0 ... 20)

```
a | prey growth rate [1/time]                    1.000
b | predation loss [1/(pred*time)]               1.000
c | predation gain [1/(prey*time)]               1.000
d | predator respiration rate [1/time]           1.000
x | prey [prey units]                            0.100
y | predators [pred units]                       0.100
Begin of simulation [time unit]                      0
End of simulation [time unit]                   20.000
Time step of computation [time unit]          1.00E-02

EQUILIBRIUM POINTS

x, y = 0, 0 /  1.000,  1.000
```

PREDATOR PREY SYSTEM 1
— y | predators (0 ... 5)

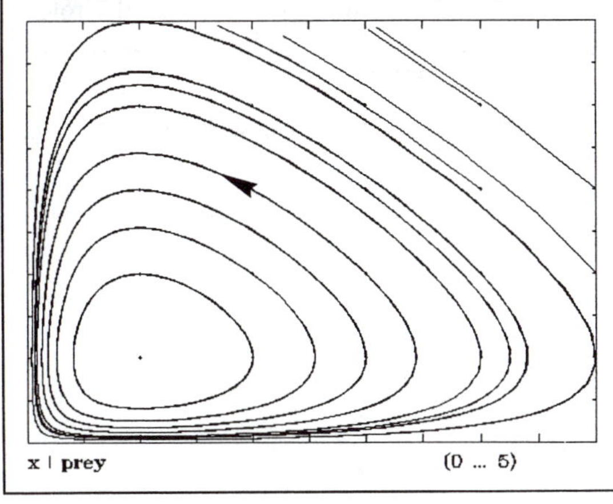

x | prey (0 ... 5)

PREDATOR-PREY SYSTEM WITH CAPACITY LIMIT M 207

1. **Description:** The model describes the dynamics of the interaction between a predator population y and a logistically growing prey population x that cannot grow beyond a capacity limit k. The behavior of the system is characterized by damped oscillations around an equilibrium point.

2. **Occurrence:** Application to all predator-prey type processes where the prey population is limited by a given capacity limit (for example, the carrying capacity of a given ecosystem which is a function of the net primary productivity).

3. **Structural characteristics:** The rate at which prey is taken depends on the prey population x as well as on the predator population y. In proportion to the non-linear predation term $(x \cdot y)$ there will be losses for x, gains for y. The prey population without predators follows logistical growth with saturation at the capacity limit k. If there is predation, the prey population at equilibrium will be less than this capacity limit.

4. **Behavioral characteristics:** Growth of the prey population is followed by a lagged growth of the predator population and later by a strong decrease of the prey population due to the large predator population. The process leads to periodically repeating oscillations of both populations. In contrast to the predator-prey system without capacity limit, the oscillations are strongly damped. Independently of the initial conditions, the system eventually approaches a stable equilibrium point.

5. **Critical parameters:** The oscillatory behavior is determined by the capacity limit k. For low k the oscillations are strongly damped. As the capacity k increases, the damping decreases and eventually disappears as k approaches infinity (this corresponds to the case without capacity limit, Model M 206). If the capacity limit is less than a certain critical value, the predator population will completely disappear.

6. **Reference run:** For a capacity limit of $k = 2$, a strongly damped oscillation with non-zero equilibrium values for both the predator and the prey population appears. If the capacity limit is reduced by one half $(k = 1)$, the predator population dies out while the prey population reaches its equilibrium value at this capacity limit.

7. **Suggestions for further study:** Investigate the role of the capacity parameter k using the option "Parameter Sensitivity" and the phase plot of y as a function of x. Examine the role of different initial values for different capacity limits k using the option "Global Analysis." Determine the influence of other parameters using the option "Parameter Sensitivity."

8. **References:** Lotka, 1925; Volterra, 1926; Richter, 1985; Wissel, 1989; Bossel, 1985 p. 91-102.

PREDATOR-PREY SYSTEM WITH CAPACITY LIMIT

M 207

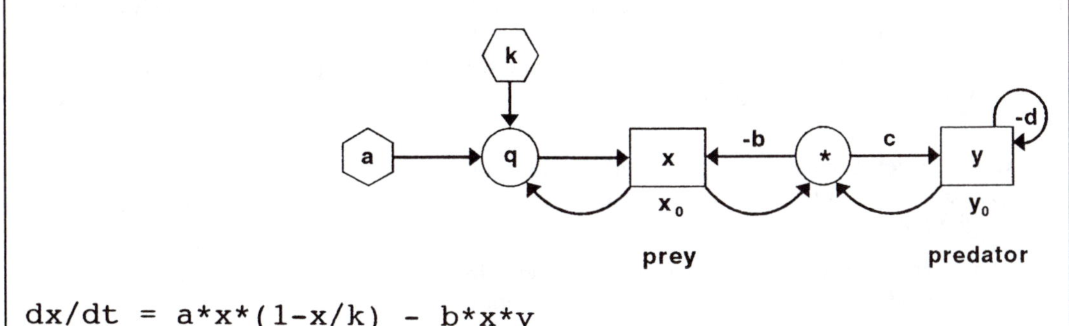


```
dx/dt = a*x*(1-x/k) - b*x*y
dy/dt = c*x*y - d*y
```

PREDATOR PREY SYSTEM 2
— x | prey (0 ... 2)
— y | predators (0 ... 2)

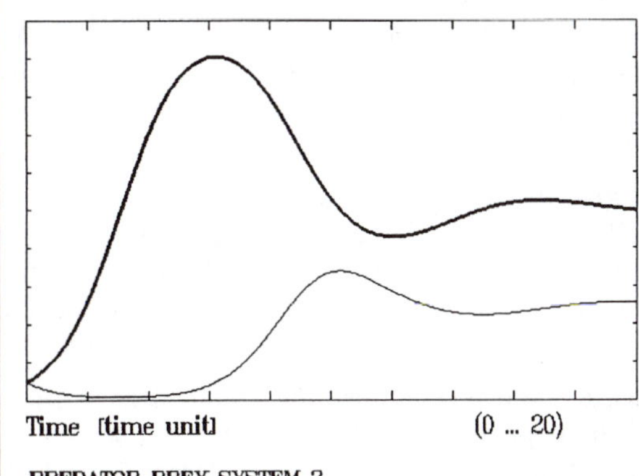

a	prey growth rate [1/time]	1.000
b	predation loss [1/(pred*time)]	1.000
c	predation gain [1/(prey*time)]	1.000
d	predator respiration rate [1/time]	1.000
k	prey carrying capacity [prey units]	2.000
x	prey [prey units]	0.100
y	predators [predator units]	0.100
Begin of simulation [time unit]		0
End of simulation [time unit]		20.000
Time step of computation [time unit]		1.00E-02

EQUILIBRIUM POINTS

x, y = 0, 0 / 2.000, 0 / 1.000, 0.500

PREDATOR PREY SYSTEM 2
— y | predators (0.010 ... 2)

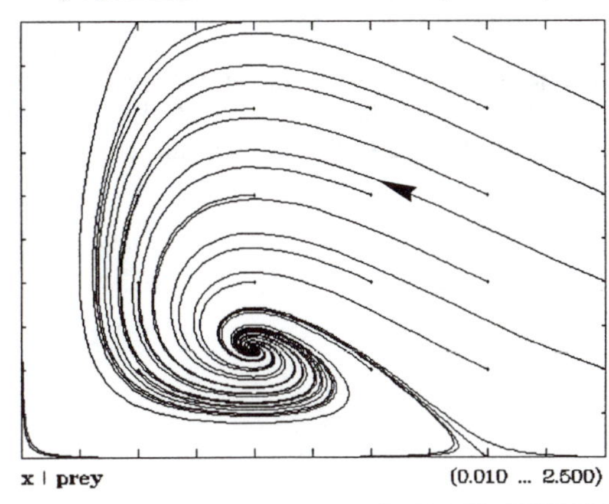

COMPETITION

M 208

1. **Description:** Competition for a common resource brings to both competitors disadvantages that are proportional to the population of competitor x and of the competitor y. The interaction term $(x \cdot y)$ is non-linear. Competition therefore impairs the development of both competitors. In the course of time that population which is at a comparative disadvantage will eventually disappear.

2. **Occurrence:** Competition between individuals, populations, and species for commonly used resources. Competition between industrial enterprises for the same market. Competition between religions, parties, and ideologies.

3. **Structural characteristics:** It is assumed here that the effect of competition $(x \cdot y)$ is proportional to the population x as well as the population y. In some competitive situations (for example, the competition for a common food supply), an additive term $(c \cdot x + d \cdot y)$ is more appropriate, where c and d are the specific rates of food consumption by x and y. The competition effect has a negative influence on both populations with specific influence rates a and b. In accordance with the conditions usually found in reality, logistic growth limitations with the capacity limits k and l have been assumed for both populations.

4. **Behavioral characteristics:** Without competition, both populations would grow to their respective capacity limit (cf. logistical growth). With competition, one population is at a disadvantage even if the initial conditions and/or the system parameters of both populations differ by only a very small amount. This eventually leads to a disappearance of the disadvantaged population. The other population then grows to its saturation limit.

5. **Critical parameters:** The ratios of the specific growth rates of the two populations (r, s) and of their competition effects (a, b) are critical: small differences will cause relative advantages that determine which population survives. Even for different system parameters, however, the initial conditions are decisive for the overall development. (This can be easily determined using the option "Global Analysis.") A stable equilibrium point exists only for one surviving population.

6. **Reference run:** For identical initial conditions for both populations and identical parameters except for the specific growth rates r and s, the population y dies out having a somewhat smaller growth rate. However, if the initial value for this population is slightly larger ($y_0 = 1.2$), population x will die out despite a somewhat more favorable specific growth rate. (Use option "Global Analysis.")

7. **Suggestions for further study:** Study the state trajectories and the disappearance of one population as a function of the initial conditions using the option "Global Analysis." Examine the dependence of the state trajectories x, y as function of the specific growth and competition rates using the option "Parameter Sensitivity."

8. **References:** Beltrami, 1987: p. 69-70.

COMPETITION

M 208

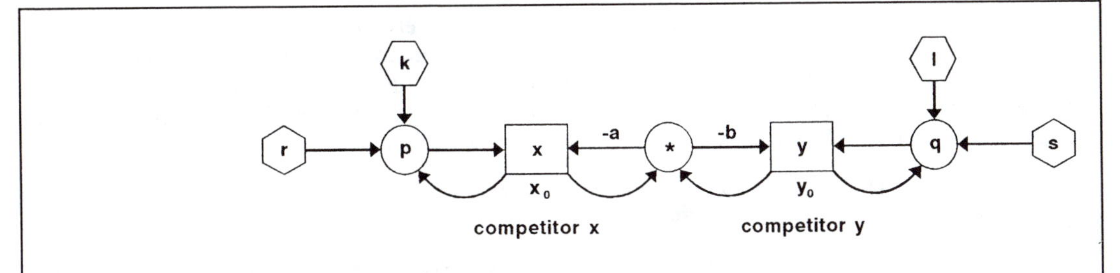


```
dx/dt = r*x*(1-x/k) - a*x*y
dy/dt = s*y*(1-y/l) - b*x*y
```

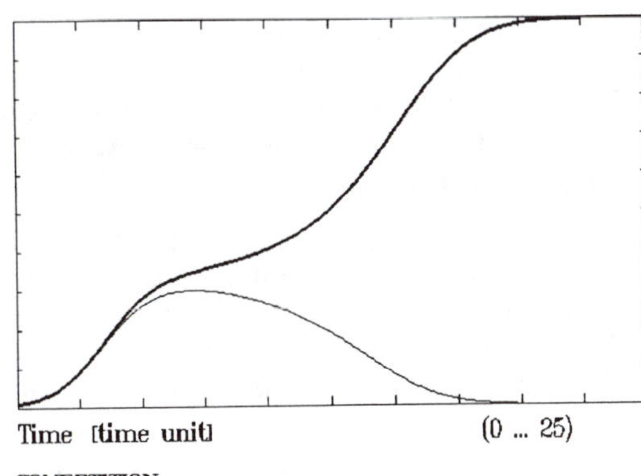

r	growth rate of X [1/time]	1.000
s	growth rate of Y [1/time]	0.990
a	compet.effect on X [1/(Yunit*time)]	1.000
b	compet.effect on Y [1/(Xunit*time)]	1.000
k	carrying capacity [X units]	2.000
l	carrying capacity [Y units]	2.000
x	population of X [X units]	2.00E-02
y	population of Y [Y units]	2.00E-02
Begin of simulation [time unit]		0
End of simulation [time unit]		25.000
Time step of computation [time unit]		2.00E-02

EQUILIBRIUM POINTS x, y =

0, 0 / 2.000, 0 / 0 , 2.000 / 0.658, 0.671

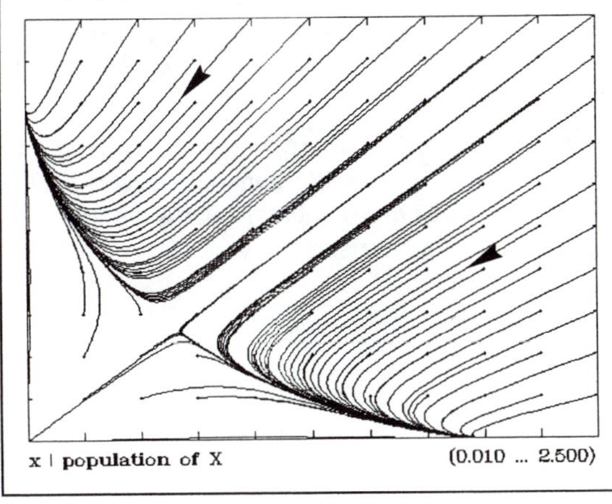

TOURISM AND ENVIRONMENTAL POLLUTION **M 209**

1. **Description:** Tourists are attracted by a region with a particularly beautiful natural environment. Even though the natural environment is able to regenerate itself up to its capacity limit (carrying capacity) if left unperturbed, it is heavily stressed and partially destroyed by the touristic infrastructure and by the accompanying pollution. This reduces the attractiveness of the area and causes a subsequent decline in the number of tourists.

2. **Occurrence:** Touristic development in coastal and mountain regions, on islands, and in "native villages." Human settlements in general.

3. **Structural characteristics:** The attractiveness of a region depends on its environmental state y. Attractiveness may be enhanced by advertisement b. The tourist influx is proportional to this attractiveness ($b \cdot y$), thereby increasing the tourist population x. The tourist population has constant losses by departing tourists at a specific rate a. The environmental destruction by tourism depends on the number of tourists and on the environmental state; it is therefore proportional to $x \cdot y$ (as in the predator-prey system) with a specific degradation rate c. If left alone, the environment would be able to regenerate itself after initial degradation with a specific regeneration rate d, eventually reaching its capacity limit k (carrying capacity).

4. **Behavioral characteristics:** Advertisement for the region ($b > 0$), together with an initially intact environment attracts tourists; the tourist population therefore increases. As the number of tourists increases, the environmental state declines. As a result, the region becomes less attractive and the tourist population also declines. The number of tourists and the environmental state finally approach an equilibrium condition where the state of the environment, however, has worsened considerably compared to the initial state of the capacity limit.

5. **Critical parameters:** The development dynamics (in particular the equilibrium state) depend very much on the amount of advertising b. More publicity will increase the tourist population but only at the cost of further degradation of the environmental situation. This specific strain c on the environment by tourism has a decisive effect on the development. The regenerative capability d of the environment also has particular significance since it determines how much tourism can be handled by the environment.

6. **Reference run:** With the parameter choice $a = c = d = k = 1$, $b = 5$, and the initial conditions $x_0 = 0.1$ and $y_0 = 1$, the system shows a rapid increase in the tourist population to almost 2 (a twentyfold increase of the initial value) and a subsequent decrease to 0.833. The environmental state (quality of the environment) decreases rapidly from an initial value of 1 to 0.167. The stable point of equilibrium is independent of the initial conditions.

7. **Suggestions for further study:** Investigate the role of the different parameters using the option "Parameter Sensitivity." Under what circumstances can strong oscillations be expected? How could a slow touristic development to a final high level of environmental quality be attained? Study the location of the equilibrium points as a function of advertisement b.

8. **References:** -

TOURISM AND ENVIRONMENTAL POLLUTION **M 209**

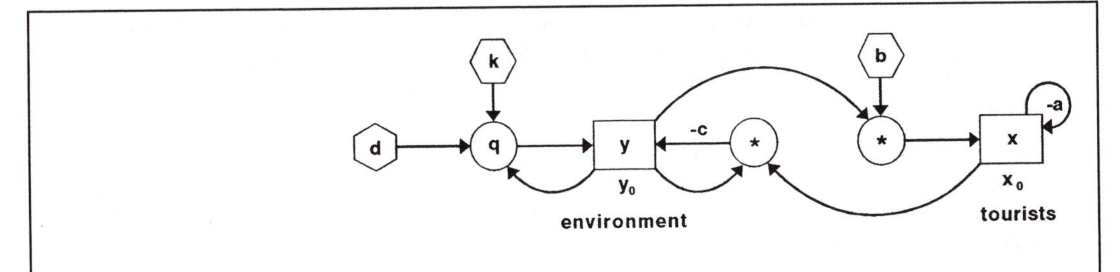

$$dx/dt = -a*x + b*y$$
$$dy/dt = d*y*(1-y/k) - c*x*y$$

TOURISM AND ENVIRONMENT
— x | tourists (0 ... 2)
— y | environment (0 ... 1)

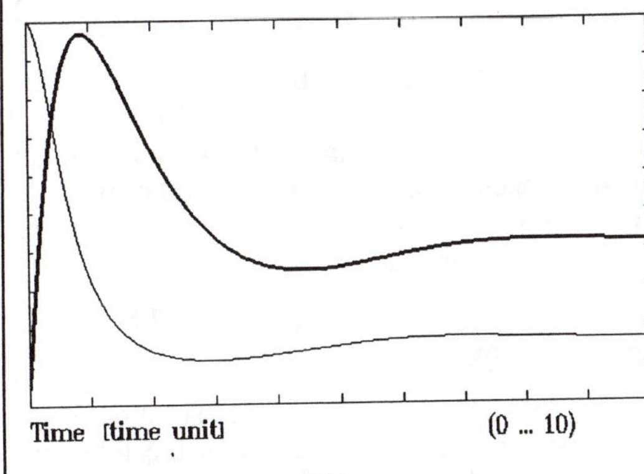

Time [time unit] (0 ... 10)

a	spec. tourist loss rate [1/time]	1.000
b	advertising effect [tour/(envir*time	5.000
c	degeneration rate [1/(tourist*time)]	1.000
d	regeneration rate [1/time]	1.000
k	carrying capacity [envir.quality]	1.000
x	tourists [tourists]	0.100
y	environment [envir.quality]	1.000
Begin of simulation [time unit]		0
End of simulation [time unit]		10.000
Time step of computation [time unit]		1.00E-02

EQUILIBRIUM POINTS

x, y = 0, 0 / 0, 1.000 / 0.833, 0.167

TOURISM AND ENVIRONMENT
— y | environment (0 ... 1)

Parameter: b | advertising effect
1 2 3 4 5

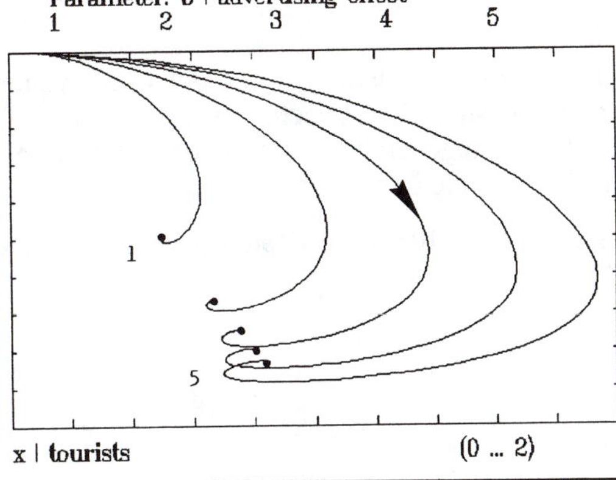

x | tourists (0 ... 2)

OVERSHOOT AND COLLAPSE

M 210

1. **Description:** The model describes the development of a population y that depends on a renewable resource x. As the resource is increasingly depleted, its ability to regenerate is also reduced until finally a recovery is no longer possible, and the resource collapses together with the population.

2. **Occurrence:** The process is typical for the (over)exploitation of renewable resources: e.g. overgrazing, deforestation, fuel wood crisis. The same temporal development caused by similar structural relationships is also found in the "global models" describing the global development of population and environment.

3. **Structural characteristics:** The growth of the consumer population is a function of its specific birthrate b, its current population level y, and the resource available per capita (x/y). Since the per capita resource effect is limited, a Michaelis-Menten saturation term is introduced. Resource use is proportional to population y and per capita resource use rate a as long as the resource level x is sufficient. If this is not the case, consumption is restricted to the currently available supply level x. As the resource becomes exhausted, the birthrate of population y will fall below the deathrate $(d \cdot y)$. Regeneration of the natural resource is described by a logistic development with a capacity limit k. For small level of the resource x, it is assumed to regenerate only slowly (regeneration rate proportional to $r \cdot x^2$, minimum regeneration $m \cdot k$).

4. **Behavioral characteristics:** Beginning with a small initial value, the population y will initially grow quickly leading to an accelerated reduction of the resource level x. The erosion of the resource base also causes a reduction of its regeneration capacity and of the resource supply rate. As a consequence, the population collapses after reaching a maximum. Together with the population, the resource breaks down completely. Because of the assumption of a minimum regeneration rate the ecosystem continues to exist on a very low level.

5. **Critical parameters:** A stable equilibrium point x, $y \geq 0$ is only possible at small birthrates. The time and severity of a collapse depend strongly on the specific birthrate b.

6. **Reference run:** The parameters of the reference run correspond roughly to those of animals on pasture land. The system collapses after about 25 years. There is no recovery in the following years; the resource remains on a very low level. If a slightly smaller birthrate is selected (0.6 instead of 0.7), the ecosystem and the animal population remain on a relatively high equilibrium level. A limit cycle appears in a narrow parameter range (here at $b = 0.63$); it marks the transition between a high and a low level of equilibrium.

7. **Suggestions for further study:** Using the option "Parameter Sensitivity" find out which values for the specific birthrate b or the "harvesting rate" d should be selected in order to attain a stabilization at a high level of y. What should be done to regenerate the resource x following a collapse? Study the transition range marked by the limit cycle in some detail.

8. **References:** Goodman, 1974: p. 377-388, Bossel, 1985: p. 103-111; Meadows, Meadows, and Randers, 1992.

OVERSHOOT AND COLLAPSE

M 210

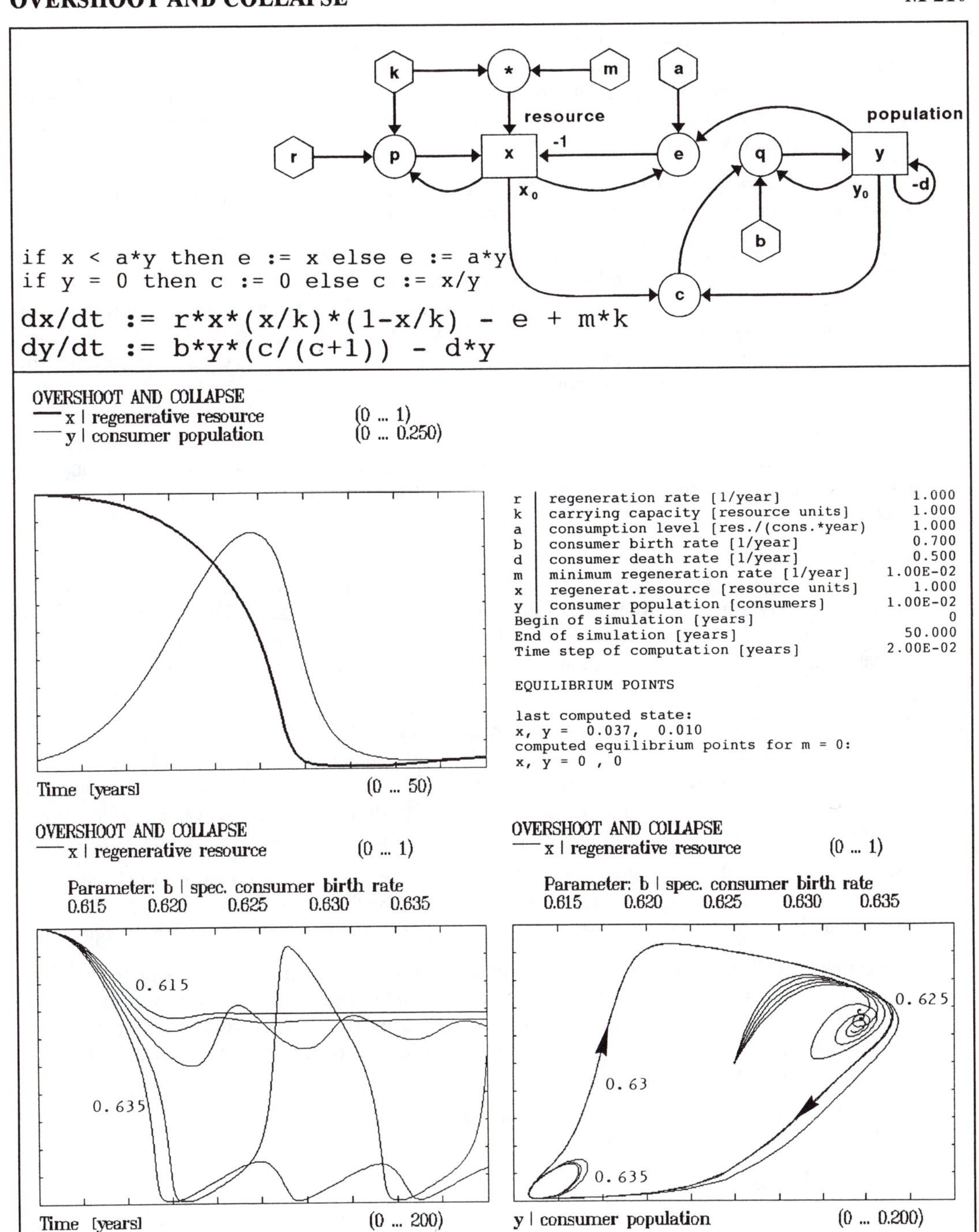

FOREST GROWTH **M 211**

1. **Description:** The leaf mass y of the leaf canopy grows up to its saturation limit k and produces photosynthate (assimilate) in proportion to the leaf mass. These products of photoproduction are partially used for respiration; the remaining surplus will turn to wood x (increment). As the respiration demand grows with the woodmass, increment will eventually stop; the energy gains then just compensate the energy losses.

2. **Occurrence:** Vegetation with wood formation as in forests or bush vegetation. Similar relationships are found in the ossification of organizations.

3. **Structural characteristics:** The leaf canopy y is assumed to grow according to logistic growth with a specific growth rate b and a saturation limit k. Saturation occurs because, as the leaf canopy becomes more dense, the light received by the lower leaf layers is reduced until the respiration losses can no longer be compensated by the small energy gains in those layers (cf. Model M 110). The production of assimilate is proportional to the leaf mass y and its specific productivity p; it may be reduced by pollution effects u(t). The efficiency r takes the respiration of leaves and fine roots into account. A further consumption of assimilate is proportional to the current wood mass (specific stemwood respiration rate s) and to leaf renewal l. The leaf mass y has losses from leaf shedding f; the woodmass x has deadwood losses with a specific deadwood loss rate d. Any surpluses remaining after subtraction of these different loss rates go into wood increment.

4. **Behavioral characteristics:** Starting with small initial values for leaf mass y and wood mass x, the leaf canopy grows to its logistic saturation limit k. The wood mass initially also grows relatively quickly since the wood-mass-specific respiration losses are relatively small. As the wood mass increases, the wood-proportional respiration losses as well as the deadwood losses increase until these losses finally attain the same value as the assimilate gains, and the increment becomes zero.

5. **Critical parameters:** The development is strongly affected by environmental pollution u(t) = q after time t. A collapse of the forest is possible if the replacement demand for leaves and fine roots, and finally also the respiration demand can no longer be covered by the energy gains. The efficiency r of the photoproduction (as ratio of net assimilation to gross leaf production) and the ratio of the specific production rate p to the stem respiration s are also critical parameters.

6. **Reference run:** With the selected default parameters, the highest increment occurs at about 20 years. Thereafter it decreases as a result of the increasing stem respiration. After year 40, when the environmental pollution q is assumed to set in, the increment is reduced even further.

7. **Suggestions for further study:** Using the option "Parameter Sensitivity," investigate the influence in particular of parameters p, r, and s. Assume different environmental influences q, and determine which pollution levels can be tolerated by the system without collapse. What kinds of environmental improvements have to take place in order to save a critically damaged system from total collapse?

8. **References:** Bossel, 1986: p. 259-288; Bossel, 1987/89: p. 245-268.

FOREST GROWTH

M 211

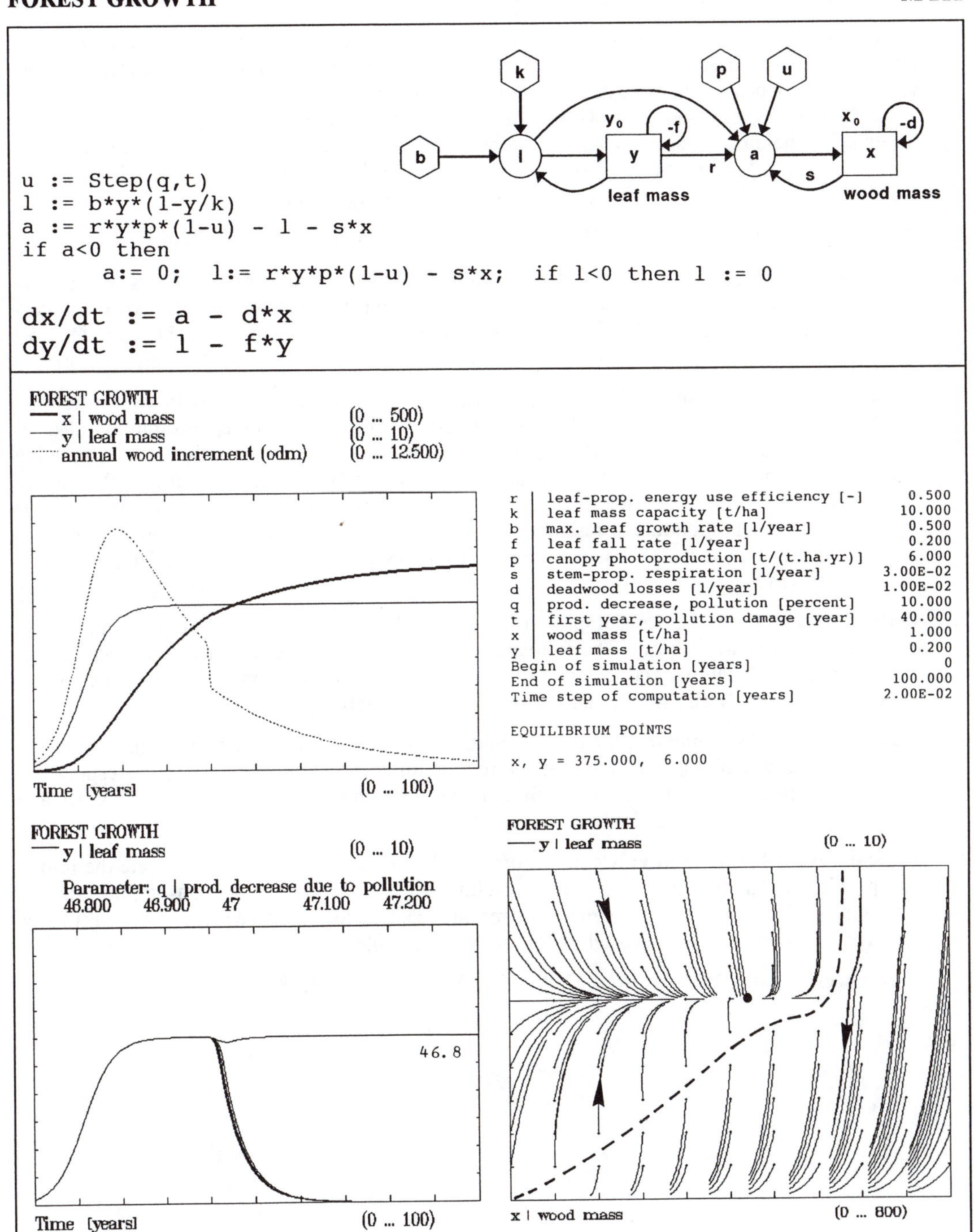

RESOURCE DISCOVERY AND DEPLETION **M 212**

1. **Description:** The discovery of resources follows a logistic development: in the beginning, when large quantities of resources have not yet been discovered, exploration is very successful and leads to an exponential increase of the discovered resources x. The discovery rate is reduced to zero when finally all resources (total amount k) have been discovered. Resource consumption depends on the amount of available resources (difference between the discovered resources x and the consumed resources y). Consumption therefore rises from an initially small value to a maximum and then again falls to zero when resources are no longer available.

2. **Occurrence:** Discovery, exploitation, and consumption of non-renewable resources.

3. **Structural characteristics:** The capacity limit of the logistic discovery process is the total available (and discoverable) amount of resources k. Consumption is proportional to the currently available amount $(x–y)$. The exploration efforts increase as the available resource volume $(x–y)$ decreases. Recycling (with the fraction r) may reduce the resource consumption.

4. **Behavioral characteristics:** The discovered supplies x increase logistically to their saturation value k. As the resource availability increases correspondingly, resource consumption also increases. This leads to a similar logistic growth of consumed supplies y with a certain time lag with respect to resource discovery. The discovery rate climbs to a maximum and then again falls to zero as resources become depleted. A similar development, again with a time lag, is found for the consumption rate of resources. Eventually, there is a point where the consumption rate becomes greater than the discovery rate. After this point in time, the available resources can only decrease.

5. **Critical parameters:** The consumption rate c and the recycling rate r are decisive for the long-term resource availability. The influence of the total volume of discoverable resources k on the resource availability and the time to depletion is smaller than one would assume since a larger resource supply also implies a steeper rate of consumption growth.

6. **Reference run:** Assuming that the maximum discovery rate d annually corresponds to 1/10 of the already discovered supplies and that the maximum consumption rate c also corresponds to 1/10 of the still available resources, maximum consumption will be reached after 55 years and total depletion of resources after about 120 years.

7. **Suggestions for further study:** Using the option "Parameter Sensitivity," investigate the role of the specific consumption rate c and the recycling rate r. For an average consumption rate c, determine by how much the "lifetime" of a resource can be increased (without recycling) if the discoverable amount k is increased tenfold and hundredfold.

8. **References:** Bossel, 1985: p. 361-377; Naill, in Meadows and Meadows 1973.

RESOURCE DISCOVERY AND DEPLETION

M 212

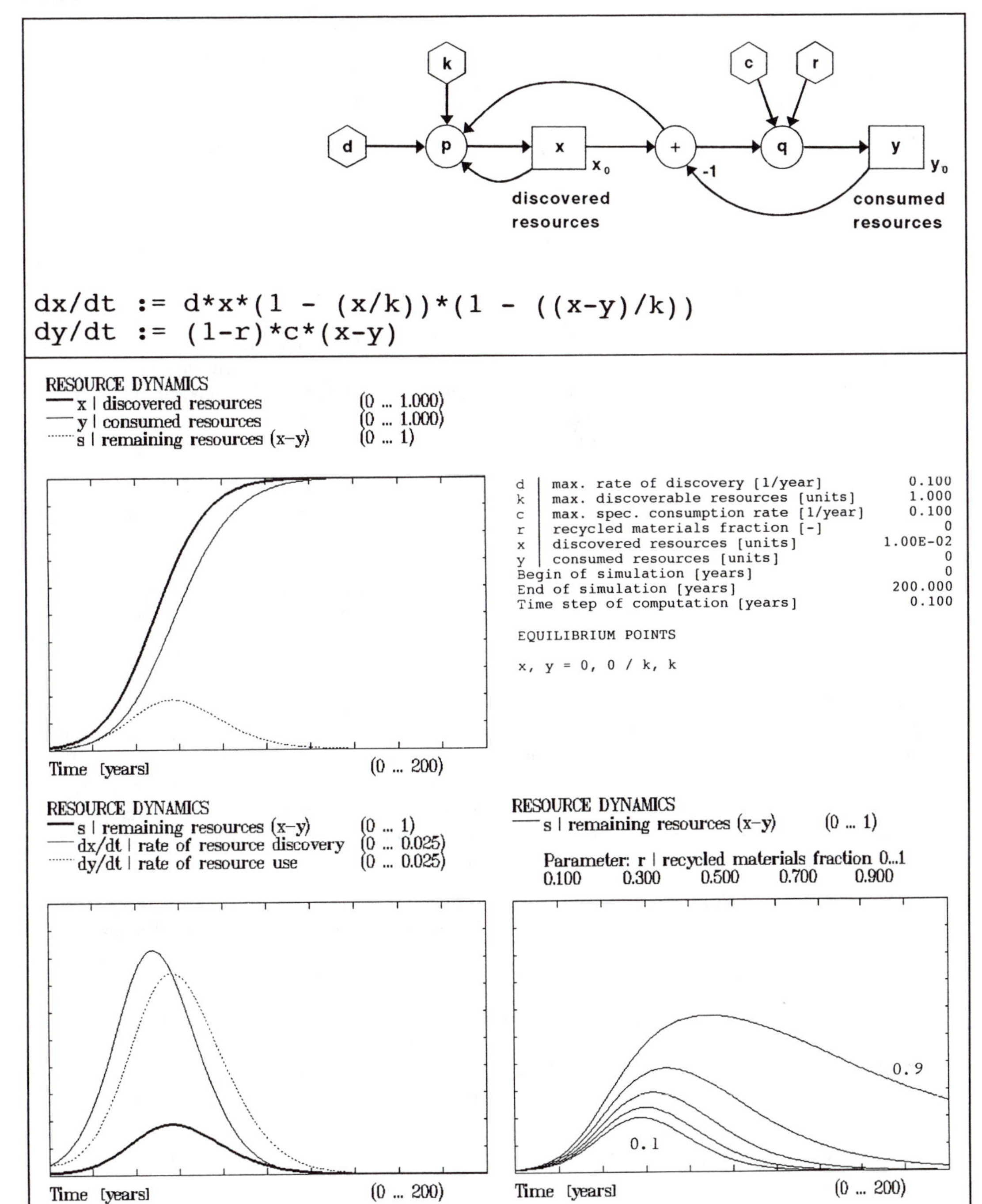

TRAGEDY OF THE COMMONS **M 213**

1. **Description:** The opportunity to exploit a communally owned renewable resource (commons) for one's own benefit may cause an individual to increase his personal gain by a small additional investment (for example, another head of cattle). Without effective (social) limits on the use of a commons, this process may lead to overexploitation and collapse.

2. **Occurrence:** Overgrazing the commons, overfishing of international fishing grounds outside of territorial waters, and fuel wood crises. State-run industries and subsidized production to "secure employment" in old, unproductive industries.

3. **Structural characteristics:** The natural renewable resource x has a logistic growth restriction with a maximum capacity k and an initial exponential growth rate r. The rate of resource consumption s is proportional to the current resource level x and the capital stock y as the means of exploitation (cattle, installations, equipment); it is also proportional to the productivity p of these means of exploitation. The capital investment rate for resource use facilities y is proportional to the deviation from the production goal g and to the net profit rate [sales income $(p \cdot m \cdot x \cdot y)$ – operating costs $(q \cdot y)$]. This expression is normalized by the maximum profit rate $(p \cdot m \cdot k)$ and multiplied by the specific investment rate i. The capital investment comes from an external source; it does not figure in the net profit.

4. **Behavioral characteristics:** At the beginning of the exploitation the (initially large) net gain causes a rapid build-up of the resource use capital, thereby increasing production quickly. As long as the production goal has not been reached, and net gains are achieved, the resource use capital y is increased further. Production therefore increases initially and with it the sales income $(p \cdot m \cdot x \cdot y)$. As production increases beyond the regeneration rate, the resource supply decreases leading to a lagged reduction of production. As resource use capital is increased, the operating expenses $(q \cdot y)$ also increase causing a decrease in the net profit. Net profit reduces to zero and finally becomes negative. In the end, the resource base is completely destroyed.

5. **Critical parameters:** If the product of carrying capacity k and specific regeneration rate r is greater than $k \cdot r = 4\ g$, i.e. if the production goal is small compared to the maximum regeneration rate of the system, then an acceptable permanent solution is possible. (The system has a total of five equilibrium points.) If the production goal is high compared to the maximum regeneration rate, the system may collapse. (It then has three equilibrium points.) The price m paid for the product plays a critical role for the system development. If it is too high, it will lead to a larger capital stock and a correspondingly larger exploitation.

6. **Reference run:** With the chosen specific rates, the system shows an initial phase of profitable exploitation over the first 60 years. This is followed by a phase of net losses; after about 100 years the resource is completely depleted. For this parameter set there is a stable equilibrium point only for very high resource capacity (> 40) and a low capital stock.

7. **Suggestions for further study:** Using the option "Parameter Sensitivity," investigate the role of the different system parameters (m, p, g, i). Assuming realistic conditions, can you find stable equilibrium points? What measures would be required in order to operate the system under these conditions? Compare the model and its results with M 214 SUSTAINABLE USE OF A RENEWABLE RESOURCE.

8. **References:** Hardin, 1968.

TRAGEDY OF THE COMMONS **M 213**

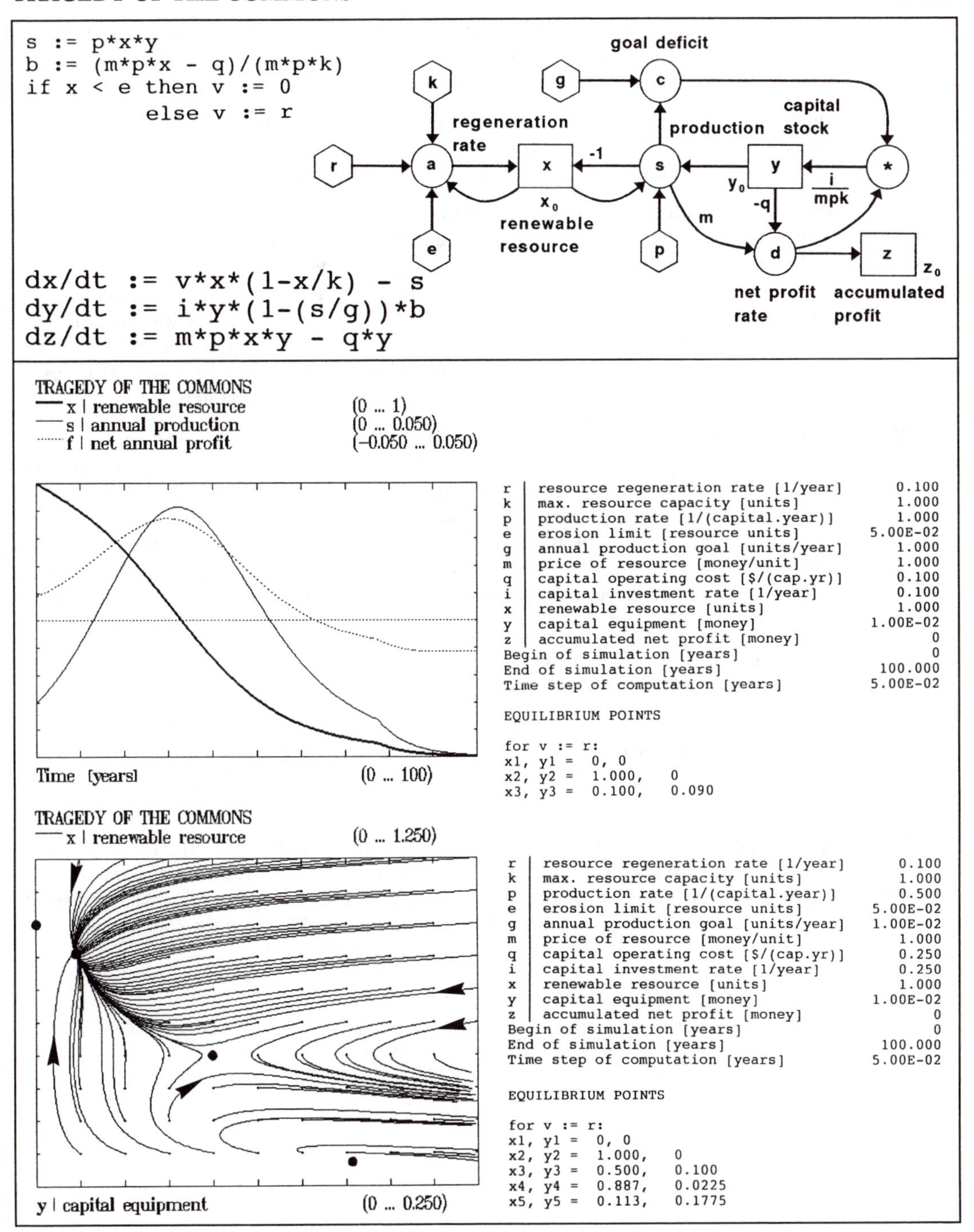


```
s := p*x*y
b := (m*p*x - q)/(m*p*k)
if x < e then v := 0
         else v := r
```

```
dx/dt := v*x*(1-x/k) - s
dy/dt := i*y*(1-(s/g))*b
dz/dt := m*p*x*y - q*y
```

r	resource regeneration rate [1/year]	0.100
k	max. resource capacity [units]	1.000
p	production rate [1/(capital.year)]	1.000
e	erosion limit [resource units]	5.00E-02
g	annual production goal [units/year]	1.000
m	price of resource [money/unit]	1.000
q	capital operating cost [$/(cap.yr)]	0.100
i	capital investment rate [1/year]	0.100
x	renewable resource [units]	1.000
y	capital equipment [money]	1.00E-02
z	accumulated net profit [money]	0
Begin of simulation [years]		0
End of simulation [years]		100.000
Time step of computation [years]		5.00E-02

EQUILIBRIUM POINTS

for v := r:
x1, y1 = 0, 0
x2, y2 = 1.000, 0
x3, y3 = 0.100, 0.090

r	resource regeneration rate [1/year]	0.100
k	max. resource capacity [units]	1.000
p	production rate [1/(capital.year)]	0.500
e	erosion limit [resource units]	5.00E-02
g	annual production goal [units/year]	1.00E-02
m	price of resource [money/unit]	1.000
q	capital operating cost [$/(cap.yr)]	0.250
i	capital investment rate [1/year]	0.250
x	renewable resource [units]	1.000
y	capital equipment [money]	1.00E-02
z	accumulated net profit [money]	0
Begin of simulation [years]		0
End of simulation [years]		100.000
Time step of computation [years]		5.00E-02

EQUILIBRIUM POINTS

for v := r:
x1, y1 = 0, 0
x2, y2 = 1.000, 0
x3, y3 = 0.500, 0.100
x4, y4 = 0.887, 0.0225
x5, y5 = 0.113, 0.1775

SUSTAINABLE USE OF A RENEWABLE RESOURCE **M 214**

1. **Description:** A logistically regenerating resource x is used sustainably, i.e. its ability to regenerate is maintained. The sustainability goal determines the maximum permissible investment in resource use capital y. Otherwise, the system is identical to Model M 213.

2. **Occurrence:** The principle of sustainability has been practiced in the forestry of some nations for several hundred years. Generally, all renewable resources should be harvested according to this principle: the utilization rate must not exceed the regeneration rate.

3. **Structural characteristics:** The model structure is mostly identical to that of Model M 213 TRAGEDY OF THE COMMONS. The major difference is that the resource level x now controls the rate of exploitation. The resource x grows according to a logistic saturation process. The investments in new resource capital stock y are proportional to the net profit rate b (sales income – maintenance costs; $m \cdot p \cdot x \cdot y - q \cdot y$) but they are now controlled in such a way that the sustainable yield is maintained at a high level. The goal for the resource level x corresponds to one half the maximum capacity k. As x falls below this value, y is also reduced; if x increases, and the net profit is positive, y is also increased by further investment.

4. **Behavioral characteristics:** Without utilization, the resource would grow logistically to its capacity limit. If resource utilization takes place, then the resource use capital y is increased until the resource x has been reduced to the value k/2. The resource level x is then maintained at this level. The system structure now leads to a constant annual production s with positive net rate of profit d = dz/dt and constant resource use capital y. There are four equilibrium points (three unstable, one stable); the trajectories in state space are relatively complex. Note that the system will collapse if the initial value for y is too high. If the erosion limit exceeds a critical value, the system also collapses.

5. **Critical parameters:** The initial value for y is a critical parameter. A stable equilibrium without collapse of the resource is possible only for an initial value y below the critical value.

6. **Reference run:** Using the same system parameters as in Model M 213 TRAGEDY OF THE COMMONS, the development stabilizes after about 70 years at an equilibrium level. Although the annual production does not reach the same maximum as in Model M 213, it is now sustainable and produces constant net profit.

7. **Suggestions for further study:** Compare the behavior with that of Model M 213 TRAGEDY OF THE COMMONS using the same parameter sets. Study the influence of the different parameters on the development and the equilibrium points of the system. Use the option "Global Analysis" to investigate how the state development depends on the initial conditions in the state space (x, y).

8. **References:** -

SUSTAINABLE USE OF A RENEWABLE RESOURCE

M 214

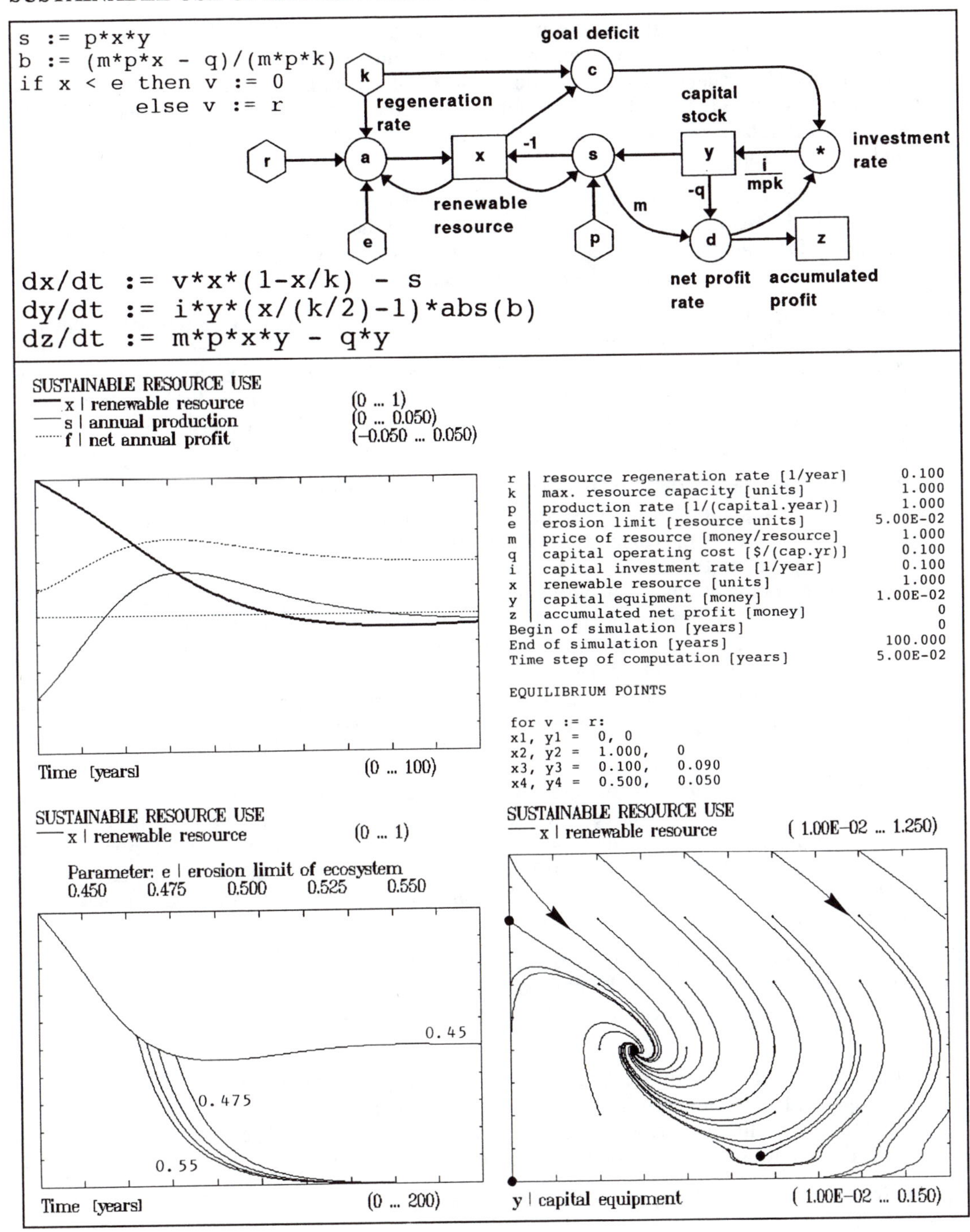

DISTURBED EQUILIBRIUM: CO_2-DYNAMICS OF THE ATMOSPHERE M 215

1. **Description:** Photosynthesis and decomposition of organic matter (litter and humus), as well as the respiration of plants and animals, remove and return to the atmosphere large amounts of carbon dioxide constantly. The combustion of fossil fuels and the destruction of forests disturb the flow equilibrium in the CO_2-exchange between the two CO_2 reservoirs, atmosphere and (living or dead) biomass; the atmospheric CO_2 level increases causing a rising greenhouse effect and climate change.

2. **Occurrence:** The phenomenon of the dynamic equilibrium of flows between two reservoirs can be found in different areas. It applies to the dynamics of the greenhouse gases in the atmosphere as well as, for example, the cycling of matter (nutrients, chemicals) in soil and groundwater and the gradual increase in the level of some compounds by pollutant inputs.

3. **Structural characteristics:** Two reservoirs with levels x and y of a substance are mutually connected by flows of the substance. A first process removes the substance at a certain rate from reservoir x and transfers it to reservoir y. The second process causes the substance to flow from reservoir y to reservoir x. Applied to the atmosphere, the system is disturbed by an additional CO_2-input from the combustion of fossil fuels which follows a logistic curve, implying eventual saturation of consumption. The model does not consider any CO_2-uptake by the ocean.

4. **Behavioral characteristics:** Beginning with an initial dynamic equilibrium of flows between the two reservoirs, the CO_2-level x in the atmosphere gradually increases by the logistically growing input f from the combustion of fossil fuels. The removal of CO_2 from the atmosphere by plants depends on the area a covered by vegetation and its average productivity p. As a result of deforestation or environmental pollution, the net primary production of the plant vegetation cover is reduced by a factor $u(t)$. In contrast to the process of photosynthesis, the decomposition and respiration processes are proportional to (living and dead) biomass y with a specific rate c. The input of CO_2 to the atmosphere from respiration and decomposition of biomass therefore remains relatively high, even if its productivity should be reduced (by pollution). In all, the inputs to the atmosphere are greater than the outflows resulting in a constant increase of the CO_2-level.

5. **Critical parameters:** The specific growth rate r, the saturation value k of the fossil fuel consumption and the deforestation scenario (intensity d, begin b, end e) are critical factors for the further development of the system. Since the CO_2-uptake is independent of x, a further CO_2-increase can only be avoided if the fossil fuel combustion and forest destruction are stopped. (In reality, some fossil fuel consumption can be compensated for by increased CO_2-uptake by the ocean.)

6. **Reference run:** The parameters chosen correspond to the historical development. The CO_2-concentration rises from about 280 ppm (parts per million) in the year 1850 to between 530 and 730 ppm in the year 2050 depending on the assumed saturation of consumption of fossil fuels (5 - 25 GtC/a).

7. **Suggestions for further study:** Use the option "Parameter Sensitivity" to investigate the further development for different scenarios of the future consumption of fossil fuels (r, k). Study the consequences of different deforestation or afforestation scenarios (d, b, e). What is the function of the decomposition rate c (average residence time $T = 1/c$)?

8. **References:** Bossel, 1985: p. 117-128.

DISTURBED EQUILIBRIUM: CO_2-DYNAMICS OF THE ATMOSPHERE **M 215**

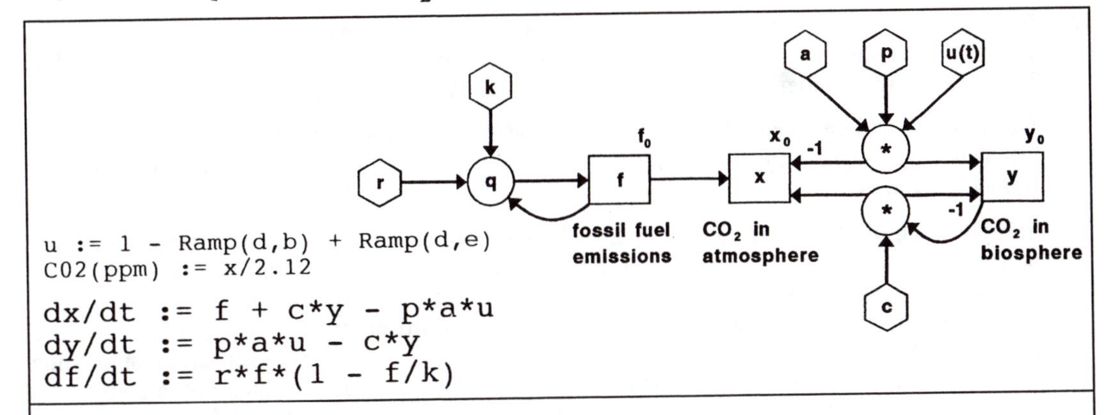

ATMOSPHERIC CO2-DYNAMICS
— y | carbon in biosphere (0 ... 3000)
— f | fossil fuel emissions (0 ... 12.500)
······ CO2 concentration in atmosphere (0 ... 1000)

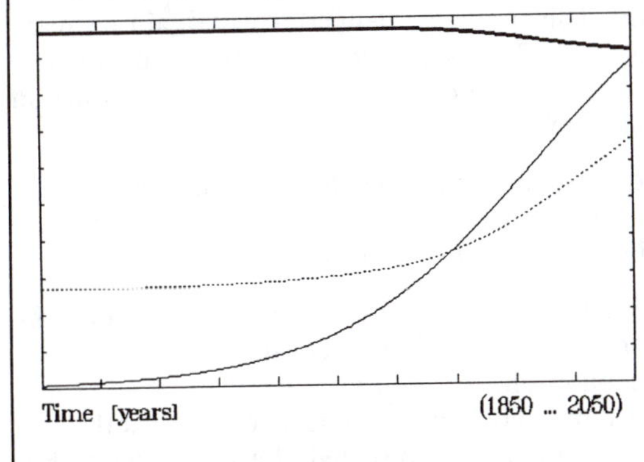

Time [years] (1850 ... 2050)

a	area, terr. ecosystems [Gkm^2]	0.145
p	net primary production [tC/(km^2.yr)	400.000
c	respiration/decomposition rate [1/yr	2.00E-02
r	fossil fuel growth rate [1/yr]	3.00E-02
k	saturation, fossil emissions [GtC/yr	15.000
d	forest destruction [pct/yr]	0.200
b	forest destruction begins, year [yr]	1970.000
e	forest destruction stops, year [yr]	2020.000
x	carbon in atmosphere [GtC]	570.000
y	carbon in biosphere [GtC]	2900.000
f	fossil fuel emissions [GtC/yr]	0.100
Begin of simulation [years]		1850.000
End of simulation [years]		2050.000
Time step of computation [years]		0.200

EQUILIBRIUM POINTS

y = p*a/c = 2900
Equilibrium only if fossil fuel emissions and forest destruction are halted.

ATMOSPHERIC CO2-DYNAMICS
— CO2 concentration in atmosphere (0 ... 800)

Parameter: k | saturation of fossil emissions
4 8 12 16 20

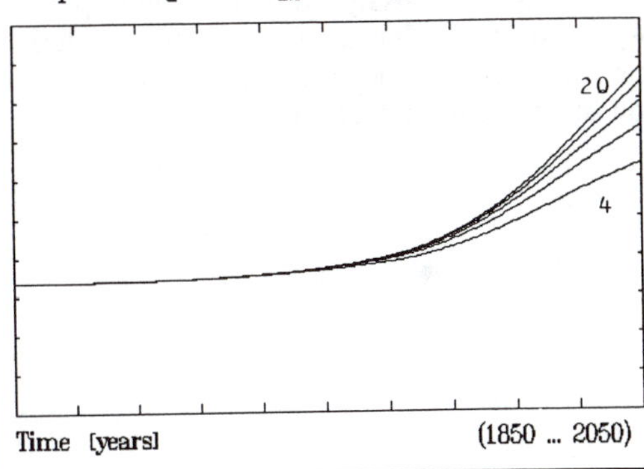

Time [years] (1850 ... 2050)

STOCKS, SALES, AND ORDERS

M 216

1. **Description:** The current stock z is determined by the time-dependent sales s(t) and the re-stocking rate r following orders. Orders are a function of the current stock and the daily sales. The delay in filling orders can lead to strong oscillations of the stock.

2. **Occurrence:** Interactions of stocks, orders, and sales in business. Market fluctuations for products having long delivery periods (plant and animal production). Demand and supply of trained personnel.

3. **Structural characteristics:** Orders o are written in reference to the current rate of sales s(t) and the current stocks z, by applying appropriate order factors a and b in an attempt to maintain a given stocking goal g. The delivery of orders is delayed by d time units. In the model the delay is realized by an exponential delay of third order. (See Model M 301 TRIPLE INTEGRATION AND EXPONENTIAL DELAY OF THIRD ORDER.) The delay causes a certain smoothing of the deliveries; over a longer period of time, the amount delivered is equal to the amount ordered but the time function of delivery is not simply a lagged time function of the orders.

4. **Behavioral characteristics:** As long as earlier orders do not show up as deliveries, the increasing stock deficit and further sales will lead to further orders which will be delivered to the warehouse only after d units of time. With these deliveries the stock increases causing a corresponding reduction in orders. Depending on the delivery delay, deliveries will again decrease after a certain period causing again more orders, etc. The system therefore has the tendency to oscillate; oscillations may be induced by random fluctuation of sales. The period of oscillation depends on the system parameters of the feedback loop.

5. **Critical parameters:** The most critical parameter which determines the period of oscillation is the delay time d. The stability or instability of the behavior is also determined by the order policy, i.e. the response to a stock deficit b and the response to the current rate of sales a.

6. **Reference run:** The normal amount of daily sales (1000 pieces) may be disturbed either by random fluctuations or by a discrete pulse. The default values of the order function are set such that the amount ordered is exactly equal to the amount sold; an additional amount of 0.125 times the stock deficit is added to the order "just to be sure." For a delivery delay of 20 days the system shows stock oscillations of roughly constant amplitude with a period of about 70 days. For shorter delivery delays, the oscillation following a single sales pulse is damped at a shorter period of oscillation.

7. **Suggestions for further study:** Use the option "Parameter Sensitivity" to investigate in particular the role of the delivery delay d, of the sales-proportional order factor a, and of the deficit-proportional order factor b. Find an order strategy which allows the smallest possible inventory while satisfying all customer requests despite (modest) random fluctuations at all times.

8. **References:** Bossel, 1987/1989: p. 191-202.

STOCKS, SALES, AND ORDERS

M 216

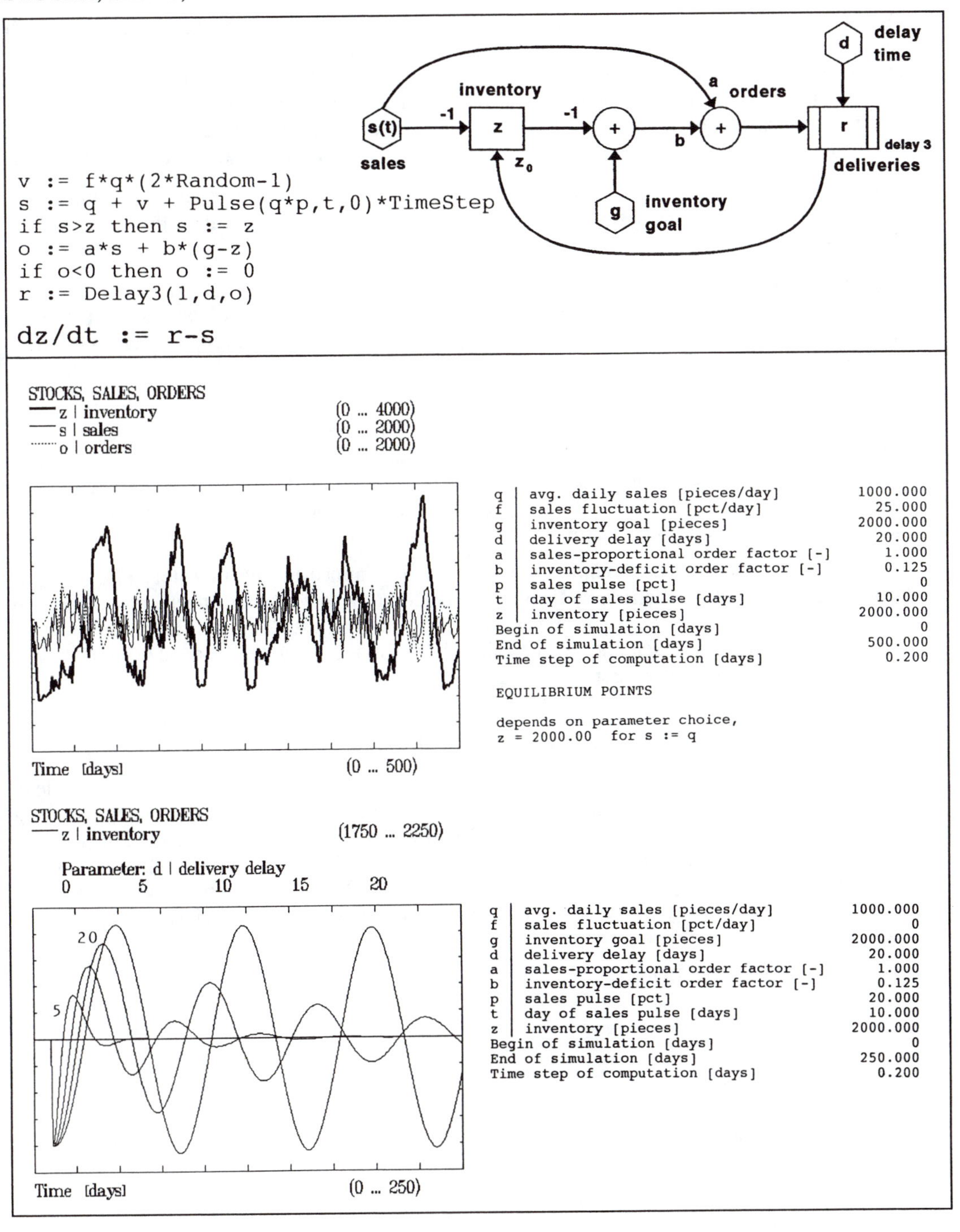

PRODUCTION CYCLE

M 217

1. **Description:** The price of a product is here assumed to be a function of the inventory y: high inventory results in lower price. Fewer products are sold if the inventory is small and the price relatively high. However, the small inventory and the corresponding high price promise future profits and therefore lead to investment activities and a higher production capacity x. The higher production capacity then again fills up the inventory leading to lower prices and a closing of some production capacity. This process may therefore generate oscillations (production cycle).

2. **Occurrence:** Production and marketing, for example, of agricultural products (commodities) (e.g. the "hog cycle"). The system structure is generally applicable to all investments controlled by market responses.

3. **Structural characteristics:** The sales rate corresponds to a normal sales rate c and a time function $u(t)$ (pulse function). It is modified in accordance to the level of supplies (the relative inventory y/g) and the price effect q (elasticity). The relative price is proportional to the relative inventory deficit $(1-(y/g))$; this term is therefore used to control the increase or decrease in production capacity x.

4. **Behavioral characteristics:** As the inventory decreases, prices rise leading to investments in new production capacity x. However, at the same time sales decrease on account of the rising prices. The new investment increases production, inventories fill up, and prices drop again. Production capacity is shut down (divested) reducing production correspondingly and producing corresponding fluctuations in the inventories and prices as well as in production capacity.

5. **Critical parameters:** The dynamics are partially determined by the inventory effect q on prices and sales and, in addition by the specific reaction a to price and inventory changes. A strong reaction (relatively large parameter a) leads to stronger oscillations. The system is linear and can be analyzed, for example, with Model M 203 LINEAR OSCILLATOR OF SECOND ORDER to determine the location of its eigenvalues and stability.

6. **Reference run:** The parameter settings result in a damped oscillation with a period of 6.5 time units. This oscillation is caused by a single sales pulse at time $t = 1$.

7. **Suggestions for further study:** Use the option "Parameter Sensitivity" to investigate the role of parameters q and a which quantify the buying reaction on one hand and the investment reaction to the price development on the other. In which parameter range do oscillations occur? Analyze the system using the Model M 203 LINEAR OSCILLATOR OF SECOND ORDER to confirm the previous results. Explain the behavior by reference to the location of the eigenvalues in the complex plane, and discuss the consequences for stability, oscillations, and the damping of oscillations of the system (cf. Sec. 7.3.10 - 7.3.15).

8. **References:** Meadows, 1970.

PRODUCTION CYCLE

M 217

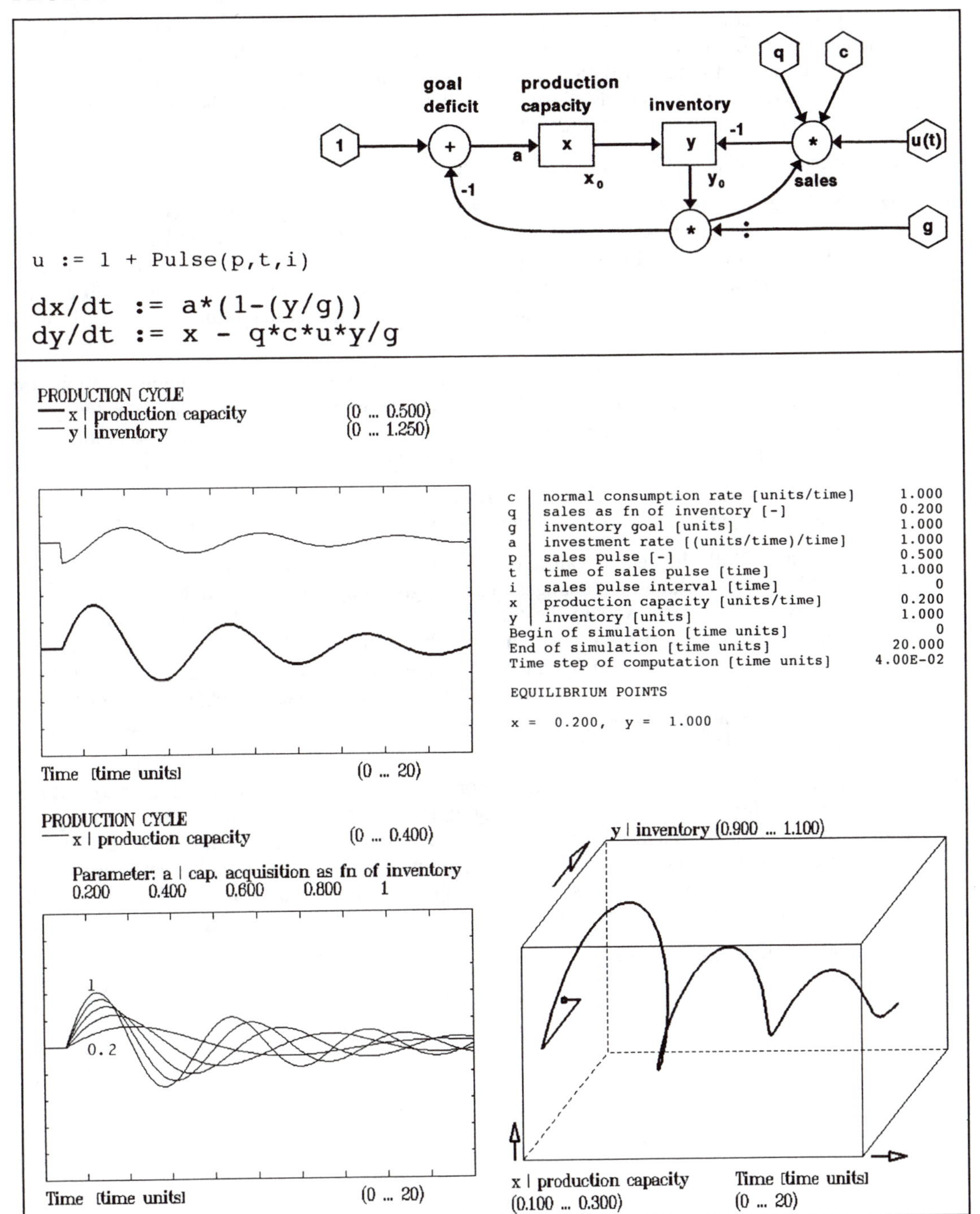

ROTATING PENDULUM

M 218

1. **Description:** A stiff, weightless rod of length r carries a point mass m at one end and is attached to a fixed frictionless bearing on the other end about which it is free to rotate in a vertical plane. If the motion is started with a high initial angular velocity y, the pendulum will rotate about its axis until air friction has slowed the motion to the point where the rotation changes to oscillation about the lower dead center where the pendulum will eventually come to rest.

2. **Occurrence:** Mechanical, gravitation-dependent pendulum.

3. **Structural characteristics:** The system with state variable y (angular velocity) and x (angle) shows a feedback loop across both state variables from which oscillations can be expected. The angular velocity is modified by damping proportional to the damping factor d and to the current angular velocity y (laminar damping). The angular acceleration component due to gravity is tangential to the pendulum trajectory and therefore changes its magnitude non-linearly with the sine of the pendulum angle x.

4. **Behavioral characteristics:** The rotational motion as well as the pendulum motion are damped. If the velocity has decreased to the point where the motion cannot pass over the upper dead center, the motion will change into damped swinging motion. Gravity and damping cause this motion to finally come to rest at the lower dead center.

5. **Critical parameters:** The damping d controls how quickly the motion comes to rest. The initial conditions, in particular the initial angular velocity, have a significant effect on the motion.

6. **Reference run:** With the default values chosen, the initial angular velocity is high enough to allow one full rotation which then changes to a strongly damped swinging motion. This motion can be observed particularly well in the three-dimensional view with the components time t, vertical position v, and horizontal position u of the pendulum. The option "Global Analysis" can be used to study the effect of different initial values on the motion. At high angular velocities, the state trajectories show rotation which slows down in the vicinity of the upper dead center (saddle). As the angular velocity decreases, the pendulum swings back and forth and finally comes to rest at the lower dead center (focus).

7. **Suggestions for further study:** Using the option "Global Analysis," determine the graph of the state trajectories for different values of damping d. How does the pendulum frequency change as a function of pendulum length and pendulum mass? Using the option "Parameter Sensitivity," study the time behavior for different damping parameters and initial angular velocities which allow at least one rotation even for strong damping. How do you explain the different equilibrium points for different values of damping?

8. **References:** Bossel, 1994: Ch. 3, 4.

ROTATING PENDULUM

M 218

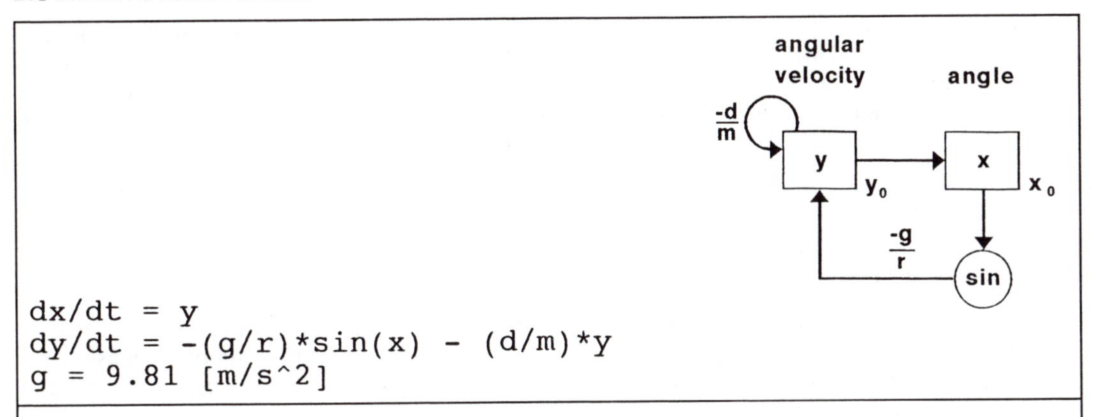

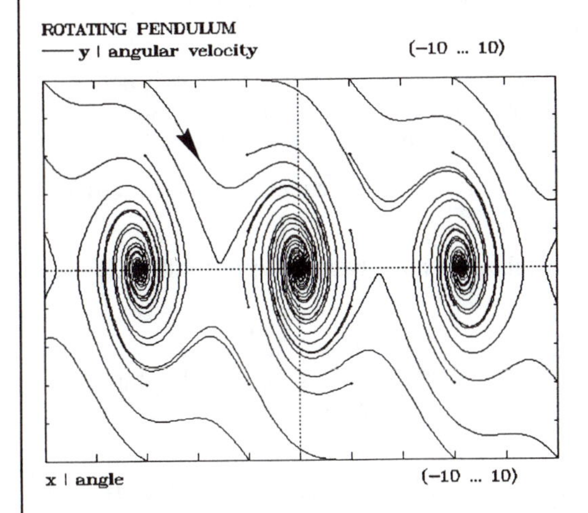

m	pendulum mass [kg]	1.000
r	pendulum radius [m]	1.000
d	viscous damping [N/(m/s)]	1.000
x	angle [radian]	0
y	angular velocity [1/sec]	10.000
Begin of simulation [seconds]		0
End of simulation [seconds]		10.000
Time step of computation [seconds]		1.00E-02

EQUILIBRIUM POINTS

stable equilibrium: x = 2*n*pi, y = 0
unstable equilibrium: x = 2*(n+1)*pi, y = 0

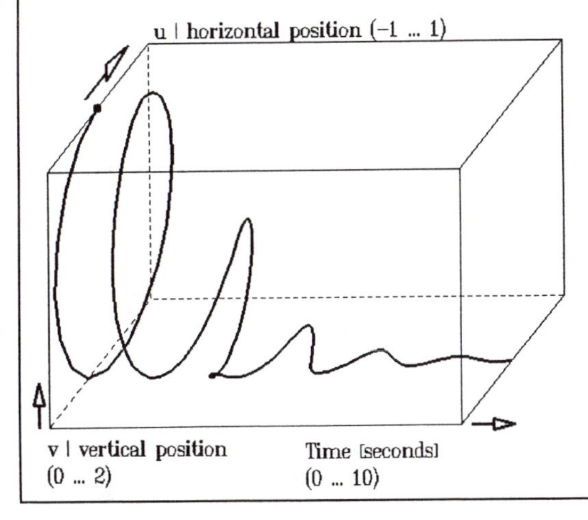

m	pendulum mass [kg]	1.000
r	pendulum radius [m]	1.000
d	viscous damping [N/(m/s)]	1.000
x	angle [radian]	3.142
y	angular velocity [1/sec]	10.000
Begin of simulation [seconds]		0
End of simulation [seconds]		10.000
Time step of computation [seconds]		1.00E-02

OSCILLATOR WITH LIMIT CYCLE (VAN DER POL) **M 219**

1. **Description:** This system consists mainly of a linear oscillator that has been modified by a non-linear structural addition producing a damping of state variable y if state variable x is large and an amplification of y if x is small. This causes the system to move very quickly into a stable oscillation of constant amplitude, independent of initial conditions. In state space (x,y), this stable oscillation appears as a closed curve (limit cycle).

2. **Occurrence:** This kind of system was initially used for the stabilization of electronic oscillations in radio tubes (triodes). It was also found to stabilize heart frequency. It can be observed in oscillations induced by fluid flows (wind-induced oscillations of structures, aerodynamic flutter), in vehicle dynamics, and in certain chemical reactions.

3. **Structural characteristics:** The direct mutual couplings of the state quantities x and y correspond to the harmonic undamped oscillator. The self-loop of y, which would be responsible for a damping for a negative sign of the coupling parameter is now modified by the current state x in such a way that for small x an amplification of y and for large x a damping of y will occur. This sign change is controlled by the term $(1-x^2)$.

4. **Behavioral characteristics:** If $x^2 > 1$, then $(1-x^2) < 0$, and damping of y will result. If, however, $x^2 < 1$, then $(1-x^2) > 0$, and y will be amplified. This causes a rapid stabilization of the oscillation on the limit cycle, and the system state continues to move on this cycle.

5. **Critical parameters:** The coupling parameter a changes the frequency of the oscillation. Oscillations can only occur if $a < 0$.

6. **Reference run:** The reference case with $a = -1$ shows a quickly stabilizing limit cycle oscillation of period 9.6.

7. **Suggestions for further study:** Use the option "Parameter Sensitivity" to study the oscillating response for different values of $a < 0$. Use the option "Global Analysis" to find the state trajectories for different initial conditions in the interior and the exterior of the limit cycle for different frequency parameters a.

8. **References:** Csaki, 1972: p. 359-362; Beltrami, 1987: p. 182-189; Bossel, 1987/89: p. 183-184; Guckenheimer and Holmes, 1983/86: p.67-82.

OSCILLATOR WITH LIMIT CYCLE (VAN DER POL) M 219

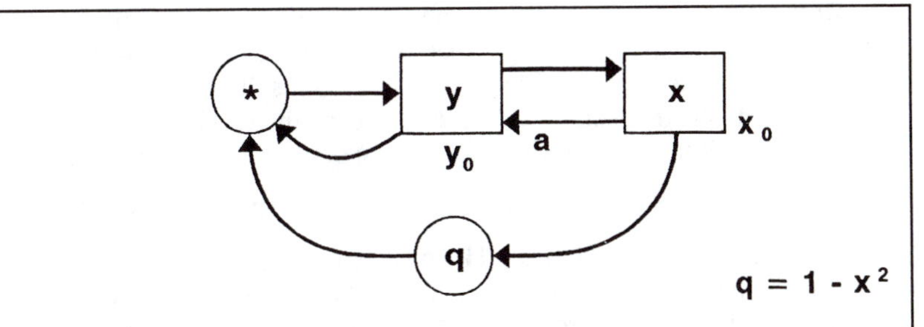


```
dx/dt = y
dy/dt = a*x + (1 - x^2)*y
```

LIMIT CYCLE OSCILLATION
— x | state1 (-5 ... 5)
— y | state2 (-5 ... 5)

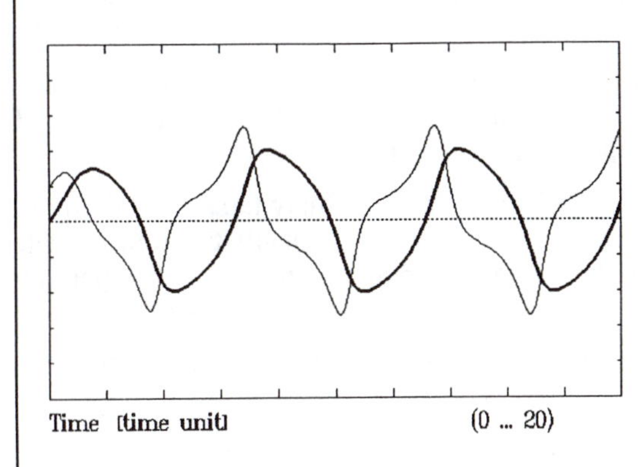

a	coupling paramter [1/time²]	-1.000
x	state1 [unit*time]	0
y	state2 [units]	1.000
Begin of simulation [time unit]		0
End of simulation [time unit]		20.000
Time step of computation [time unit]		1.00E-02

EQUILIBRIUM POINTS

stable limit cycle around
unstable equilibrium point at x = 0, y = 0

LIMIT CYCLE OSCILLATION
— y | state2 (-5 ... 5)

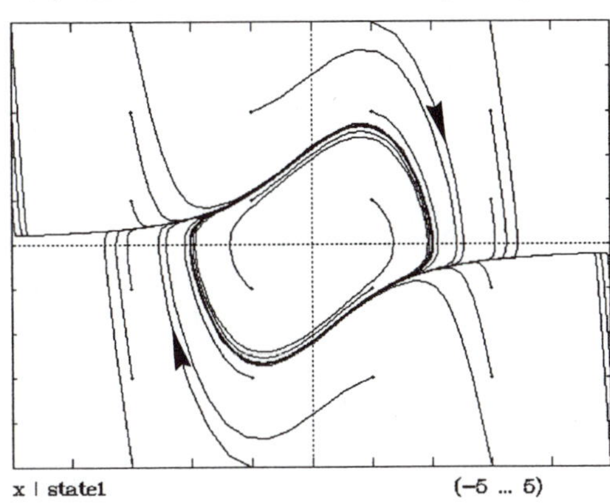

BISTABLE OSCILLATOR

M 220

1. **Description:** The core of the system is again a linear oscillator which has been modified by an additional negative coupling of x^3 to y. Depending on initial conditions, the system trajectories will end at one of two stable equilibrium points. The system can be physically realized by a leaf spring mounted between two permanent magnets. The initial conditions determine at which of the two magnets (equilibrium points) the system will come to rest.

2. **Occurrence:** Electronic flip-flop circuit; steel spring between two permanent magnets.

3. **Structural characteristics:** The basic structure of the system is that of a linear oscillator with exponential damping (self-loop) d of state variable y. In addition, there is a restoring force proportional to the third power of the displacement x (as is the case, for example, in the bending of beams).

4. **Behavioral characteristics:** From the system equations follow two stable equilibrium points at $x = 1$ and $x = -1$ as well as an unstable equilibrium point at $x = 0$. The free motion ends at one of the stable equilibrium points; the resting point depends on the initial conditions. The damping of the state variable y depends on the velocity itself (y) and the damping constant d.

5. **Critical parameters:** Very different trajectories are obtained for different values of damping d. The initial state has a critical influence on the final state.

6. **Reference run:** A damping of $d = 1$ results in a strongly damped oscillation which comes to rest either at $x = 1$ or $x = -1$ and $y = 0$ depending on the initial conditions. The totality of the state trajectories is best analyzed using the option "Global Analysis."

7. **Suggestions for further study:** Using the option "Global Analysis," study the state trajectories for different values of damping $0 < d < 1$. Use the option "Parameter Sensitivity" to investigate which time behavior will result from different choices for damping parameters and initial values.

8. **References:** DeRusso, Roy and Close, 1965: p. 483-488; Guckenheimer and Holmes, 1983/86: p. 82-91; Bossel, 1987/89: p. 187-188.

BISTABLE OSCILLATOR

M 220

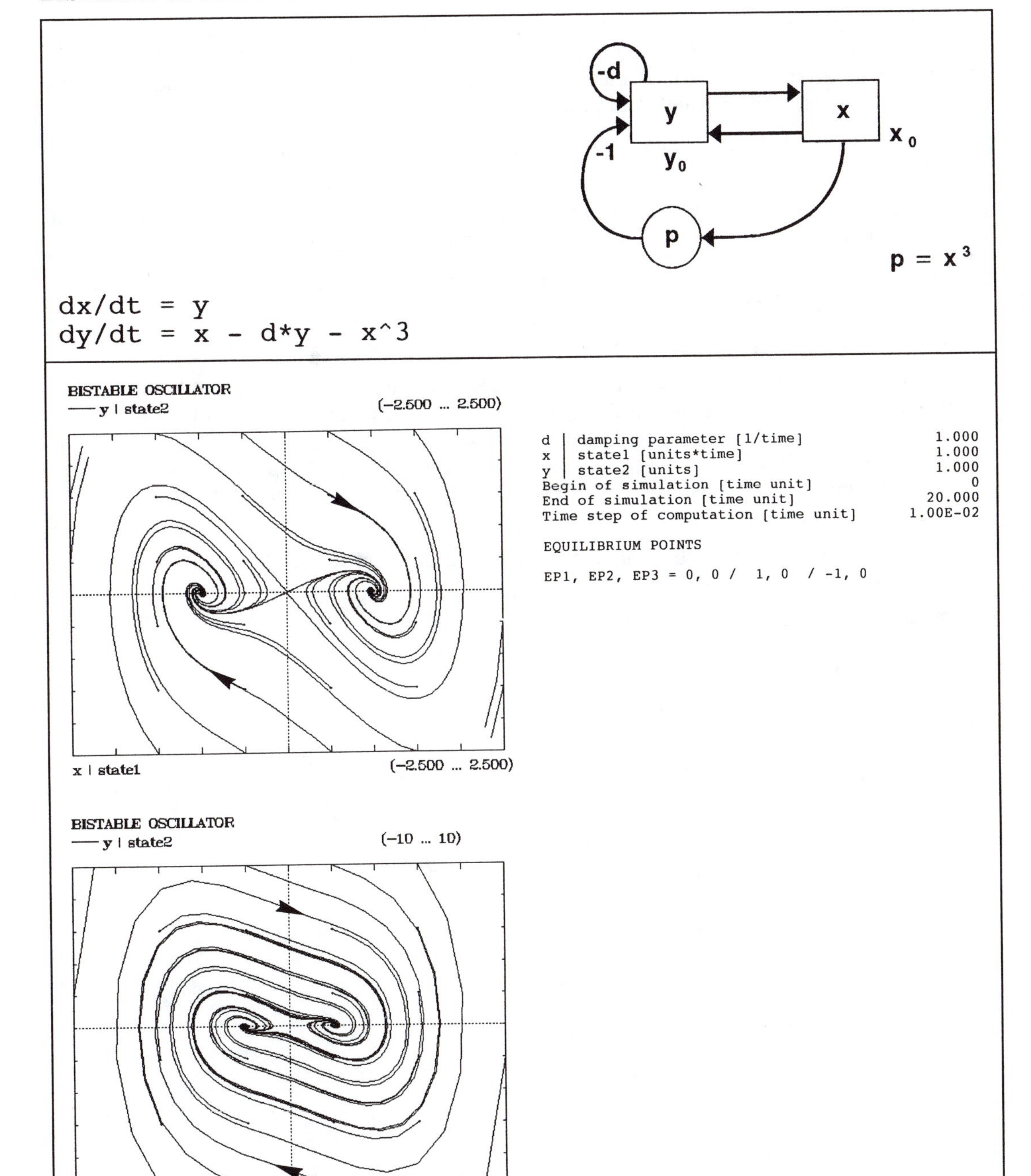

CHAOTIC BISTABLE OSCILLATOR M 221

1. **Description:** The system is identical to the bistable oscillator (M 220) except for a sinusoidal input with frequency w and amplitude factor q. This now results in chaotic motion. The system can be physically realized by a leaf spring between two permanent magnets where the entire system is now moved back and forth in sinusoidal motion.
2. **Occurrence:** Bistable oscillator with sinusoidal forcing; periodic forced motion between two attractors.
3. **Structural characteristics:** The basic structure is that of the linear damped oscillator with a sinusoidal input. In addition, there is a non-linear restoring force (proportional to x^3).
4. **Behavioral characteristics:** Depending on damping d, frequency w, and amplitude q, chaotic motion will result with different shapes of the attracting regions.
5. **Critical parameters:** Forcing frequency w, forcing amplitude q, and damping factor d.
6. **Reference run:** With the default values, the motion first moves around one of the equilibrium points while the amplitude increases. At some point it jumps into the vicinity of the other equilibrium point, rotates around it, and eventually jumps back to the neighborhood of the first equilibrium point. This motion is best studied in state space (x, y). Amplification of the motion and the jumping of the trajectory to the other region of attraction are also obvious if x and y are plotted as function of time.
7. **Suggestions for further study:** Study the motion for different values of w, q and d in state space (x, y).
8. **References:** Guckenheimer and Holmes, 1983/86: p. 82-91.

CHAOTIC BISTABLE OSCILLATOR

M 221

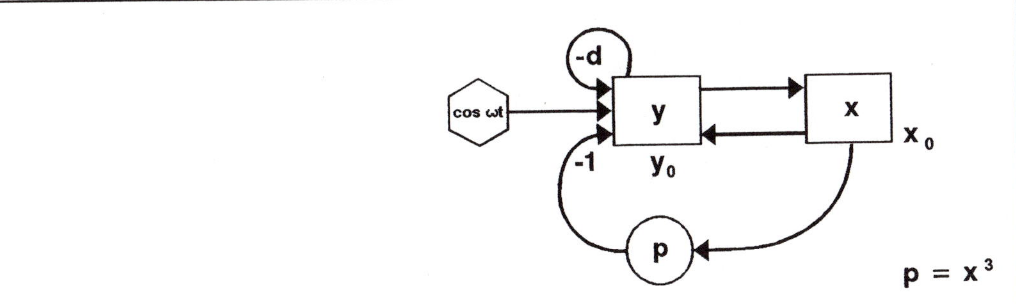


```
dx/dt = y
dy/dt = x - x^3 - d*y + q*cos(w*time)
```

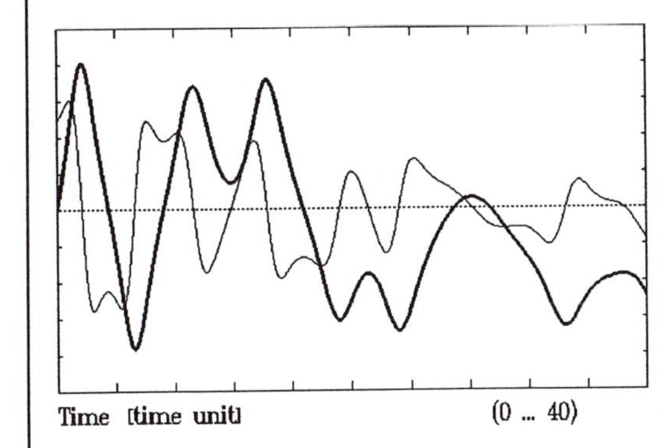

d	damping parameter [1/time]	0.250
q	amplitude, sine forcing [units/time]	0.300
w	frequency, sine forcing [1/time]	1.000
x	state1 [units*time]	0
y	state2 [units]	1.000
Begin of simulation [time unit]		0
End of simulation [time unit]		40.000
Time step of computation [time unit]		1.00E-02

EQUILIBRIUM POINTS

chaotic attractor
equilibrium points of unforced system:
x, y = 0, 1 / 0, -1 (stable), 0, 0 (unstable)

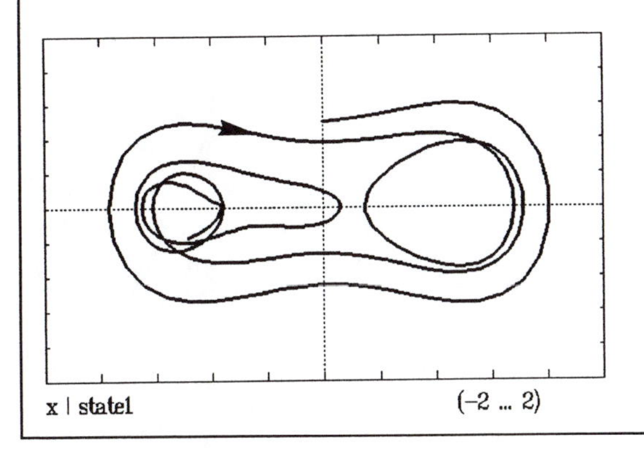

6.3 Dynamic Systems with Three or Four State Variables

301 TRIPLE INTEGRATION AND EXPONENTIAL DELAY
302 POPULATION DYNAMICS WITH THREE GENERATIONS
303 LINEAR OSCILLATOR OF THIRD ORDER
304 MINIWORLD: POPULATION, CONSUMPTION, POLLUTION
305 PREDATOR WITH TWO PREY POPULATIONS
306 TWO PREDATORS WITH ONE PREY POPULATION
307 BIRDS, INSECTS, AND FOREST
308 NUTRIENT CYCLING AND PLANT COMPETITION
309 CHAOTIC ATTRACTOR (RÖSSLER)
310 HEAT, WEATHER, AND CHAOS (LORENZ SYSTEM)
311 COUPLED DYNAMOS AND CHAOS
312 BALANCING AN INVERTED PENDULUM

TRIPLE INTEGRATION AND EXPONENTIAL DELAY M 301

1. **Description:** In this system the input function u(t) is integrated three times in succession with respect to time. For a feedback parameter $a = 0$, the numerical integration corresponds to the analytical integration formulae. If the magnitude of the negative feedback $|a| > 0$, then there will be a delaying effect at each of the integrators (cf. Models M 103 and M 201 EXPONENTIAL DELAY of first and second order). The triple delayed integration leads to an exponential delay of third order which is often used in simulation models to delay and smooth functions (SIMPAS function: Delay3).

2. **Occurrence:** Numerous physical processes and exponential delays of third order.

3. **Structural characteristics:** Case $a = 0$: The first integrator z integrates the input signal u(t) over time. The output of the second integrator y is the time integral of the input signal z. The output of the third integrator x is a time integral of the input signal y.

   Case $a > 0$: If a negative feedback loop exists on all three integrators, there will be a state-proportional loss rate at each of the integrators. If u(t) is a step function with $u = \text{const}$ for $t > t_0$, the state z will increase to the point where the exponential loss rate is exactly equal to the input signal u at which point it remains constant. This constant output signal z now becomes the input to integrator y where it again undergoes the same process. Again, there will be a constant output signal y after some time. The same process repeats at integrator x.

4. **Behavioral characteristics:** With zero feedback ($a = 0$) the process corresponds to the triple application of integration rules. For $a > 0$, a corresponding delay effect occurs at each integrator with a total delay time of $3/a$.

5. **Critical parameters:** For strong feedback there will be a small signal delay; for weak feedback the delay is much longer. For a constant signal u , we obtain the following equilibrium values at the three integrators: $z^* = u/a$, $y^* = u/a^2$, and $x^* = u/a^3$.

6. **Reference run:** The default values of the model correspond to a triple integration of a unit step function beginning at $t = 1$ with a damping parameter $a = 1$. The initial response of the integrators is as follows: z is proportional to t, y is proportional to t^2, and x is proportional to t^3. After some time, each of the integrators saturates at its equilibrium value. The input signal is delayed three times by a time delay $(1/a)$. The total delay is $T = 3/a$.

7. **Suggestions for further study:** Investigate the dependence of the equilibrium values of the three integrators on the feedback parameter a. Use the model with $a = 0$ to integrate the different test functions (pulse, step, ramp, sin) one, two, and three times over time; compare the results with the analytical integration formulae. Use non-zero feedback values a, and investigate the response of the third order delay to the different test functions.

8. **References:** -

TRIPLE INTEGRATION AND EXPONENTIAL DELAY **M 301**

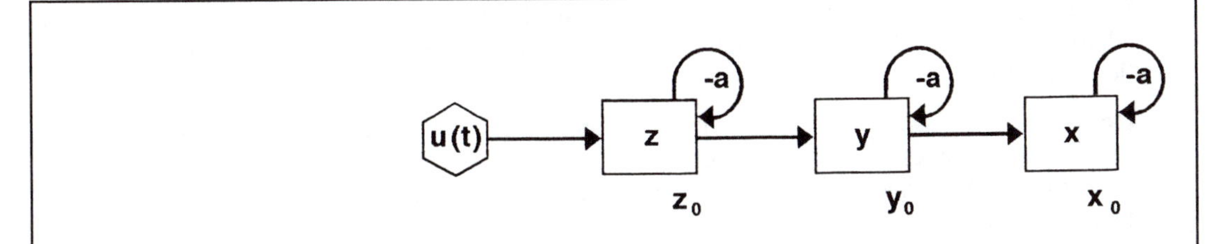


```
u := Pulse(p,tp,i) + Step(s,ts) + Ramp(r,tr) + q*sin(2*pi*w*time)
dx/dt := y - a*x
dy/dt := z - a*y
dz/dt := u - a*z
```

TRIPLE INTEGRATION
x | state1 (0 ... 1.250)
y | state2 (0 ... 1.250)
z | state3 (0 ... 1.250)
u | exogenous change rate (0 ... 1.250)

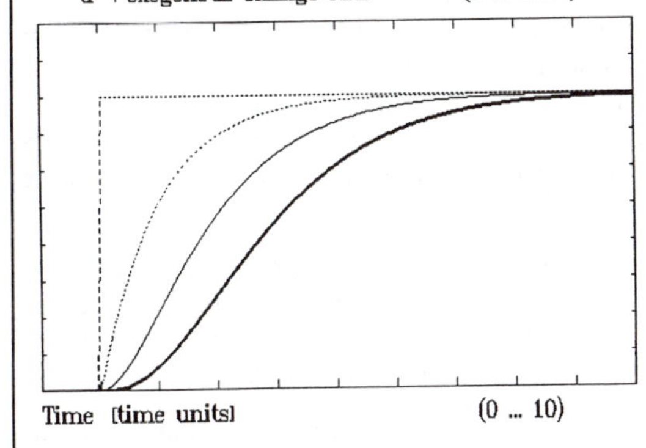

a	decay rate [1/time]	1.000
p	PULSE volume [units]	0
tp	initial pulse time [time]	1.000
i	pulse interval [time]	1.000
s	STEP height [units/time]	1.000
ts	step time [time]	1.000
r	RAMP slope [units/time^2]	0
tr	ramp time [time]	1.000
q	SINE wave amplitude [units/time]	0
w	sine wave frequency [rad/time]	0.100
x	state1 [units*time^2]	0
y	state2 [units*time]	0
z	state3 [units]	0
Begin of simulation [time units]		0
End of simulation [time units]		10.000
Time step of computation [time units]		2.00E-02

EQUILIBRIUM POINTS

depend on input; x = y = z = 0 for autonomous system
z = u/a, y = u/(a*a), x = u/(a*a*a) for u = const

TRIPLE INTEGRATION
x | state1 (0 ... 5)

Parameter: a | spec. decay rate
0.600 0.700 0.800 0.900 1

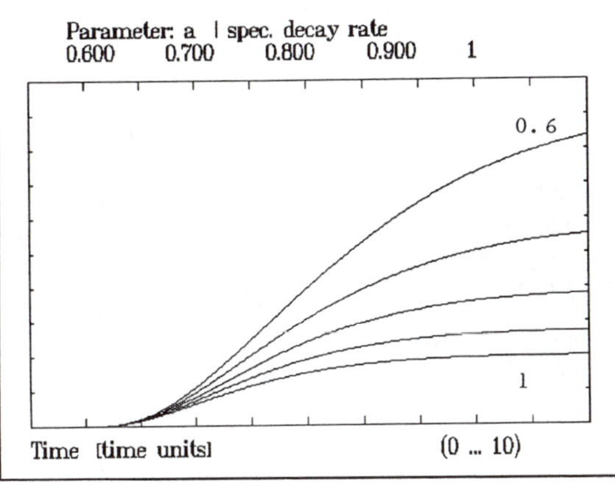

POPULATION DYNAMICS WITH THREE GENERATIONS M 302

1. **Description:** The model describes the population dynamics of three generations: children z, adults y, and old people x. In each age class, outflows (deaths), inflows by births, and transitions from the younger age class are considered. Only adults "produce" babies.

2. **Occurrence:** The model describes human population development, and more generally also population dynamics of animals and plants. A similar approach is used, for example, to describe an insect population (egg, caterpillar, pupa, butterfly) or forest dynamics (tree height or diameter classes).

3. **Structural characteristics:** The transition rate from one age class to the next is proportional to the "donor" population and inversely proportional to the residence time in the age class. Each age class has losses by (age class specific) mortality (p, q, r) and gains by transition from the younger age class. The number of births is proportional to the number of women $(= y/2)$ and the number of children $u(t)$ they have during their residence time n as adults: births = $(y/2) \cdot u(t)/n$.

4. **Behavioral characteristics:** As a consequence of the residence times in the different age classes (m in z, n in y), a delay is associated with each age class. The model therefore can produce population "waves" (baby booms and their long-term consequences) as a result of fertility fluctuations (number of births). In particular, birth control will only have a significant effect after about one generation when the number of potential parents has been reduced by the earlier control efforts. A large number of children leads to many more births a generation later. In order to quickly stabilize a population it may therefore be necessary to reduce the number of children per family below the replacement value (of roughly 2.3 children per family) for a limited time, for example, by only allowing one child per family.

5. **Critical parameters:** The most critical parameter of the population development is the fertility, i.e. the number of children per woman. In the model it can be described as a function of time using the scenario parameters a, b, c, and d. Because of the delay of roughly one generation, an early anticipatory response is necessary to deal effectively with population problems.

6. **Reference run:** The default parameters correspond to a (fictitious) country of initially 100 million inhabitants and high fertility (five children per woman) up to the year 2000. Thereafter, the number of children per family is reduced linearly to 2.3 in the year 2050. Despite these rigorous measures of population control the population climbs to 260 million people in about 60 years. The percentage of children decreases in the population while the percentage of old people increases strongly.

7. **Suggestions for further study:** Investigate different scenarios of population development by choosing appropriate parameters (a, b, c, d). Use the option "Parameter Sensitivity" to find the influence of time d on the population development, if at this time the replacement value of 2.3 children per family is to be reached. Use the model to compute the population development in selected developing and industrial countries. (The necessary statistical information can be found, for example, in current issues of *State of the World* (Brown since 1980 or similar sources).

8. **References:** Meadows et. al., 1974; Bossel, 1985/87: p. 84-91.

POPULATION DYNAMICS WITH THREE GENERATIONS

M 302

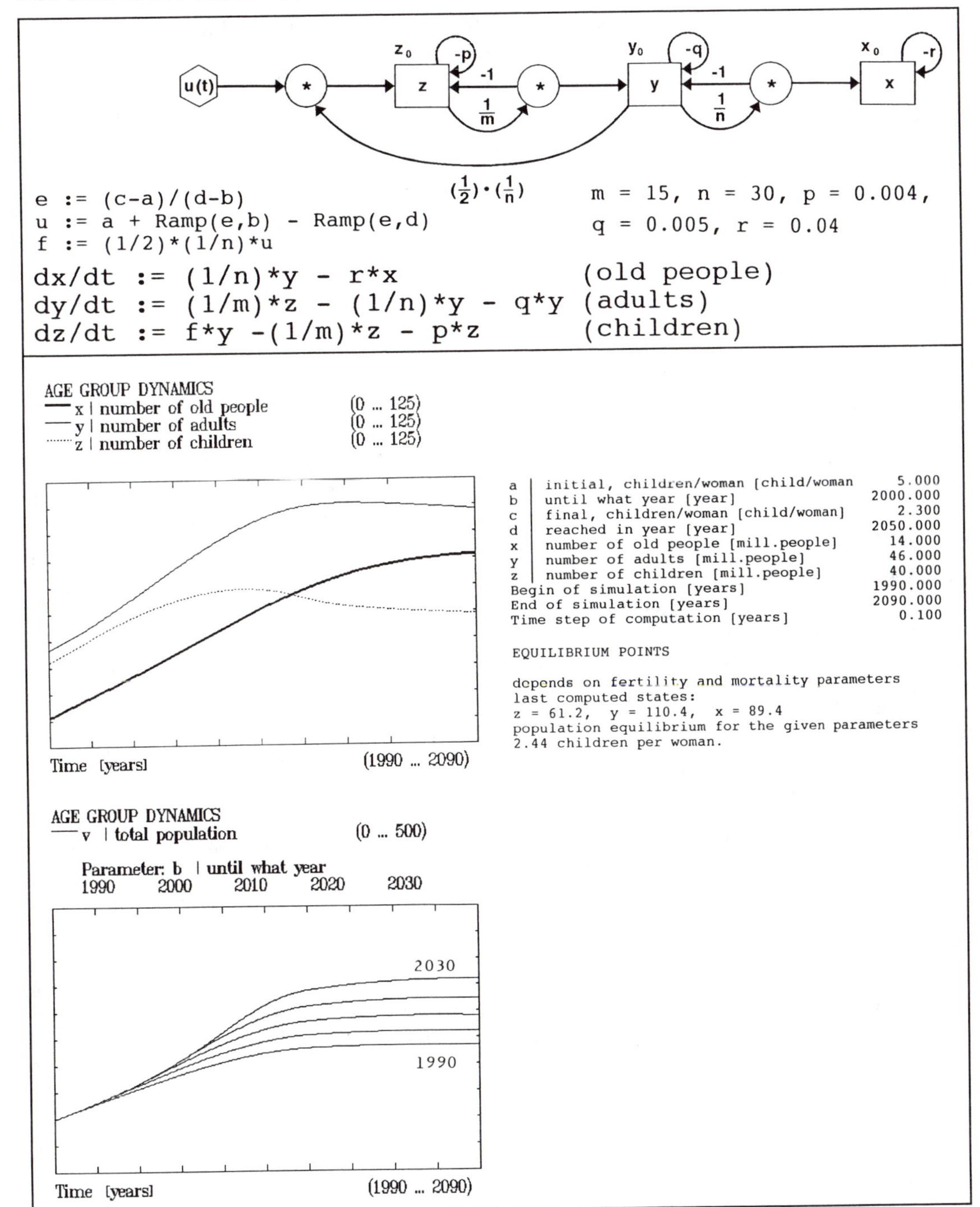

LINEAR OSCILLATOR OF THIRD ORDER

M 303

1. **Description:** In this system the output of the first integrator is integrated a second and a third time; the outputs of all three integrators are fed back to the first one. This is a linear dynamic system of third order. The behavioral possibilities are identical to that of the linear oscillator of second order (damped and undamped, periodic and aperiodic behavior).

2. **Occurrence:** Technical systems, in particular control systems.

3. **Structural characteristics:** The system structure shown here corresponds to the standard form of linear dynamic systems, with system matrix **A**:

$$\mathbf{A} = \begin{bmatrix} 0 & 1 & 0 \\ 0 & 0 & 1 \\ a & b & c \end{bmatrix}$$

This system matrix has the characteristic polynomial

$$-\lambda^3 + c\cdot\lambda^2 + b\cdot\lambda + a = 0.$$

The eigenvalues of the system follow from this. All other linear systems of third order can be reduced to this form (cf. Sec. 7.3.8). In the standard form, all state variables are fed back to the first state variable z of the integration chain. In addition, an exogenous input function $u(t)$ may act as a forcing function.

4. **Behavioral characteristics:** The three state variables correspond to the three eigenvalues and three characteristic modes of type $e^{\lambda t}$ or $e^{\sigma t} \cdot (\sin \omega t)$. Here σ is the real part of the (complex) eigenvalue: $\sigma = \mathrm{Re}(\lambda)$. The general solution is a combination of the possible periodic, aperiodic, damped, or undamped solutions. The system behavior is stable only if the real part of the eigenvalue is negative, i.e. for $\mathrm{Re}(\lambda) < 0$. The location of eigenvalues in the complex plane determines the behavior of the system (as for the oscillator of second order).

5. **Critical parameters:** The parameter c determines mainly the damping; the parameters a and b determine the oscillation and its frequency.

6. **Reference run:** The default values for the parameters $(a = -1,\ b = -1,\ c = -2)$ result in a strongly damped oscillation following a unit impulse at time $t = 1$ to the system at rest. From the time plot of the response, one obtains a period of oscillation of approximately 6.3. The output signals of the second and third integrator are delayed with respect to the preceding integrator.

7. **Suggestions for further study:** Use the option "Parameter Sensitivity" to search for parameters with a more strongly damped solution. Find the parameter combination where the oscillations disappear. How can frequency and period of the oscillation be changed by parameter changes? Find a parameter set for a fully aperiodic damped motion (corresponding to three purely real and negative eigenvalues).

8. **References:** Books on control system theory.

LINEAR OSCILLATOR OF THIRD ORDER

M 303

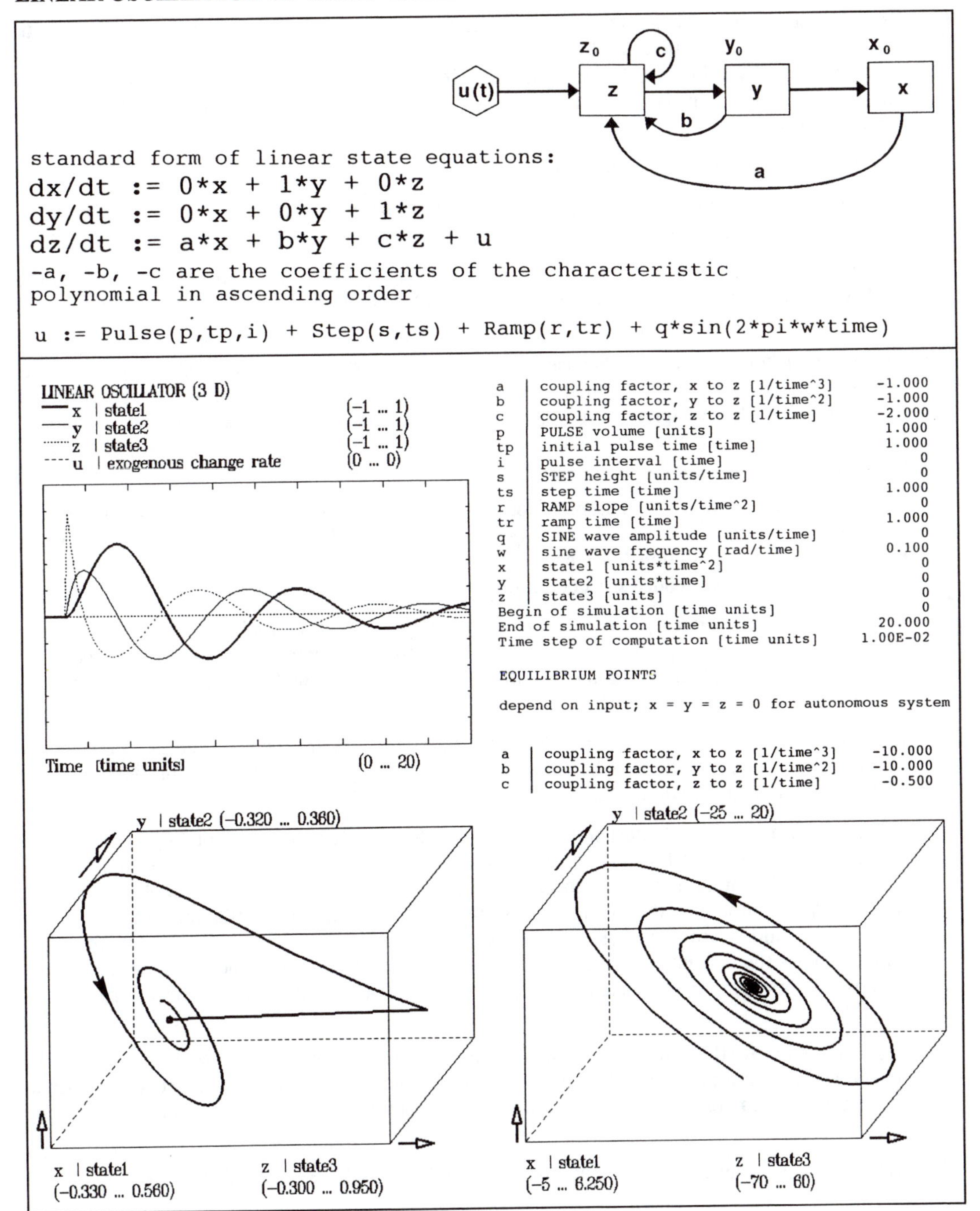

MINIWORLD: POPULATION, CONSUMPTION, AND POLLUTION M 304

1. **Description:** This highly aggregated model attempts to combine several important processes to describe mutual interactions and the resulting developments: population development, consumption growth, and environmental pollution. It exhibits the response dynamics also found in other global models, in particular the process of rapid growth and subsequent collapse of the three state variables.

2. **Occurrence:** Regional and global development.

3. **Structural characteristics:** The model is composed of three submodels describing population development (x), environmental pollution (y), and the development of consumption (z).

   Population development x: Births $(b \cdot x)$ are influenced by the level of consumption z and the quality of the environment $(q = m/y)$, where m is a threshold level. Deaths $(d \cdot x)$ are affected by environmental pollution y.

   Consumption development z: Per capita consumption (as a measure of material throughput) follows a process of saturation which is influenced by the level of consumption z (proportional to the capital stock per capita), a normal growth rate c, the environmental pollution y, and efforts to control consumption k. A larger value of k means a stronger damping of the consumption growth rate.

   Environmental pollution y: The pollution rate $(e \cdot x \cdot z)$ is proportional to population and the level of consumption. The rate at which pollution is broken down by natural processes $(a \cdot y)$ is proportional to the pollution level but will not increase beyond a critical level (here $= a \cdot m$) when the system is overloaded with pollution.

4. **Behavioral characteristics:** An initial increase in population and consumption leads to (delayed) environmental pollution which eventually strongly reduces the population level. At later stages the system shows damped oscillations; eventually it will come to rest at an equilibrium point.

5. **Critical parameters:** The most important control parameters are the birthrate b and the control of consumption growth k. The parameter k has a very significant effect on the long-term development and on the equilibrium point.

6. **Reference run:** Reasonable assumptions are used for the default values of the specific rates: specific rate of pollution breakdown $a = 0.1$, specific rate of pollution by consumption $e = 0.02$, specific birthrate $b = 0.03$, specific deathrate $d = 0.01$, pollution threshold $m = 1$, and consumption control $k = 0.1$ (saturation at 10 times the initial per capita consumption level). Since x, y, and z are relative state variables, the initial value is chosen as 1. For these parameter values the system shows strongly damped oscillations with a period of 120 years. A first collapse of population takes place after about 40 years as a consequence of a steeply rising pollution level.

7. **Suggestions for further study:** Use the option "Parameter Sensitivity" to study the role of the consumption control parameter k. Examine the influence of the other parameters on the development of a system, on minima and maxima, on oscillations, and on the location of equilibrium values.

8. **References:** Meadows, Meadows, and Randers, 1992; Bossel, 1994: Ch. 2, 5.

MINIWORLD: POPULATION, CONSUMPTION, AND POLLUTION M 304

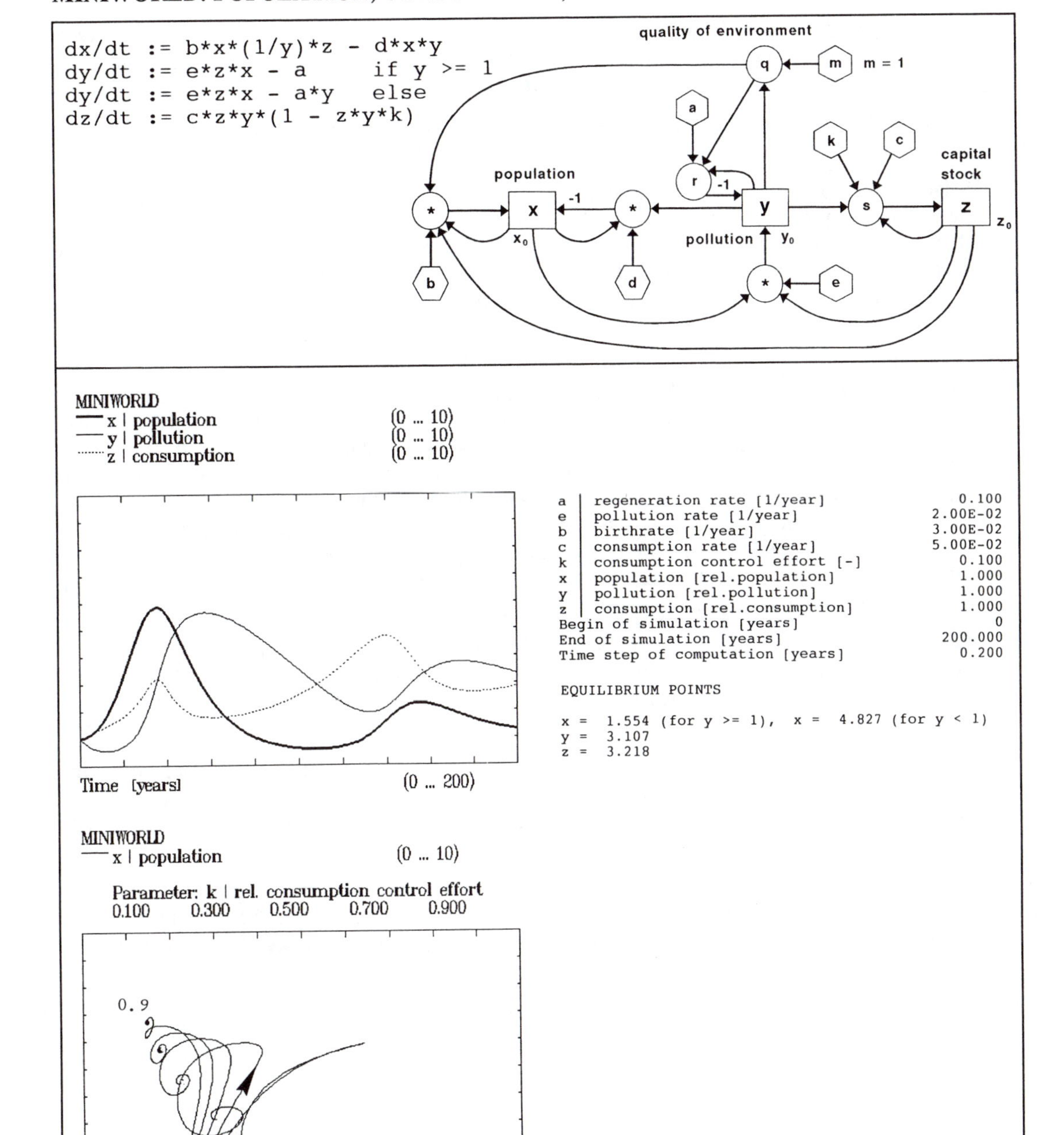

PREDATOR WITH TWO PREY POPULATIONS M 305

1. **Description:** In this model a predator z has two sources of food: prey x and prey y. One prey population is slightly disadvantaged compared to the other prey population (for example, a lower net growth rate) and will eventually disappear.

2. **Occurrence:** Predator-prey dependencies in ecosystems; social exploitation processes, for example, those involving different ethnic groups.

3. **Structural characteristics:** The predator z is coupled to the two prey populations x and y by way of two predator-prey couplings. The structure of each of these couplings corresponds to that of the simple predator-prey system (cf. Model M 206). The prey populations grow with specific net growth rates a and b. The predator has a specific respiration rate c. The relative losses of the prey populations x and y from predation are given by specific loss parameters d and e; the corresponding relative gains of the predator population from predation of x or y are given by f and g.

4. **Behavioral characteristics:** The system again exhibits the oscillations typical of predator-prey systems. However, the disadvantaged prey (having a lower growth rate or a relatively higher loss rate) is affected much more strongly than the less disadvantaged prey. Since the predator's existence is not dependent on the more disadvantaged prey, this prey population does not have an inherent protection (by a feedback loop affecting the predator), and it will therefore eventually disappear. More generally, in a system having a number of prey populations as potential food source for a predator, all prey populations except the last remaining one will eventually disappear (according to this model). (In the real world organisms appear to avoid this fate by differentiation and niche specialization.)

5. **Critical parameters:** All other parameters being equal, the difference between the growth parameters a and b (relative growth rates) of the two prey populations is critical. The prey population that either regenerates more slowly or is more strongly affected by predation will eventually die out.

6. **Reference run:** The default parameter settings for both prey populations are equal except for the relative rates of growth. The net growth rate a of population x is slightly less than for population y (with a growth rate b); it will therefore die out. The predator-prey oscillation has a period of about 70 time units.

7. **Suggestions for further study:** Using the option "Parameter Sensitivity," study the development of the system for different parameter combinations. Investigate whether and how the development, especially the equilibrium state, depends on the initial values of the three populations.

8. **References:** -

PREDATOR WITH TWO PREY POPULATIONS

M 305

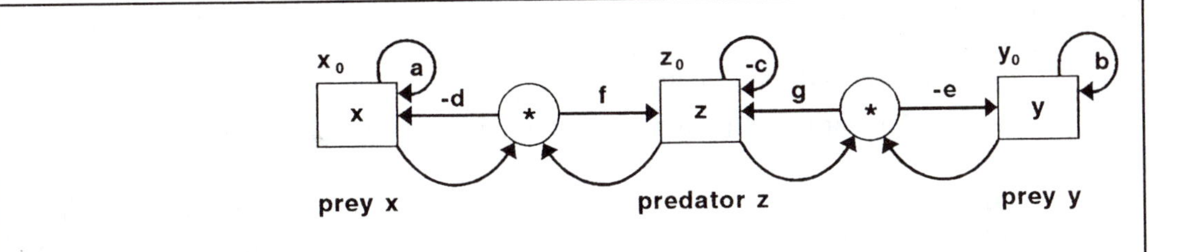


```
dx/dt := a*x - d*x*z
dy/dt := b*y - e*y*z
dz/dt := f*x*z + g*y*z - c*z
```

1 PREDATOR, 2 PREYS
x | prey1 population (0 ... 2.500)
y | prey2 population (0 ... 2.500)
z | predator population (0 ... 2.500)

Time [time units] (0 ... 200)

a	growth rate of preyX [1/time]	0.100
b	growth rate of preyY [1/time]	0.120
c	respiration rate, predator [1/time]	0.100
d	pred. loss, preyX [1/(pred*time)]	0.100
e	pred. loss, preyY [1/(pred*time)]	0.100
f	pred. gain, preyX [1/(preyX*time)]	0.100
g	pred. gain, preyY [1/(preyY*time)]	0.100
x	preyX population [preyX units]	1.000
y	preyY population [preyY units]	1.000
z	predator population [predator units]	1.000
Begin of simulation [time units]		0
End of simulation [time units]		200.000
Time step of computation [time units]		0.100

EQUILIBRIUM POINTS

x, y, z = 0 , 0 , 0
0 , 1.000 , 1.200
1.000 , 0 , 1.000

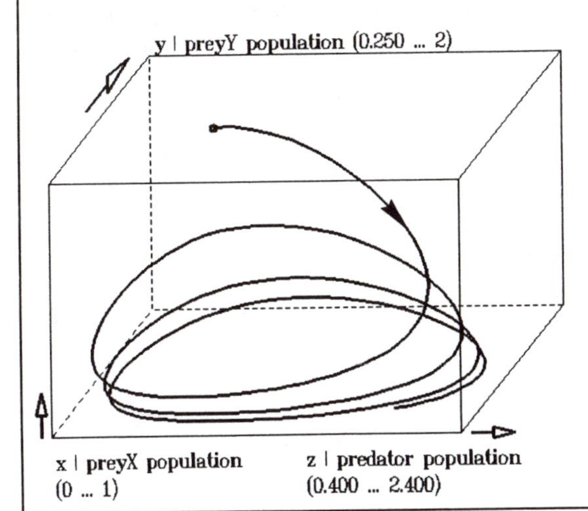

TWO PREDATORS WITH ONE PREY POPULATION M 306

1. **Description:** Two predator populations x and y depend on a common food source, the population z. In this system the more disadvantaged predator (for example, the one with the lower net growth rate) will eventually disappear.

2. **Occurrence:** Predator-prey dependencies in ecosystems; social exploitation with dependence on a common resource.

3. **Structural characteristics:** The system with one prey and two predator populations again has two predator-prey couplings $(x \cdot z, y \cdot z)$. These predator-prey couplings are identical to those of the normal predator-prey system (Model M 206). Predators x and y have specific respiration rates (a, b); the prey z has a specific net growth rate c. The predation gains of the predators are characterized by parameters d and e, the predation losses of the prey by parameters f and g.

4. **Behavioral characteristics:** As expected, the system shows the typical predator-prey oscillations for the three populations. Since the predator populations are both dependent on the prey population, it cannot disappear: overexploitation of the prey would reduce the predator populations to a level which would protect the prey from dying out. However, the predator with a relative disadvantage (for example, as a result of a higher respiration rate) receives relatively less energy to maintain and increase its biomass even if the initial populations x and y are the same. In the long run, the disadvantaged predator will therefore die out.

5. **Critical parameters:** The differences between the gain and loss rates of the competing predators are critical: the predator with the higher rate of respiration (a, b) or the lower predation gain rate (e, d) loses in the long run.

6. **Reference run:** The default values of the parameter sets for both predators are the same except for the specific respiration rates. The respiration rate of predator x is slightly higher $(a = 0.12, b = 0.1)$. Population x therefore dies out in the long run. The system shows oscillations with a period of about 70 time units.

7. **Suggestions for further study:** Use the option "Parameter Sensitivity" to study the development for different parameter combinations. Using the option "Global Analysis" determine whether and how the behavior (the equilibrium solution in particular) depends on the initial conditions.

8. **References:** -

TWO PREDATORS WITH ONE PREY POPULATION

M 306

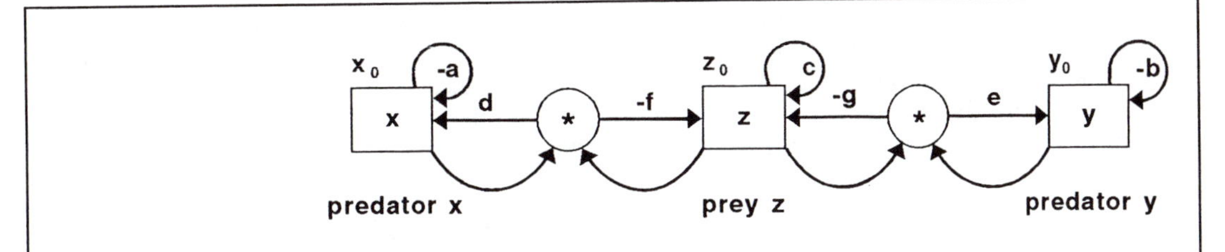


```
dx/dt := - a*x + d*x*z
dy/dt := - b*y + e*y*z
dz/dt := - f*x*z - g*y*z + c*z
```

1 PREY, 2 PREDATORS
x | predator1 population (0 ... 2.500)
y | predator2 population (0 ... 2.500)
z | prey population (0 ... 2.500)

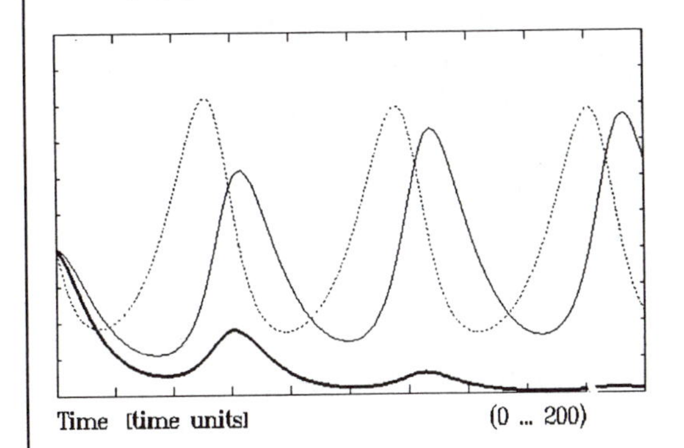

a	respiration rate of predX [1/time]	0.120
b	respiration rate of predY [1/time]	0.100
c	growth rate of prey [1/time]	0.100
d	pred. gain, predX [1/(prey*time)]	0.100
e	pred. gain, predY [1/(prey*time)]	0.100
f	prey loss, predX [1/(predX*time)]	0.100
g	prey loss, predY [1/(predY*time)]	0.100
x	predatorX population [predX units]	1.000
y	predatorY population [predY units]	1.000
z	prey population [prey units]	1.000
Begin of simulation [time units]		0
End of simulation [time units]		200.000
Time step of computation [time units]		0.100

EQUILIBRIUM POINTS

```
x, y, z =  0 , 0 , 0
           0 ,  1.000 ,  1.000
           1.000 , 0 ,   1.200
```

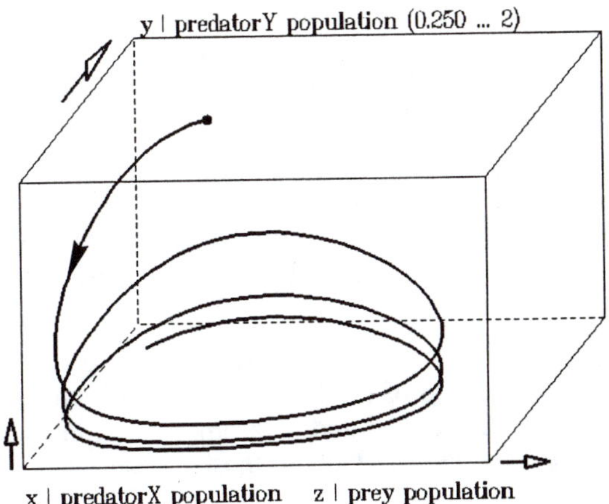

BIRDS, INSECTS, AND FOREST

M 307

1. **Description:** A region having a maximum biomass capacity k (carrying capacity) is partially covered by forest x, and partially by grassland (k–x). The bird population z needs the forest for nesting; it feeds on the insect population y. Insects need the forest as a source of food and grassland for the development of insect larvae. As the forest disappears, the conditions improve for the insects; at a certain stage the insect population explodes and destroys the remaining forest.

2. **Occurrence:** Ecological equilibrium in a complex system of three trophic levels. Dynamics of ecosystems during pest outbreaks and under environmental pollution.

3. **Structural characteristics:** The model is composed of three submodels: forest, insects, and birds.

   Forest: The structure of this model is that of a logistically growing state variable. Forest biomass x is affected by the cutting of trees (provided as scenario u(t)) and by insect damage q. The feeding rate of insects depends on the insect population and—by way of a Michaelis-Menten saturation with the half-saturation constant d—on the current forest area.

   Insects: The basic structure is again that for logistic growth of a population. In this case the capacity is variable and depends on the forest cover x and the grassland (k – x). The insect population has losses s from predation by birds. These losses are proportional to the bird population; they also depend—with a half-saturation constant e—on the insect population.

   Birds: This population also has the fundamental structure of logistic growth with a variable capacity that in this case depends on forest area x and insect population y.

4. **Behavioral characteristics:** If the forest area is large enough, birds and insects are able to survive in small populations. As the forest biomass declines, the conditions for the insect population improve significantly (note the effect of the term $x \cdot (k-x)$) resulting in an explosion of the insect population which either destroys the forest completely or only partially and for a limited time. The basic behavior corresponds to that of growth and subsequent collapse (MODEL M 210 OVERSHOOT AND COLLAPSE).

5. **Critical parameters:** The loss of forest biomass by deforestation has a decisive effect on the further development and in particular on the possibility of collapse. Breakdown will occur if the forest cover drops below a certain critical value. The cutting strategy (strength h and duration t) therefore has a crucial influence on the development.

6. **Reference run:** If (for the given parameter set) there is no deforestation, a mild insect calamity will develop after seven years but it cannot cause any collapse of the system. If, however, the deforestation rate is increased to 5 percent per year, a strong insect outbreak and subsequent breakdown of the forest as well as the insect and bird populations will occur after about seven years.

7. **Suggestions for further study:** Using the option "Parameter Sensitivity," study the consequences of different forest use scenarios. Investigate the role of the initial conditions using the option "Global Analysis." Use the option "Parameter Sensitivity" and a variation of parameters within reasonable bounds to determine which of the parameters has the most critical effect on the system.

8. **References:** Richter, 1985: p. 91-98.

BIRDS, INSECTS, AND FOREST **M 307**

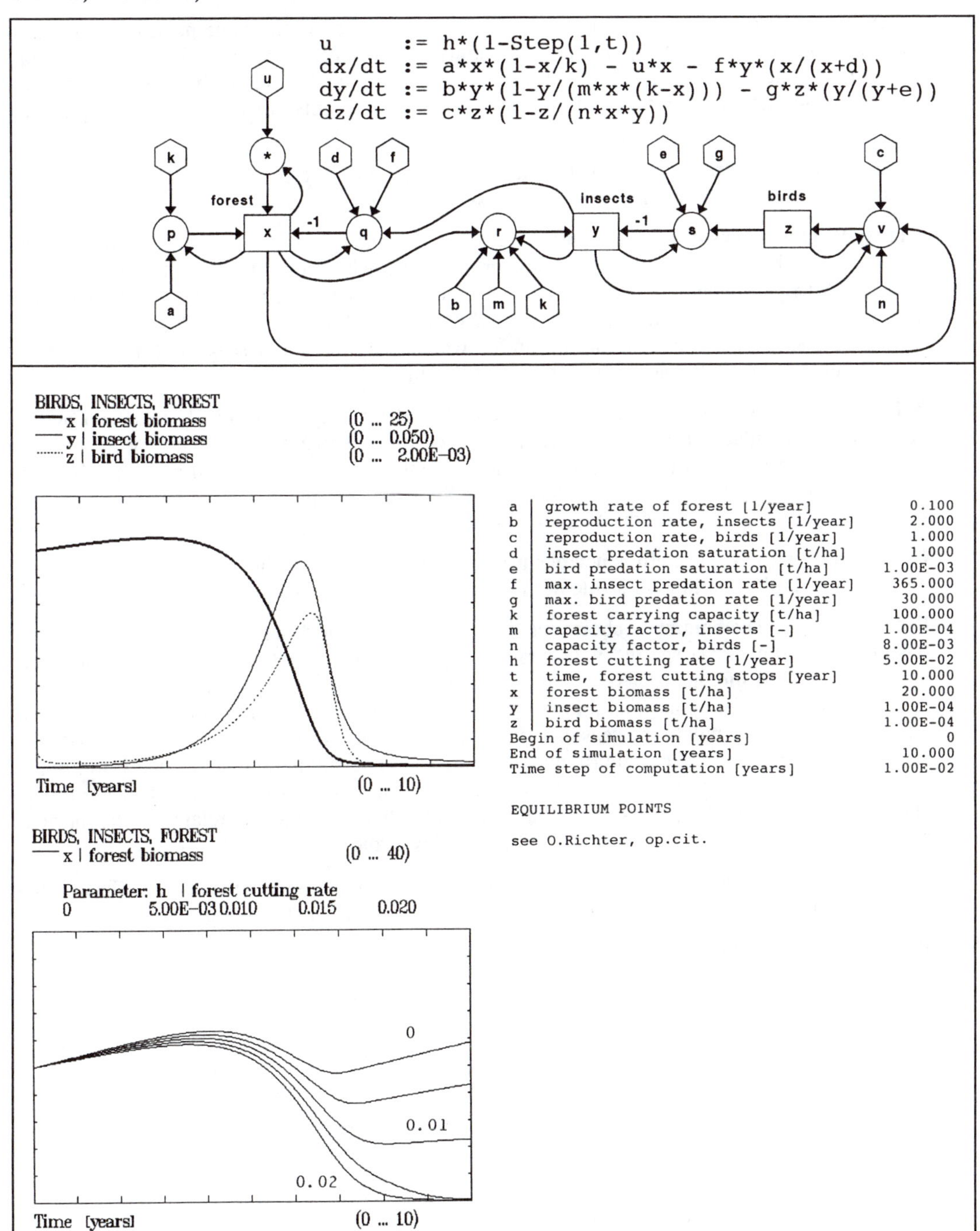

a	growth rate of forest [1/year]	0.100
b	reproduction rate, insects [1/year]	2.000
c	reproduction rate, birds [1/year]	1.000
d	insect predation saturation [t/ha]	1.000
e	bird predation saturation [t/ha]	1.00E-03
f	max. insect predation rate [1/year]	365.000
g	max. bird predation rate [1/year]	30.000
k	forest carrying capacity [t/ha]	100.000
m	capacity factor, insects [-]	1.00E-04
n	capacity factor, birds [-]	8.00E-03
h	forest cutting rate [1/year]	5.00E-02
t	time, forest cutting stops [year]	10.000
x	forest biomass [t/ha]	20.000
y	insect biomass [t/ha]	1.00E-04
z	bird biomass [t/ha]	1.00E-04
Begin of simulation [years]		0
End of simulation [years]		10.000
Time step of computation [years]		1.00E-02

EQUILIBRIUM POINTS

see O.Richter, op.cit.

NUTRIENT CYCLING AND PLANT COMPETITION **M 308**

1. **Description:** Two plant populations x and y grow together on the same piece of land, making use of the same nutrient supply n. This supply is constantly replenished by the mineralization of plant litter l (leaf and fine root litter, etc.). The model describes both the nutrient recycling in a terrestrial ecosystem and the competition for nutrients of two plant species growing on it.

2. **Occurrence:** Terrestrial ecosystems in general, vegetation in forests, fields, and meadows. The same basic structure also applies to aquatic ecosystems.

3. **Structural characteristics:** The plant populations x and y grow as a function of nutrient supply n with the specific growth rates a and b which depend on nutrient saturation (n/(n+k), Michaelis-Menten) and may change with the seasons (f(t)). Both plant populations drop their litter at specific rates c and d, adding it to the litter supply l. This supply is mineralized at a specific mineralization rate m·f(t) which is a function of seasonal time. The mineralized nutrient enters the nutrient supply n at this rate.

4. **Behavioral characteristics:** An initially available nutrient supply is quickly taken up by the plants and accumulates in their biomass. The litter supply increases corresponding to the specific litter loss and mortality rates c and d; the nutrient supply is replenished by mineralization (with specific rate m) of the litter supply. In the long run, the plant population that is able to store more nutrients and to produce less litter fall will dominate.

5. **Critical parameters:** The efficiency of nutrient accumulation, i.e. the litter loss and mortality rates c and d play an important role in the competition process between the plant populations.

6. **Reference run:** The parameter choice for the reference run corresponds to a pioneer species x (r-strategist) and a climax species y (k-strategist). The plant population x has a rapid nutrient uptake and a short lifetime (0.5 years); the plant population y on the other hand is characterized by a slow nutrient uptake and a long lifetime (100 years). A sine function with a period of one year is used to simulate the seasonal fluctuations of nutrient uptake during the growth period and of the mineralization rate (temperature effect). The simulation shows that initially the pioneer plant dominates but that it will eventually be surpassed by the climax species. In the end, the pioneer plant disappears. Note that this competitive advantage is related to the nutrient supply only; differences in light availability do not play a role in this model.

7. **Suggestions for further study:** Use the option "Parameter Sensitivity" to study the effect of different parameter choices for the system development (in particular the parameters a, b, c, and d.

8. **References:** Jørgenson, 1992.

NUTRIENT CYCLING AND PLANT COMPETITION

M 308

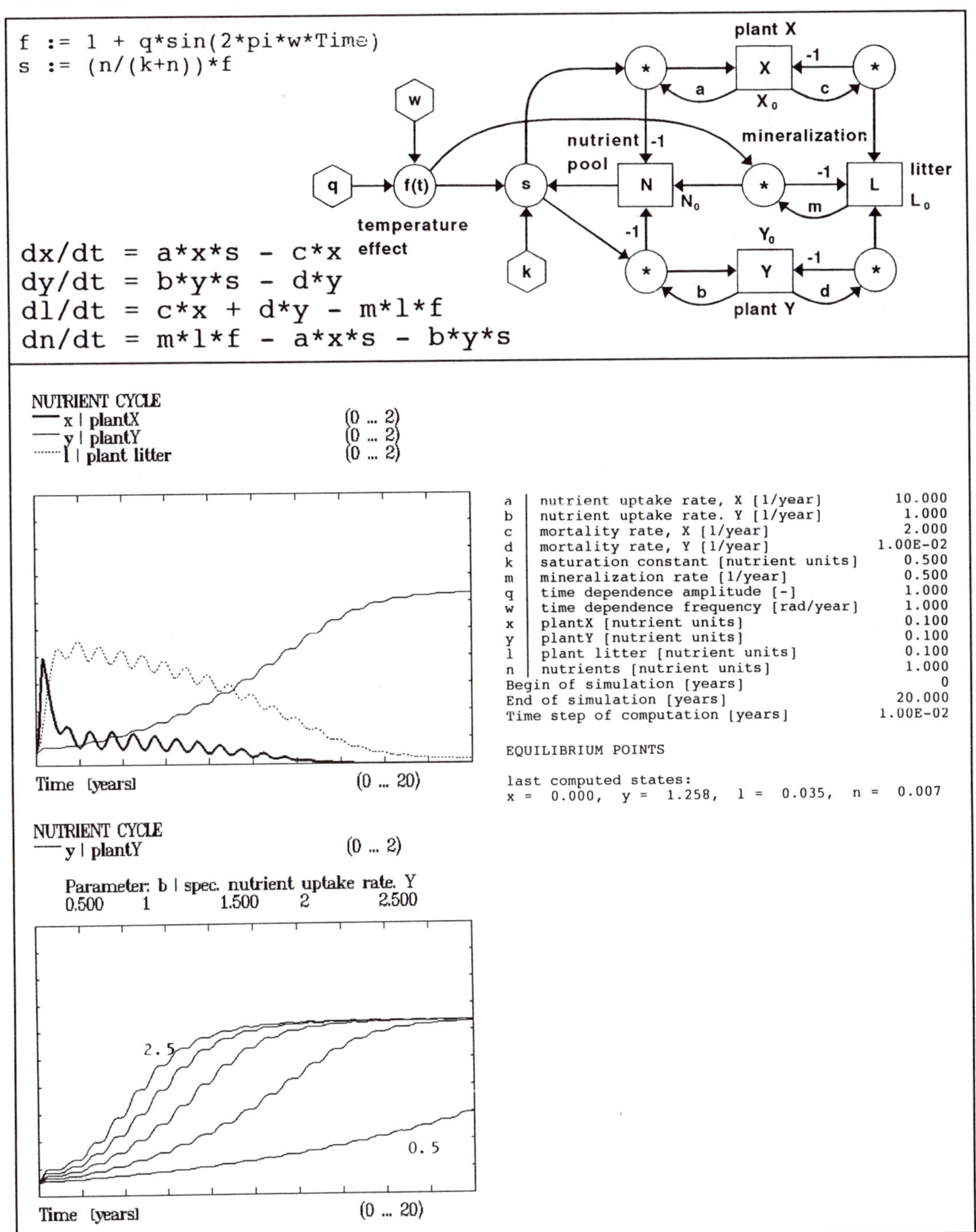

CHAOTIC ATTRACTOR (RÖSSLER)

M 309

1. **Description:** The system which consists at its core of a linear oscillator with a non-linear structure modification, produces a stable limit cycle with increasing period doublings for an increasing parameter c. As c increases further, chaotic behavior develops.

2. **Occurrence:** Mathematical construction (Rössler, 1976).

3. **Structural characteristics:** If considered in isolation, the links between the state variables x and y correspond to those of the linear oscillator. The state variable z changes x; the rate of change of z itself is determined by the difference $(x - c)$ which may produce a sign change in the rate of change of z.

4. **Behavioral characteristics:** As a result of the sign change of the rate $(x-c)\cdot z$, the system exhibits motion on a limit cycle. As c increases, the periods of the limit cycle are doubled. If c increases beyond a certain value $c > c_{\infty}$, the system exhibits chaos.

5. **Critical parameters:** The parameter c determines the period of oscillation and the appearance of chaos.

6. **Reference run:** The default parameters produce chaotic behavior. The time plot shows different oscillation periods in the course of the simulation. The shape of the Rössler attractor is best studied in three-dimensional state space (x, y, z).

7. **Suggestions for further study:** Examine the time behavior and the shape of the attractor as a function of the parameter c using the time plot and the three-dimensional display. Study the influence of other parameters.

8. **References:** Rössler, 1976; Jetschke, 1989: p. 136-138; Thompson and Stewart, 1986: p. 235-253.

CHAOTIC ATTRACTOR (RÖSSLER) M 309

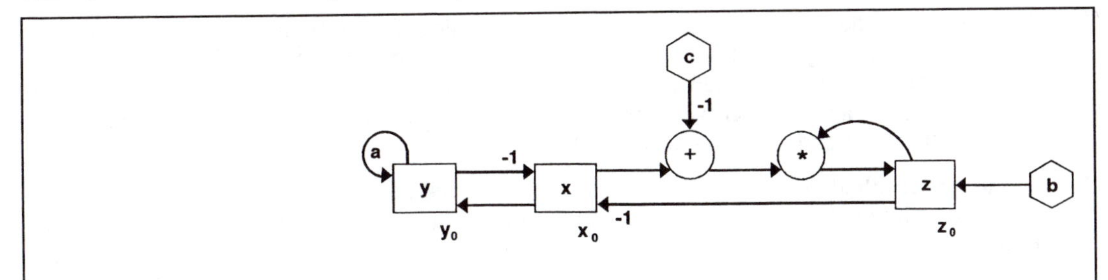


```
dx/dt := - y - z
dy/dt := x + a*y
dz/dt := b + (x - c)*z
```

CHAOTIC ATTRACTOR
x (−30 ... 30)
y (−48 ... 12)
z (−12 ... 48)

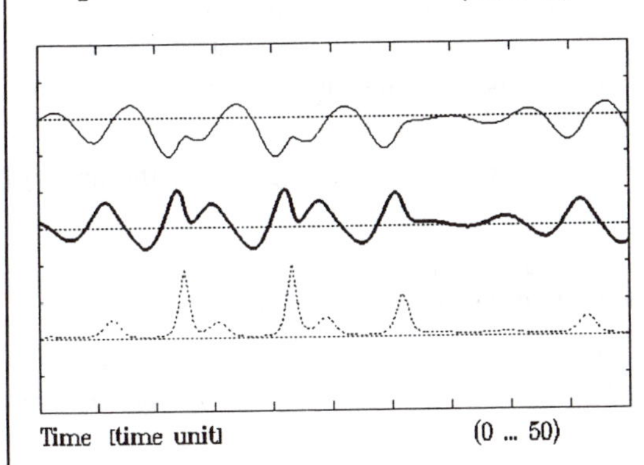

Time [time unit] (0 ... 50)

a [-]	0.550
b [-]	2.000
c [-]	4.000
x [-]	1.000
y [-]	0
z [-]	0
Begin of simulation [time unit]	0
End of simulation [time unit]	50.000
Time step of computation [time unit]	1.00E-02

EQUILIBRIUM POINTS

chaotic attractor around x = 0, y = 0, z = 0

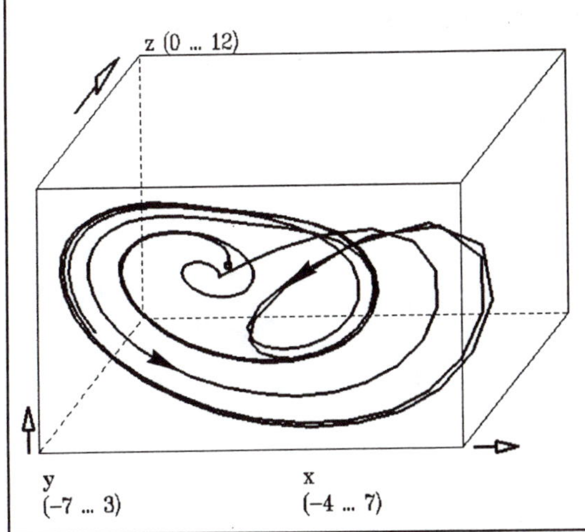

HEAT, WEATHER, AND CHAOS (LORENZ SYSTEM) M 310

1. **Description:** The Lorenz system is an approximate description of the hydro-thermodynamic equations for the coupling of heat convection and heat conduction in fluid flows, in particular for the description of Bénard cell flows. The state variable x describes the velocity profile, the state variables y and z describe the temperature distribution. In a certain parameter range (see default values), the system shows chaotic behavior.

2. **Occurrence:** Fluid flows over a heated surface where cellular flow patterns may develop under certain conditions (Bénard cells). These cellular flows may occur in meteorology, on the surface of the sun, in flat ponds, or in fluid flows under cooled surfaces. Other chaotic systems (laser, or dynamo, cf. Model M 311 COUPLED DYNAMOS AND CHAOS) show a very similar structure.

3. **Structural characteristics:** Non-linearities $(x \cdot z,\ x \cdot y)$; possible sign reversal for the rate $dx/dt = a \cdot (y - x)$. As a general rule, however, chaotic behavior cannot be recognized from the system structure alone.

4. **Behavioral characteristics:** The state trajectory encircles one of two attraction centers but is captured after an unpredictable number of rotations by the other region of attraction. This leads to a typical butterfly shape of the attractor. Note that these state trajectories cannot cross at the same point in space since the deterministic motion would otherwise have to be repeated. Cross-over points can therefore only have identical values for two, never three, state coordinates. Viewing the state trajectories in three-dimensional state space from different directions confirms the observation that trajectories will never pass through the same point.

5. **Critical parameters:** The parameter b (here $b = 28$) is particularly critical for the onset of chaos.

6. **Reference run:** The default parameter setting uses the standard parameters of the Lorenz system. This produces the "Lorenz butterfly" in the three-dimensional state space presentation.

7. **Suggestions for further study:** Change parameter b and determine the range of b in which chaos appears. Study the influence of other parameters on the system behavior, in particular on the onset of chaos.

8. **References:** Jetschke, 1989: p. 130-136; Thompson and Stewart, 1986: p. 212-234.

HEAT, WEATHER, AND CHAOS (LORENZ-SYSTEM) M 310

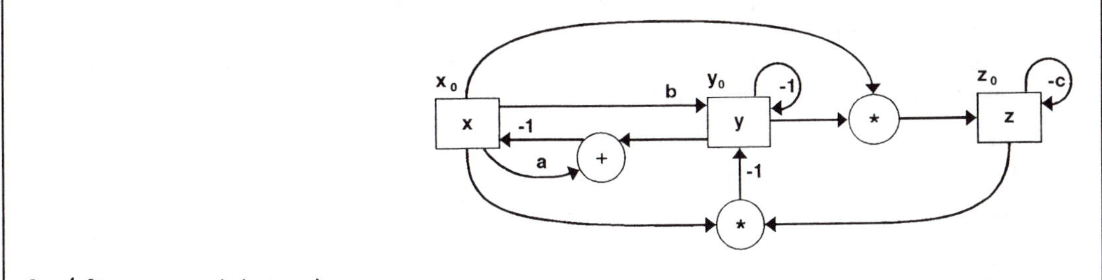


```
dx/dt := a*(y-x)
dy/dt := -x*z + b*x -y
dz/dt :=  x*y - c*z
```

HEAT, WEATHER, CHAOS

x	(-90 ... 90)
y	(-150 ... 30)
z	(0 ... 180)

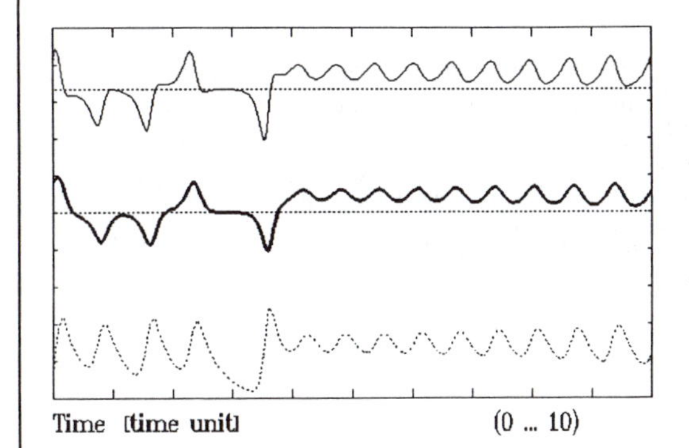

a [1/time]	10.000
b [1/time]	28.000
c [1/time]	2.667
x [-]	15.000
y [-]	15.000
z [-]	15.000
Begin of simulation [time unit]	0
End of simulation [time unit]	10.000
Time step of computation [time unit]	1.00E-02

EQUILIBRIUM POINTS

chaotic attractor
x = 0, y = 0, z = 0
x, y, z = 8.485 , 8.485 , 27.000
-8.485 , -8.485 , 27.000

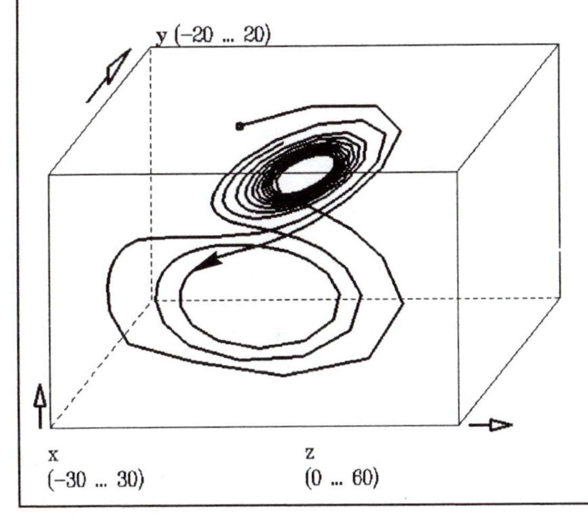

COUPLED DYNAMOS AND CHAOS

M 311

1. **Description:** Two identical dynamos are coupled to each other such that the current of one dynamo excites the magnetic field of the other. The state variables x and y represent the current in both circuits. The angular velocity of rotation of dynamo x corresponds to the state variable z. This system exhibits chaotic behavior.
2. **Occurrence:** Coupled dynamos. The Lorenz system of a heated fluid and the chaotic laser system show similar system structure.
3. **Structural characteristics:** The structural connections between the state variables x and y are symmetric. The difference of the angular velocities of the two dynamos $(\omega_x - \omega_y) = \text{const}$; for this reason only $\omega_x = z$ is considered. The parameter c corresponds to the difference between the two angular velocities. The forcing rate of the system is 1. The system has three non-linear couplings $x \cdot z$, $y \cdot z$, and $x \cdot y$.
4. **Behavioral characteristics:** The state trajectory encircles one of two unstable equilibrium points several times. After an unpredictable number of revolutions it jumps to a trajectory around the other point of equilibrium. The system shows bistable behavior.
5. **Critical parameters:** The parameters a and c determine the time behavior and the shape of the attractor.
6. **Reference run:** The parameters of the default setting result in unpredictable motion from one region of attraction to the other with several subsequent rotations around the corresponding equilibrium point. These rotations correspond to oscillations in the time plot.
7. **Suggestions for further study:** Investigate the role of the parameters a and b on the shape of the attractor using the three-dimensional presentation in state space.
8. **References:** Beltrami, 1987: p. 214 - 218.

COUPLED DYNAMOS AND CHAOS

M 311

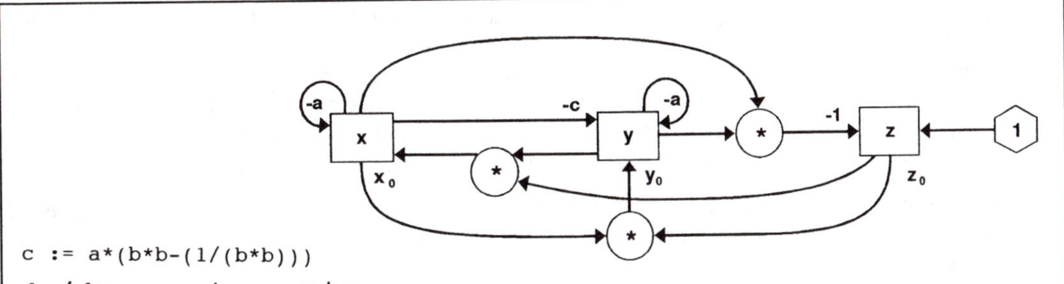


```
c := a*(b*b-(1/(b*b)))
dx/dt := z*y - a*x
dy/dt := (z - c)*x - a*y
dz/dt := 1 - x*y
```

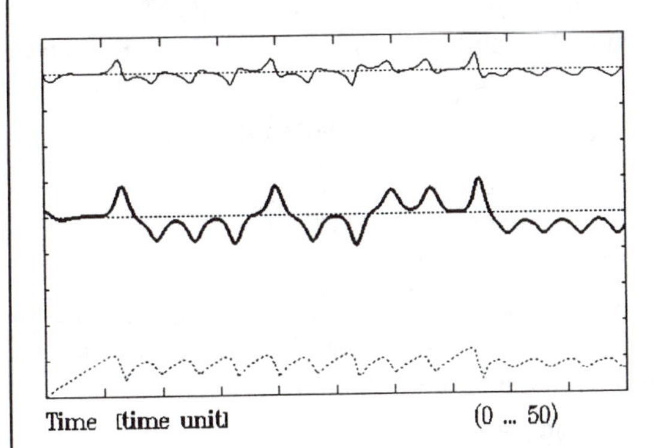


```
a | parameter 1 [-]                          1.000
b | parameter 2 [-]                          2.000
x | current in circuit 1 [-]                 1.000
y | current in circuit 2 [-]                 0
z | angular velocity [-]                     0
Begin of simulation [time unit]              0
End of simulation [time unit]               50.000
Time step of computation [time unit]      1.00E-02

EQUILIBRIUM POINTS

chaotic attractor
x, y, z =  2.000 ,  0.500 , 4.000
          -2.000 , -0.500 , 4.000
```

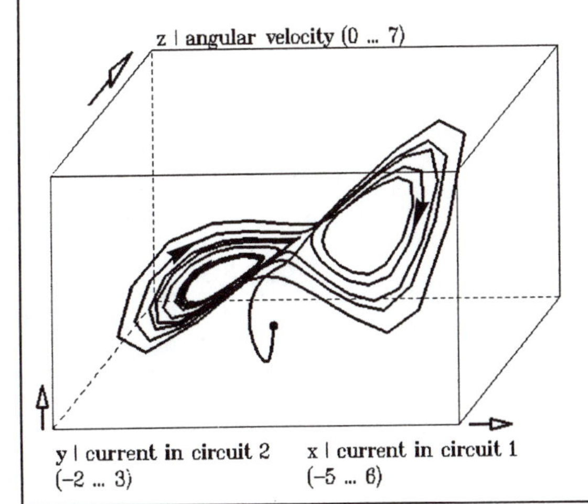

BALANCING AN INVERTED PENDULUM

M 312

1. **Description:** The inverted pendulum is stabilized by moving its point of rotation back and forth to "catch" the falling pendulum and bring it back to an upright position. This requires that the falling motion (angle and angular velocity) is sensed; these measurements are used to compute the required motion of the point of rotation. Similarly, the deviation from the horizontal reference position is sensed; these measurements of position and velocity are used to determine the necessary acceleration of the point of rotation to return it to the reference position.

2. **Occurrence:** Stabilizing the inverted pendulum (or a broom); stabilizing a rocket at take-off. Generally: an example for demonstrating that unstable systems can be stabilized by proper control measures.

3. **Structural characteristics:** The angle y, the angular velocity q, the position x, and the velocity v are used to define a control function u as a linear combination of these state variables and appropriate control parameters e, f, h, and k. A corresponding time-varying control force is coupled back to the angular acceleration and the horizontal acceleration to obtain accelerations of angle and position that stabilize the pendulum. A randomly fluctuating perturbation s(t) acts as a disturbance.

4. **Behavioral characteristics:** For a suitable choice of control parameters, the falling and swinging motion of the inverted pendulum is quickly stabilized, and the pendulum is brought back to its initial vertical (unstable) position.

5. **Critical parameters:** The parameters f and e (concerning the angular motion) are especially critical for the stabilization process.

6. **Reference run:** The set of default values for the parameters (100/30/3/10) is close to "optimal" and leads to a quick stabilization even for random disturbances.

7. **Suggestions for further study:** Use the option "Parameter Sensitivity" to study the behavior for different parameter settings. Find the limits of stability for the different control parameters e, f, h, and k. Begin by stabilizing the rotation, and then the horizontal motion.

8. **References:** Bossel, 1987/89: p. 203-217; Bossel, 1994: Ch. 5.4.

BALANCING AN INVERTED PENDULUM

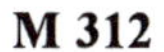
M 312

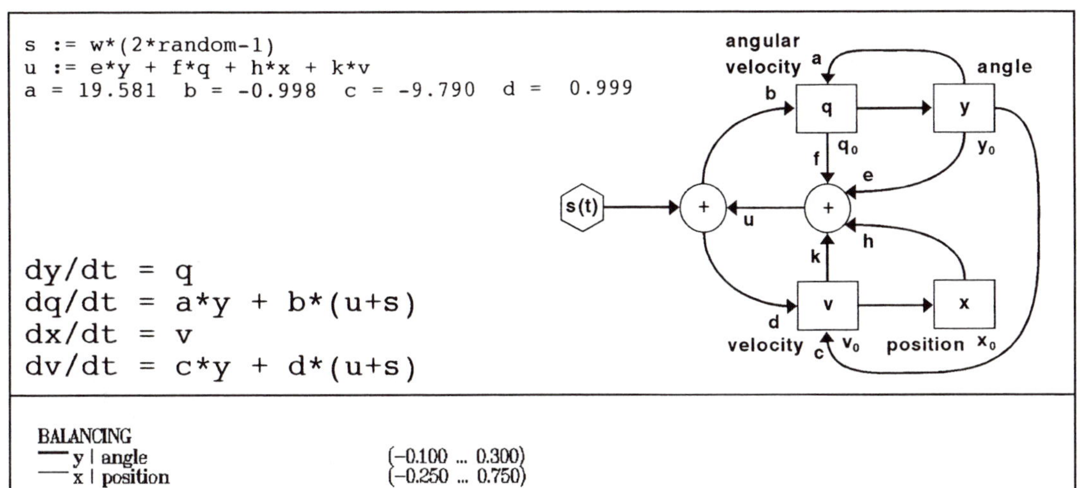


```
s := w*(2*random-1)
u := e*y + f*q + h*x + k*v
a = 19.581   b = -0.998   c = -9.790   d =   0.999
```

```
dy/dt = q
dq/dt = a*y + b*(u+s)
dx/dt = v
dv/dt = c*y + d*(u+s)
```

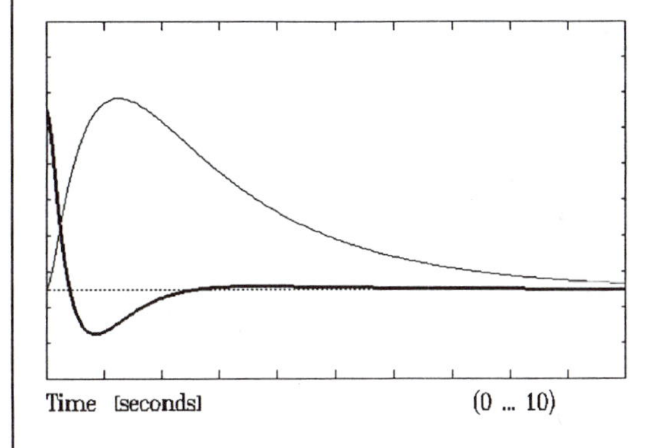

e	angle feedback param [N/rad]	100.000
f	ang.velocity feedback par [N/(rad/s)	30.000
h	position feedback param [N/m]	3.000
k	velocity feedback param [N/(m/sec)]	10.000
m	pendulum mass [kg]	1.000
w	perturbation amplitude [N]	0
y	angle [rad]	0.200
q	angular velocity [rad/sec]	0
x	position [m]	0
v	velocity [m/sec]	0
Begin of simulation [seconds]		0
End of simulation [seconds]		10.000
Time step of computation [seconds]		4.00E-02

EQUILIBRIUM POINTS

x = 0, y = 0 for appropriate feedback parameters

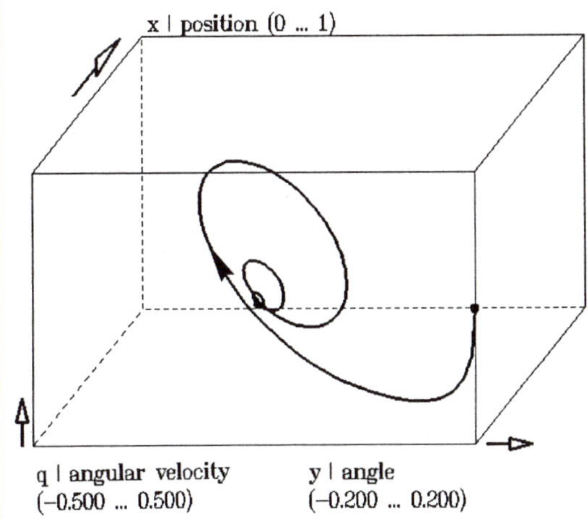

e	angle feedback param [N/rad]	50.000
f	ang.velocity feedback par [N/(rad/s)	30.000
h	position feedback param [N/m]	3.000
k	velocity feedback param [N/(m/sec)]	10.000
m	pendulum mass [kg]	1.000
w	perturbation amplitude [N]	0
y	angle [rad]	0.200
q	angular velocity [rad/sec]	0
x	position [m]	0
v	velocity [m/sec]	0
Begin of simulation [seconds]		0
End of simulation [seconds]		10.000
Time step of computation [seconds]		4.00E-02

CHAPTER 7

MATHEMATICAL SYSTEMS ANALYSIS

7.0 Introduction

An individual simulation run describes the dynamics of a model system under specific conditions. More general information about the system can be obtained from a large number of simulations in which important parameters are changed over a certain range of conditions. Because of the large number of possible parameter combinations, plausible parameter sets have to be formulated as coherent and consistent "scenarios."

Even systematic simulation studies retain this trial-and-error, groping-in-the-dark feature. It would therefore be desirable to obtain information about the complete spectrum of possible system states and system behavior directly and analytically from the state equations. This is possible for linear systems for which a well-developed set of analytical tools is readily available. However, most of the interesting systems found in reality are non-linear, and for these systems the analytical approach is available only to a very limited extent.

Nevertheless, all possibilities for analytical study should be used in examining a non-linear system. In particular, this means finding the equilibrium points or other regions of attraction and investigating the system behavior in the neighborhood of these points or regions. In such studies, (local) linearization of non-linear systems and the detailed analysis of the linear substitute system using the methods of linear systems analysis play an important role.

In Section 7.1 of this chapter, system concepts will be explained, state equations for continuous and discrete systems will be developed, the linearization of non-linear state equations will be discussed, and the conditions for equilibrium of the different system types will be defined.

Since further analysis requires some elementary knowledge of vector and matrix algebra, of eigenvalues and eigenvectors of system matrices, etc., the corresponding concepts are summarized in Section 7.2. Section 7.3 deals in particular with the unforced (autonomous) solution of the linear dynamic system, the reformulation of the system representation using a change of vector basis, and the behavioral response and stability of the unforced system. In Section 7.4, the superposition principle is used to determine the dynamics of forced (non-autonomous) behavior of linear dynamic systems with periodic and aperiodic inputs. Section 7.5 deals with behavior and stability of non-linear systems.

The presentation is quite condensed and limited to essentials. It may suffice as a summary for those who are familiar with the concepts. For those who are new to the subject, it may serve as an introduction to a field with which numerous texts have dealt in detail. (See bibliography.)

7.1 State Equations of Dynamic Systems

7.1.1 System concepts

A system S exists in a system environment U. It consists of elements E_p which are connected to each other by their influences w_{pq} (from element E_p on element E_q) (Fig. 7.1). A (hypothetical) system boundary G separates the elements of the system from other elements in the system environment. There are influences (inputs) u_i from the system environment on the system. Observable system behavior is described by behavior variables v_j (output). The state of the system is described by state variables z_n. State variables are storage or memory variables (stocks, levels) of the system. Not every system element corresponds to a state variable. The observable behavior is a result of influences from the environment and the dynamics of the state variables.

7.1.2 System quantities as vectors

Vectors serve as a compact representation of quantities having several components. For system studies, system inputs (influences of the system environment), system outputs (system behavior), and system states are combined in corresponding vectors. Vectors are denoted by bold lower-case letters, matrices by bold capital letters.

Input vector:

$$\mathbf{u} = \begin{bmatrix} u_1 \\ u_2 \\ \dots \\ u_i \end{bmatrix}$$

Output vector:

$$\mathbf{v} = \begin{bmatrix} v_1 \\ v_2 \\ \dots \\ v_j \end{bmatrix}$$

State vector:

$$\mathbf{z} = \begin{bmatrix} z_1 \\ z_2 \\ \dots \\ z_n \end{bmatrix}$$

Rate vector (time rate of change of states):

$$\mathbf{z}' = d\mathbf{z} / dt = \begin{bmatrix} z'_1 \\ z'_2 \\ \dots \\ z'_n \end{bmatrix} = \begin{bmatrix} dz_1 / dt \\ dz_2 / dt \\ \dots \\ dz_n / dt \end{bmatrix}$$

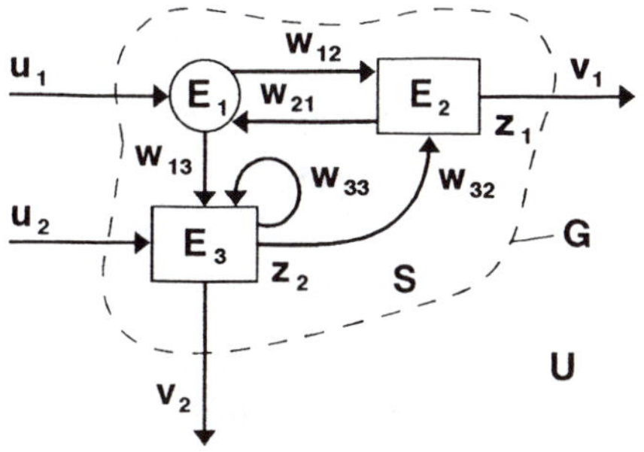

Fig. 7.1: System concepts: System S, system environment U, system boundary G, inputs from the system environment u_i, output from the system to the environment (behavior) v_j, system elements E_p, state variables z_n, influence w_{pq} of one element on another element.

7.1.3 General state and behavior equations

In the general case, the system state $\mathbf{z}$ and system behavior $\mathbf{v}$ are functions of the environmental influence $\mathbf{u}$, the system state $\mathbf{z}$, and possibly of time t (e.g. time-dependent parameter change):

$$\mathbf{z}(t) = \mathbf{f}^{o}(\mathbf{z}(t), \mathbf{u}(t), t) \qquad \text{(state equation)}$$

$$\mathbf{v}(t) = \mathbf{g}(\mathbf{z}(t), \mathbf{u}(t), t) \qquad \text{(behavior equation).}$$

The simultaneous condition for $\mathbf{z}$ (as a function of $\mathbf{z}$) has to be resolved by using previous state information.

A **time-discrete system** is defined at discrete intervals of time

$t = 0 \cdot \Delta T,\ 1 \cdot \Delta T,\ 2 \cdot \Delta T, \ldots k \cdot \Delta T \ldots$

where ΔT represents the chosen (constant) size of the time step. A particular point in time is defined by the corresponding time index k.

If the current system state is a result of the conditions of the previous time step, the system equations for a time-discrete system can be written as

$$\begin{aligned} \mathbf{z}(k+1) &= \mathbf{f}(\mathbf{z}(k), \mathbf{u}(k), k) && \text{(state equation)} \\ \mathbf{v}(k) &= \mathbf{g}(\mathbf{z}(k), \mathbf{u}(k), k) && \text{(behavior equation).} \end{aligned} \tag{7.1}$$

Using another common notation for the time index, the system equations of the time-discrete system can be written as

$$\begin{aligned} \mathbf{z}_{k+1} &= \mathbf{f}(\mathbf{z}_k, \mathbf{u}_k, k) \\ \mathbf{v}_k &= \mathbf{g}(\mathbf{z}_k, \mathbf{u}_k, k) . \end{aligned}$$

In a **time-continuous system** the rate of change of the state variables ("rate" for short) $d\mathbf{z}/dt = \mathbf{z}'$ can be determined from the conditions at time t; the new state then follows from the previous (or initial) state and integration of the rate of change with respect to time.

In this case the system equations are:

$$\begin{aligned} \mathbf{z}'(t) &= \mathbf{f}(\mathbf{z}(t), \mathbf{u}(t), t) && \text{(state equation)} \\ \mathbf{v}(t) &= \mathbf{g}(\mathbf{z}(t), \mathbf{u}(t), t) && \text{(behavior equation).} \end{aligned} \tag{7.2}$$

For the discrete system (Eq. 7.1) as well as the continuous system (Eq. 7.2) the behavior variables $\mathbf{v}(t)$ can be computed directly from algebraic (or logical) expressions.

7.1.4 General system diagram for dynamic systems

The general system equations (Eq. 7.1, 7.2) for discrete or continuous systems can be repesented in the form of a general system diagram (Fig. 7.2).

The box for the state variable $\mathbf{z}$ represents:

1. for the discrete system: storage of the current state vector $\mathbf{z}(k+1)$ as defined by state function $\mathbf{f}$;
2. for the continuous system: time integration of the current rate of change $\mathbf{z}'(t) = d\mathbf{z}/dt$ as defined by $\mathbf{f}$.

In both cases the current value of the state variable $\mathbf{z}$ can be found in the box.

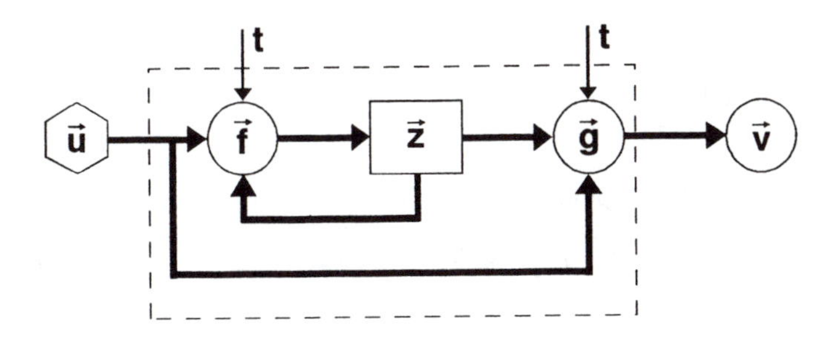

Fig. 7.2: General diagram for dynamic systems.

7.1.5 State computation

The state equation of the discrete system (Eq. 7.1) is a first order difference equation. It already represents the prescription for the state computation.

The state equation of the continuous system (Eq. 7.2) represents the computation of the rate of change **z'**(t) as an algebraic expression. This rate of change must be integrated over time, subject to the initial conditions $\mathbf{z}(t_0)$, in order to obtain **z**(t). Analytical integration is possible only for linear systems (see below) and for a few special non-linear systems. However, numerical integration is always possible and can be easily done for *all* time-continuous systems.

7.1.6 Numerical integration of the state equation

Euler-Cauchy integration: Using a time step of integration ΔT, the new state is computed from the previous state and the rate of change $(d\mathbf{z}/dt) = \mathbf{f}$ as determined (from Eq. 7.2) by using the previous state. (In the following, $\mathbf{f}(\mathbf{z}, \mathbf{u}, t)$ is condensed to $\mathbf{f}(\mathbf{z}, t)$.)

$$\mathbf{z}(t+\Delta T) = \mathbf{z}(t) + \mathbf{f}(\mathbf{z}, t) \cdot \Delta T\,. \tag{7.3}$$

The computational error of this procedure is of order $O(\Delta T)^2$. For this reason, the procedure normally requires a very small step size which may lead to truncation errors and long computation times.

Runge-Kutta procedure (fourth order): This procedure results in a much smaller computational error of order $O(\Delta T)^5$ and therefore allows a larger step size ΔT. However, the state function **f** has to be computed four times per computation step. Each intermediate vector quantity $\mathbf{k}_i$ consists of n components k_{ij}, i.e. $\mathbf{k}_i = (k_{i1}, k_{i2}, \dots, k_{in})$.

$$\mathbf{k}_1 = \Delta T \cdot \mathbf{f}(\mathbf{z}, t)$$
$$\mathbf{k}_2 = \Delta T \cdot \mathbf{f}(\mathbf{z} + \mathbf{k}_1/2,\ t + \Delta T/2)$$
$$\mathbf{k}_3 = \Delta T \cdot \mathbf{f}(\mathbf{z} + \mathbf{k}_2/2,\ t + \Delta T/2)$$
$$\mathbf{k}_4 = \Delta T \cdot \mathbf{f}(\mathbf{z} + \mathbf{k}_3,\quad t + \Delta T)\ . \tag{7.4a}$$

The result of the integration is computed from

$$\mathbf{z}(t+\Delta t) = \mathbf{z}(t) + \mathbf{k}_1/6 + \mathbf{k}_2/3 + \mathbf{k}_3/3 + \mathbf{k}_4/6\ . \tag{7.4b}$$

7.1.7 Transformation to first order state equations

For the analytical as well as the numerical treatment of the state equations, the first order form of the difference equation (Eq. 7.1) or differential equation (Eq. 7.2) is of particular relevance. An ordinary difference or differential equation of n-th order can always be transformed to n equations of first order.

7.1.8 Transformation of an n-th order differential equation

Let the original differential equation for the dependent variable y be a function of y itself as well as its time derivatives dy^i/dt^i up to the highest order n.

1. Introduce n new state variables z_i, the first of which corresponds to the original quantity y.

2. The other (n–1) state variables correspond to the time derivatives of y up to the n-th order.

3. Solve the original differential equation for the highest derivative. The resulting expression contains only derivatives of lower order, or the z_i respectively.

4. In this way, obtain a system of n differential equations of first order of the form

$$d\mathbf{z}/dt = \mathbf{z}' = \mathbf{f}(\mathbf{z}, \mathbf{u}, t)\ .$$

The following scheme summarizes the procedure.

state equations (7.5)

$$
\begin{array}{lcllcl}
y & = & & & & z_1 \\
dy/dt & = & z'_1 & = & & z_2 \\
d^2y/dt^2 & = & z'_2 & = & & z_3 \\
 & & \ldots & & & \\
d^{n-1}y/dt^{n-1} & = & z'_{n-1} & = & & z_n \\
d^ny/dt^n & = & z'_n & = & & f(z_1, z_2, \ldots z_n, \mathbf{u}, t)
\end{array}
$$

Example: original (non-linear) system

$$y'' + y^2\, y' + ky = u$$

Solving for the highest derivative:

$$y'' = u - y^2\, y' - ky$$

Renaming $y = z_1$, we obtain the state equations

$$
\begin{array}{lclcl}
y' & = & z'_1 & = & z_2 \\
y'' & = & z'_2 & = & u - z_1{}^2\, z_2 - k \cdot z_1
\end{array}
$$

7.1.9 Transformation of an n-th order difference equation

Let the original difference equation for the dependent variable y be a function of y for time indexes ranging from the lowest time index (k–i) to the highest time index (k+j). The total range of time steps is $(k+j) - (k-i) = n$.

1. Solve the difference equation for the variable y with the highest time index.
2. Shift all indexes by the same number of steps such that the term with the highest time index becomes $y(k+1)$.
3. Introduce n new state variables z_i where

state equations (7.6)

$$
\begin{array}{lclcl}
y(k+1-n) & = & & & z_1(k) \\
y(k+2-n) & = & z_1(k+1) & = & z_2(k) \\
 & & \ldots & & \\
y(k) & = & z_{n-1}(k+1) & = & z_n(k) \\
y(k+1) & = & z_n(k+1) & = & f(z_1(k), z_2(k), \ldots, z_n(k), u\,(k), k)
\end{array}
$$

4. The state equations for a discrete, time-variant, non-linear dynamic system can then be written:

   $$\mathbf{z}(k+1) = \mathbf{f}(\mathbf{z}(k), \mathbf{u}(k)) .$$

Example: Non-linear difference equation with time-variant parameter

$$k \cdot y(k+2) \cdot y(k+1) = [y(k) \cdot y(k-1)]^{1/2}$$

Order: $(k+2) - (k-1) = 2+1 = 3$

Highest index: $k+2$

Solving for y with the highest index:

$$y(k+2) = [y(k) \cdot y(k-1)]^{1/2} / (k \cdot y(k+1))$$

Renaming (index shift by -1)

$$y(k+1) = [y(k-1) \cdot y(k-2)]^{1/2} / ((k-1) \cdot y(k))$$

New state variables:

					state equations
$y(k-2)$	=		=	$z_1(k)$	
$y(k-1)$	=	$z_1(k+1)$	=	$z_2(k)$	
$y(k)$	=	$z_2(k+1)$	=	$z_3(k)$	
$y(k+1)$	=	$z_3(k+1)$	=	$[z_2(k) \cdot z_1(k)]^{1/2} / ((k-1) \cdot z_3(k))$	

The original non-linear difference equation of third order is now converted to a system of three difference equations of first order for the state variables z_1, z_2, and z_3. The non-linearity and time-variance of the original system are conserved in the third equation.

7.1.10 State equation and system dynamics

The behavior function $\mathbf{g}$ (in Eq. 7.1, 7.2) merely represents an algebraic transformation of state variables $\mathbf{z}$ and system inputs $\mathbf{u}$ to observable behavior (output) $\mathbf{v}$. This means that the system's characteristic dynamics depend entirely on the state function $\mathbf{f}$ which determines the actual current system state $\mathbf{z}$:

$$\mathbf{z}(k+1) = \mathbf{f}(\mathbf{z}(k), \mathbf{u}(k), k) \quad \text{(discrete system)} \quad (7.1a)$$
$$\mathbf{z}'(t) \quad = \mathbf{f}(\mathbf{z}(t), \mathbf{u}(t), t) \quad \text{(continuous system).} \quad (7.2a)$$

Given an appropriate state function **f**, particular state values **z** can be realized by providing an appropriate input **u** from the system environment to the system. Changing **f** and/or operating on **u** are therefore two possible approaches to changing a system's dynamics and controlling system behavior.

Before this task can be tackled, the autonomous behavior (eigendynamics, characteristic dynamics) of the unforced system with time-invariant parameters must be analyzed, i.e. the behavior of the system with zero or time-invariant inputs ($\mathbf{u} = \mathbf{0}$ or $\mathbf{u} = \mathbf{u}_c$). In this case the observed system dynamics are entirely due to the internal dynamics of the system.

For autonomous systems with time-invariant parameters the state equations simplify to

$$\mathbf{z}(k+1) = \mathbf{f}(\mathbf{z}(k)) \quad \text{(discrete system)}$$
$$\mathbf{z}'(t) \quad = \mathbf{f}(\mathbf{z}(t)) \quad \text{(continuous system).}$$

7.1.11 Linearization of the state equation

In general, the vector state equation

$$\mathbf{z}'(t) = \mathbf{f}(\mathbf{z}, \mathbf{u}, t)$$

is non-linear and can therefore not be treated by analytical methods except in a few special cases.

By contrast, the linear vector state equation of the form

$$\mathbf{z}'(t) = \mathbf{A}(t)\,\mathbf{z}(t) + \mathbf{B}(t)\,\mathbf{u}(t)$$

can easily be dealt with analytically.

Since the behavior of non-linear systems is often of interest only in a limited state region and for a narrow set of initial conditions and input functions, it makes sense to linearize the original non-linear system in the region of interest and to study its behavior using a locally valid linear substitute system.

A **linear term** is a term of first order (proportionality) in the state variable z_i or its time derivatives. The linearity requirement does not apply to functions of the independent variable t which may have any structure.

Linear terms in differential equations of order N therefore have the general form

$$a \cdot f(t) \cdot z,\ a \cdot f(t) \cdot \left(\frac{dz}{dt}\right),\ \ldots a \cdot f(t) \cdot \left(\frac{d^n z}{dt^n}\right); \quad n = 1, 2, \ldots, N\ .$$

A **linear differential equation** is a differential equation consisting of a **sum of linear terms**. All other differential equations are non-linear.

Linear approximation: Linearization is particularly simple and admissible if a. interest focuses on system behavior in the vicinity of a certain reference state or on a limited state region and if b. the non-linearities of the system are limited to (a few) functions which can be approximated by linear relationships in the region of interest without affecting essential properties of the system. The linear dependence can often be obtained by graphical approximation, in particular for empirical functions.

7.1.12 Perturbation approach

This approach considers the effects of small perturbations $\mathbf{\Delta z}$ from a reference state $\mathbf{z}_0$. The state is then given by

$$\mathbf{z} = \mathbf{z}_0 + \mathbf{\Delta z}\ .$$

This approach is particularly useful for the analysis of system behavior in the vicinity of an equilibrium point $(\mathbf{z}_0 = \mathbf{z}^*)$ of a non-linear system.

Replacing the original state vector $\mathbf{z}$ by $\mathbf{z}_0 + \mathbf{\Delta z}$ and performing the non-linear operations as prescribed by the system of differential equations results in a. terms which are linear in $\mathbf{\Delta z}$ and b. other terms of higher order in $\mathbf{\Delta z}$. If terms of second order (quadratic) and higher order are neglected, an expression remains which is linear in $\mathbf{\Delta z}$. Obviously, this approximation is permissible only for small $\mathbf{\Delta z}$. Often it is possible to subtract the equation for the reference state from the linearized expression. The remaining expression is the linear state equation for the **state perturbation** $\mathbf{\Delta z}$. The subsequent analysis then deals with this linear system of **perturbation differential equations**.

In the following, the approach is demonstrated for the analysis of a (non-linear) predator-prey system in the vicinity of an equilibrium point.

Example: Predator-prey system (non-linear)

$$x' = a_1 x + a_2 x y$$
$$y' = b_1 y + b_2 x y\ .$$

The rates of change disappear at the equilibrium point x_0, y_0:

$$x' = 0 = a_1 x_0 + a_2 x_0 y_0$$
$$y' = 0 = b_1 y_0 + b_2 x_0 y_0\ .$$

In the neighborhood of the equilibrium point we have:

$$x = x_0 + \Delta x$$
$$y = y_0 + \Delta y\,.$$

Introducing these expressions into the differential equations yields

$$x' = a_1 \cdot (x_0 + \Delta x) + a_2 \cdot (x_0 + \Delta x) \cdot (y_0 + \Delta y)$$
$$y' = b_1 \cdot (y_0 + \Delta y) + b_2 \cdot (x_0 + \Delta x) \cdot (y_0 + \Delta y)\,.$$

and after multiplication:

$$x' = a_1\, x_0 + a_2\, x_0\, y_0 + a_1\, \Delta x + a_2\, y_0\, \Delta x + a_2\, x_0\, \Delta y + a_2\, \Delta x\, \Delta y$$
$$y' = b_1\, y_0 + b_2\, x_0\, y_0 + b_1\, \Delta y + b_2\, y_0\, \Delta x + b_2\, x_0\, \Delta y + b_2\, \Delta x\, \Delta y\,.$$

As a result of the equilibrium condition, the first two terms on the right hand side are zero. In the neighborhood of the equilibrium point, the last term is very small and can be neglected. Since

$$x' = \frac{dx}{dt} = \frac{d(x_o + \Delta x)}{dt} = \frac{d(\Delta x)}{dt} = \Delta x'$$

and, similarly, $y' = \Delta y'$, the system reduces to a **linear system** (perturbation differential equation):

$$\Delta x' = (a_1 + a_2\, y_0)\, \Delta x + (a_2\, x_0)\ \Delta y$$
$$\Delta y' = (b_2\, y_0)\, \Delta x + (b_1 + b_2\, x_0)\ \Delta y\,.$$

This can be expressed as a linear vector state equation

$$\Delta \mathbf{z}' = \mathbf{A}\, \Delta\, \mathbf{z}\,,$$

with **A** being the system matrix of the substitute linear system, valid at and in the neighborhood of the equilibrium point

$$\mathbf{A} = \begin{bmatrix} a_1 + a_2\, y_0 & a_2\, x_0 \\ b_2\, y_0 & b_1 + b_2\, x_0 \end{bmatrix}.$$

7.1.13 Approximation by Taylor expansion

If a non-linear relationship $f(\mathbf{z}) = f\,(z_1, z_2, \ldots z_n)$ is provided as an analytical expression, or if it can be described analytically (e.g. as an approximation of an empirical function), linearization is possible by developing a Taylor expansion around the reference point $\mathbf{z} = \mathbf{a} = (a_1, a_2, \ldots a_n)$. The general form of the Taylor expansion is given by:

$$
\begin{aligned}
f(z_1, z_2, \ldots z_n) &= f(a_1, a_2, \ldots a_n) + \sum_{i=1}^{n} \left(\frac{\partial f}{\partial z_i} \right)_{\mathbf{a}} \cdot (z_i - a_i) \\
&+ \frac{1}{2!} \sum_{i=1}^{n} \sum_{j=1}^{n} \left(\frac{\partial^2 f}{\partial z_i \partial z_j} \right)_{\mathbf{a}} \cdot (z_i - a_i)(z_j - a_j) . \\
&+ \text{terms of higher order}
\end{aligned}
$$

If terms of second and higher order are neglected, only linear relationships remain. Higher order terms may become significant under special circumstances (for example, if all first order terms are zero).

Even for complex functions, this method will yield simple linear approximations (see list in Sec. 7.5.5). In this list the reference point $z = a$ was normalized to "1." The table also makes clear why—in the neighborhood of a reference point—a simple influence diagram (containing only summations and multiplicative parameters) may provide a valid approximation to a complex system (cf. Sec. 7.5.5 on pulse dynamics).

7.1.14 Linearization of the state equation: Jacobi matrix

The linearization approach is not restricted to a reference state; it can also be applied in the neighborhood of a reference trajectory of the system state. In this case the Taylor expansion of the full non-linear system must be developed around the reference trajectory. This reference trajectory is a given development of the system state from which the actual system state is assumed to deviate only minimally. Linearization then again leads to perturbation differential equations describing the state perturbation from the reference trajectory. These perturbation equations are linear and can therefore be analyzed using the methods of linear systems analysis. This approach is particulary well-suited for dealing with control problems and processes. If the control system correctly solves its control task (i.e. if the perturbation differential equation is stable), it will return the system state to the neighborhood of the reference trajectory despite disturbances. This means that under certain circumstances, Taylor expansion around a reference trajectory and analysis of the resulting substitute linear differential system may provide a valid description of the dynamics of a non-linear system even under non-equilibrium conditions and in a larger region of state space.

Linearization along a reference trajectory is based on the state equation

$$\mathbf{z}' = \mathbf{f}(\mathbf{z}, \mathbf{u}) \qquad \text{or} \qquad \mathbf{z}(k+1) = \mathbf{f}(\mathbf{z}(k)), \mathbf{u}(k)) .$$

The following derivation is for the continuous system.

Let $\mathbf{z}_0(t)$ be the (prescribed) reference state vector (reference trajectory) and $\mathbf{u}_0(t)$ the (prescribed) reference input vector (control vector). If the input is exactly equal to $\mathbf{u}_0$, then the system state is exactly $\mathbf{z}_0$, i. e. the state equation is satisfied:

$$\mathbf{z}'_0 = \mathbf{f}(\mathbf{z}_0, \mathbf{u}_0)\,.$$

Let the actual state and the actual control input be slightly perturbed from the reference conditions:

$$\mathbf{z} = \mathbf{z}_0 + \Delta\mathbf{z}$$
$$\mathbf{u} = \mathbf{u}_0 + \Delta\mathbf{u}\,.$$

$\Delta\mathbf{z}$ and $\Delta\mathbf{u}$ are the perturbations from the reference state vector and the reference input vector.

State vector $\mathbf{z}$ and input vector $\mathbf{u}$ must satisfy the non-linear state equation:

$$d(\mathbf{z}_0 + \Delta\mathbf{z})/dt = \mathbf{z}'_0 + \Delta\mathbf{z}' = \mathbf{f}(\mathbf{z}_0 + \Delta\mathbf{z}, \mathbf{u}_0 + \Delta\mathbf{u})\,.$$

Since the perturbations $\Delta\mathbf{z}$ and $\Delta\mathbf{u}$ are assumed to remain small, each component of this equation can be written as a Taylor expansion:

$$\frac{d(z_{i0} + \Delta z_i)}{dt} \approx f_i(\mathbf{z}_0, \mathbf{u}_0) + \frac{\partial f_i}{\partial z_1}\Delta z_i + \ldots + \frac{\partial f_i}{\partial z_n}\Delta z_n$$
$$+ \frac{\partial f_i}{\partial u_1}\Delta u_i + \ldots + \frac{\partial f_i}{\partial u_m}\Delta u_m$$

The derivatives have to be taken along the reference trajectory. (All partial derivatives are assumed to exist; higher order terms are neglected.)

Along a reference trajectory, $z'_{i0} = f_i(\mathbf{z}_0, \mathbf{u}_0)$. Subtracting this reference condition from the previous set of equations yields the following system of linear differential equations for the state perturbations:

$$\frac{d(\Delta z_i)}{dt} \approx \left(\frac{\partial f_i}{\partial z_1}\right)_0 \cdot \Delta z_1 \ldots + \left(\frac{\partial f_i}{\partial z_i}\right)_0 \cdot \Delta z_i + \ldots + \left(\frac{\partial f_i}{\partial z_n}\right)_0 \cdot \Delta z_n \qquad i = 1, 2, \ldots n$$
$$+ \left(\frac{\partial f_i}{\partial u_1}\right)_0 \cdot \Delta u_1 \ldots + \left(\frac{\partial f_i}{\partial u_j}\right)_0 \cdot \Delta u_j + \ldots + \left(\frac{\partial f_i}{\partial u_m}\right)_0 \cdot \Delta u_m \qquad j = 1, 2, \ldots m$$

This system can be written as a vector equation after definition of the following Jacobi matrices:

$$\mathbf{A} = \begin{bmatrix} \frac{\partial f_1}{\partial z_1} & \dots & \frac{\partial f_1}{\partial z_n} \\ \dots & \dots & \dots \\ \frac{\partial f_n}{\partial z_1} & \dots & \frac{\partial f_n}{\partial z_n} \end{bmatrix}_0 = \left(\frac{\partial \mathbf{f}}{\partial \mathbf{z}}\right)_0 \tag{7.7a}$$

$$\mathbf{B} = \begin{bmatrix} \frac{\partial f_1}{\partial u_1} & \dots & \frac{\partial f_1}{\partial u_m} \\ \dots & \dots & \dots \\ \frac{\partial f_n}{\partial u_1} & \dots & \frac{\partial f_n}{\partial u_m} \end{bmatrix}_0 = \left(\frac{\partial \mathbf{f}}{\partial \mathbf{u}}\right)_0 \tag{7.7b}$$

The index "$_0$" implies that the partial differentials must be evaluated along the reference trajectory (or at the reference point).

The approximating system of linear differential equations for the state perturbations of a continuous system can now be written using the Jacobi matrices:

$$\mathbf{\Delta z'} = \mathbf{A}\ \mathbf{\Delta z} + \mathbf{B}\ \mathbf{\Delta u}\ .$$

Similarly, one obtains for the discrete system

$$\mathbf{\Delta z}(k+1) = \mathbf{A}\ \mathbf{\Delta z}(k) + \mathbf{B}\ \mathbf{\Delta u}(k)\ .$$

In general, these equations will be time-variant, i.e. $\mathbf{A}(t)$ and $\mathbf{B}(t)$ may be functions of time.

The linear perturbation differential equations are obviously also valid for a constant reference state (e.g. equilibrium point) with $\mathbf{z}_0 = \text{const}$ and $\mathbf{u}_0 = \text{const}$.

In many cases, non-linear conditions apply over a wide range of the state region of interest and cannot be adequately dealt with by a single linearization. In such cases piecewise linearization can often be used. When passing from one region to another, system parameters must then be changed. In computer simulations this can easily be accomplished by corresponding program statements.

7.1.15 Equilibrium points

Equilibrium points (stationary points, fixed points) are the natural resting points of systems. Knowledge of their locations is important for the assessment of system behavior. At these points in state space the rates of change of the state variables are zero. Equilibrium points are also singular points where the direction of state trajectories is undefined since trajectories from all directions may end (or begin) here. Equilibrium points may be stable or unstable (see 7.3.7).

A state vector $\mathbf{z}^*$ characterizes an **equilibrium point** if for a constant input $\mathbf{u}^*$ the system will remain in state $\mathbf{z}^*$ for all future times once it has reached that state. This applies to all linear and non-linear systems.

Generally the equilibrium points of time-invariant systems with a constant input $\mathbf{u}^*$ follow from the condition

$$\mathbf{0} = \mathbf{f}(\mathbf{z}^*, \mathbf{u}^*) \qquad \text{(continuous system)} \qquad (7.8)$$

$$\mathbf{z}^* = \mathbf{f}(\mathbf{z}^*, \mathbf{u}^*) \qquad \text{(discrete system)} . \qquad (7.9)$$

In both cases $\mathbf{z}^*$ is determined as the solution vector of the algebraic system of Equations 7.8 or 7.9, respectively.

7.1.16 Equilibrium points of non-linear systems

For the **continuous system**

$$\mathbf{z}'(t) = \mathbf{f}(\mathbf{z}(t), t)$$

the equilibrium points follow from the condition $d\mathbf{z}/dt = \mathbf{z}' = \mathbf{0}$, i.e.

$$\mathbf{f}(\mathbf{z}^*, t) = \mathbf{0} .$$

For the **time-invariant continuous system**

$$\mathbf{z}'(t) = \mathbf{f}(\mathbf{z}(t))$$

the time dependence in the condition for the equilibrium points disappears, and

$$\mathbf{f}(\mathbf{z}^*) = \mathbf{0} .$$

For the **discrete time-variant system**

$$\mathbf{z}(k+1) = \mathbf{f}(\mathbf{z}(k), k)$$

the equilibrium points follow from the condition

$$\mathbf{z}^* = \mathbf{f}(\mathbf{z}^*, k) \ \text{ for all } k .$$

If the **system** is **time-invariant**

$$\mathbf{z}(k+1) = \mathbf{f}(\mathbf{z}(k))$$

then the condition for the equilibrium points becomes

$$\mathbf{z}^* = \mathbf{f}(\mathbf{z}^*) .$$

For a non-linear system, the equilibrium points are found as solutions of non-linear algebraic equations. These systems may therefore have several (even infinitely many) equilibrium points. Any distribution in state space is possible .

7.1.17 Equilibrium points of continuous linear systems

Consider the unforced (free, autonomous) linear system:

$$\mathbf{z}'(t) = \mathbf{A}\,\mathbf{z}(t)\,.$$

For $\mathbf{z} = \mathbf{0}$ this yields $\mathbf{z}' = \mathbf{0}$: this state vector is therefore an equilibrium point of the system. Obviously the origin $\mathbf{z} = \mathbf{0}$ is always an equilibrium point of autonomous continuous systems. In addition, it is the only equilibrium point of such a system unless the system matrix $\mathbf{A}$ is singular, i.e. $\mathbf{0}$ appears as an eigenvector of $\mathbf{A}$. In this case other equilibrium points exist.

For a forced (non-autonomous) system with constant input $\mathbf{u}^*$

$$\mathbf{z}'(t) = \mathbf{A}\,\mathbf{z}(t) + \mathbf{B}\,\mathbf{u}(t)$$

the equilibrium condition is

$$\mathbf{0} = \mathbf{A}\,\mathbf{z}^* + \mathbf{B}\,\mathbf{u}^*\,.$$

If $\mathbf{A}$ is non-singular, the equilibrium point is defined by

$$\mathbf{z}^* = -\mathbf{A}^{-1}\,\mathbf{B}\,\mathbf{u}^*\,.$$

If $\mathbf{A}$ is singular, the existence of equilibrium points is undetermined.

7.1.18 Equilibrium points of discrete linear systems

The unforced (free, autonomous) linear system

$$\mathbf{z}(k+1) = \mathbf{A}\,\mathbf{z}(k)$$

always has $\mathbf{z}^* = \mathbf{0}$ as an equilibrium point. Since at this point the condition

$$\mathbf{z}^* = \mathbf{A}\,\mathbf{z}^*$$

applies, the eigenvector $\mathbf{z}^*$ corresponding to the eigenvalue $\lambda = 1$ is also an equilibrium point. If the system does not have an eigenvalue $\lambda = 1$, then $\mathbf{z}^* = \mathbf{0}$ is the only equilibrium point.

For the forced (non-autonomous) system

$$\mathbf{z}(k+1) = \mathbf{A}\,\mathbf{z}(k) + \mathbf{B}\,\mathbf{u}(k)$$

the equilibrium point must satisfy the condition

$$\mathbf{z}^* = \mathbf{A}\,\mathbf{z}^* + \mathbf{B}\,\mathbf{u}^*\,.$$

If $\lambda = 1$ is not an eigenvalue of the system matrix $\mathbf{A}$, then the matrix $[\mathbf{I} - \mathbf{A}]$ is non-singular, and a unique solution for the equilibrium point exists:

$$\mathbf{z}^* = [\mathbf{I} - \mathbf{A}]^{-1}\, \mathbf{B}\, \mathbf{u}^* .$$

If $\lambda = 1$ is an eigenvalue, then either no equilibrium point exists or an infinite number exist. This possibility has little practical relevance since even slight changes of the coefficients of **A** will result in an eigenvalue which differs from 1, yielding a unique equilibrium point.

7.2 Matrix Operations for Linear Dynamic Systems

7.2.1 Operations with matrices and vectors

The simplest form of an autonomous time-continuous system is obtained if the rates of change $z'_i = dz_i/dt$ are linearly dependent on the states z_i :

$$\begin{aligned}
dz_1/dt &= z'_1 = a_{11}\, z_1 + a_{12}\, z_2 + \ldots + a_{1n}\, z_n \\
dz_2/dt &= z'_2 = a_{21}\, z_1 + a_{22}\, z_2 + \ldots + a_{2n}\, z_n \\
\ldots &= \ldots\ldots \\
dz_n/dt &= z'_n = a_{1n}\, z_1 + a_{n2}\, z_2 + \ldots + a_{nn}\, z_n .
\end{aligned} \tag{7.10}$$

Here the a_{ij} are constant or time-dependent system parameters quantifying the contribution of state z_j to the rate of change z'_i.

Using column vectors for the system state

$$\mathbf{z} = \begin{bmatrix} z_1 \\ z_2 \\ \cdots \\ z_n \end{bmatrix}$$

and for the rates of change of the system state

$$\frac{d\mathbf{z}}{dt} = \mathbf{z}' = \begin{bmatrix} z'_1 \\ z'_2 \\ \cdots \\ z'_n \end{bmatrix}$$

and employing the system matrix

$$\mathbf{A} = \begin{bmatrix} a_{11} & a_{12} & \cdots & a_{1n} \\ a_{21} & a_{22} & \cdots & a_{2n} \\ \cdots & \cdots & \cdots & \cdots \\ a_{n1} & a_{n2} & \cdots & a_{nn} \end{bmatrix},$$

the state equations (Eq. 7.10) can be written in much more compact form

$$dz/dt = \mathbf{z}' = \mathbf{A}\,\mathbf{z}\,. \tag{7.11}$$

The rules of linear algebra apply to all operations with matrices and vectors.

Matrix addition is defined only for matrices of the same dimension (m rows, n columns).

If $\mathbf{A} = [a_{ij}]$, $\mathbf{B} = [b_{ij}]$, $\mathbf{C} = [c_{ij}]$ and $\mathbf{C} = \mathbf{A} + \mathbf{B}$, then the elements of $\mathbf{C}$ are determined by

$$c_{ij} = a_{ij} + b_{ij}\,,$$

i.e. elements having the same index are added up individually.

Scalar multiplication of matrices. Multiplication of a matrix $\mathbf{A} = [a_{ij}]$ by a scalar α (real or complex number) is defined by

$$\alpha\mathbf{A} = [\alpha\, a_{ij}]\,,$$

i.e. each matrix element must be multiplied by the factor α.

Matrix multiplication. The matrix product $\mathbf{C} = \mathbf{AB}$ is defined only if the number of columns of $\mathbf{A}$ agrees with the number of rows of $\mathbf{B}$.

If $\mathbf{A}$ is an (m·n) matrix and $\mathbf{B}$ an (n·p) matrix, the matrix $\mathbf{C} = \mathbf{A}\,\mathbf{B}$ is defined as an (m·p) matrix with elements

$$c_{ik} = \sum_{j=1}^{n} a_{ij}\, b_{jk}\,,$$

From this follows in particular for the product $\mathbf{A}\,\mathbf{z}$ (in the state equation 7.11) the column vector $\mathbf{y}$ whose elements are exactly equal to the right hand side of Equation 7.10.

$$y_i = \sum_{j=1}^{n} a_{ij}\, z_j\,.$$

The definition of the **determinant** of a quadratic (n·n) matrix $\mathbf{A}$ is useful for the solution of linear systems of equations. The determinant is a scalar number. It is defined by

$$\det\mathbf{A} = \sum_{j=1}^{n} a_{ij}C_{ij} = \sum_{i=1}^{n} a_{ij}C_{ij}\,.$$

Here the **cofactor** C_{ij} of element a_{ij} is defined by

$$C_{ij} = (-1)^{i+j} M_{ij}\,,$$

where the **minor** M_{ij} is the determinant of the matrix obtained from $\mathbf{A}$ by deleting the i-th row and the j-th column. By using this Laplace decomposition, an n-th order determinant can be computed from a combination of determinants of order (n-1).

Unless the determinant of a quadratic matrix $\mathbf{A}$ is equal to zero, the (n·n) Matrix $\mathbf{A}$ possesses an (n·n) **inverse matrix** $\mathbf{A}^{-1}$. The product of both matrices is the **identity matrix** $\mathbf{I}$:

$$\mathbf{A}\,\mathbf{A}^{-1} = \mathbf{I}$$

where

$$\mathbf{I} = \begin{bmatrix} 1 & 0 & . & 0 \\ 0 & 1 & . & 0 \\ . & . & . & . \\ 0 & 0 & . & 1 \end{bmatrix} .$$

Denoting the elements of the inverse matrix $\mathbf{A}^{-1}$ by a_{ij}^{-1}, the inverse can be computed from

$$\mathbf{A}^{-1} = [a_{ij}^{-1}] = [C_{ji}] / \det \mathbf{A}$$

where the cofactor $C_{ji} = (-1)^{j+i}\ M_{ji}$ is defined in terms of the corresponding minor.

7.2.2 Eigenvalues, eigenvectors, and the characteristic equation

Eigenvalues λ_i and **eigenvectors** $\mathbf{e}_i$ of a (symmetric) matrix $\mathbf{A}$ are determined by the **eigenvector equation**

$$\mathbf{A}\,\mathbf{e} = \lambda\,\mathbf{e}$$

which can also be written as

$$[\mathbf{A} - \lambda\mathbf{I}]\,\mathbf{e} = \mathbf{0} . \tag{7.12}$$

This homogeneous equation has a non-trivial solution $\mathbf{e} \neq \mathbf{0}$ only if the matrix is singular, i.e. if

$$\det[\mathbf{A} - \lambda\mathbf{I}] = 0 . \tag{7.13}$$

This equation is called the **characteristic equation** of $\mathbf{A}$. Expansion of the determinant leads to the **characteristic polynomial**

$$p(\lambda) = \lambda^n + a_{n-1}\,\lambda^{n-1} + a_{n-2}\,\lambda^{n-2} + \ldots + a_1\lambda + a_0 = 0 . \tag{7.14}$$

Here n is the dimension of the matrix **A**. The n roots of this polynomial of order n are the **eigenvalues** of the matrix **A**.

The polynomial may also be written as:

$$p(\lambda) = (\lambda - \lambda_1)(\lambda - \lambda_2) \ldots (\lambda - \lambda_n) = 0 .$$

The equation is obviously satisfied for each one of the n eigenvalues λ_i. If the coefficients a_{ij} of the matrix **A** are real numbers, the eigenvalues λ_i may be real or complex, i.e.

$$\lambda_i = \sigma_i + j\omega_i$$

with real part $Re(\lambda_i) = \sigma_i$ and imaginary part $Im(\lambda_i) = \omega_i$.

Substituting the individual eigenvalues into the eigenvector equation (Eq. 7.12) and performing the corresponding matrix multiplications yields the n eigenvectors $\mathbf{e}_i$, respectively their n components e_{ij}

$$\mathbf{e}_1 = \begin{bmatrix} e_{11} \\ e_{12} \\ \ldots \\ e_{1n} \end{bmatrix}, \quad \mathbf{e}_2 = \begin{bmatrix} e_{21} \\ e_{22} \\ \ldots \\ e_{2n} \end{bmatrix}, \quad \ldots, \quad \mathbf{e}_n = \begin{bmatrix} e_{n1} \\ e_{n2} \\ \ldots \\ e_{nn} \end{bmatrix} .$$

The matrix composed of the n eigenvectors $\mathbf{e}_i$ is referred to as **modal matrix** or **eigenvector matrix**:

$$\mathbf{M} = [\mathbf{e}_1 \; \mathbf{e}_2 \; \ldots \; \mathbf{e}_n] .$$

A quadratic (n·n) matrix **A** with distinct eigenvalues $\lambda_1, \lambda_2, \ldots \lambda_n$ (i.e. all eigenvalues are different) and corresponding eigenvectors $\mathbf{e}_1, \mathbf{e}_2, \ldots \mathbf{e}_n$ can be transformed into the (diagonal) eigenvalue matrix $\mathbf{\Lambda}$ by a transformation using the modal matrix **M:**

$$\mathbf{\Lambda} = \mathbf{M}^{-1} \mathbf{A} \mathbf{M}$$

or

$$\mathbf{A} = \mathbf{M} \mathbf{\Lambda} \mathbf{M}^{-1} ,$$

where $\mathbf{\Lambda}$ is the **eigenvalue matrix**

$$\mathbf{\Lambda} = \begin{bmatrix} \lambda_1 & 0 & \cdots & 0 \\ 0 & \lambda_2 & \cdots & 0 \\ \ldots & \ldots & \ldots & \ldots \\ 0 & 0 & \cdots & \lambda_n \end{bmatrix}$$

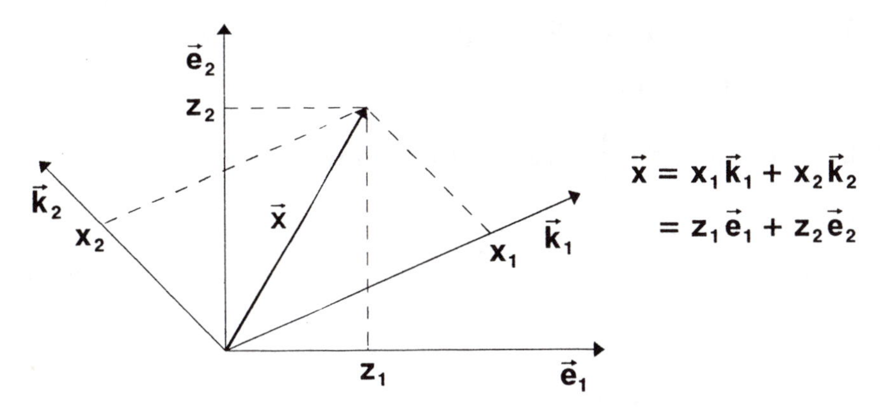

Fig. 7.3: Basis transformation of a vector. In the orthogonal basis $(\mathbf{e}_1, \mathbf{e}_2)$ the vector $\mathbf{x}$ has coordinates z_1, z_2 ; in the non-orthogonal basis $(\mathbf{k}_1, \mathbf{k}_2)$ it has coordinates x_1, x_2.

7.2.3 Transformation of basis

A vector $\mathbf{x}$ is determined by the direction of the basis vectors chosen for its description (coordinate system) and the length of the projection of the original vector on each of the basis vectors (Fig. 7.3). If the eigenvectors $\mathbf{e}_i$ are used as basis vectors, a vector $\mathbf{x}$ can be expressed as

$$\mathbf{x} = z_1 \mathbf{e}_1 + z_2 \mathbf{e}_2 + \ldots + z_n \mathbf{e}_n = \begin{bmatrix} \mathbf{e}_1 \, \mathbf{e}_2 \, \ldots \mathbf{e}_n \end{bmatrix} \begin{bmatrix} z_1 \\ z_2 \\ \ldots \\ z_n \end{bmatrix}$$

or

$$\mathbf{x} = \mathbf{M}\,\mathbf{z}\,.$$

Here the z_i are the lengths (with positive or negative signs) of the $\mathbf{x}$ vector in the new basis of eigenvectors $\mathbf{e}_i$.

The components z_i in the eigenvector basis can therefore be determined from the components x_i of the original basis by

$$\mathbf{z} = \mathbf{M}^{-1}\,\mathbf{x}\,.$$

These operations with eigenvalues and eigenvectors of the system matrix $\mathbf{A}$ are important for the analysis of linear time-discrete and time-continuous systems.

7.3 Behavior and Stability of Linear Systems in Unforced Motion

7.3.1 Form of the general solution of the state equation

The complete solution of the state equation of **linear systems**

$$\mathbf{z}' = \mathbf{A}\,\mathbf{z} + \mathbf{B}\,\mathbf{u} \qquad \text{(continuous system)} \qquad (7.15)$$

or

$$\mathbf{z}(k+1) = \mathbf{A}\,\mathbf{z}(k) + \mathbf{B}\,\mathbf{u}(k) \qquad \text{(discrete system)} \qquad (7.16)$$

consists of two very different components

$$\mathbf{z} = \mathbf{z}_h + \mathbf{z}_p\,.$$

The first component $\mathbf{z}_h$, the homogeneous (free, unforced) solution of the autonomous (unforced) system

$$\mathbf{z}' = \mathbf{A}\,\mathbf{z}$$

or

$$\mathbf{z}(k+1) = \mathbf{A}\,\mathbf{z}(k)$$

depends only on the initial values of the state variables $\mathbf{z}_0$ or $\mathbf{z}(t_0)$ and is independent of the input vector $\mathbf{u}$. The free solution describes that part of system behavior that is due entirely to the initial conditions, uninfluenced by inputs from outside of the system. In stable systems, the influence of the initial conditions must disappear after some time. Eventually, the current system state is therefore no longer determined by the initial conditions.

By contrast, the second component of the solution, the particular (forced) solution $\mathbf{z}_p$ is independent of the initial conditions and depends exclusively on the time development of the input vector $\mathbf{u}$ in the period from t_0 through t. Since the effect of initial conditions eventually disappears in stable systems, the forced solution eventually dominates (provided such inputs are acting on the system).

7.3.2 Linear dynamic systems

The simplest form of the state equation is the linear form where the rate of change of the state variables is determined by a linear combination of the current states:

$$\mathbf{z}(k+1) = \mathbf{A}\,\mathbf{z}(k) \qquad \text{(discrete system)} \qquad (7.17)$$

$$\mathbf{z}'(t) = \mathbf{A}\,\mathbf{z}(t) \qquad \text{(continuous system).} \qquad (7.18)$$

$\mathbf{A}$ is the **system matrix** of the linear system. In a time-invariant system, all elements a_{ij} of this matrix are constants. If the system matrix is non-singular, i.e. if its determinant is not equal to zero:

$$\det \mathbf{A} \neq 0$$

then as a result of Conditions 7.8 and 7.9, the only equilibrium point of the autonomous linear dynamic system is given by the state

$$\mathbf{z} = \mathbf{0}\,.$$

7.3.3 Solution for autonomous time-invariant discrete system

Introducing the initial value $\mathbf{z}(0)$ into the recursive state equation (Eq. 7.17)

$$\mathbf{z}(k+1) = \mathbf{A}\,\mathbf{z}(k)$$

leads to the result

$$\begin{aligned} \mathbf{z}(1) &= \mathbf{A}\,\mathbf{z}(0) \\ \mathbf{z}(2) &= \mathbf{A}\,\mathbf{z}(1) = \mathbf{A}\cdot\mathbf{A}\,\mathbf{z}(0) = \mathbf{A}^2\,\mathbf{z}(0) \\ &\dots \\ \mathbf{z}(k) &= \mathbf{A}^k\,\mathbf{z}(0) \qquad \text{(discrete system)} \end{aligned} \tag{7.19}$$

$\mathbf{A}^k$ is called the state transition matrix of the linear discrete system.

7.3.4 Solution using the diagonal eigenvalue matrix

The state transition matrix $\mathbf{A}^k$ can be expressed in terms of the eigenvalues λ_n and the eigenvectors $\mathbf{e}_n$ of the system matrix $\mathbf{A}$, respectively by the modal matrix composed of the eigenvectors $\mathbf{e}_n$

$$\mathbf{M} = [\mathbf{e}_1\ \mathbf{e}_2 \dots \mathbf{e}_n]$$

and by its inverse $\mathbf{M}^{-1}$ as

$$\mathbf{A}^k = \mathbf{M}\,\mathbf{\Lambda}^k\,\mathbf{M}^{-1}$$

where (for discrete eigenvalues) the eigenvalue matrix is given by

$$\mathbf{\Lambda} = \begin{bmatrix} \lambda_1 & 0 & \cdots & 0 \\ 0 & \lambda_2 & \cdots & 0 \\ \cdots & \cdots & \cdots & \cdots \\ 0 & 0 & \cdots & \lambda_n \end{bmatrix}.$$

Using this result, the solution of the homogeneous system $\mathbf{z}(k+1) = \mathbf{A}\,\mathbf{z}(k)$ can then be expressed as

$$\mathbf{z}(k) = \mathbf{M}\,\mathbf{\Lambda}^k\,\mathbf{M}^{-1} \cdot \mathbf{z}(0)\,. \tag{7.20}$$

7.3.5 Solution for the autonomous time-invariant continuous system

The homogeneous (autonomous) vector state equation is given by Equation 7.18:

$$\mathbf{z}' = \mathbf{A}\,\mathbf{z} \qquad \text{with} \qquad \mathbf{z}(0) = \mathbf{z}_0\,.$$

Assume that the solution vector $\mathbf{z}$ can be represented by a vector power series in t, whose coefficients are given by column vectors $\mathbf{a}_i$:

$$\mathbf{z} = \mathbf{a}_0 + \mathbf{a}_1\,t + \mathbf{a}_2\,t^2 + \ldots + \mathbf{a}_n\,t^n + \ldots$$

By differentiating with respect to t we obtain the time rate of change $\mathbf{z}'$. Introducing the power series for $\mathbf{z}$ and $\mathbf{z}'$ into the homogeneous differential equation Equation 7.18 yields:

$$\mathbf{a}_1 + 2\mathbf{a}_2\,t + 3\mathbf{a}_3\,t^2 + \ldots = \mathbf{A}\,(\mathbf{a}_0 + \mathbf{a}_1\,t + \mathbf{a}_2\,t^2 + \ldots)\,.$$

Comparing coefficients on both sides of the equation, one obtains

$$\mathbf{a}_1 = \mathbf{A}\,\mathbf{a}_0$$
$$\mathbf{a}_2 = \mathbf{A}\,\mathbf{a}_1/2 = \mathbf{A}\,\mathbf{A}\,\mathbf{a}_0/2 = \mathbf{A}^2\,\mathbf{a}_0/2$$
$$\ldots\ldots$$
$$\mathbf{a}_n = \mathbf{A}^n\,\mathbf{a}_0/n!$$

where the coefficient vector $\mathbf{a}_0$ is given by the initial condition $\mathbf{z}_0 = \mathbf{a}_0$. The expression for the solution vector $\mathbf{z}$ now becomes

$$\mathbf{z} = \mathbf{z}_0 + \mathbf{A}\,\mathbf{z}_0\,t + (\mathbf{A}^2/2)\,\mathbf{z}_0\,t^2 + \ldots + (\mathbf{A}^n/n!)\,\mathbf{z}_0\,t^n + \ldots$$
$$= (\mathbf{I} + \mathbf{A}\,t + (\mathbf{A}^2/2)\,t^2 + \ldots + (\mathbf{A}^n/n!)\,t^n + \ldots)\,\mathbf{z}_0\,.$$

The expansion within the parentheses is abbreviated as $e^{\mathbf{A}t}$, called **matrix exponential** or **transition matrix** $\mathbf{\Phi}(t)$. The solution of the homogeneous (autonomous) vector differential equation can then be written as

$$\mathbf{z}(t) = e^{\mathbf{A}t}\,\mathbf{z}_0 = \mathbf{\Phi}(t)\,\mathbf{z}_0\,. \tag{7.21}$$

The transition matrix $e^{\mathbf{A}t} = \mathbf{\Phi}(t)$ is a linear transformation (a quadratic matrix of dimension $n \cdot n$) which carries the initial state $\mathbf{z}_0$ into the new system state $\mathbf{z}(t)$.

7.3.6 Solution using the diagonal matrix exponential

The matrix exponential $e^{\mathbf{A}t}$ can be expressed in terms of the eigenvalues λ_n and eigenvectors $\mathbf{e}_n$ of the matrix $\mathbf{A}$, i.e. by the modal matrix composed of the $\mathbf{e}_n$

$$\mathbf{M} = [\mathbf{e}_1\ \mathbf{e}_2 \dots \mathbf{e}_n]$$

and its inverse matrix $\mathbf{M}^{-1}$:

$$e^{\mathbf{A}t} = \mathbf{M}\, e^{\Lambda t}\, \mathbf{M}^{-1} \qquad \text{where}$$

$$e^{\Lambda t} = \begin{bmatrix} e^{\lambda_1 t} & 0 & \cdots & 0 \\ 0 & e^{\lambda_2 t} & \cdots & 0 \\ \cdots & \cdots & \cdots & \cdots \\ 0 & 0 & \cdots & e^{\lambda_n t} \end{bmatrix}.$$

Using this result, the solution of the homogenous system may be written as

$$\mathbf{z}(t) = \mathbf{M}\, e^{\Lambda t}\, \mathbf{M}^{-1} \cdot \mathbf{z}_0 \,. \tag{7.22}$$

Derivation of $e^{\mathbf{A}t} = \mathbf{M}\, e^{\Lambda t}\, \mathbf{M}^{-1}$ using $\mathbf{A}^k = \mathbf{M}\Lambda^k\mathbf{M}^{-1}$:

$$\begin{aligned} e^{\mathbf{A}t} &= \mathbf{I} + \mathbf{A}\cdot t + \mathbf{A}^2 \cdot t^2/2! + \dots \\ &= \mathbf{I} + (\mathbf{M}\Lambda\mathbf{M}^{-1})\cdot t + (\mathbf{M}\Lambda^2\mathbf{M}^{-1})\cdot t^2/2! + \cdots \\ &= \mathbf{M}(\mathbf{I} + \Lambda t + \Lambda^2 t^2/2! + \dots)\mathbf{M}^{-1} \\ e^{\mathbf{A}t} &= \mathbf{M}\, e^{\Lambda t}\, \mathbf{M}^{-1} \end{aligned}$$

7.3.7 Stability of linear systems

A linear time-invariant system with constant input $\mathbf{u}$ is asymptotically **stable** if, for arbitrary initial conditions $\mathbf{z}_0$ or $\mathbf{z}(0)$, the state vector $\mathbf{z}(t)$ or $\mathbf{z}(k)$ approaches the equilibrium state $\mathbf{z}^*$ as time progresses.

Statements about stability as a function of the eigenvalues λ_i of the system matrix $\mathbf{A}$ follow directly from the solutions (Eq. 7.20, 7.22).

The homogeneous **discrete linear system** is **stable** if the absolute value of all eigenvalues is less than 1 (i.e. the eigenvalues are inside the unit circle in the complex eigenvalue plane)

$$|\lambda_i| < 1 \,. \tag{7.23}$$

The homogeneous **continuous linear system** ist **stable** if all eigenvalues have a negative real part (i.e. all eigenvalues are located in the left half-plane of the complex eigenvalue plane),

$$\mathrm{Re}(\lambda_i) < 0 . \tag{7.24}$$

The stability property of a linear system is a direct function of the system matrix **A**. It is independent of the forcing of the system.

A forced system is stable with respect to a set $U = \{u(t)\}$ of input signals if the resulting state vector remains bounded for all inputs belonging to this set. For example, an **un**damped oscillatory system $(\mathrm{Re}(\lambda_i) = 0)$ is unstable if excited by an input at its resonance frequency.

7.3.8 General form, standard form, and normal form: transformations

Let $\mathbf{A}_x$ be the (n·n) system matrix in the general form

$$\mathbf{A}_x = [a_{ij}] .$$

Let $\mathbf{x}$ be the state vector. The state equation of the linear homogeneous system is

$$\mathbf{x}' = \mathbf{A}_x \mathbf{x} \qquad \text{(continuous system)}$$
$$\mathbf{x}(k+1) = \mathbf{A}_x \mathbf{x}(k) \qquad \text{(discrete system).}$$

Let $\mathbf{A}_x$ have discrete eigenvalues λ_i and the corresponding modal matrix $\mathbf{M}_x$. Let $\mathbf{z}$ be the state vector of the diagonalized system with system matrix $\mathbf{A}_z = \mathbf{\Lambda}$ (consisting of the discrete eigenvalues λ_i of the system matrix $\mathbf{A}_x$).

$$\mathbf{z}' = \mathbf{\Lambda}\, \mathbf{z} \qquad \text{or} \qquad \mathbf{z}(k+1) = \mathbf{\Lambda}\, \mathbf{z}(k) .$$

The state vector $\mathbf{x}$ can then be determined (see Sec. 7.2.3) from

$$\mathbf{x} = \mathbf{M}_x \mathbf{z}$$

and the state vector $\mathbf{z}$ from

$$\mathbf{z} = \mathbf{M}_x^{-1} \mathbf{x} .$$

For a system with matrix $\mathbf{A}_y$, the same eigenvalues λ_i, and the state equation

$$\mathbf{y}' = \mathbf{A}_y \mathbf{y} \qquad \text{or} \qquad \mathbf{y}(k+1) = \mathbf{A}_y \mathbf{y}(k),$$

the state vector $\mathbf{y}$ can be expressed as

$$\mathbf{y} = \mathbf{M}_y \mathbf{z}$$

and the state vector $\mathbf{z}$ as

$$\mathbf{z} = \mathbf{M}_y^{-1} \mathbf{y} .$$

From this one obtains for the transformation of $\mathbf{x}$ into the state vector $\mathbf{y}$:

$$\mathbf{x} = \mathbf{M}_x \mathbf{M}_y^{-1} \mathbf{y}$$

and

$$\mathbf{y} = \mathbf{M}_y \mathbf{M}_x^{-1} \mathbf{x} .$$

The transformation therefore requires the modal matrices $\mathbf{M}_x$ and $\mathbf{M}_y$ of both systems or the eigenvectors of $\mathbf{A}_x$ and $\mathbf{A}_y$ which follow from the corresponding eigenvector equations:

$$[\mathbf{A}_x - \lambda \mathbf{I}]\, \mathbf{e}_x = \mathbf{0} \qquad \text{and} \qquad [\mathbf{A}_y - \lambda \mathbf{I}]\, \mathbf{e}_y = \mathbf{0} .$$

The **general form** of the $(n \cdot n)$ system matrix has an arbitrary distribution of real coefficients

$$\mathbf{A}_x = [a_{ij}] .$$

The **normal form** of the system matrix corresponds to the diagonal eigenvalue matrix $\mathbf{\Lambda}$ with eigenvalues λ_i on its main diagonal:

$$\mathbf{A}_z = \mathbf{\Lambda} = \begin{bmatrix} \lambda_1 & 0 & \cdots & 0 \\ 0 & \lambda_2 & \cdots & 0 \\ \cdots & \cdots & \cdots & \cdots \\ 0 & 0 & \cdots & \lambda_n \end{bmatrix} .$$

The **standard form** of the system matrix can be obtained from the characteristic polynomial of the system $\mathbf{A}_x$:

$$p(\lambda) = \lambda^n + a_{n-1} \lambda^{n-1} + a_{n-2} \lambda^{n-2} + \ldots + a_1 \lambda + a_0 = 0$$

by arranging the coefficients as follows:

$$\mathbf{A}_y = \begin{bmatrix} 0 & 1 & 0 & \cdots & 0 & 0 \\ 0 & 0 & 1 & \cdots & 0 & 0 \\ \cdots & \cdots & \cdots & \cdots & \cdots & \cdots \\ 0 & 0 & 0 & \cdots & 0 & 1 \\ -a_0 & -a_1 & -a_2 & \cdots & -a_{n-2} & -a_{n-1} \end{bmatrix} . \qquad (7.25)$$

$\mathbf{A}_x$, $\mathbf{A}_y$ and $\mathbf{A}_z$ obviously possess an identical characteristic polynomial and therefore the same eigenvalues and identical dynamics and stability.

The practical relevance of the transformation from the general form to the standard form or to the normal form lies in a very significant reduction of the number of possible linkages between n state variables, i.e. a simplification of the state equations as well as the corresponding system diagrams:

	Number of possible linkages:
General form:	n^2
Standard form:	$(2n)-1$
Normal form:	n

The advantage of the more efficient description in the behaviorally equivalent standard form or normal form comes at the price of having to recompute the results from the standard form (state vector **y**) or the normal form (state vector **z**) to the general form (state vector **x**) of the original system using the transformation matrices (output matrices) $\mathbf{C}_y$ and $\mathbf{C}_z$:

$$\mathbf{x} = \mathbf{C}_y \mathbf{y} \qquad\qquad \mathbf{x} = \mathbf{C}_z \mathbf{z}$$

where $\qquad\qquad$ where

$$\mathbf{C}_y = \mathbf{M}_x \mathbf{M}_y^{-1} \qquad\qquad \mathbf{C}_z = \mathbf{M}_x .$$

The procedure is demonstrated in the following example. Note the structurally different form of the three systems with state vectors **x**, **y**, and **z** which produce identical behavior (x_1, x_2), however.

7.3.9 Behaviorally equivalent systems: example

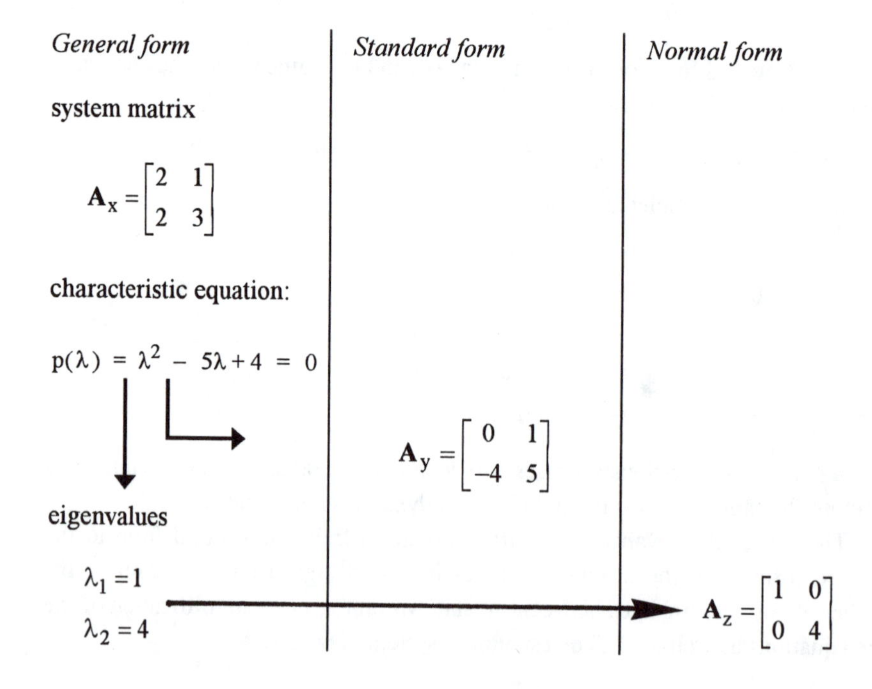

Modal matrices $\mathbf{M} = \left[\mathbf{e}_1\,\mathbf{e}_2 \ldots \mathbf{e}_n\right]$ from

$$[\mathbf{A}_x - \lambda\mathbf{I}]\,\mathbf{e}_x = \mathbf{0} \qquad [\mathbf{A}_y - \lambda\mathbf{I}]\,\mathbf{e}_y = \mathbf{0} \qquad [\mathbf{A}_z - \lambda\mathbf{I}]\,\mathbf{e}_z = \mathbf{0}$$

$$\mathbf{M}_x = \begin{bmatrix} 1 & 1 \\ -1 & 2 \end{bmatrix} \qquad \mathbf{M}_y = \begin{bmatrix} 1 & 1 \\ 1 & 4 \end{bmatrix} \qquad \mathbf{M}_z = \begin{bmatrix} 1 & 0 \\ 0 & 1 \end{bmatrix}$$

inverted:

$$\mathbf{M}_x^{-1} = \tfrac{1}{3}\begin{bmatrix} 2 & -1 \\ 1 & 1 \end{bmatrix} \qquad \mathbf{M}_y^{-1} = \tfrac{1}{3}\begin{bmatrix} 4 & -1 \\ -1 & 1 \end{bmatrix} \qquad \mathbf{M}_z^{-1} = \begin{bmatrix} 1 & 0 \\ 0 & 1 \end{bmatrix}$$

Transformation of the state vectors $\mathbf{y}$ and $\mathbf{z}$ to system output (= $\mathbf{x}$)

$$\mathbf{x} = \mathbf{C}_x\,\mathbf{x} \qquad \mathbf{x} = \mathbf{C}_y\,\mathbf{y} \qquad \mathbf{x} = \mathbf{C}_z\,\mathbf{z}$$

with output matrix

$$\mathbf{C}_x = \mathbf{I} = \begin{bmatrix} 1 & 0 \\ 0 & 1 \end{bmatrix} \qquad \mathbf{C}_y = \mathbf{M}_x\mathbf{M}_y^{-1} = \begin{bmatrix} 1 & 0 \\ -2 & 1 \end{bmatrix} \qquad \mathbf{C}_z = \mathbf{M}_x = \begin{bmatrix} 1 & 1 \\ -1 & 2 \end{bmatrix}$$

inverted:

$$\mathbf{C}_x^{-1} = \begin{bmatrix} 1 & 0 \\ 0 & 1 \end{bmatrix} \qquad \mathbf{C}_y^{-1} = \begin{bmatrix} 1 & 0 \\ 2 & 1 \end{bmatrix} \qquad \mathbf{C}_z^{-1} = \begin{bmatrix} 2 & -1 \\ 1 & 1 \end{bmatrix}\cdot\tfrac{1}{3}$$

and therefore

$$\mathbf{y} = \mathbf{C}_y^{-1}\mathbf{x} \qquad \mathbf{z} = \mathbf{C}_z^{-1}\mathbf{x}.$$

The state $\mathbf{x}$ follows from each of the three systems by

$$\begin{aligned} x_1 &= x_1 \\ x_2 &= x_2 \end{aligned} \qquad \begin{aligned} x_1 &= 1\cdot y_1 + 0\cdot y_2 \\ x_2 &= -2\cdot y_1 + 1\cdot y_2 \end{aligned} \qquad \begin{aligned} x_1 &= 1\cdot z_1 + 1\cdot z_2 \\ x_2 &= -1\cdot z_1 + 2\cdot z_2. \end{aligned}$$

The initial values for each of the system representations have to be determined from $\mathbf{x}_0$ by the corresponding transformation.

$x_{10} = x_{10}$	$x_{10} = 1 \cdot x_{10} + 0 \cdot x_{20}$	$z_{10} = \frac{2}{3} \cdot x_{10} - \frac{1}{3} \cdot x_{20}$
$x_{20} = x_{20}$	$y_{20} = 2 \cdot x_{10} + 1 \cdot x_{20}$	$z_{20} = \frac{1}{3} \cdot x_{10} + \frac{1}{3} \cdot x_{20}$.

With these results the simulation diagrams can be drawn (Fig. 7.4). Corresponding simulations produce identical results for $\mathbf{x}$. If considered as "black boxes," these three systems are completely identical. For the same initial conditions, they produce identical behavior.

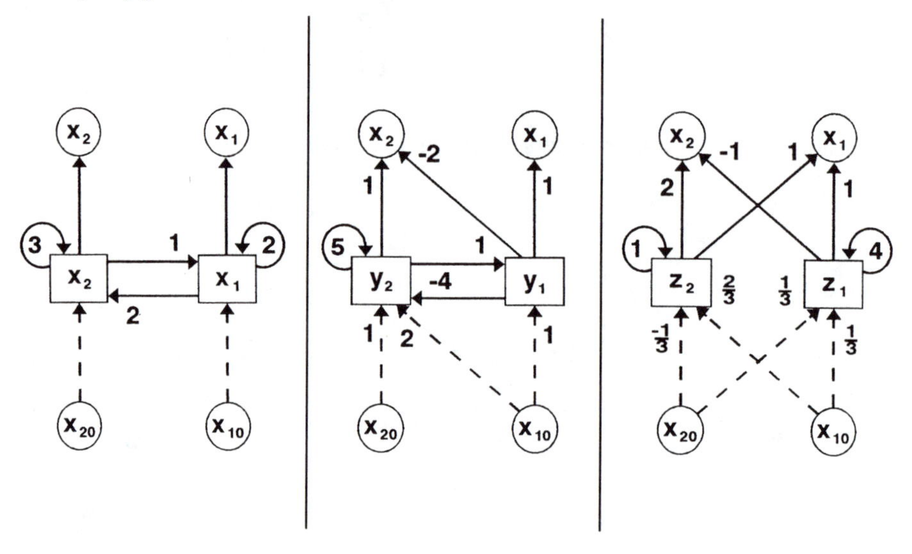

Fig. 7.4: Three systems with identical behavior (x_1, x_2) but different state variables and system structure. The system matrix has general form (left), standard form (center), and normal form (right). Eigenvalues of all three systems are identical.

7.3.10 Behavioral modes of linear systems

The characteristic equation (Equation 7.13)

$$\det[\mathbf{A} - \lambda \mathrm{I}] = 0$$

of the (n·n) matrix $\mathbf{A}$ with real coefficients a_{ij}, or the corresponding characteristic polynomial (Eq. 7.14)

$$p(\lambda) = \lambda^n + a_{n-1}\,\lambda^{n-1} + a_{n-2}\,\lambda^{n-2} + \ldots + a_1\,\lambda + a_0 = 0$$

has n eigenvalues λ_i, which may be

1. real or
2. complex (real part + imaginary part).

Complex roots (Case 2) can only appear in conjugate pairs. The following behavioral modes are possible:

7.3.11 Continuous systems

In the general case, eigenvalues are complex conjugates:

$$\lambda = \sigma \pm j\omega .$$

Eigenvalues are real if the imaginary component (ω) vanishes.

1. **Real eigenvalue** $\lambda = \sigma$

The only behavioral mode possible (eigenmode) is $e^{\sigma t}$. This permits exponential growth ($\sigma > 0$), constant value ($\sigma = 0$), or exponential decay of the state variable ($\sigma < 0$). The system is unstable if one of the eigenvalues is positive ($\lambda_i > 0$).

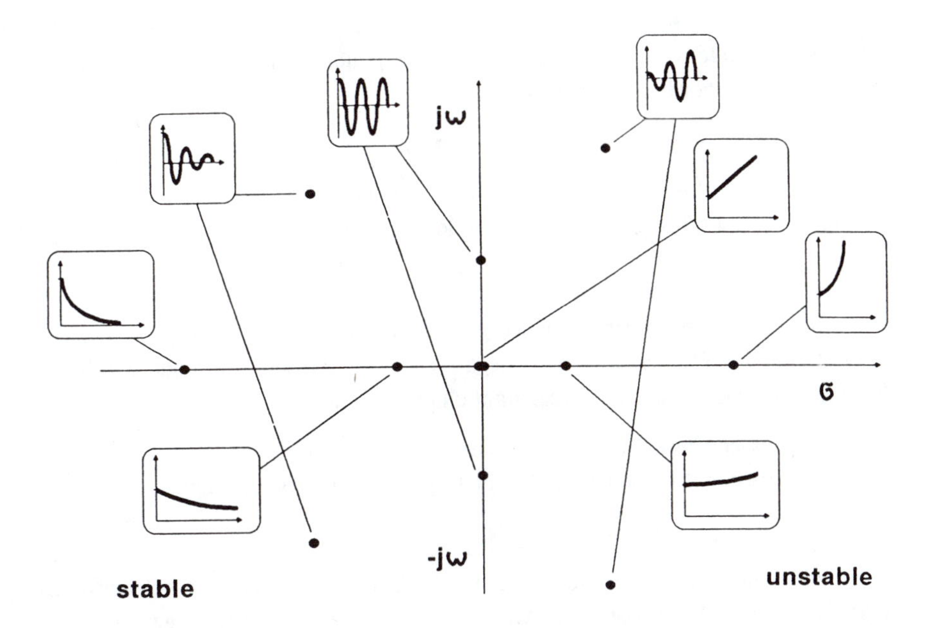

Fig. 7.5: Behavioral modes of a linear system as a function of the location of its eigenvalues in the complex plane (root locus).

2. **Complex-conjugate eigenvalue pair** $\lambda_{1,2} = \sigma \pm j\omega$
A complex-conjugate eigenvalue pair with (identical) real part σ and conjugate imaginary parts $(+j\omega)$ and $(-j\omega)$ corresponds to the complex behavioral modes

$$e^{(\sigma+j\omega)t}, \; e^{(\sigma-j\omega)t} .$$

Using Euler's formula, they can be rewritten as the characteristic modes (eigenmodes)

$$e^{\sigma t} (A \sin \omega t + B \cos \omega t) .$$

Terms of this nature describe an oscillation with frequency ω and exponential damping $(\sigma < 0)$ or amplification $(\sigma > 0)$. The frequency of oscillation is therefore specified by the imaginary part ω of the eigenvalue while the rate of damping or amplification is determined by the real part σ. The special case $\sigma = 0$ corresponds to an undamped oscillation. These behavioral possibilities of linear continuous system are shown in Figure 7.5 as a function of the location of the eigenvalues in the complex plane.

7.3.12 Discrete systems

Here also the eigenvalues are complex in the general case. They are best expressed in polar coordinates (radius r, angle θ):

$$\lambda = \sigma + j\omega = r\, e^{j\theta} = r (\cos \theta + j \sin \theta).$$

1. **Real eigenvalue:** $\lambda = r, \quad \theta = 0$
The corresponding characteristic mode (eigenmode) is

$$r^k .$$

The terms of this geometric sequence grow in absolute value if $|r| > 1$ and decrease if $|r| < 1$. If the eigenvalue is negative $(r < 0)$, the solution mode will be an alternating geometric sequence (constant change of sign).

2. **Complex-conjugate eigenvalue pair:** $\lambda = r (\cos \theta \pm j \sin \theta)$
The corresponding behavioral mode (eigenmode) is of the form

$$r^k (A \sin \theta k + B \cos \theta k)$$

i.e. an oscillation over the time index k whose amplitude will geometrically increase (or decrease) in accordance with the absolute value r of the eigenvalue.

7.3.13 Stability behavior of a two-dimensional linear system

Behavior and stability of a linear system are determined by its system matrix **A**, respectively by its eigenvalues λ_i. Since complex eigenvalues can appear in all linear systems of order two and higher and since roots of another type are not possible if the system matrix has real coefficients, two-dimensional linear systems exhibit all the possible behavioral modes of linear systems of any order. Their analysis is therefore of general relevance.

Possible behavioral modes can be categorized as function of the coefficients of the system matrix **A** of the two-dimensional continuous system in terms of source, sink, node, saddle, center, and focus (cf. Model M 203 LINEAR OSCILLATOR OF SECOND ORDER in the systems zoo).

Let $\mathbf{A} = \begin{bmatrix} a & b \\ c & d \end{bmatrix}$ and $p = a+d$, $q = ad - bc$. Then:

source, unstable:	if $4q = p^2$ and $p > 0$
sink, stable:	if $4q = p^2$ and $p \le 0$
focus, unstable:	if $4q > p^2$ and $p > 0$
center, marginally stable:	if $4q > p^2$ and $p = 0$
focus, stable:	if $4q > p^2$ and $p < 0$
line source, unstable:	if $q = 0$ and $p \ge 0$
line sink, stable:	if $q = 0$ and $p < 0$
saddle, unstable:	if $q < 0$
node, unstable:	if $q > 0$ and $4q < p^2$ and $p > 0$
node, stable:	if $q > 0$ and $4q < p^2$ and $p \le 0$.

Corresponding behavioral regions are shown in Figure 7.6 as function of p and q.

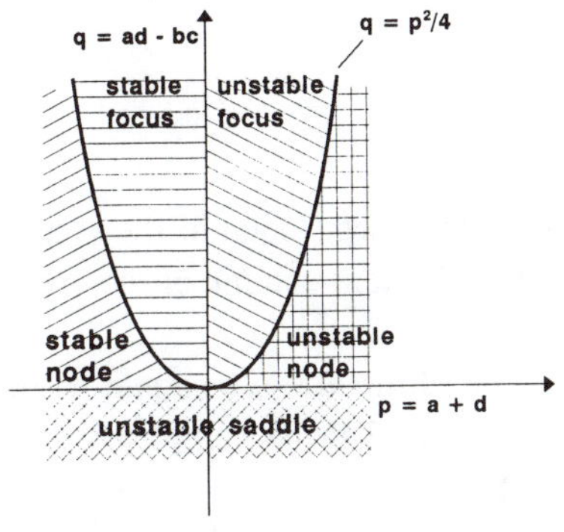

Fig. 7.6: Dependence of behavior and stability of a linear system with two state variables on its system parameters $p = a + d$ and $q = ad - bc$.

7.3.14 Stability test for linear systems

For a **continuous system** the stability condition is:

$$\mathrm{Re}(\lambda) < 0\ .$$

For a **discrete system** it is:

$$|\lambda| < 1\ .$$

If the characteristic equation of the system matrix **A** of a **continuous system** is available in the form of Equation 7.14,

$$\lambda^n + a_{n-1}\,\lambda^{n-1} + \ldots + a_1\,\lambda + a_0 = 0$$

then it is possible to determine the number of unstable eigenvalues of the system by constructing Routh's array and applying the **Routh criterion**. The coefficients a_i of the characteristic equation are arranged according to the following scheme:

$$\begin{array}{l|llll}
n & a_n & a_{n-2} & a_{n-4} & \ldots \\
n-1 & a_{n-1} & a_{n-3} & a_{n-5} & \ldots \\
n-2 & b_1 & b_2 & b_3 & \ldots \\
n-3 & c_1 & c_2 & & \ldots \\
\ldots & \ldots & & & \\
1 & f_1 & & & \\
0 & g_1 & & &
\end{array}$$

where (7.26)

$$\begin{aligned}
b_1 &= \left(1/a_{n-1}\right)\cdot\left(a_{n-1}\,a_{n-2} - a_n\,a_{n-3}\right) \\
b_2 &= \left(1/a_{n-1}\right)\cdot\left(a_{n-1}\,a_{n-4} - a_n\,a_{n-5}\right) \\
&\ldots \\
c_2 &= 1/b_1\cdot\left(b_1\,a_{n-5} - a_{n-1}\,b_3\right) \\
c_2 &= 1/b_1\cdot\left(b_1\,a_{n-5} - a_{n-1}\,b_3\right) \\
&\ldots
\end{aligned}$$

Routh criterion: The number of unstable eigenvalues (roots) of the system is equal to the number of sign changes in the first coefficient column of the array. (A single sign change already means instability.)

If a zero appears in the first coefficient column of the array, it should be replaced by a small positive quantity ε. The array can then be completed. The Routh test is then applied under the condition $\varepsilon \to 0$.

The criterion can be used to determine the magnitude and/or sign of system parameters guaranteeing a stable system.

The Routh criterion can also be used for **discrete systems**. However, in this case we must account for the fact that the stability region corresponds to the area inside the unit circle in the complex z-plane.

The transformation

$$\lambda = \frac{z+1}{z-1} \quad \text{or} \quad z = \frac{\lambda+1}{\lambda-1}$$

maps the interior of the unit circle in the complex z-plane onto the left half-plane in the λ-plane. If this mapping is applied to the characteristic equation of a discrete system

$$A_n z^n + A_{n-1} z^{n-1} + \ldots + A_1 z + A_0 = 0$$

then the transformation $z = (\lambda+1)/(\lambda-1)$ yields a polynomial in the form of Equation 7.14 and of the same order in λ. After construction of the Routh array for this polynomial, the Routh criterion can be applied. Each sign reversal in the first column indicates an unstable eigenvalue outside of the unit circle in the z-plane.

7.3.15 Remarks on the behavior of linear continuous systems

1. The differential equation of a second order linear oscillator

$$\frac{d^2y}{dt^2} + 2\xi\omega_0\frac{dy}{dt} + \omega_0^2\, y = 0$$

can also be written in the standard form as two first order differential equations:

$$x'_1 = x_2$$
$$x'_2 = -\omega_0^2\, x_1 - 2\,\xi\,\omega_0\, x_2$$

or

$$\mathbf{x}' = \mathbf{A}\,\mathbf{x}$$

with the system matrix

$$\mathbf{A} = \begin{bmatrix} 0 & 1 \\ -\omega_0^2 & -2\xi\omega_0 \end{bmatrix}.$$

The characteristic equation follows directly from this standard form of the system matrix (cf. Eq. 7.25):

$$\lambda^2 + 2\,\xi\,\omega_0\,\lambda + \omega_0^2 = 0\,.$$

This quadratic equation has the solutions

$$\lambda_{1,2} = -\xi\,\omega_0 \pm \omega_0\sqrt{\xi^2 - 1}\,.$$

Here ω_0 is the **natural (characteristic) frequency** of the undamped system. In the absence of damping $(\xi = 0)$, the unforced response of the system (homogeneous solution) will be a harmonic oscillation of frequency ω_0. If the system is damped $(\xi > 0)$, a gradual increase in damping will initially result in oscillations whose frequency is modified by the damping factor. As the damping factor increases to $\xi = 1$, the oscillations disappear completely. The value $\xi = 1$ is therefore called critical damping. For $\xi > 1$ (supercritical damping), the eigenvalues have no imaginary parts; oscillation is therefore no longer possible. Oscillations can therefore occur only in the subcritically damped region.

2. A linear unforced system of second or higher order may start to oscillate with certain frequencies in response to given initial conditions (initial displacement) or a given aperiodic input (pulse, step). Excitation by a periodic input is therefore not always required for oscillation. The frequencies of oscillation are a function of the coefficients of the differential equation or of the system matrix **A**, respectively. These frequencies correspond to the characteristic frequencies of the system, modified by its damping.
3. A linear system of order n has n eigenvalues, among them at most $n/2$ complex-conjugate eigenvalue pairs. This means that an n-th order system can have at most $n/2$ (n even) or $(n-1)/2$ characteristic frequencies (n odd).
4. An amplification of the response, and likewise instability, occurs if there exists a single eigenvalue with a positive real part. Conversely, a decaying response (stability) is possible only if the real parts of all eigenvalues are negative. If the real part of one or several eigenvalues vanishes, the system will have marginal stability. In the absence of further perturbations, the amplitude of oscillation of a harmonic oscillator (zero damping) will therefore remain constant with time.
5. In the long run, the eigenvalue with the smallest damping will dominate the free motion. For the assessment of the long-term behavior of the system it therefore suffices to consider only this eigenvalue.
6. Linear systems of order higher than $n = 2$ cannot have qualitatively different behavior than linear systems of first and second order. The study of the behavior of systems of second order therefore has fundamental relevance.
7. The location of eigenvalues (root locus) in the complex plane provides information about general behavior and stability of the system. Often the root locus graph is sufficient for an assessment of system behavior (cf. Fig. 7.5).

8. The stability and general behavior of linear systems depend *only* on the system matrix $\mathbf{A}$, i.e. on the coefficients of the differential equation. In contrast to non-linear systems, the behavioral characteristic of linear systems is independent of initial values or input functions.

7.4 Behavior of Linear Systems in Forced Motion

7.4.1 Linear system and the superposition principle

For a linear continuous (or discrete) system, the state equation (Eq. 7.1, 7.2)

$$\mathbf{z}' = \mathbf{f}(\mathbf{z}, \mathbf{u}, t) \qquad \text{or} \qquad \mathbf{z}(k+1) = \mathbf{f}(\mathbf{z}(k), \mathbf{u}(k), k)$$

simplifies to Equation 7.15 or 7.16:

$$\mathbf{z}' = \mathbf{A}(t)\,\mathbf{z} + \mathbf{B}(t)\,\mathbf{u} \qquad \text{or} \qquad \mathbf{z}(k+1) = \mathbf{A}(k)\,\mathbf{z}(k) + \mathbf{B}(k)\,\mathbf{u}(k),$$

and the behavior equation (output equation)

$$\mathbf{v} = \mathbf{g}(\mathbf{z}, \mathbf{u}, t) \qquad \text{or} \qquad \mathbf{v}(k) = \mathbf{g}(\mathbf{z}(k), \mathbf{u}(k), k)$$

simplifies to

$$\mathbf{v} = \mathbf{C}(t)\,\mathbf{z} + \mathbf{D}(t)\,\mathbf{u} \qquad \text{or} \qquad \mathbf{v}(k) = \mathbf{C}(k)\,\mathbf{z}(k) + \mathbf{D}(k)\,\mathbf{u}(k)\,.$$

For a time-invariant system, the state and output equations reduce to

$$\mathbf{z}' = \mathbf{A}\,\mathbf{z} + \mathbf{B}\,\mathbf{u} \qquad \text{or} \qquad \mathbf{z}(k+1) = \mathbf{A}\,\mathbf{z}(k) + \mathbf{B}\,\mathbf{u}(k)$$

$$\mathbf{v} = \mathbf{C}\,\mathbf{z} + \mathbf{D}\,\mathbf{u} \qquad \text{or} \qquad \mathbf{v}(k) = \mathbf{C}\,\mathbf{z}(k) + \mathbf{D}\,\mathbf{u}(k)\,. \qquad (7.27)$$

Superposition principle: The response $\mathbf{z}(t)$ of a linear system to several simultaneous inputs $\mathbf{u}_1(t), \mathbf{u}_2(t), \dots, \mathbf{u}_n(t)$ is equal to the sum of the responses to the individual inputs, i.e. if $\mathbf{z}_i(t)$ is the system response to $\mathbf{u}_i(t)$, then

$$\mathbf{z}(t) = \sum_{i=1}^{n} \mathbf{z}_i(t) \qquad \text{or} \qquad \mathbf{z}(k) = \sum_{i=1}^{n} \mathbf{z}_i(k)\,.$$

The superposition principle is valid also for a time-variant system matrix $\mathbf{A}(t)$ as well as for a time-variant input matrix $\mathbf{B}(t)$.

Since an arbitrary input function $\mathbf{u}(t)$ can be approximated by a sum of elementary functions (e.g. pulse, step, ramp, sine functions), the superposition principle allows the computation of the response of a linear system to an arbitrary input by summation of the individual responses to the elementary components of the input signal.

7.4.2 Representation of aperiodic input functions

The **impulse function** $\delta(t)$ is of particular significance for system studies. It is defined by the following properties:

$$\delta(t-a) = 0 \text{ for } t \neq a$$

$$\int_{a-\varepsilon}^{a+\varepsilon} \delta(t-a)dt = 1 \text{ with } \varepsilon > 0$$

and therefore

$$\int_{-\infty}^{t} \delta(t-a)dt = \sigma(t-a) .$$

The integral of the unit impulse function over a time interval, including the point in time $t = a$ at which the impulse occurs, is therefore the unit step function $\sigma(t-a)$; the step occurs at the time of the impulse $t = a$.

Using the impulse function, an arbitrary input function $u(t)$ can be approximated by the sum of discrete impulses

$$u(t) \approx \sum_{i=1}^{n} u(t_i) \cdot \Delta t \cdot \delta(t-t_i).$$

At times t_i and at time intervals Δt, impulses of strength $u(t_i) \cdot \Delta t$ (pulse area) are input to the system. Although the signal itself is quite different from the original input signal, the approximation by the sum of impulse functions has (approximately) the same **effect** on the system as the original function. We are therefore **not** dealing with an approximation of the function u(t) itself but rather of its effect on the system.

For a "continuing" input function $u(t)$ with $u(t) = 0$ for $t < 0$ (or $k < 0$), we then obtain

$$u(t) \approx \sum_{k=0}^{\infty} [u(k\Delta\tau)] \cdot \delta(t-k\Delta\tau) \cdot \Delta\tau$$

in a discrete representation and

$$u(t) = \int_{0}^{\infty} u(\tau) \cdot \delta(t-\tau) \cdot d\tau$$

in a continuous representation after taking the limit $\Delta\tau \rightarrow d\tau \rightarrow 0$.

7.4.3 Representation of periodic input functions

System inputs are often of periodic nature, i.e. an input function of a certain shape is repeated after a certain time period. In this case the periodic function can be represented by a sum of sine or cosine oscillations (or phase-shifted sine oscillations) of different frequencies. This Fourier series approximation can be written as

$$u(t) = A_0 + \sum_{n=1}^{\infty}\left(A_n \cos n\omega t + B_n \sin n\omega t\right)$$

or, equivalently

$$u(t) = A_0 + \sum_{n=1}^{\infty} C_n \sin\left(n\omega t + \varphi_n\right) .$$

The initially unknown coefficients A_0 , A_n , and B_n have to be determined by computing the following integrals over the time period T:

$$A_0 = \frac{1}{T}\int_{-T/2}^{+T/2} u(\tau)\, d\tau$$

$$A_n = \frac{2}{T}\int_{-T/2}^{+T/2} u(\tau) \cos n\omega\tau \;\, d\tau$$

$$B_n = \frac{2}{T}\int_{-T/2}^{+T/2} u(\tau) \sin n\omega\tau \;\, d\tau .$$

C_n and φ_n follow from

$$C_n = \sqrt{A_n^{\,2} + B_n^{\,2}}$$

$$\varphi_n = \arctan\left(\frac{B_n}{A_n}\right)$$

$$\omega = \frac{2\pi}{T} .$$

Evaluation of the integrals can be done analytically, by numerical approximation, or even graphically. Normally, a few terms of the Fourier series are sufficient for a satisfactory approximation of a function. More terms have to be included if greater accuracy is required.

By expanding the interval from T to ∞ it is also possible to approximate aperiodic functions by Fourier transformations.

7.4.4 Solution for the forced linear time-invariant system

Application of the superposition principle to the state equation of the forced linear discrete time-invariant system (Eq. 7.27b)

$$\mathbf{z}(k+1) = \mathbf{A}\,\mathbf{z}(k) + \mathbf{B}\,\mathbf{u}(k)$$

with input $\mathbf{u}(k)$ yields the solution

$$\mathbf{z}(k) = \mathbf{A}^k\,\mathbf{z}(0) + \sum_{j=0}^{k-1} \mathbf{A}^{k-j-1}\,\mathbf{B}\,\mathbf{u}(j). \tag{7.28}$$

The forced linear continuous time-invariant system (7.27a)

$$\mathbf{z}' = \mathbf{A}\,\mathbf{z} + \mathbf{B}\,\mathbf{u}$$

with input $\mathbf{u}(t)$ has the solution

$$\mathbf{z}(t) = e^{\mathbf{A}t} \cdot \mathbf{z}_0 + e^{\mathbf{A}t} \cdot \int_0^t e^{-\mathbf{A}\tau}\,\mathbf{B}\,\mathbf{u}(\tau)\,d\tau$$

$$= e^{\mathbf{A}t} \cdot \mathbf{z}_0 + \int_0^t e^{\mathbf{A}(t-\tau)}\,\mathbf{B}\,\mathbf{u}(\tau)\,d\tau. \tag{7.29}$$

This result can be interpreted as the response to a sum of individual impulses $\mathbf{u}$ at the different points in time τ.

The first term ($e^{\mathbf{A}t}\,\mathbf{z}_0$) represents the free (autonomous) system behavior resulting from the initial condition $\mathbf{z}_0$. The integrand of the second term can be interpreted as follows: the input function $\mathbf{u}(t)$ is represented by individual pulses of magnitude $\mathbf{u}(\tau)$ and duration $\Delta\tau$. Application of pulse $\mathbf{u}(\tau)\cdot\Delta\tau$ at time $t = \tau$ and transformation by input matrix $\mathbf{B}$ changes the system state by $\Delta\mathbf{z} = \mathbf{B}\,\mathbf{u}(\tau)\cdot\Delta\tau$. Beginning at time τ (where $(t-\tau) = 0$), the individual impulse applied at time $t = \tau$ begins to decay according to the state transition matrix $e^{\mathbf{A}t}$, i.e. the system response to the individual impulse is

$$e^{\mathbf{A}(t-\tau)}\,\mathbf{B}\,\mathbf{u}(\tau)\,\Delta\tau.$$

Summation of the responses to all individual impulses using the superposition principle up to time t, and integration in the limit $\Delta\tau \to d\tau \to 0$ leads to the Expression 7.29. Expression 7.28 for the discrete system is analogous.

7.4.5 Diagonalization of the system and decoupling of the characteristic modes

In the study of forced behavior, we again obtain an especially simple system representation if a suitable transformation is used to replace the original system matrix

A by an equivalent matrix having non-zero coefficients on the main diagonal only. This diagonalization is possible if the system matrix **A** has distinct eigenvalues. If repeated eigenvalues are involved, transformation to the Jordan canonical form is still possible where the eigenvalues again appear on the main diagonal, and the only other non-zero elements are a certain number of unit elements ("1") on the superdiagonal. Here we only consider the case where the eigenvalues of the system matrix **A** of the discrete or continuous system are distinct.

Since the approach for discrete and continuous systems does not differ in principle, the procedure is developed in parallel in the following. The point of departure are the system equations for the time-invariant discrete and continuous systems.

Discrete system	***Continuous system***
$\mathbf{x}(k+1) = \mathbf{A}\,\mathbf{x}(k) + \mathbf{B}\,\mathbf{u}(k)$ $\mathbf{v}(k) = \mathbf{C}\,\mathbf{x}(k) + \mathbf{D}\,\mathbf{u}(k)$	$\mathbf{x}' = \mathbf{A}\,\mathbf{x} + \mathbf{B}\,\mathbf{u}$ $\mathbf{v} = \mathbf{C}\,\mathbf{x} + \mathbf{D}\,\mathbf{u}$
General solution:	
$\mathbf{x}(k) = \mathbf{A}^k\,\mathbf{x}(0) + \sum_{j=0}^{k-1} \mathbf{A}^{k-j-1}\,\mathbf{B}\,\mathbf{u}(j)$	$\mathbf{x}(t) = e^{\mathbf{A}t}\mathbf{x}_0 + e^{\mathbf{A}t}\int_0^t e^{-\mathbf{A}\tau}\,\mathbf{B}\,\mathbf{u}(\tau)\,d\tau$
$\mathbf{v}(k) = \mathbf{C}\,\mathbf{x} + \mathbf{D}\,\mathbf{u}$	$\mathbf{v}(t) = \mathbf{C}\,\mathbf{x} + \mathbf{D}\,\mathbf{u}$

Procedure:

1. Determine the eigenvalues of **A**. If the eigenvalues are not distinct, find out whether a distinct set of eigenvalues can be produced by admissible changes in the system matrix.

2. Determine the eigenvectors of the system matrix **A**.

3. Construct the modal matrix **M** of eigenvectors.

4. Transform the original state variable **x** :

Discrete	Continuous
$\mathbf{x}(k) = \mathbf{M}\,\mathbf{z}(k)$	$\mathbf{x}(t) = \mathbf{M}\,\mathbf{z}(t)$

5. Transform the original homogeneous system

$\mathbf{x}(k+1) = \mathbf{A}\,\mathbf{x}(k)$	$\mathbf{x}'(t) = \mathbf{A}\,\mathbf{x}(t)$
$\mathbf{M}\,\mathbf{z}(k+1) = \mathbf{A}\,\mathbf{M}\,\mathbf{z}(k)$	$\mathbf{M}\,\mathbf{z}'(t) = \mathbf{A}\,\mathbf{M}\,\mathbf{z}(t)$ or
$\mathbf{z}(k+1) = \mathbf{M}^{-1}\,\mathbf{A}\,\mathbf{M}\,\mathbf{z}(k)$	$\mathbf{z}'(t) = \mathbf{M}^{-1}\,\mathbf{A}\,\mathbf{M}\,\mathbf{z}(t)$.
Since $\mathbf{M}^{-1}\,\mathbf{A}\,\mathbf{M} = \mathbf{\Lambda}$,	
we obtain the normal form	
$\mathbf{z}(k+1) = \mathbf{\Lambda}\,\mathbf{z}(k)$	$\mathbf{z}'(t) = \mathbf{\Lambda}\,\mathbf{z}(t)$.

6. The transition matrix is obtained from

$\mathbf{A}^k = \mathbf{M}\,\mathbf{\Lambda}^k\,\mathbf{M}^{-1}$	$e^{\mathbf{A}t} = \mathbf{M}e^{\lambda t}\,\mathbf{M}^{-1}$ where
$\mathbf{\Lambda} = \begin{bmatrix} \lambda_1 & 0 & \cdots & 0 \\ 0 & \lambda_2 & \cdots & 0 \\ \cdots & \cdots & \cdots & \cdots \\ 0 & 0 & \cdots & \lambda_n \end{bmatrix}$	$e^{\Lambda t} = \begin{bmatrix} e^{\lambda_1 t} & 0 & \cdots & 0 \\ 0 & e^{\lambda_2 t} & \cdots & \cdots \\ \cdots & \cdots & \cdots & \cdots \\ 0 & \cdots & \cdots & e^{\lambda_n t} \end{bmatrix}$.

7. The original system is now represented by n decoupled systems.

8. The transformation converts the state equations in standard form to the **state equations in normal form**:

$\mathbf{z}(k+1) = \mathbf{\Lambda}\,\mathbf{z}(k) + \mathbf{B}_n\,\mathbf{u}(k)$	$\mathbf{z}' = \mathbf{\Lambda}\,\mathbf{z} + \mathbf{B}_n\,\mathbf{u}$
$\mathbf{v}(k) = \mathbf{C}_n\,\mathbf{z}(k) + \mathbf{D}_n\,\mathbf{u}(k)$	$\mathbf{v} = \mathbf{C}_n\,\mathbf{z} + \mathbf{D}_n\,\mathbf{u}$

where

$$\mathbf{\Lambda} = \mathbf{M}^{-1}\,\mathbf{A}\,\mathbf{M}$$
$$\mathbf{B}_n = \mathbf{M}^{-1}\,\mathbf{B}$$
$$\mathbf{C}_n = \mathbf{C}\,\mathbf{M}$$
$$\mathbf{D}_n = \mathbf{D}\,.$$

9. The **general solution** of the normal form is given by (7.30, 7.31)

$$\mathbf{z}(k) = \boldsymbol{\Lambda}^k \mathbf{z}(0) + \sum_{j=0}^{k-1} \boldsymbol{\Lambda}^{k-j-1} \mathbf{B}_n \mathbf{u}(j) \quad \Bigg| \quad \mathbf{z}(t) = e^{\Lambda t} \mathbf{z}_0 + e^{\Lambda t} \cdot \int_0^t e^{-\Lambda \tau} \mathbf{B}_n \mathbf{u}(\tau)\, d\tau$$

10. For the **reverse transformation** of the $\mathbf{z}$-vector into the $\mathbf{x}$-vector use (cf. Point 4.)

$$\mathbf{x}(k) = \mathbf{M}\,\mathbf{z}(k) \quad \Bigg| \quad \mathbf{x}(t) = \mathbf{M}\,\mathbf{z}(t)\ .$$

7.4.6 Linear system response to periodic input functions (frequency response)

The previous sections were concerned with the transition behavior of linear systems in response to aperiodic inputs. This behavior is controlled by the transition matrix $\boldsymbol{\Phi} = e^{\mathbf{A}t}$, i.e. by the system matrix $\mathbf{A}$. Behavioral dynamics and stability can therefore be deduced from the unforced motion (homogeneous solution).

It became evident that for low values of damping, oscillations at the characteristic frequencies could develop, even if the input function itself did not contain oscillations. It can be expected that these characteristic oscillations will amplify if they are excited by a periodic input having a frequency close to the characteristic frequency. In this section, we shall therefore deal with the system response to oscillating inputs. Because of the possible resonance processes, it is to be expected that the system response will strongly depend on the forcing frequency of the input. This dependence is called the **frequency response** of the system.

The fact that an arbitrary periodic signal can be approximated by a Fourier series containing contributions by different frequencies is of particular relevance for the investigation of frequency response. The superposition principle for linear systems guarantees that system response to arbitrary periodic input signals can be determined by summing the individual responses to the individual components of the input signal. Since arbitrary periodic signals can be approximated by a sum of sine waves of different frequencies, it suffices to investigate the frequency response of a system with respect to sine waves covering the whole range of possible frequencies. For an n-th order system of differential equations

$$\frac{d^n x}{dt^n} + a_{n-1} \frac{d^{n-1} x}{dt^{n-1}} + \ldots + a_0 x = u(t) \tag{7.32}$$

or its n state equations in the standard form ($\mathbf{b}$ = column vector $(0, 0 \ldots, 1)$)

$$\mathbf{x}' = \mathbf{A}\,\mathbf{x} + \mathbf{B}\,\mathbf{u} = \mathbf{A}\,\mathbf{x} + \mathbf{b}\,u(t)\,.$$

We obtain for the sinusoidal input signal

$$u(t) = U \cos \omega t$$

the output signal

$$x(t) = |G|\, U \cos(\omega t + \varphi)$$

where the complex **frequency transfer function** = frequency response $G(j\omega)$

$$G(j\omega) = |G(\omega)|\, e^{i\varphi(\omega)} = 1/[(j\omega)^n + a_{n-1}(j\omega)^{n-1} + \ldots + a_0] = 1/[\det(j\omega\mathbf{I} - \mathbf{A})]$$

This expression can be written down directly using the coefficients of the characteristic polynomial or of the differential equation (Eq. 7.32), respectively.

The **gain** expresses the amplitude ratio of output signal to input signal. It is equal to the absolute value of the frequency response. In terms of the real and imaginary parts of $G(j\omega)$:

$$|G(j\omega)| = \sqrt{(\mathrm{Re}\,G)^2 + (\mathrm{Im}\,G^2)}\;.$$

The **phase** expresses the phase shift of output to input signal. It is given by

$$\varphi = \arctan \frac{\mathrm{Im}\,G(j\omega)}{\mathrm{Re}\,G(j\omega)}\;.$$

The geometric relationships are shown in Figure 7.7.

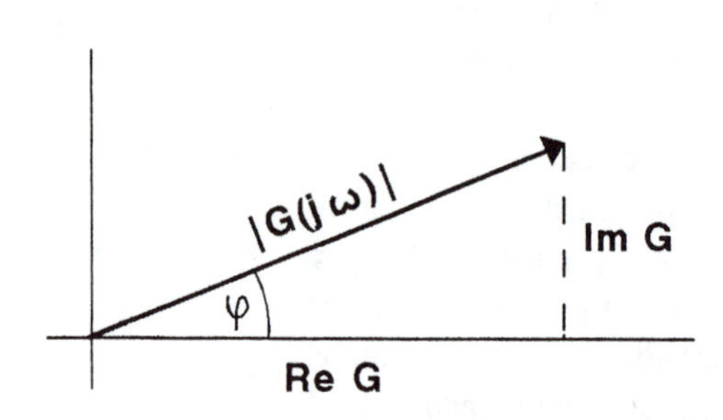

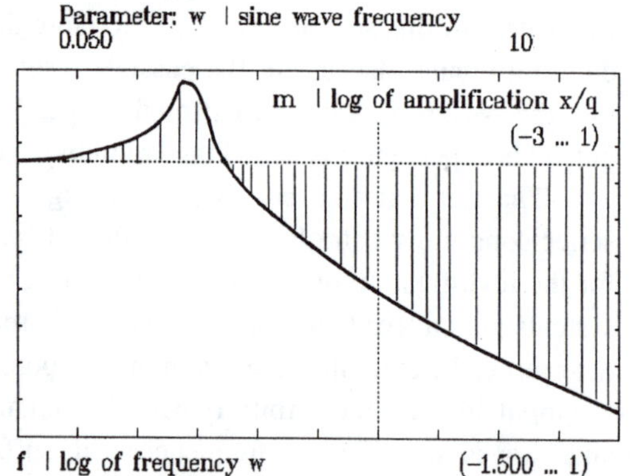

Fig. 7.7 (left): Magnitude (gain) and phase angle (phase) of the complex frequency transfer function $G(j\omega)$.

Fig. 7.8 (right): Frequency response of a linear oscillator of 2nd order (Bode plot): Plot of logarithm of amplitude of state variable x as function of the logarithm of the frequency of the forcing oscillation $u = \sin(2\pi\omega t)$. (Model M203 with $a = 0$, $b = 1$, $c = -1$, $d = -0.1$.)

The **frequency response** $G(j\omega)$ characterizes the behavior of a linear system in response to the frequency of the harmonic input signal. The plot of the magnitude of the frequency response $|G(\omega)|$ as function of frequency is called **amplitude response** while the plot of the phase angle $\varphi(\omega)$ as function of frequency is termed **phase response**. The amplitude of the output signal is obtained by multiplying input amplitude by amplitude response; its phase shift follows from the phase angle.

The analysis shows that excitation by an harmonic oscillation causes the system output to follow this oscillation with a frequency-dependent phase shift φ. The amplitude of the output oscillation $|G|\ |U|$ also depends on the input frequency ω. Amplitude ratio and phase shift are determined by the coefficients of the differential equation or more generally by the elements of the system matrix **A**. As an example, the analysis of the frequency reponse of a second order oscillator using Model M 203 of the systems zoo in Figure 7.8 clearly shows resonance near the characteristic frequency ω_0 of the system. The smaller the damping, the stronger the resonance. For vanishing damping ($\xi = 0$), excitation at the resonance frequency results in infinitely large resonance amplitude.

A linear oscillator of second order shows the following behavior with respect to the phase angle: at low frequencies, the system can still follow the input oscillation, and the phase angle therefore lags the input oscillation by a small amount only. As the frequency of excitation increases, the phase angle also increases (more so for stronger damping) until the phase lag reaches exactly 90 degrees at the point of resonance (where excitation frequency = characteristic frequency). As the excitation frequency increases further, the phase lag increases further and finally reaches 180 degrees at very high frequencies.

In order to find the system response to an arbitrary periodic input signal (which can be represented by a sum of sine waves of different frequencies using Fourier analysis), the frequency response must be known for the whole range of frequencies. In practical applications the experimental determination of the frequency response is of particular significance. Often, the systems to be investigated are extremely complex or difficult to describe in mathematical terms. The method employs a signal generator which excites the system at different input frequencies. The resulting complex output signal, i.e. amplitude and phase, is measured and related to the input signal. Using these measurements, diagrams similar to Figure 7.8 are drawn (Bode plots, see below). Even for complex systems, these frequency response curves can often be approximated by the response curves of much simpler linear systems whose parameters can be deduced from the plots. Extensive analytical investigations to better understand the behavior of the real system can then be made using the simpler model system, and a corresponding control system can be developed to correct unstable or undesirable system behavior.

7.4.7 Representation of frequency response

The **frequency response characteristic** (Bode plot) is particularly well-suited for plotting the results of experimental frequency response measurements, and for determining the type and parameters of the system equation of the real system or a simpler model system approximating its behavior. The Bode plot uses logarithmic scales for both amplitude and frequency (cf. Fig 7.8). In this plot, the frequency response characteristic can be approximated by a succession of straight lines with break points at the characteristic frequencies of the system where the phase angle also changes relatively quickly. These break frequències and the slope of the straight lines between the break points correspond directly to the characteristic equation of the system which can therefore be reconstructed from the measured response. The procedure is described in detail in the control systems literature.

The **frequency response locus** (Nyquist diagram) is a plot of the complex frequency transfer function $G(j\omega)$ in the complex plane. In this diagram, the amplitude response (gain) for a given frequency is represented by the length of the radius vector from the coordinate origin while the angle of this vector with the positive real axis is equal to the corresponding phase angle φ. The locus of gain and phase as a function of increasing frequency traces a curve in the complex plane which begins at location (1, 0) for frequency 0 (constant signal). It ends at the origin (0, 0) as the frequency becomes infinite since the system can follow frequencies above its highest characteristic frequency only with vanishingly small amplitude (cf. Fig. 7.8). The Nyquist plot is of particular relevance for stability studies. The procedure is described in detail in the control systems literature.

7.5 Behavior and Stability of Non-linear Systems

7.5.1 Stability of non-linear systems

The stability of non-linear systems differs greatly from that of linear systems:

1. The stability of a non-linear system generally depends on the input $\mathbf{u}$. For example, the unforced (autonomous) system may be stable but may become unstable under certain input conditions (e.g. step function). On the other hand, it may be possible to stabilize an unstable non-linear system by a proper input function.
2. The stability of an unforced non-linear system generally depends on the initial state $\mathbf{z}_0$. This is a result of the fact that a non-linear system often has several equlibrium points, each of which may show different stability behav-

ior. The location of the initial state with respect to the equilibrium points then determines the system's trajectory into, or away from, an equilibrium point, i.e. stability or instability.

For a non-linear system, the stability definition has to be more restricted since it will no longer be valid for the entire state space, as for linear systems. The stability properties can only be used to describe behavior in the vicinity of a given equilibrium point. The system state may remain in the neighborhood of the equilibrium point but it may also increasingly move away from it. Behavior is called stable if the state remains in the vicinity of the equilibrium point or moves towards it as time increases. Behavior is unstable if the state moves away from the equilibrium point as time progresses.

For non-linear systems, stability therefore always refers to behavior in the neighborhood of a given equilibrium point. For the purpose of stability analysis it is then permissible to replace the full non-linear system description by a simpler approximation valid in this neighborhood. Often a linear approximation is sufficient; it is obtained by linearization of the non-linear system near the equilibrium point (c.f. Sec. 7.1.11 – 7.1.14).

7.5.2 Attractors of non-linear systems

While linear systems (with non-singular matrix **A**) exhibit identical behavior and stability characteristics in all of state space and can only have a single equilibrium point, additional complex phenomena can be found for non-linear systems. Furthermore, their behavior and stability depend on system input and initial state.

Equilibrium points

A non-linear system may have **several equilibrium points**. In the immediate neighborhood of such a point, behavior and stability of the non-linear system will generally not be different from that of its linear approximation. For topological reasons, the existence of several equilibrium points may imply lines or surfaces of separation (separatrix) on which the singular points are located and which divide the state space into regions having different behavioral characteristics. If the system state is originally on one side of the surface of separation, it cannot possibly move to the other side in free (unforced) motion.

For example, Model M 213 TRAGEDY OF THE COMMONS may have up to five equilibrium points of which only two are stable. Model M 220 BISTABLE OSCILLATOR has surfaces of separation which wind around the three equilibrium points in spirals.

Limit cycles

Occasionally, non-linear systems may have limit cycles. Limit cycles are closed state trajectories which can never be crossed in free motion and which therefore divide the state space into different behavioral regions.

An example is the van der Pol oscillator (Model M 219 of the systems zoo)

$$z'_1 = z_2$$
$$z'_2 = (1 - z_1^2)\, z_2 - z_1.$$

Here, the limit cycle is caused by the non-linear damping function $(1 - z_1^2)$. For small absolute values of z_1, it causes positive feedback (i.e. amplification); for large values it causes negative feedback (i.e. damping). This drives the system into an intermediate stable periodic state, the limit cycle. The inner and outer state trajectories all move towards and onto the limit cycle. Limit cycles may be stable, unstable, or semistable. (The van der Pol cycle is stable.) Model M 210 OVERSHOOT AND COLLAPSE also has a limit cycle in a narrow parameter range.

A criterion for limit cycles can be derived from the damping function $(1 - z_1^2)$ of the van der Pol oscillator: a limit cycle can be expected if the function is symmetric and causes amplification for some state values and damping for others.

Tori

In analogy to the limit cycles of two-dimensional systems, systems having more than two state variables may have surfaces which "capture" the motion of the system. These attractors are called **torus**. For example, in a given three-dimensional system the motion may be captured by a ring. This surface corresponds to two independent stable motions whose frequencies correspond to the motion on the small radius and on the large radius of the ring. Similar higher-dimensional attractors may be found in systems of higher order. They are the result of superposition of two or more oscillations of the limit cycle type.

Chaotic attractors

For most real systems, neighboring state trajectories remain close to each other as time progresses. These systems are therefore predictable. The future development can be determined from the initial conditions, and this development is not very sensitive to (measurement) errors in the initial conditions.

However, there is another class of systems that have surfaces of attraction on which the system state will be located after some time (of free motion) while exact prediction of the location is not possible. This is caused by exponential divergence of state trajectories on the attractor surface. Since the trajectories cannot leave this surface, it must allow a "folding back" of the trajectories. Two trajectories with almost identical initial conditions may be on very different paths after a short time.

Systems with these **strange** or **chaotic attractors** no longer have any predictability: the final state could be anywhere on the attractor. There is therefore no longer any direct relationship between the past (initial condition) and the future, even though these systems are completely deterministic in their formulation.

The constant stretching and folding of the state trajectories of a chaotic attractor can be compared to the rolling out and folding of a chunk of dough: here a spoonful of flour ("initial condition") is distributed all over the dough (the "chaotic attractor") after several foldings and stretchings with the rolling pin. The final location of a grain of flour does not say anything about its initial location.

The existence of chaos in dynamic systems means that there are fundamental limits to computability if the system has a chaotic attractor. However, this does not imply complete arbitrariness of future system states since these have to be located on the attractor. It is then the task of systems analysis to find these surfaces.

Examples for chaotic systems in the systems zoo are: Model M 221 CHAOTIC BISTABLE OSCILLATOR, M 309 CHAOTIC ATTRACTOR (RÖSSLER), M 310 HEAT, WEATHER AND CHAOS (LORENZ SYSTEM), and M 311 COUPLED DYNAMOS AND CHAOS. The folding and mixing of the state trajectories can be studied especially well in the Rössler attractor.

7.5.3 Structural dynamics of systems

Often systems are encountered in practice that switch to a different mode upon reaching certain system states, thereby qualitatively changing their behavior. This switching may come about by changing of parameters, interruption or activation of certain structural connections, or the connection or disconnection of whole subsystems. Plants and animals are equipped with such mechanisms in order to cope with stress situations (e.g. water stress or predation).

From a systems theoretical point of view, this switching means a change of the state equations: as a function of system state, one set of state equations is replaced by another. We refer to this as structural change or structural dynamics. In simulations, the switching process is easily implemented by inclusion of proper logical conditions or of functions allowing stepwise changes.

In the behavior and stability analysis of these systems, one has to keep in mind the fact that the change of state equations also implies a change of equilibrium points and of the corresponding stability conditions.

Switching properties, and changes of structure and behavior, can be found, for example, in Models M 210 OVERSHOOT AND COLLAPSE, M 211 FOREST GROWTH, M 213 TRAGEDY OF THE COMMONS, M 214 SUSTAINABLE USE OF A RENEWABLE RESOURCE, M 304 MINIWORLD.

7.5.4 Comparison of linear and non-linear dynamic systems

Linear systems	Non-linear systems
The amplitude of the system response is exactly proportional to the amplitude of the input signal (or the initial condition). The character (shape and development) is independent of the amplitude of the input signal and initial conditions.	A change in the amplitude of an input signal (or of an initial condition) will in general not result in a proportional change of the system response. The character of the system response (shape and development) may change drastically with initial conditions and amplitude of input signal.
The superposition principle holds: the system response to complex periodic or aperiodic inputs can be computed as the sum of responses to individual elementary functions (i.e. sine, pulse).	The superposition principle does not apply. The system response cannot be computed as the sum of system responses to individual input signals.
The equilibrium state is independent of the initial conditions.	The equilibrium state depends on the initial conditions.
A single attractor (equilibrium point) of the autonomous system at $\mathbf{z} = \mathbf{0}$.	Several attractors of different types are possible: equilibrium points, limit cycles, tori, and chaotic attractors.
No qualitative change of behavior over the entire state space.	Qualitative change of behavior possible (bifurcation, structural dynamics).
Stability is a system property which is independent of sign and absolute value of system inputs or initial conditions.	Stability may depend on sign and absolute value of system inputs and/or initial conditions.
Quasi-stationary output of the forced system (after decay of the influence of initial conditions) contains only frequencies identical to input frequencies.	Quasi-stationary output may contain higher harmonic and subharmonic frequencies in response to a given input frequency.

Resonance may appear at certain input frequencies. Continuous changes of input frequency result in continuous changes of gain (amplitude ratio) and phase (phase shift).	Sudden changes of amplitude and phase are possible near resonance. Behavior depends on the magnitude of input oscillations.
	Limit cycle is possible: independently of inputs or initial conditions, a stable oscillation of constant amplitude and frequency develops even in the absence of oscillating inputs.

7.5.5 Perturbation dynamics of the influence structure

The influence diagram shows how effects (impacts, influences) are passed on in the (model) system from one element to the next. It is therefore tempting to use this system sketch to deduce information about possible system behavior. This can be done by qualitative assessments, numerical investigations, logical deduction, or mathematical analysis of the influence graph. For example, negative feedback loops in the influence diagram suggest damped, stable motion; positive feedbacks suggest amplification of initial disturbances and instability.

It may also be possible to compute the propagation of small perturbations (pulses) through the system, even if the system elements are only vaguely known, and to identify from numerical studies of perturbation dynamics critical system parameters and critical structural connections.

The influence diagram does not distinguish between different types of system elements (state variables, rates, intermediate variables). All are treated as identical holding elements which store information and pass it on to the neighboring nodes at the next instant. The influence diagram then reduces to a linear system whose dynamics and stability can be studied using well-known methods of linear systems analysis (cf. Sec. 7.3 – 7.4). In this investigation, the eigenvalues (characteristic values) of the system matrix play an important role; they lead to information concerning system behavior and the stability of the influence graph.

For the (rare) case where the real system can be fully and validly approximated by a set of linear difference or differential equations, the influence graph represents an adequate simulation diagram of the system. In this case, it is possible to go beyond qualitative analysis and to derive quantitatively valid conclusions about system behavior. In particular, this is the case if (local) linearization of a system is possible at an equilibrium point or near a reference trajectory (cf.

Sec. 7.1.11 ff.). In all other cases, attempts to use the influence diagram of nonlinear systems to derive valid information about system behavior and stability are bound to fail. The conclusions are usually totally misleading. There is no alternative to correctly representing elements and relationships with their idiosyncracies—and this means building a full dynamic simulation model.

The influence diagram describes the influence structure of a system without differentiating the elements. For the purpose of determining how perturbations propagate in the influence structure, we can either assume a discrete formulation where the discrete state of a node is a function of the discrete states of neighboring nodes; or a continuous formulation where the rate of change at a node depends on the states of neighboring nodes. In the discrete case, the nodes are represented by holding elements (unit delays) that hold information for one time interval Δt. In the continuous case, the nodes are represented by integrators. In both cases, only additive combinations of the inputs at the nodes are permitted. Of the possible formulations, three are particularly relevant for investigating perturbation dynamics.

In the **discrete process**, states $\mathbf{x}_k$ at the nodes change at discrete intervals as a function of states at neighboring nodes where the system matrix $\mathbf{A}$ describes the coefficients of the linear transformation:

$$\mathbf{x}_{k+1} = \mathbf{A}\,\mathbf{x}_k \,.$$

The process is (state) stable if all eigenvalues of $\mathbf{A}$ have absolute value less than 1:

$$|\lambda_i| < 1 \,.$$

In the **pulse process**, pulses $\mathbf{p}_k = (\mathbf{x}_{k+1} - \mathbf{x}_k)$ at the nodes change at discrete intervals as a function of the pulses at neighboring nodes. The system matrix $\mathbf{A}$ now describes this linear transformation of pulses:

$$\mathbf{p}_{k+1} = \mathbf{A}\,\mathbf{p}_k \,.$$

The process is (pulse)stable if all eigenvalues of $\mathbf{A}$ have absolute value equal or less than 1:

$$|\lambda_i| \leq 1 \,.$$

Note that a system may be pulse-stable but state-unstable (for example, if the pulse remains at a constant value, the state will grow indefinitely).

In the **continuous process**, the state $\mathbf{x}$ changes continuously at the rate of change $d\mathbf{x}/dt$ which follows as a linear state transformation with system matrix $\mathbf{A}$:

$$d\mathbf{x}/dt = \mathbf{A}\,\mathbf{x} \,.$$

This process is stable if the real parts of all eigenvalues of $\mathbf{A}$ are negative:

$$\mathrm{Re}(\lambda_i) < 0 \,.$$

Marginal stability is obtained for $\mathrm{Re}(\lambda_i) = 0$.

Care must always be used in the application of linear models (as, for example, influence graphs) since these are not able to describe behaviorally relevant (non-linear) system characteristics. The influence graph is therefore only an initial step in the model building process which requires as the next essential step the determination of the specific properties of the system elements and their functional connections and their linkage in the simulation model.

For small perturbations ($x \ll 1$, $y \ll 1$) from reference state ($x^* = 1$, $y^* = 1$) the following linear approximations hold:

$$
\begin{aligned}
(1 \pm x)^n &\approx 1 \pm nx \\
(1 \pm x)^2 &\approx 1 \pm 2x \\
(1 \pm x)^{0.5} &\approx 1 \pm x/2 \\
1/(1+x) &\approx 1 - x \\
e^x &\approx 1 + x \\
\ln(1+x) &\approx x \\
\cos x &\approx 1 \\
\sin x &\approx x \\
(1+x)(1+y) &\approx 1 + x + y, \text{ etc.}
\end{aligned}
$$

7.6 Summary of Important Results

Where applicable, mathematical systems theory can produce much more general conclusions about a given system than simulations. This chapter presents a collection of the most important mathematical concepts. The point of departure of mathematical analysis are the system equations developed in the modeling process, i.e. the state equations for the state variables $\mathbf{z}$ and the behavior equations for the output variables $\mathbf{v}$ as functions of input variables $\mathbf{u}$ and state variables $\mathbf{z}$.

The most important results are summarized in the following compilation; it is restricted to time-invariant systems.

1. The general form of the **system equations** is for continuous systems

$$
\begin{aligned}
d\mathbf{z}/dt &= \mathbf{f}(\mathbf{z}(t), \mathbf{u}(t)) \\
\mathbf{v}(t) &= \mathbf{g}(\mathbf{z}(t), \mathbf{u}(t))
\end{aligned}
$$

and for discrete systems

$$
\begin{aligned}
\mathbf{z}(k+1) &= \mathbf{f}(\mathbf{z}(k), \mathbf{u}(k)) \\
\mathbf{v}(k) &= \mathbf{g}(\mathbf{z}(k), \mathbf{u}(k)) .
\end{aligned}
$$

2. The **equilibrium conditions** at the equilibrium state $\mathbf{z^*}$ (for constant input $\mathbf{u^*}$) are for the continuous system

   $$\mathbf{0} = \mathbf{f}(\mathbf{z^*}, \mathbf{u^*})$$

   and for the discrete system

   $$\mathbf{z^*} = \mathbf{f}(\mathbf{z^*}, \mathbf{u^*}) .$$

3. The **system equations of linear systems** are for the continuous system:

   $$d\mathbf{z}/dt = \mathbf{A}\,\mathbf{z} + \mathbf{B}\,\mathbf{u}$$
   $$\mathbf{v} \quad = \mathbf{C}\,\mathbf{z} + \mathbf{D}\,\mathbf{u}$$

   and for the discrete system

   $$\mathbf{z}(k+1) = \mathbf{A}\,\mathbf{z}(k) + \mathbf{B}\,\mathbf{u}(k)$$
   $$\mathbf{v}(k) \quad = \mathbf{C}\,\mathbf{z}(k) + \mathbf{D}\,\mathbf{u}(k) .$$

4. **Linearization:** In general, non-linear systems can be linearized in the neighborhood of their equilibrium points $\mathbf{z^*}$. For $\mathbf{u} = \mathbf{0}$ and the definition

   $$\Delta\mathbf{z} = \mathbf{z} - \mathbf{z^*}$$

   one obtains linear state equations for the perturbation state $\Delta\mathbf{z}$. For the continuous system

   $$\Delta\mathbf{z}' = d(\Delta\mathbf{z})/dt = \mathbf{A}\,\Delta\mathbf{z}$$

   and for the discrete system

   $$\Delta\mathbf{z}(k+1) = \mathbf{A}\,\Delta\mathbf{z}(k) .$$

   Here the system matrix $\mathbf{A}$ is the **Jacobi matrix** which has to be evaluated at the equilibrium point $\mathbf{z^*}$.

   $$\mathbf{A} = \mathbf{J} = \begin{bmatrix} \frac{\partial f_1}{\partial z_1} & \cdots & \frac{\partial f_1}{\partial z_n} \\ \cdots & \cdots & \cdots \\ \frac{\partial f_n}{\partial z_1} & \cdots & \frac{\partial f_n}{\partial z_n} \end{bmatrix}_{\mathbf{z^*}}$$

5. The **superposition principle** applies to linear systems: the total system response is composed of the sum of individual responses to initial conditions and to the individual components of the input signal.

6. **Analytical solutions** can be developed for (forced and unforced) linear dynamic systems. For the continuous system one obtains

$$\mathbf{z}(t) = e^{\mathbf{A}t}\,\mathbf{z}_0 + e^{\mathbf{A}t}\cdot\int_0^t e^{-\mathbf{A}\tau}\,\mathbf{B}\,\mathbf{u}(\tau)\,d\tau$$

and for the discrete system

$$\mathbf{z}(k) = \mathbf{A}^k\,\mathbf{z}(0) + \sum_{j=0}^{k-1}\mathbf{A}^{k-j-1}\,\mathbf{B}\,\mathbf{u}(j)\,.$$

The system matrix $\mathbf{A}$ determines the properties and characteristic dynamics of the system in forced and unforced motion.

7. The n **eigenvalues** λ_i of the (n·n) system matrix $\mathbf{A}$ follow from the characteristic equation

   $\det\,[\mathbf{A} - \lambda\mathbf{I}] = 0\,.$

8. The n **eigenvectors** $\mathbf{e}_i$ of the system matrix $\mathbf{A}$ are determined from the eigenvector equation written for each of the n eigenvalues λ_i :

   $[\mathbf{A} - \lambda_i\mathbf{I}]\,\mathbf{e} = \mathbf{0}\,.$

   The eigenvector matrix $\mathbf{M}$ (modal matrix) consists of the n eigenvectors (column vectors) $\mathbf{e}_i$.

9. The system matrix $\mathbf{A}$ can be transformed into the diagonal **eigenvalue matrix** $\mathbf{\Lambda}$ with the help of the modal matrix $\mathbf{M}$:

   $\mathbf{\Lambda} = \mathbf{M}^{-1}\,\mathbf{A}\,\mathbf{M}$
   $\mathbf{A} = \mathbf{M}\,\mathbf{\Lambda}\,\mathbf{M}^{-1}\,.$

   If the eigenvalues are distinct, they appear on the main diagonal of $\mathbf{\Lambda}$; all other coefficients are equal to zero.

10. Replacing $\mathbf{A}$ by $\mathbf{M}\,\mathbf{\Lambda}\,\mathbf{M}^{-1}$ in the solutions of linear systems (Point 6) leads to **stability conditions** for linear systems: a continuous system is stable if the real parts of all eigenvalues of $\mathbf{A}$ are negative. A discrete system is stable if the absolute values of all eigenvalues of $\mathbf{A}$ are less than 1.

11. Without having to determine the eigenvalues, the **Routh criterion** allows finding the number of unstable eigenvalues of the system and determining the **conditions for instability** as a function of system parameters. The Routh criterion for continuous systems can also be applied to discrete systems after applying the transformation $\lambda = (z+1)/(z-1)$.

12. **Different system formulations** of same dimension n (i.e. different $\mathbf{A}$) may have the same eigenvalues and therefore **identical behavior**: the choice of state variables is not unique.

13. The **general form** of the system matrix (with n^2 possible structural connections) can be transformed by using the coefficients of the characteristic polynomial into the much simpler **standard form** (with (2n-1) possible connections). If the **normal form** is used, the state variables are completely **decoupled** (for discrete eigenvalues), and the eigenvalues appear as feedback parameters of the self-loops (n possible connections). Using the modal matrices, the state variables of a given system can be transformed into the state variables of equivalent systems.

14. If excited by a periodic function, the response of a linear system depends on the frequency of the input signal (frequency response). Resonance may occur. From the **frequency response** one can draw conclusions about the structure (state equation) of the system.

15. Non-linear systems differ fundamentally from linear systems. In particular, they may have several equilibrium points, or other regions of attraction. **Linearization** can therefore only be applied **locally**.

16. **Influence diagrams** are derived using the implicit assumption of **small perturbations** from a reference state. Since there is no differentiation between system elements—all are of the same type—and only additive connections are permitted, influence diagrams are linear representations of systems.

17. The information contained in **influence diagrams** of the form "system element z_i has an effect on system element z_j" can be interpreted in different ways, all of which lead to **linear system formulations** that can be used for perturbation analysis:

    a. "The state of z_j follows directly from the state of z_i." This can be formulated as a discrete system

    $\mathbf{z}(k+1) = \mathbf{A}\, \mathbf{z}(k)$

    where $\mathbf{A}$ is the system matrix of the influence diagram.

    b. "The change of state of z_j (pulse at z_j) is a direct result of the change of state at z_i (pulse at z_i)." This formulation is formally equivalent to (a):

    $\mathbf{p}(k+1) = \mathbf{A}\, \mathbf{p}(k)$.

    A pulse is defined by $\mathbf{p}(k+1) = \mathbf{z}(k+1) - \mathbf{z}(k)$. The study of the pulse dynamics of influence diagrams is based on this formulation.

    c. "The rate of change of state z_j follows directly from the state z_i." This formulation leads to a system of linear differential equations of first order:

    $dz/dt = \mathbf{A}\, \mathbf{z}$.

BIBLIOGRAPHY

Aris, R. 1978: *Mathematical Modelling Techniques.* Pitman, London / San Francisco.

Ashby, W. R. 1956: *An Introduction to Cybernetics.* John Wiley, New York.

Banks, S. P. 1986: *Control Systems Engineering.* Prenticc Hall, Englewood Cliffs NJ.

Beltrami, E. 1987: *Mathematics for Dynamic Modeling.* Academic Press Orlando FL / London.

Bender, E. A. 1978: *An Introduction to Mathematical Modeling.* John Wiley, New York.

Bennett, R. J., Chorley, R. J. 1978: *Environmental Systems: Philosophy, Analysis and Control.* Methuen, London.

Berg, E., Kuhlmann, F. 1993: *Systemanalyse und Simulation für Agrarwissenschaftler und Biologen - Methoden und PASCAL-Programme zur Modellierung dynamischer Systeme.* Verlag Eugen Ulmer, Stuttgart.

Bertalanffy, L. von 1969: *General System Theory: Foundations, Development, Applications.* G. Braziller Publishing, New York.

Bocklisch, St. F. 1987: *Prozeßanalyse mit unscharfen Verfahren.* VEB Verlag Technik, Berlin.

Bossel, H. 1978: *Bürgerinitiativen entwerfen die Zukunft - Neue Leitbilder, neue Werte, 30 Szenarien.* Fischer, Frankfurt/M.

Bossel, H. 1985: *Umweltdynamik.* TeWi Verlag, München.

Bossel, H. 1986: *Dynamics of forest dieback - systems analysis and simulation.* Ecological Modelling 34 (259-288).

Bossel, H. 1987/1989/1992: *Simulation dynamischer Systeme - Grundwissen, Methoden, Programme.* Vieweg Braunschweig/Wiesbaden.

Bossel, H. 1990: *Umweltwissen - Daten, Fakten, Zusammenhänge.* Springer Berlin/Heidelberg/New York.

Bossel, H. 1992: *Modellbildung und Simulation - Konzepte, Verfahren, Modelle und Simulationsprogramme.* Vieweg Verlag, Braunschweig/Wiesbaden. 2nd ed. 1994.

Bossel, H. 1994: *Modeling and Simulation.* A K Peters, Wellesley MA.

Bossel, H., Bruenig, E.F. 1991: *Natural Resource Systems Analysis.* Deutsche Stiftung für Internationale Entwicklung (DSE), Feldafing.

Bossel, H., Hornung, B. R., Müller-Reißmann, K.-F. 1989: *Wissensdynamik mit DEDUC.* Vieweg Braunschweig/Wiesbaden.

Bossel, H., Strobel, M. 1978: *Experiments with an 'intelligent' world model.* Futures vol. 10, no. 3, June (191-212).

Casti, J.L. 1979: *Connectivity, Complexity, and Catastrophe in Large-Scale Systems.* Wiley, Chichester.

Cellier, F. E. 1991: *Continuous System Modeling.* Springer Verlag New York / Berlin / Heidelberg.

Clocksin, W.F., Mellish, C.S. 1984: *Programming in Prolog.* Springer Verlag, Berlin.

Close, C. M., Frederick, D. K. 1978: *Modeling and Analysis of Dynamic Systems.* Houghton Mifflin, Boston MA.

Csaki, F. 1972: *Modern Control Theories - Nonlinear, Optimal and Adaptive Systems.* Akademiai Kiado, Budapest.

Csaki, F. 1973: *Die Zustandsraummethode in der Regelungstechnik.* Akademiai Kiado, Budapest.

DeRusso, P.M., Roy, R. J., Close, C. M. 1965: *State Variables for Engineers.* Wiley, New York NY.

DiStefano, J. J., Stubberud, A. R., Williams, I. J. 1967: *Feedback and Control Systems.* Schaum, New York NY.

Dorf, R. C. 1989: *Modern Control Systems,* 5th ed. Addison-Wesley, Reading, Mass.

Ester, J. 1987: *Systemanalyse und mehrkriterielle Entscheidung.* VEB Verlag Technik, Berlin.

Fishwick, P. A., Luker, P. A. (eds) 1991: *Qualitative Simulation, Modeling, and Analysis.* Springer Verlag, New York / Berlin / Heidelberg.

Flood, R.L., Carson, E.R. 1988: *Dealing With Complexity - An Introduction to the Theory and Practice of Systems Science.* Plenum, New York.

Forrester, J. W. 1961: *Industrial Dynamics.* MIT Press, Cambridge MA.

Forrester, J. W. 1968: *Principles of Systems.* Wright-Allen Press, Cambridge MA.

Forrester, J. W. 1970: *World Dynamics.* MIT Press, Cambridge MA. (Productivity Press, Cambridge MA).

France, J., Thornley, J. H. M. 1984: *Mathematical Models in Agriculture.* Butterworths, London.

Frost, R.A. 1986: *Introduction to Knowledge Base Systems.* Collins, London.

Gleick, J. 1987: *Chaos - Making a New Science.* Viking, New York.

Glisson, T.H. 1985: *Introduction to System Analysis.* McGraw-Hill New York.

Göldner, K. 1981 (Bd.1), 1983 (Bd.2), mit S. Kubik 1983 (Bd.3): *Mathematische Grundlagen der Systemtheorie.* Fachbuchverlag Leipzig und Verlag Harri Deutsch, Thun und Frankfurt/M.

Goodman, M. R. 1974: *Study Notes in System Dynamics.* Wright-Allen Press, Cambridge MA.

Gopal, M. 1987: *Modern Control System Theory.* John Wiley, Singapore/New York.

Gottwald, S. 1993: *Fuzzy Sets and Fuzzy Logic - The Foundations of Application, from a Mathematical Point of View.* Vieweg, Wiesbaden, and Teknea, Toulouse.

Guckenheimer, J., Holmes, P. 1983/1986: *Nonlinear Oscillations, Dynamical Systems, and Bifurcations of Vector Fields.* Springer New York / Berlin / Heidelberg / Tokyo.

Hardin, G. 1968: *The tragedy of the commons.* Science, vol. 162 (1243-1248).

Holland, J.H. 1975: *Adaptation in Natural and Artificial Systems.* University of Michigan, Ann Arbor.

Hudetz, W. 1977: *Construction of dynamic system models using interactive computer graphics.* In: H. Bossel (ed.): *Concepts and Tools of Computer-assisted Policy Analysis.* Birkhäuser Basel (266-291).

Ivachnenko, A. G., Müller, J.-A. 1984: *Selbstorganisation von Vorhersagemodellen.* VEB Verlag Technik, Berlin.

Jetschke, G. 1989: *Mathematik der Selbstorganisation.* Deutscher Verlag der Wissenschaften, Berlin.

Jørgenson, S.E. 1992: *Exergy and ecology.* Ecol. Modelling 63 (185-214).

Kheir, N. A. (ed) 1988: *Systems Modeling and Computer Simulation.* Marcel Dekker, New York.

Kirkpatrick, S. et al. 1983: *Optimization by Simulated Annealing.* Science 220, 671-680.

Kosko, B. 1992: *Neural Networks and Fuzzy Systems.* Prentice Hall International, Englewood Cliffs N.J.

Kosmol, P. 1991: *Optimierung und Approximation.* Walter de Gruyter, Berlin and New York.

Kreß, D. 1987: *Angewandte Systemtheorie - Signale und lineare Systeme.* S. 80-155 in Philippow 1987.

Law, A. M., Kelton, W. D. 1990: *Simulation Modeling and Analysis,* 2nd ed. McGraw-Hill, New York.

Lotka, H.J. 1925: *Elements of Physical Biology.* Williams and Wilkins, Baltimore.

Luenberger, D. G. 1979: *Introduction to Dynamic Systems - Theory, Models, and Applications.* John Wiley, New York.

May, R.M. (ed.) 1976: *Theoretical Ecology: Principles and Applications.* Blackwell, Oxford.

Meadows, D.L. 1970: *Dynamics of Commodity Production Cycles.* Wright-Allen Press (Productivity Press), Cambridge MA.

Meadows, D. H., Meadows, D. L., Randers, J. 1992: *Beyond the Limits.* Chelsea Green, Post Mills VT.

Meadows, D. H., Meadows, D. L., Randers, J., Behrens, W. W. III 1972: *The Limits to Growth.* Potomac Associates, Washington.

Meadows, D.L., Meadows, D.H. 1973: *Towards Global Equilibrium.* Wright-Allen Press (Productivity Press), Cambridge MA.

Meadows, D.L. et al. 1974: *The Dynamics of Growth in a Finite World.* Wright-Allen Press (Productivity Press), Cambridge MA.

Meadows, D., Richardson, J., Bruckmann, G. 1982: *Groping in the Dark - The First Decade of Global Modelling.* Wiley, Chichester 1982.

Metzler, W. 1987: *Dynamische Systeme in der Ökologie.* Teubner, Stuttgart.

Negoita, C.V. 1985: *Expert Systems and Fuzzy Systems.* Benjamin/Cummings, Menlo Park, CA.

Nicolis, G., Prigogine, I. 1989: *Exploring Complexity.* W. H. Freeman, New York.

Page, B. 1991: *Diskrete Simulation.* Springer Verlag, Berlin.

Palm III, W. J. 1983: *Modeling, Analysis, and Control of Dynamic Systems.* John Wiley, New York.

Penning de Vries, F. W. T., Jansen, D. M., ten Berge, H. F. M., Bakema, A. 1989: *Simulation of Ecophysiological Processes of Growth in Several Annual Crops.* Pudoc, Wageningen.

Penning de Vries, F. W. T., van Laar, H. H. (eds) 1982: *Simulation of Plant Growth and Crop Production.* Pudoc, Wageningen.

Peschel, M. 1992: *Rechnergestützte Analyse regelungstechnischer Systeme.* Akademie Verlag, Berlin.

Peschel, M., Riedel, C. 1976: *Polyoptimierung - eine Entscheidungshilfe für ingenieurtechnische Kompromißlösungen.* Verlag Technik, Berlin.

Philippow, E. (Hg.) 1987: *Taschenbuch Elektrotechnik, Band 2: Grundlagen der Informationstechnik,* Verlag Technik, Berlin.

Pichler, F. 1975: *Mathematische Systemtheorie.* Walter de Gruyter, Berlin.

Piefke, F. 1991: *Simulation mit dem Personalcomputer.* Hüthig Buch Verlag Heidelberg.

Press, W. H., Flannery, B. P., Teukolsky, S. A., Vetterling, W. T. 1986: *Numerical Recipes - The Art of Scientific Computing.* Cambridge University Press, Cambridge.

Puccia, C. J., Levins, R. 1985: *Qualitative Modeling of Complex Systems.* Harvard University Press, Cambridge MA / London.

Puppe, F. 1991: *Einführung in Expertensysteme.* Springer Verlag, Berlin, 2nd ed.

Reinisch, K. 1974: *Kybernetische Grundlagen und Beschreibung kontinuierlicher Systeme.* Verlag Technik, Berlin.

Rheingold, H. 1992: *Virtual Reality.* Mandarin, London.

Richmond, B., Peterson, S., 1992: *An Introduction to Systems Thinking.* High Performance Systems, Hanover NH.

Richter, C. 1988: *Optimierungsverfahren und BASIC-Programme.* Akademie Verlag, Berlin 1988.

Richter, O. 1985: *Simulation des Verhaltens ökologischer Systeme* - Mathematische Methoden und Modelle. VCH Weinheim.

Richter, O., Söndgerath, D. 1990: *Parameter Estimation in Ecology.* VCH Weinheim.

Roberts, N. 1983: *Introduction to Computer Simulation - A System Dynamic Modeling Approach.* Addison-Wesley, Reading MA 1983.

Rössler, O. E. 1976: *Different types of chaos in two simple differential equations.* Z. Naturf. 31a (1664-1670).

Schmidt, G. 1982: *Grundlagen der Regelungstechnik.* Springer, Berlin.

Schwefel, H. P. 1977: *Numerische Optimierung von Computer-Modellen mittels der Evolutionsstrategie.* Birkhäuser, Basel.

Schwefel, H.P. 1981: *Numerical Optimization of Computer Models.* Wiley, Chichester.

Smith, J. M. 1987: *Mathematical Modeling and Digital Simulation for Engineers and Scientists.* John Wiley, New York.

Spriet, J. A., Vansteenkiste, G. C. 1982: *Computer-Aided Modeling and Simulation.* Academic Press, London.

STELLA II: *High Performance Systems,* 45 Lyme Road, Hanover NH.

Steuer, R.E. 1986: *Multiple Criteria Optimization: Theory, Computation, and Application.* Wiley, New York.

Stewart, I. 1989: *Does God Play Dice? - The Mathematics of Chaos.* Blackwell, Cambridge MA.

Stöcker, H. 1993: *Taschenbuch mathematischer Formeln und moderner Verfahren.* Verlag Harri Deutsch, Frankfurt/M, 2nd ed.

Thompson, J. M. T., Stewart, H. B. 1986: *Nonlinear Dynamics and Chaos.* John Wiley, Chichester / New York.

Unbehauen R., 1983 (4.Aufl): *Systemtheorie.* Oldenbourg, München.

Volterra, V. 1926: *Variazioni e fluttuazioni del numero d'individui in specie animali conviventi.* Mem. Accad. Lincei 6 (31-113).

Wissel, C. 1989: *Theoretische Ökologie.* Springer Berlin/Heidelberg/New York.

Wunsch, G. 1987: *Allgemeine Systemtheorie.* S. 9-79 in Philippow 1987.

Zeigler, B. P. 1976: *Theory of Modeling and Simulation.* John Wiley, New York.

Zimmermann, H.-J. 1991: *Fuzzy Set Theory and Its Applications.* Kluwer Academic Publishers, Boston/Dordrecht/London.

APPENDIX

PROGRAM LISTINGS

Prog. 2.1 GLOBSIM.PAS
TurboPascal simulation program for a primitive global model

```
Program Globsim;
(* H.Bossel: Modellbildung und Simulation 910512 *)

uses crt,graph;
var
   N,P,C,Q,NRate,PRate,CRate,a,b,k,d,e,f,g,dt,t,final,xscale,yscale:real;
   answer,Nstr,Pstr,Cstr,fstr,gstr,finstr: string;
   more: boolean;
   graphDriver,graphMode,sx,sy,xnull,ynull: integer;
begin
   more := true;
   while (more) do
   begin
      clrscr;
      writeln ('DYNAMICS OF A NONLINEAR MINI-WORLD MODEL');
      writeln;
      writeln ('control parameter g (1)');
      readln (g);
      writeln ('control parameter f (0.1)');
      readln (f);
      writeln ('simulation period [years] (200)');
      readln (final);
      graphDriver := detect;
      initGraph (graphDriver, graphMode, '');
      sx := (getMaxX+1) div 100;
      sy := (getMaxY+1) div 100;
      xnull := round(0*sx);                {origin}
      ynull := round(90*sy);
      line (xnull+round(100*sx),ynull,xnull,ynull);
      moveTo(xnull,ynull);
      xscale := 100/final;                 {scaling}
      yscale := 20;
      t := 0;                              {parameters}
      a := 0.1;  b := 0.03;  k := 0.05;  d := 0.01;  e := 0.02;
      dt := 0.2;
      N := 1;                              {initial values}
      P := 1;
      C := 1;

      while t<final do                     {begin of time loop}
      begin
         t := t+dt;
         lineTo (round(t*xscale*sx), round(ynull-P*yscale));
         circle (round(t*xscale*sx), round(ynull-N*yscale),1);
         circle (round(t*xscale*sx), round(ynull-C*yscale),3);
         moveTo (round(t*xscale*sx), round(ynull-P*yscale));
         Q := 1/P;
         NRate := b*N*g*Q*C - d*N*P;       {state equations (rates)}
```

```
        if Q>=1 then PRate := e*C*N - a*P
           else PRate := e*C*N - a*P*Q;
        CRate := k*C*P*(1-(C*P*f));
        N := N+NRate*dt;                {numerical integration}
        P := P+PRate*dt;
        C := C+CRate*dt;
        if (N+P+C)>1000 then t:=final;  {end of time loop}
     end;
     str(N:6:3,Nstr); str(P:6:3,Pstr); str(C:6:3,Cstr);
     str(g:3:2,gstr); str(f:3:2,fstr); str(final:5:0,finstr);
     OutTextXY(xnull,ynull+10,'N = '+Nstr+', P = '+Pstr+', C = '+Cstr);
     OutTextXY(xnull,ynull+20,'g = '+gstr+', f = '+fstr+', period = '+finstr);
     readln; closeGraph;
     writeln ('repeat? (j/n)');
     readln (answer);
     if (answer='n') or (answer='N') then
        more := false;
  end;
end.
```

Prog. 4.1 MODLSTYL.TXT
Programming template for SIMPAS model units MODEL.PAS

```
UNIT MODEL;

INTERFACE
    uses Base, Crt, Graph;
    procedure InitialInfo;
    procedure ModelEqs;
    procedure Summary;

IMPLEMENTATION
    var
        data declarations

    procedure InitialInfo;
    begin
        Title       := 'model title';
        Description := 'brief description';
        TimeUnit    := 'simulation time unit';
        Author      := 'name, date, references';

        StateVariable[1] := 'label: (initial value) [dimension]';
        (other state variables)
        ParamQuestion[1] := 'label: (default value) [dimension]';
        (other system parameters)
        ScenaQuestion[1] := 'label: (default value) [dimension]';
        (other scenario parameters)
        TableFunction[1] := 'output variable y : input variable x'
        + '/x1,y1/x2,y2/ ... /xn,yn//';
        (other table functions)
        OutVarText[1]    := 'output label: [dimension]';
        (other output variables in addition to the state variables)
        Start    := begin of simulation;
        Final    := end of simulation;
        TimeStep := timestep of computation;
        (other initialization data, if necessary)
    end;

    procedure ModelEqs;
    begin
        All model equations are located in this procedure. The following
        assignments must appear in the procedure:
        at the beginning of the procedure (for T=START):
        if Time = Start then
        begin
            name of system parameter in the program   := ParamAnswer[1];
            (other system parameters)
            name of scenario parameter in the program := ScenaAnswer[1];
            (other scenario parameters)
        end;
```

```
        (at the beginning of the procedure, for all T):
        name of state variable in the program     := State[1];
        (other state variables)
        (at the end of the procedure, for all T):
        Rate[1] := rate of change of state variable;
        (other rates of change)
        OutVariable[1] := quantity to be stored as output;
        (other output quantities)
    end;

    procedure Summary;
    begin
        (procedure supplied by user for additional presentation of results)
    end;
end.
```

Prog. 4.2 PREDATOR.MOD
Sample SIMPAS model unit of a predator-prey system

```
UNIT MODEL;
(* PREDATOR.MOD, H.Bossel 901227 *)

INTERFACE
   uses Base,Crt;
   procedure InitialInfo;
   procedure ModelEqs;
   procedure Summary;


IMPLEMENTATION

   const
      specFoxRespiration = 0.2;
      specRabbitGrowth   = 0.08;
      KillRisk           = 0.002;
      specFoxGain        = 0.2;

   var
      Rabbits, Foxes, OrigCapacity, Capacity, FreeCapacity, Meetings,
      RabbitGrowth, RabbitKills, FoxRespiration, FoxGains,
      CapacityReduction, SwitchTime: real;

   procedure InitialInfo;
   begin
      Title       := 'PREDATOR-PREY SYSTEM';
      Description := 'mutual dependence of predator and prey';
      TimeUnit    := 'weeks';
      Author      := 'H.Bossel (901227)';

      StateVariable[1] := 'rabbits: (50) [no.]';
      StateVariable[2] := 'foxes: (2) [no.]';
      ParamQuestion[1] := 'orig.carr.capacity: (2000) [no.of rabbits]';
      ScenaQuestion[1] := 'time of carr.cap. reduction: (200) [weeks]';
      ScenaQuestion[2] := 'percent of carr.cap. reduction: (75) [%]';
      OutVarText[1]    := 'free capacity: [no. of rabbits]';
      Start            := 0;
      Final            := 500;
      TimeStep         := 0.5;
   end;

   function CapacityFactor(Time:real):real;
   begin
      CapacityFactor := 1;
      if (Time>SwitchTime) then
         CapacityFactor := (100-CapacityReduction)/100;
   end;
```

```
   procedure ModelEqs;
   begin
      Rabbits      := State[1];
      Foxes        := State[2];
      OrigCapacity := ParamAnswer[1];
      SwitchTime   := ScenaAnswer[1];
      CapacityReduction := ScenaAnswer[2];
      Capacity     := OrigCapacity*CapacityFactor(Time);
      FreeCapacity := Capacity - Rabbits;   (* negative if overcrowding *)
      RabbitGrowth :=(1/OrigCapacity)*FreeCapacity*specRabbitGrowth*Rabbits;
      FoxRespiration := specFoxRespiration*Foxes;
      Meetings     := Rabbits*Foxes;
      RabbitKills  := Meetings*KillRisk;
      FoxGains     := RabbitKills*specFoxGain;
      OutVariable[1] := FreeCapacity;
      Rate[1]      := RabbitGrowth-RabbitKills;
      Rate[2]      := FoxGains-FoxRespiration;
   end;

   procedure Summary; begin end;
end.
```

Prog. 4.3 FCTNDEMO.MOD
Model unit for demonstration of SIMPAS functions and simulation

```
UNIT MODEL;
(* FCTNDEMO.MOD H.Bossel: Modellbildung und Simulation 920112 *)

INTERFACE
   uses Base,Crt,Graph;
   procedure InitialInfo;
   procedure ModelEqs;
   procedure Summary;

IMPLEMENTATION
   var
      DelT, X, Y1, Y3, IntX, IntY1, IntY3: real; Select: byte;

   procedure InitialInfo;
   begin
      Title       := 'SIMPAS FUNCTIONS';
      Description := 'Demonstration of SIMPAS functions which can be used '
     +'as test inputs or for other purposes: pulse function, step function, '
     +'ramp function, sine function, table function, exponential delays of '
     +'1st and 2nd order.';
      TimeUnit    := 'time unit';
      Author      := 'H.Bossel: Modellbildung und Simulation 920112';
      StateVariable[1] := 'integral of X, no delay: (0) [-]';
      StateVariable[2] := 'integral of X with Delay1: (0) [-]';
      StateVariable[3] := 'integral of X with Delay3: (0) [-]';
      ParamQuestion[1] := '1-Pulse,2-Step,3-Ramp,4-Sine,5-Tbf: (5) [-]';
      ScenaQuestion[1] := 'delay time: (1) [-]';
      TableFunction[1] := 'X : time'
         + '/0,0/1,0/1.2,1/1.5,0.2/2,0/2.5,-0.2/2.8,-1/3,0/4,0//';
      OutVarText[1]    := 'signal: [-]';
      OutVarText[2]    := 'signal with Delay1: [-]';
      OutVarText[3]    := 'signal with Delay3: [-]';
      Start            := 0;
      Final            := 5;
      TimeStep         := 0.01;
   end;

   procedure ModelEqs;
   begin
      if Time=Start then
      begin
         Select := round(ParamAnswer[1]);
         DelT   := ScenaAnswer[1];
      end;
      IntX := State[1]; IntY1 := State[2]; IntY3 := State[3];
      if Select=1 then X := Pulse(1,1,0) + Pulse(-1,2,0);
      if Select=2 then X := Step(1,1) - Step(2,2) + Step(1,3);
      if Select=3 then X := Ramp(1,1) - Ramp(2,1.5) + Ramp(2,2.5) - Ramp(1,3);
```

```
      if Select=4 then
      begin
         if (Time>=1) and (Time<=3) then
         X := Sin(pi*(Time-1)) else X := 0;
      end;
      if Select=5 then X := Tbf(1,Time);
      Y1 := Delay1(1,DelT,X); Y3 := Delay3(1,DelT,X);
      OutVariable[1] := X; OutVariable[2] := Y1;OutVariable[3] := Y3;
      Rate[1] := X;
      Rate[2] := Y1;
      Rate[3] := Y3;
   end;

   procedure Summary; begin end;

end.
```

Prog. 4.4 PENDULUM.MOD
SIMPAS model unit for rotating pendulum dynamics

```
UNIT MODEL;
(* PENDULUM.MOD H.Bossel: Modellbildung und Simulation 910612 *)

INTERFACE
   uses Base,Crt,Graph;
   procedure InitialInfo;
   procedure ModelEqs;
   procedure Summary;


IMPLEMENTATION

   var
      angle, angular_velocity, angular_acceleration,
      mass, radius, gravitation, damping, velocity,
      gravitational_force, damping_force, acceleration_force,
      acceleration, horizontal, vertical: real;

   procedure InitialInfo;
   begin
      Title       := 'ROTATING PENDULUM';
      Description
      := 'Nonlinear damped oscillation of pendulum in circular motion';
      TimeUnit    := 'seconds';
      Author      := 'H.Bossel: Modellbildung und Simulation 910612';

      StateVariable[1] := 'angle: (0) [radian]';
      StateVariable[2] := 'angular_velocity: (10) [1/s]';
      ParamQuestion[1] := 'mass: (1) [kg]';
      ParamQuestion[2] := 'radius: (1) [m]';
      ScenaQuestion[1] := 'damping: (1) [N/(m/s)]';
      OutVarText[1]    := 'vertical position: [m]';
      OutVarText[2]    := 'horizontal position: [m]';

      Start            := 0;
      Final            := 10;
      TimeStep         := 0.02;
   end;

   procedure ModelEqs;
   begin
      if Time=Start then
      begin
         mass       := ParamAnswer[1];
         radius     := ParamAnswer[2];
         damping    := ScenaAnswer[1];
         gravitation:= 9.81;
      end;
```

```
      angle                := State[1];
      angular_velocity     := State[2];

      velocity             := radius * angular_velocity;
      gravitational_force  := mass * gravitation * sin(angle);
      damping_force        := damping * velocity;
      acceleration_force   := - gravitational_force - damping_force;
      acceleration         := acceleration_force / mass;
      angular_acceleration := acceleration / radius;
      horizontal           := radius * sin(angle);
      vertical             := radius * (1 - cos(angle));
      OutVariable[1]       := vertical;
      OutVariable[2]       := horizontal;

      Rate[1] := angular_velocity;
      Rate[2] := angular_acceleration;
   end;

   procedure Summary; begin end;

end.
```

Prog. 4.5 FISHERY.MOD
SIMPAS model unit for fishery dynamics

```
UNIT MODEL;
(* Fishery.mod H.Bossel: Modellbildung und Simulation 910612 *)

INTERFACE
   uses Base,Crt,Graph;
   procedure InitialInfo;
   procedure ModelEqs;
   procedure Summary;

IMPLEMENTATION

   var
      fish, boats, area, fish_carry_capacity, max_fish_capacity,
      max_fish_growth_rate, operating_costs, boat_price,
      boat_lifetime, decommission_rate, max_fish_catch, fish_price,
      investment_fraction_boats, fish_density, fish_growth_rate,
      catch_potential, fish_catch, catch_proceeds, boat_maintenance,
      net_income, catch_chance, investment_capital_boats, purchase_boats,
      decommission_boats, max_boat_number: real;
      p, i, q, o, d, k, f, a, c, e, fish_eq_pt, boat_eq_pt: real;
      fish_locator: boolean;

   procedure InitialInfo;
   begin
      Title       := 'FISHERY DYNAMICS';
      Description
         := 'Fish population and fishing fleet dynamics as function '
            +'of management, ecological and economic parameters.';
      TimeUnit    := 'years';
      Author      := 'H.Bossel: Modellbildung und Simulation 910612';

      StateVariable[1]    := 'fish: (5000) [t fish]';
      StateVariable[2]    := 'boats: (100) [boats]';
      area                := 100; {km_}
      fish_carry_capacity := 100; {t fish/km_}
      max_fish_capacity   := area * fish_carry_capacity; {t fish}
      catch_chance        := 0.8;

      ParamQuestion[1] := 'max spec fish growth rate: (1) [1/year]';
      ParamQuestion[2] := 'spec maintenance costs: (50000) [$/(boat.year)]';
      ParamQuestion[3] := 'boat price: (100000) [$/boat]';
      ParamQuestion[4] := 'boat lifetime: (15) [year]';
      ScenaQuestion[1] := 'max spec fish catch: (100) [t fish/(boat.year)]';
      ScenaQuestion[2] := 'fish price: (1000) [$/t fish]';
      ScenaQuestion[3] := 'investment fraction boats: (0.5) [-]';
      ScenaQuestion[4] := 'fish_locator (0/1): (0) [-]';
      ScenaQuestion[5] := 'max number of boats: (1000) [boats]';
      OutVarText[1]    := 'fish_catch: [t fish/year]';
```

```
   Start    := 0;
   Final    := 20;
   TimeStep := 0.02;
end;

procedure ModelEqs;
begin
   if Time=Start then
   begin
      max_fish_growth_rate       := ParamAnswer[1];
      operating_costs            := ParamAnswer[2];
      boat_price                 := ParamAnswer[3];
      boat_lifetime              := ParamAnswer[4];
      decommission_rate          := 1/boat_lifetime;
      max_fish_catch             := ScenaAnswer[1];
      fish_price                 := ScenaAnswer[2];
      investment_fraction_boats  := ScenaAnswer[3];
      if ScenaAnswer[4] = 1 then
               fish_locator := true
          else fish_locator := false;
      max_boat_number            := ScenaAnswer[5];
   end;

   if State[1]<0 then State[1] := 0;
   fish  := State[1];
   boats := State[2];

   fish_density      := fish / max_fish_capacity;
   fish_growth_rate  := max_fish_growth_rate * fish * (1 - fish_density);
   catch_potential   := max_fish_catch * boats;
   if not fish_locator then
      fish_catch     := catch_potential * fish_density;
   if fish_locator then
      if fish>0 then
         fish_catch  := catch_potential * catch_chance
      else fish_catch := 0;
   catch_proceeds     := fish_price * fish_catch;
   boat_maintenance   := operating_costs * boats;
   net_income         := catch_proceeds - boat_maintenance;
   investment_capital_boats := investment_fraction_boats * net_income;
   purchase_boats     := investment_capital_boats / boat_price;
   if boats>max_boat_number then
      purchase_boats  := 0;
   decommission_boats := decommission_rate * boats;

   Rate[1] := fish_growth_rate - fish_catch;
   Rate[2] := purchase_boats - decommission_boats;
   OutVariable[1] := fish_catch;
end;
```

```
   procedure Summary;
   begin
      p := fish_price;
      i := investment_fraction_boats;
      q := boat_price;
      o := operating_costs;
      d := decommission_rate;
      k := max_fish_capacity;
      f := max_fish_catch;
      a := max_fish_growth_rate;
      c := p*i/q;
      e := o*i/q+d;
      fish_eq_pt := k*e/(f*c);
      boat_eq_pt  := (k*a/f)*(1-e/(f*c));
      if boat_eq_pt <= 0 then
      begin
         fish_eq_pt  := k;
         boat_eq_pt  := 0;
      end;
      ClrScr;
      writeln;
      writeln (title);
      writeln;
      writeln ('equilibrium points:');
      writeln;
      writeln ('fish [t fish]  :  ', fish_eq_pt:6:0);
      writeln ('boats  [boats] :  ', boat_eq_pt:6:0);
      readln;
   end;

end.
```

Prog. 5.1 WORLDSIM.MOD
SIMPAS model unit for the Miniworld global model

```
UNIT MODEL;
(* Worldsim.mod H.Bossel: Modellbildung und Simulation 910813 *)

INTERFACE
   uses Base,Crt,Graph;
   procedure InitialInfo;
   procedure ModelEqs;
   procedure Summary;


IMPLEMENTATION

   var
      population, capital, pollution, absorption_rate, birth_rate,
      capital_growth_rate, death_rate, pollution_rate, birth_control,
      consumption_control, quality, births, deaths, degradation, threshold,
      regeneration, conservation, capital_growth, renewal, consumption: real;
      RunCase: byte;

   procedure InitialInfo;
   begin
      Title       := 'MINIWORLD';
      Description
          := 'Dynamics resulting from the interaction of population ,'
            +'consumption, pollution.';
      TimeUnit    := 'years';
      Author      := 'H.Bossel 910512, 910813';

      StateVariable[1] := 'population: (1) [rel]';
      StateVariable[2] := 'capital : (1) [rel]';
      StateVariable[3] := 'pollution: (1) [rel]';

      ParamQuestion[1] := '1-all, 2-popul, 3-pollution, 4-capital : (1) [-]';
      ParamQuestion[2] := 'pollution absorption rate: (0.1) [1/year]';
      ParamQuestion[3] := 'pollution rate: (0.02) [1/year]';
      ScenaQuestion[1] := 'birth rate: (0.03) [1/year]';
      ScenaQuestion[2] := 'capital growth rate: (0.05) [1/year]';
      ScenaQuestion[3] := 'birth control: (1.0) [-]';
      ScenaQuestion[4] := 'consumption control: (0.1) [-]';
      death_rate       := 0.01;

      Start            := 0;
      Final            := 100;
      TimeStep         := 0.2;
   end;
```

```
procedure ModelEqs;
begin
   if Time=Start then
   begin
      RunCase            := round(ParamAnswer[1]);
      absorption_rate    := ParamAnswer[2];
      pollution_rate     := ParamAnswer[3];
      birth_rate         := ScenaAnswer[1];
      capital_growth_rate := ScenaAnswer[2];
      birth_control      := ScenaAnswer[3];
      consumption_control := ScenaAnswer[4];
      threshold          := 1;
   end;
   population := State[1];
   capital    := State[2];
   pollution  := State[3];
case RunCase of
1: begin
      quality      := threshold/pollution;
      births       := birth_rate*population*quality*capital*birth_control;
      deaths       := death_rate*population*pollution;
      degradation  := pollution_rate*population*capital;
      conservation := 1 - consumption_control*pollution*capital;
      capital_growth:= capital_growth_rate*capital*pollution*conservation;
      if quality <  1 then renewal := quality else renewal := 1;
      regeneration := absorption_rate*pollution*renewal;
      consumption  := capital;
   end;
2: begin
      births         := birth_rate*population;
      deaths         := death_rate*population;
      capital_growth := 0; degradation := 0;  regeneration := 0;
   end;
3: begin
     births := 0; deaths := 0; capital_growth := 0;
      quality      := threshold/pollution;
      if quality < 1 then renewal := quality else renewal := 1;
      degradation  := pollution_rate;
      regeneration := absorption_rate*pollution*renewal;
   end;
4: begin
      births := 0; deaths := 0; degradation := 0; regeneration := 0;
      conservation   := 1 - consumption_control*capital;
      capital_growth := capital_growth_rate*capital*conservation;
   end;
end;
   Rate[1] := births - deaths;
   Rate[2] := capital_growth;
   Rate[3] := degradation - regeneration;
end;

procedure Summary; begin end;  end.
```

Prog. 5.2 WORLDVAL.MOD
SIMPAS model unit for the Miniworld with orientor assessment

```
UNIT MODEL;
(* Worldval.mod, H.Bossel: Modellbildung und Simulation 910813 *)

INTERFACE
   uses Base,Crt,Graph;
   procedure InitialInfo;
   procedure ModelEqs;
   procedure Summary;

IMPLEMENTATION
   var
      population, capital, pollution, absorption_rate, birth_rate,
      capital_growth_rate, death_rate, pollution_rate, birth_control,
      consumption_control, quality, births, deaths, degradation,
      regeneration, conservation, capital_growth, renewal, consumption: real;
      EXIS1, EXIS2, EFFI1, EFFI2, FREE1, FREE2, FREE3, SECU1, SECU2,
      ADAP1, ADAP2, ADAP3, ADAP4, SOLI1, EXIS, EFFI, FREE, SECU, ADAP,
      SOLI, SATIS: real;

   procedure InitialInfo;
   begin
      Title       := 'MINIWORLD WITH ORIENTATION';
      Description := 'Dynamics of population, consumption, and pollution '
      +'are mapped on the system orientors to allow assessment of overall '
      +'system performance.';
      TimeUnit    := 'years';
      Author      := 'H.Bossel 910512, 910813';
      StateVariable[1]  := 'population: (1) [rel]';
      StateVariable[2]  := 'capital : (1) [rel]';
      StateVariable[3]  := 'pollution: (1) [rel]';
      ParamQuestion[1]  := 'pollution absorption rate: (0.1) [1/Jahr]';
      ParamQuestion[2]  := 'pollution rate: (0.02) [1/Jahr]';
      ScenaQuestion[1]  := 'birth rate: (0.03) [1/Jahr]';
      ScenaQuestion[2]  := 'capital growth rate: (0.05) [1/Jahr]';
      ScenaQuestion[3]  := 'birth control: (1.0) [-]';
      ScenaQuestion[4]  := 'consumption control: (0.1) [-]';
      death_rate        := 0.01;
      {table functions for orientation:}
      TableFunction[1]  := 'EXIS1 : population /0,0/0.1,0/0.2,1/1,1//';
      TableFunction[2]  := 'EXIS2 : quality /0,0/0.1,0/0.2,1/1,1//';
      TableFunction[3]  := 'EFFI1: regeneration_degradation'
                           +'/0,0/0.95,0/1.5,1/2,1//';
      TableFunction[4]  := 'EFFI2 : quality_capital /0,0/0.4,0/1,1/2,1//';
      TableFunction[5]  := 'FREE1 : consumption /0,0/0.8,0/2,1/3,1//';
      TableFunction[6]  := 'FREE2 : quality /0,0/0.5,0/1,1/2,1//';
      TableFunction[7]  := 'FREE3 : deaths_population'
                           +'/0,0/0.01,1/0.03,0/0.05,0//';
```

```
    TableFunction[8]  := 'SECU1 : deaths_births'
                         +'/0,0/0.9,0/1,1/1.1,0/2,0//';
    TableFunction[9]:=   'SECU2 : regeneration_degradation'
                         +'/0,0/0.95,0/1.2,1/2,1//';
    TableFunction[10] := 'ADAP1 : quality /0,0/0.5,0/1.5,1/2,1//';
    TableFunction[11] := 'ADAP2 : population /0,1/1.5,1/4,0/5,0//';
    TableFunction[12] := 'ADAP3 : capital_growth'
                         +'/0,1/0.01,1/0.02,0/0.03,0//';
    TableFunction[13] := 'ADAP4 : capital /0,0/0.5,0/2,1/3,1//';
    TableFunction[14] := 'SOLI1 : population /0,0/0.5,0/1,1/2,1//';
    OutVarText[1]     := 'orientor satisfaction, total: [-]';
    OutVarText[2]     := 'existence: [-]';
    OutVarText[3]     := 'effectiveness: [-]';
    OutVarText[4]     := 'freedom of action: [-]';
    OutVarText[5]     := 'security: [-]';
    OutVarText[6]     := 'adaptivity: [-]';
    OutVarText[7]     := 'consideration: [-]';
    Start             := 0;
    Final             := 100;
    TimeStep          := 0.2;
 end;

 procedure ModelEqs;
 begin
    if Time=Start then
    begin
       absorption_rate     := ParamAnswer[1];
       pollution_rate      := ParamAnswer[2];
       birth_rate          := ScenaAnswer[1];
       capital_growth_rate := ScenaAnswer[2];
       birth_control       := ScenaAnswer[3];
       consumption_control := ScenaAnswer[4];
       threshold           := 1;
    end;
    population  := State[1];
    capital     := State[2];
    pollution   := State[3];

    quality        := threshold/pollution ;
    births         := birth_rate*population*quality*capital*birth_control;
    deaths         := death_rate*population*pollution;
    degradation    := pollution_rate*population*capital ;
    conservation   := 1 - consumption_control*pollution*capital ;
    capital_growth := capital_growth_rate*capital*pollution*conservation;
    if quality <  1 then renewal := quality;
    if quality >= 1 then renewal := 1;
    regeneration   := absorption_rate*pollution *renewal;
    consumption    := capital;

    Rate[1]      := births - deaths;
    Rate[2]      := growth;
    Rate[3]      := degradation - regeneration;
```

```
    {orientor module:}
    EXIS1 := Tbf(1,population);
    EXIS2 := Tbf(2,quality);
    EFFI1 := Tbf(3,regeneration/degradation);
    EFFI2 := Tbf(4,quality/capital);
    FREE1 := Tbf(5,consumption);
    FREE2 := Tbf(6,quality);
    FREE3 := Tbf(7,deaths/population);
    SECU1 := Tbf(8,deaths/births);
    SECU2 := Tbf(9,regeneration/degradation);
    ADAP1 := Tbf(10,quality);
    ADAP2 := Tbf(11,population);
    ADAP3 := Tbf(12,capital_growth);
    ADAP4 := Tbf(13,capital);
    SOLI1 := Tbf(14,quality);
    if (EXIS1*EXIS2>0)
      then EXIS := (EXIS1+EXIS2)/2 else EXIS := 0;
    if (EFFI1*EFFI2>0)
      then EFFI := (EFFI1+EFFI2)/2 else EFFI := 0;
    if (FREE1*FREE2*FREE3>0)
      then FREE := (FREE1+FREE2+FREE3)/3 else FREE := 0;
    if (SECU1*SECU2>0)
      then SECU := (SECU1+SECU2)/2 else SECU := 0;
    if (ADAP1*ADAP2*ADAP3*ADAP4>0)
      then ADAP := (ADAP1+ADAP2+ADAP3+ADAP4)/4 else ADAP := 0;
    if (SOLI1>0)
      then SOLI := SOLI1 else SOLI := 0;
    SATIS := (EXIS+EFFI+FREE+SECU+ADAP+SOLI)/6;
    Outvariable[1] := SATIS;
    Outvariable[2] := EXIS;
    Outvariable[3] := EFFI;
    Outvariable[4] := FREE;
    Outvariable[5] := SECU;
    Outvariable[6] := ADAP;
    Outvariable[7] := SOLI;
  end;

  procedure Summary; begin end;

end.
```

Prog. 5.3 FISHOPT.MOD
SIMPAS model unit of fishery dynamics with quality criteria

```
UNIT MODEL;
(* Fishopt.mod H.Bossel: Modellbildung und Simulation 910612, 910814 *)

INTERFACE
   uses Base,Crt,Graph;
   procedure InitialInfo;
   procedure ModelEqs;
   procedure Summary;

IMPLEMENTATION
   var
      fish, boats, area, spec_fish_capacity, max_fish_capacity,
      max_spec_fish_growth_rate, spec_maintenance_costs, boat_price,
      boat_lifetime, spec_decommissioning_rate, max_spec_fish_catch,
      fish_price, investment_fraction_boats, fish_density, fish_growth_rate,
      catch_potential, fish_catch, catch_proceeds, boat_maintenance, net_income,
      catch_chance, investment_capital_boats, purchase_boats, decommission_boats,
      max_boat_number, catch_weight, profit_weight, profit_rate, rel_fish_catch,
      rel_profit_rate, quality: real;
      p, i, q, o, d, k, f, a, c, e, fish_eq_pt, boat_eq_pt: real;
      fish_locator: boolean;

   procedure InitialInfo;
   begin
      Title        := 'FISHERY OPTIMIZATION';
      Description
         := 'Dynamics of fish population, fishing fleet, and economic '
           +'returns. Use of weighted criteria to allow optimization.';
      TimeUnit     := 'years';
      Author       := 'H.Bossel: Modellbildung und Simulation 910612, 910814';
      StateVariable[1]    := 'fish: (5000) [t fish]';
      StateVariable[2]    := 'boats: (100) [boats]';
      area                := 100; {km_}
      spec_fish_capacity  := 100; {t fish/km_}
      max_fish_capacity   := area * spec_fish_capacity; {t fish}
      catch_chance        := 0.8;
      ParamQuestion[1] := 'max spec fish growth rate: (1) [1/year]';
      ParamQuestion[2] := 'spec maintenance costs: (50000) [$/(boat.year)]';
      ParamQuestion[3] := 'boat price: (100000) [$/boat]';
      ParamQuestion[4] := 'boat lifetime: (15) [yeare]';
      ScenaQuestion[1] := 'max spec fish catch: (100) [t fish/(boat.year)]';
      ScenaQuestion[2] := 'fish price: (1000) [$/t fish]';
      ScenaQuestion[3] := 'investment fraction boats: (0.5) [-]';
      ScenaQuestion[4] := 'fish locator (0/1): (0) [-]';
      ScenaQuestion[5] := 'max number of boats: (1000) [boats]';
      ScenaQuestion[6] := 'catch weight: (0) [-]';
      ScenaQuestion[7] := 'profit weight: (1) [-]';
```

```
   OutVarText[1]    := 'fish catch: [t fish/year]';
   OutVarText[2]    := 'profit rate: [$/year]';
   OutVarText[3]    := 'quality index: [-]';
   Start            := 0;
   Final            := 20;
   TimeStep         := 0.02;
end;

procedure ModelEqs;
begin
   if Time=Start then
   begin
      max_spec_fish_growth_rate := ParamAnswer[1];
      spec_maintenance_costs    := ParamAnswer[2];
      boat_price                := ParamAnswer[3];
      boat_lifetime             := ParamAnswer[4];
      spec_decommissioning_rate := 1/boat_lifetime;
      max_spec_fish_catch       := ScenaAnswer[1];
      fish_price                := ScenaAnswer[2];
      investment_fraction_boats := ScenaAnswer[3];
      if ScenaAnswer[4] = 1 then
               fish_locator := true
          else fish_locator := false;
      max_boat_number := ScenaAnswer[5];
      catch_weight    := ScenaAnswer[6];
      profit_weight   := ScenaAnswer[7];
   end;
   if State[1]<0 then State[1] := 0;
   fish  := State[1];
   boats := State[2];
   fish_density      := fish / max_fish_capacity;
   fish_growth_rate
        := max_spec_fish_growth_rate * fish * (1 - fish_density);
   catch_potential    := max_spec_fish_catch * boats;
   if not fish_locator then
      fish_catch      := catch_potential * fish_density;
   if fish_locator then
      if fish>0 then
         fish_catch   := catch_potential * catch_chance
      else fish_catch:= 0;
   catch_proceeds     := fish_price * fish_catch;
   boat_maintenance   := spec_maintenance_costs * boats;
   net_income         := catch_proceeds - boat_maintenance;
   investment_capital_boats := investment_fraction_boats * net_income;
   purchase_boats     := investment_capital_boats / boat_price;
   if boats>max_boat_number then
      purchase_boats := 0;
   decommission_boats:= spec_decommissioning_rate * boats;
   profit_rate        := net_income - investment_capital_boats;
   rel_fish_catch     := fish_catch/max_fish_capacity;
   rel_profit_rate    := profit_rate/(fish_price*max_fish_capacity);
```

```
      quality
        := ((catch_weight*rel_fish_catch)+(profit_weight*rel_profit_rate))
           *100/(catch_weight+profit_weight);
      Rate[1]        := fish_growth_rate - fish_catch;
      Rate[2]        := purchase_boats - decommission_boats;
      OutVariable[1] := fish_catch;
      OutVariable[2] := profit_rate;
      OutVariable[3] := quality;
   end;

   procedure Summary;
   begin
      p := fish_price;
      i := investment_fraction_boats;
      q := boat_price;
      o := spec_maintenance_costs;
      d := spec_decommissioning_rate;
      k := max_fish_capacity;
      f := max_spec_fish_catch;
      a := max_spec_fish_growth_rate;
      c := p*i/q;
      e := o*i/q+d;
      fish_eq_pt := k*e/(f*c);
      boat_eq_pt := (k*a/f)*(1-e/(f*c));
      if boat_eq_pt <= 0 then
      begin
         fish_eq_pt := k;
         boat_eq_pt := 0;
      end;
      ClrScr;
      writeln;
      writeln (title);
      writeln;
      writeln ('equilibrium points:');
      writeln;
      writeln ('fish [t fish] :  ', fish_eq_pt:6:0);
      writeln ('boats [boats] :  ', boat_eq_pt:6:0);
      readln;
   end;

end.
```

Prog. 5.4 BALANCER.MOD
SIMPAS model unit for balancing control of inverted pendulum

```
UNIT MODEL;
(Balancer.mod, H.Bossel: Systemdynamik 1987, Modellbildung und Simulation 910824)

INTERFACE
   uses Base,Crt,Graph;
   procedure InitialInfo;
   procedure ModelEqs;
   procedure Summary;

IMPLEMENTATION
   var
      a, aa, amp, ar, b, c, d, g, ii, kw, kwt, kx, kxt, l, mp, mw, rp,
      s, u, w, wt, wtt, x, xt, xtt: real;

   procedure InitialInfo;
   begin
      Title       := 'BALANCING CONTROL';
      Description
          := 'Stabilization of an inverted pendulum by feedback control';
      TimeUnit    := 'seconds';
      Author      := 'H.Bossel: Systemdynamik 1987, 910824';
      StateVariable[1] := 'angle: (0.2) [rad]';
      StateVariable[2] := 'angular velocity: (0) [rad/sec]';
      StateVariable[3] := 'position: (0) [m]';
      StateVariable[4] := 'velocity: (0) [m/sec]';
      StateVariable[5] := 'control work: (0) [Nm]';
      ParamQuestion[1] := 'angle feedback factor: (100) [N/rad]';
      ParamQuestion[2] := 'angular velocity feedback factor:(30)[N/(rad/sec)]';
      ParamQuestion[3] := 'position feedback factor: (3) [N/m]';
      ParamQuestion[4] := 'velocity feedback factor: (10) [N/(m/sec)]';
      ParamQuestion[5] := 'pendulum mass: (1) [kg]';
      ScenaQuestion[1] := 'disturbance amplitude: (0) [N]';
      OutVarText[1]    := 'control force: [N]';
      OutVarText[2]    := 'disturbance force: [N]';
      OutVarText[3]    := 'control power: [Nm/sec]';
      Start            := 0;
      Final            := 10;
      TimeStep         := 0.01;
   end;

   procedure ModelEqs;
   begin
      if Time=Start then
      begin
         g   := 9.81;  {gravitation constant, m/sec_}
         l   := 1; {pendulum length, m}
         mw  := 1; {carriage mass, kg}
         rp  := 0.05; {radius of pendulum mass, m}
```

```
        kw  := ParamAnswer[1];
        kwt := ParamAnswer[2];
        kx  := ParamAnswer[3];
        kxt := ParamAnswer[4];
        mp  := ParamAnswer[5];
        amp := ScenaAnswer[1];
        {computation of system constants:}
        ii  := 2*mp*rp*rp/5;  {spherical pendulum mass}
        aa  := ii*(mp+mw)+mp*mw*l*l;
        a   := g*(mp+mw)*mp*l/aa;
        b   := -mp*l/aa;
        c   := -g*mp*mp*l*l/aa;
        d   := (ii+mp*l*l)/aa;
        randomize;
     end;
     w   := State[1];
     wt  := State[2];
     x   := State[3];
     xt  := State[4];
     ar  := State[5];
     s   := amp*(2*random-1);  {disturbance function}
     u   := kw*w + kwt*wt + kx*x + kxt*xt; {control function}
     wtt := a*w + b*(u+s);
     xtt := c*w + d*(u+s);
     Outvariable [1] := u;
     Outvariable [2] := s;
     Outvariable [3] := abs(u*xt);
     Rate[1] := wt;
     Rate[2] := wtt;
     Rate[3] := xt;
     Rate[4] := xtt;
     Rate[5] := abs(u*xt);
  end;

  procedure Summary;
  var
     lx, ly: real; info, kwstr, kwtstr, kxstr, kxtstr, arstr: string;
     i, n, sx, sy, xw, yw, xp, yp, rr: integer;
     GraphMode: integer;
  begin
     GraphMode := GetGraphMode;
     SetGraphMode(Graphmode);
     sx := (GetMaxX+1) div 100;
     sy := (GetMaxY+1) div 100;
     line (0*sx, 90*sy, 100*sx, 90*sy);
     n := 0;
     repeat
        line (n*sx*100 div 20, 90*sy, n*sx*100 div 20, 93*sy);
        n := n+1;
     until n=21;
     line (10*sx*100 div 20, 90*sy, 10*sx*100 div 20, 96*sy);
     str(kw:3:0,kwstr); str(kwt:3:0,kwtstr);
```

```
    str(kx:3:0,kxstr); str(kxt:3:0,kxtstr);
    info := ' control function: u = ' +kwstr +'*w + ' +kwtstr +'*dw/dt + '
         +kxstr +'*x + '+ kxtstr+ '*dx/dt';
    OutTextXY(0*sx,5*sy,info);
    str(ar:6:3,arstr);
    info := ' control work [Nm]  = ' + arstr;
    OutTextXY(0*sx,10*sy,info);
    repeat
       i := 0;
       repeat
          i := i+1;
          w := Result[2,i];
          x := Result[4,i];
          lx := l*sin(w);
          ly := l*cos(w);
          xw := 50*sx + round(x*50*sx);  {carriage position}
          yw := 85*sy;
          xp := xw + round(lx*50*sx); {position of pendulum mass}
          yp := yw - round(ly*66*sy);
          rr := round(rp*sx*50);
          SetColor(15);
          Rectangle (xw-5*sx,yw,xw+5*sx,yw+4*sy);
          Circle (xw,yw,4);
          Line (xw,yw,xp,yp);
          Circle(xp,yp,rr);
          Delay (15);
          SetColor(0);
          Rectangle (xw-5*sx,yw,xw+5*sx,yw+4*sy);
          Circle (xw,yw,4);
          Line (xw,yw,xp,yp);
          Circle(xp,yp,rr);
       until i=250;
    until KeyPressed;
    readln;
    RestoreCrtMode;
  end;

end.
```

Notes on the computer programs on the accompanying diskette

The diskette contains

1. the systems zoo with 50 simulation models (SIMZOO.EXE),
2. all simulation models discussed and documented in Chapters 2 through 5 in TurboPascal source code (*.PAS or *.MOD),
3. two SIMPAS program patterns (MODLSTYL.TXT and PREDATOR.MOD),
4. the SIMPAS simulation software (SIMPAS.PAS, BASE.TPU, SIMUL.TPU).

All programs can be executed on IBM-DOS personal computers (and compatible computers).

The systems zoo program SIMZOO.EXE already contains all common graphics interfaces and it can therefore be executed on all PC's without further preparation (call from DOS: SIMZOO). (TurboPascal is not required for this). The 50 different simulation models are accessible from the menus which appear after starting the program. The systems zoo uses SIMPAS as its simulation environment; the operation of the simulation models and the use of the different options is therefore identical to the simulations shown in Chapters 4 and 5. Documentations of the parameters and the simulation results in tabular or graphical form as they appear on the screen can be printed out if a suitable printer is connected to the computer.

The simulation models discussed in Chapters 4 and 5 are also contained in SIMZOO; using this program, they can be executed immediately without first compiling them using SIMPAS and TurboPascal.

The simulation models characterized by *.PAS can be run directly under TurboPascal without using SIMPAS.

The simulation models characterized by *.MOD are model units for SIMPAS. Before using SIMPAS, a model unit must be saved as MODEL.PAS. It can then be compiled together with SIMPAS.PAS, BASE.TPU, and SIMUL.TPU under TurboPascal to obtain an executable simulation program (SIMPAS.EXE).

In compiling a SIMPAS model, make sure that the SIMPAS-TPU's used correspond to the TurboPascal version used. The different SIMPAS-TPU's are found on the diskette in the corresponding directories SIMPAS50, SIMPAS55, SIMPAS60, and SIMPAS70 corresponding to TurboPascal 5.0, 5.5, 6.0, and 7.0. If you are working with TurboPascal 5.5 for example, you may only use BASE.TPU and SIMUL.TPU from the directory SIMPAS55. Note that the SIMPAS version on the diskette is restricted to a maximum of ten state variables.

The use of the SIMPAS simulation shell is explained in detail in Chapter 4 (Secs. 4-1 and 4-2).

INDEX